Black Holes, Gravitational Radiation and the Universe

Fundamental Theories of Physics

An International Book Series on The Fundamental Theories of Physics: Their Clarification, Development and Application

Volume 100

Black Holes, Gravitational Radiation and the Universe

Essays in Honor of C.V. Vishveshwara

edited by

Bala R. Iyer

Raman Research Institute,
Bangalore, India

and

Biplab Bhawal

TAMA Project,
National Astronomical Observatory,
Osawa, Mitaka, Tokyo, Japan

SPRINGER-SCIENCE+BUSINESS MEDIA, B.V.

A C.I.P. Catalogue record for this book is available from the Library of Congress.

ISBN 978-90-481-5121-9 ISBN 978-94-017-0934-7 (eBook)
DOI 10.1007/978-94-017-0934-7

Printed on acid-free paper

TABLE OF CONTENTS

HIGHER DIMENSION
KILLING
YANO
COLLECTED
• BLACK HOLES FOR BED TIME
• COSMIC PICTURE BOOK
FRENET SERRET SPINS
FERMI WALKER RIDES
GHOST VILLE 13
• BASE CAMP
• BLACK HOLE HUNT
EINSTEIN SOLUTION
UNDER COVER ORGANISATION
• SCHWARZSCHILD STABLE
• KERR WHIRPOOL

PREFACE

Our esteemed colleague C. V. Vishveshwara, popularly known as Vishu, turned sixty on 6th March 1998. His colleagues and well wishers felt that it would be appropriate to celebrate the occasion by bringing out a volume in his honour. Those of us who have had the good fortune to know Vishu, know that he is unique, in a class by himself. Having been given the privilege to be the volume's editors, we felt that we should attempt something different in this endeavour.

Vishu is one of the well known relativists from India whose pioneering contributions to the studies of black holes is universally recognised. He was a student of Charles Misner. His Ph. D. thesis on the stability of the Schwarzschild black hole, coordinate invariant characterisation of the stationary limit and event horizon for Kerr black holes and subsequent seminal work on quasi-normal modes of black holes have passed on to become the starting points for detailed mathematical investigations on the nature of black holes. He later worked on other aspects related to black holes and compact objects. Many of these topics have matured over the last thirty years. New facets have also developed and become current areas of vigorous research interest. No longer are black holes, ultracompact objects or event horizons mere idealisations of mathematical physicists but concrete entities that astrophysicists detect, measure and look for. Astrophysical evidence is mounting up steadily for black holes. The laser interferometric gravitational wave experiments, LIGO and VIRGO, under way in the USA and Europe respectively, should be completed by 2001 and could provide our first direct detection of gravitational waves from binary black hole systems. There are new concrete proposals for quantum theories of gravity with deep implications for the microstructure of black holes.

We felt that the time was just right to assess the progress made in these various topics of black hole research via short reviews by respective experts. Thus instead of being a collection of many different topics the volume would weave separate threads into the majestic black hole tapestry. We have attempted to carefully plan the volume with a view of making it uniform and useful. This we felt would give Vishu great pleasure since it could have a longer term pedagogic value and prove to be useful to the relativity community in general, especially fresh graduate students, fresh post docs and other research scientists. This attempt is our personal tribute to Vishu on behalf of all those numerous colleagues, students and lay persons who have been touched by the magic realism of his popular talks, humour and deep erudition. We hope the volume leads the reader along a seemingly lazy innocuous trail that began in the sixties, through today, to the future. It is an attempt to offer a grand panoramic view of black hole physics before the new millennium.

We were gratified with the warm response we received from all the contributors to our request to write for the volume. Some of them have known Vishu personally for a long time, others just via his papers. It was gratifying to find the international relativity community as a whole, still like one huge extended family bound by wonderful emotions and a sense of common endeavour, regardless of age or any other divisive factors. We thank our contributing colleagues for their critical and thorough assessments of the areas of their speciality and their participation in The Fest at the GR15 at IUCCA, Pune, in December 1997. A word of explanation is in order in regard to two articles by Abhay Ashtekar in the book. While the book was in its final stages the exciting work on implications of nonperturbative quantum gravity for black hole entropy appeared and we felt that this must be reviewed in the volume. Though Abhay had already sent us his contribution to the Fest earlier, we are happy that Abhay and Krasnov graciously agreed to our suggestion to write the review at very short notice.

It is a pleasure to thank G. Manjunatha for his meticulous, enthusiastic and uncomplaining assistence over the last few months, in preparing this volume for publication. We thank B. S. Gujjarappa for the caricature of Vishu that appears in these proceedings and Raju Verghese for producing the photographic prints in the volume. We thank A. Ratnakar for his assistence in checking the publication list of C. V. Vishveshwara, and Sai Iyer and A. Gopakumar for help with preparing some articles for production. We acknowledge the cooperation of our publishers Kluwer Academic Publishers, The Netherlands and in particular Dr. David Larner and Margaret Deignan in bringing out this volume. We would like to express our appreciation of Kluwer's attempts to publish highly specialised books and serve this relatively small academic community.

Bala Iyer

Biplab Bhawal

BIBLIOGRAPHICAL NOTES

The quotations in the book are from the following sources:

1. C. Behr, E. L. Schucking, C. V. Vishveshwara and W. Wallace, *Kinematics of Relativistic Ejection*, Astronom. Journ., 81, 147 (1976).
2. Thibault Damour, *The problem of motion in Newtonian and Einsteinian gravity*, in '300 years of Gravitation', eds. S. W. Hawking and W. Israel (CUP, Cambridge, 1987).
3. Robert H. Dicke in *The Theoretical Significance of Experimental Relativity* (1964).
4. F. A. E. Pirani, *Gravitational radiation*, in 'Gravitation: An introduction to current research', ed. L. Witten, (Wiley, 1962).
5. B. Rossi, Preface to *X-ray Astronomy*, ed. R. Giaconni and H. Gursky, Dordrecht, Reidel (1974).
6. Kip Thorne, *Black holes and time warps*, W. W. Norton and Co. (1994).
7. C. V. Vishveshwara, *Black Holes for Bedtime*, in 'Gravitation, Quanta and the Universe', eds. A. R. Prasanna, J. V. Narlikar and C. V. Vishveshwara (Wiley Eastern, 1980),
8. C. V. Vishveshwara, *The Engelbert Experience: Pathways from the past*, in 'On Einstein's Path: Essays in Honor of Engelbert Schucking', eds. Alex Harvey (Springer-Verlag, 1998).
9. C. V. Vishveshwara, *After the Fall: From Adam and Eve to Apples and Elevators*, in 'Gravitational and Cosmology: At the Turn of the Millenium', eds. N. K. Dadhich and J. V. Narlikar (IUCAA, Pune, 1998).
10. John Wheeler, in *Black holes and time warps*, Kip Thorne, W. W. Norton and Co. (1994).
11. Clifford M. Will in *Was Einstein right?* (Oxford, 1986)

1. THE BLACK HOLE EQUILIBRIUM PROBLEM

B. CARTER
Département d'Astrophysique Relativiste et de Cosmologie,
Centre National de la Recherche Scientifique,
Observatoire de Paris, 92195 Meudon, France.

1. Introduction

The subject to be dealt with here is the progress that has been achieved so far towards a definitive reply to the question of what equilibrium states are possible for an isolated black hole, taking the qualification "isolated" to mean that it is a global solution of the Einstein vacuum equations or their electrovac generalisation in a 4-dimensional asymptotically flat space-time background. This article is dedicated to C.V. Vishveshwara, who was one of the first to appreciate the importance of this problem, and who played an important role in persuading others to take the problem seriously as something of potential astrophysical relevance by providing the first convincing proof [1] that in at least one case, namely that of the Schwarzschild solution, such an equilibrium state can be stable. The concise overview given here is a shortened version of a more technically detailed account recently presented at the 8th Marcel Grossmann meeting [2].

In recent years there has been a considerable resurgence of mathematical work on black hole equilibrium states, but most of it has been concerned with more or less exotic generalisations of Einstein's theory to allow for inclusion of various scalar and other (e.g. Yang Mills type) fields such as those occuring in low energy limits of superstring theory. Another active research area that is also beyond the scope the present discussion is the astrophysically motivated subject of black hole equilibrium states with surrounding rings of matter, such as would result from accretion from external sources.

One reason why there has not been so much recent work on the more fundamental problem of the vacuum equilibrium states of a black hole in ordinary four dimensional Einstein theory is the widespread belief that the solution was already shown long ago to be just what was predicted by

B. R. Iyer and B. Bhawal (eds.), Black Holes, Gravitational Radiation and the Universe, 1–16.

my original 1967 conjecture [3, 4], i.e. that it is completely provided by the subset of Kerr solutions [5] for which $a^2 \leq M^2$ where $a = J/M$ is the ratio of the angular momentum J to the mass M. This belief rapidly gained general acceptance when – following the example of Israel's earlier 1967 work [6] providing strong evidence that (as has since been confirmed) the only strictly static (not just stationary) solutions were given by the special Schwarzschild ($J = 0$) case – I was able [7] in 1971 to obtain a line of argument that provided rather overwhelming, though by no means absolutely watertight, mathematical evidence to the effect that the most general solution is indeed included in the Kerr family.

Many people supposed that the conversion of the original plausibility argument into an utterly unassailable mathematical proof was merely a physically insignificant technical formality, that could be left as an exercise for the amusement of obsessively rigoristic pure mathematicians. However a more fundamental reason for the slowness of subsequent progress was that – as was soon shown by those, starting with Hawking [8, 9], who took the question seriously after all – the mathematical work needed to deal with the various apparently small technical gaps and loose ends in the comparitively simple 1971 argument [7, 10] is harder than might have been hoped. Thus despite the very considerable efforts of many people – of whom some of the most notable after Hawking [8, 9, 11] have been Robinson [12, 13, 14], followed (for the electrovac generalisation) by Bunting [15, 16] and Mazur [17, 18]), and most recently (since my last comprehensive review of the subject [19]) Wald and his collaborators [20, 21, 22, 23] – there still remains a lot that needs to be done before we shall have what could be considered a mathematically definitive solution even of the pure vacuum problem, not to mention the more formidable challenge of its electrovac generalisation.

To provide a historical background, the next section recalls some of the relevant results (culminating in Israel's theorem) obtained during the prolonged period of general confusion that I refer to as the "preclassical phase", prior to the introduction of the term "black hole" and to the general recognition that the disciples of Ginzburg and Zel'dovich had been right [24] in arguing that what it represents is a generic phenomenon – not just an unstable artefact of spherical symmetry as many people, including Israel himself [25], had speculated. Section 3 then describes the rapid progress made during what I refer to as the "classical phase" – the beginning of what Israel [25] has referred to as the "age of enlightenment"– immediately following the definitive formulation of the concept of a "black hole" (in terms of the "outer past event horizon" in an asymptotically flat background) so that the corresponding equilibrium state problem could at last be posed in a mathematically well defined form. Section 4 describes the substantial though slower progress that has been made in what I refer to as the "postclassical phase", that began when the main stream of work on black holes had been diverted to quantum aspects following the discovery of the Hawking effect [26].

The final section draws attention to the mathematically challenging (even if physically less important) problems that still remain to be tackled: these include not only the questions concerning the technically awkward degenerate limit case and the assumptions about spherical topology and analyticity that have been discussed by Chrusciel [27] and also, in a very extensive and up to date review, by Heusler [28], but also the largely neglected question of the assumption of causality, i.e. the absence of closed timelike curves.

2. The preclassical phase (1915-67)

What I refer to as the preclassical phase in the development of black hole theory is the period of unsystematic accumulation of more or less relevant results prior to the actual use of the term "black hole". This period began with the discovery in 1916 by Schwarzschild [29] of his famous asymptotically flat vacuum solution, whose outer region is strictly static, with a hypersurface orthogonal timelike Killing vector whose striking feature is that its magnitude tends to zero on what was at first interpreted as a spacetime singularity, but was later recognised to be interpretable as a regular boundary admitting a smooth extension to an inner region where the Killing vector becomes spacelike. This preclassical phase culminated in Israel's 1967 discovery [6] of a mathematical argument to the effect that the Schwarzschild solution is uniquely characterised by these particular properties, i.e. the original spherical example is the only example. The significance of this discovery was the subject of an intense debate that precipitated the transition to the "classical phase", inaugurating what Israel [25] has termed the "age of enlightenment", which dawned when the preceeding confusion at last gave way to a clear concensus. Thrilling eyewitness accounts of the turbulent evolution of ideas in the "golden age" of rapid progress, during the transition from the preclassical to the classical phase, have been given by Israel himself [25] and from a different point of view by Thorne [24]

A historical account of the more dilatory fumbling in the early years of the preclassical phase has been given by Eisenstaedt [30]. Due to a renaissance of interest (following the observational discovery of quasars) progress was much more rapid during what Thorne [24] has referred to as the "golden age", which began during the last half dozen years of the preclassical phase and continued through what I call the classical phase . It was during the late "golden" period of the pre-classical phase, roughly from 1961 to 1967, that results of importance for vacuum black hole theory began to be obtained without reliance on a presupposition of spherical symmetry. The most important of these results were of course Kerr's 1963 discovery [5] of the family of stationary asymptotically flat vacuum solutions characterised by a degenerate "type D" Weyl tensor, and the 1967 Israel theorem [6] referred to above.

As well as these two specific discoveries, the most significant development during this final "golden" period of the preclassical phase was the animated debate – under the leadership of Ginzburg and Zel'dovich [31] in what was then the Soviet Union, of Wheeler and later on Thorne [32] in the United States, and of Sciama and Penrose [33] in Britain – from which the definitive conceptual machinery and technical jargon of black hole theory finally emerged. Prior to the discovery of the Kerr solution [5], when the only example considered was that of Schwarzschild, it had not been thought necessary to distinguish what Wheeler later termed the "ergosphere" – where the Killing vector generating the stationary symmetry of the exterior ceases to be timelike – from the "outer past event horizon" bounding what Wheeler later termed the "black hole" region, from which no future timelike trajectory can escape to the asymptotically flat exterior. In its original version, Israel's 1967 theorem [6] (as well as its electrovac generalisation [36]) was effectively formulated in terms of an "infinite redshift surface" that was effectively taken to be what in strict terminology was really the "ergosphere" rather than the "outer past event horizon": this meant that the significance of the theorem for the theory for what were to be called a "black holes" was not clear until it was understood that – as shown in my thesis [3, 34] and pointed out independently about the same time by Vishveshwara [35] – subject to the condition of strict staticity postulated by Israel (but not in the more general stationary case exemplified by the non-spherical Kerr [5] solutions) that the "outer past event horizon" actually will coincide with the ergosphere.

One of the first to appreciate the distinction between (what would come to be known as) the horizon and the ergosphere, and to recognise the members of the relevant ($a^2 \leq M^2$) Kerr subset as prototypes of what would come to be known as black hole solutions was Boyer [37, 38, 39]. However at the time of the first detailed geometrical investigations of the Kerr solutions [40, 38] it was assumed that this was just particularly simple case within what might turn out to be a much more extensive category. However the publication of the 1967 Israel theorem [6] – which went much of the way towards proving that the Schwarzschild [29] solution is the only strictly static example – immediately leads to the question of whether the Kerr solutions might not be similarly unique.

The explicit formulation of this suggestion came later to be loosely referred to in the singular as the "Israel-Carter conjecture", but there were originally not one but two distinct versions. The stronger version – suggested by the manner in which the Israel theorem was originally formulated – conjectured that the relevant Kerr subfamily might be the only stationary solutions that are well behaved outside and on a regular "infinite redshift surface" – a potentially ambiguous term that in the context of the original version [6] of Israel's theorem effectively meant what was later to be termed an "ergosurface" rather than an "event horizon". The weaker version, first written unambiguously in my 1967 thesis [3, 4], conjectured that the relevant Kerr subfamily might be the only stationary solutions that are well behaved all the way in to a regular black hole horizon, not just outside the ergosphere.

Work by Bardeen [41] and others on the effects of stationary orbiting matter rings (which can occur outside the horizon but inside the ergosphere of an approximately Kerr background) soon made it evident that strong version of the conjecture is definitely wrong, no matter how liberally one interprets the rather vague qualifications "regular" and "well behaved". On the other hand the upshot of the work to be described in the following sections is to confirm the validity of my weaker version, as expressed in terms of the horizon rather than the ergosphere. It is nevertheless to be remarked that, as was pointed out by Hartle and Hawking [42], the generalisation of this conjecture from the Kerr pure vacuum solutions [5] to the Kerr-Newman electrovac solutions [43] is not valid, since the solutions due to Papapetrou [44] and Majumdar [45] provide counter-examples. It is also to be emphasised that, as will be discussed in the final section, the question still remains entirely open, even in the pure vacuum case, if the interpretation of the qualification "well behaved" is relaxed so as to permit causality violation outside the horizon of the kind that is actually observed [4] to occur in the inner regions of the Kerr examples.

3. The classical phase (1968-75)

The "classical phase" in the development of black hole theory began when the appropriate conceptual framework and terminology became available, facilitating clear formulation of the relevant mathematical problems, whose solutions could then be sought by systematic research programs, not just by haphazard approach of the"unenlightened" preceeding period. The relevant notions had already began to become clear to a small number of specialists – notably Wheeler's associates, including Thorne, Misner, and their respective students Price and Vishveshwara, in the United States, and Penrose's associates, including Hawking and myself in Britain – during the period of accelerated activity at the end of what I call the preclassical phase.

Once the black hole scenario had been mathematically defined, it became urgent to settle the question of its potential physical relevance, which requires that it should occur not just as an unstable special case (which was the implication that Israel was at first inclined to draw from his theorem [25]) but as a generic phenomenon as Zel'dovich and his collaborators had been claiming [24]).

The strongest conceivable confirmation of the general validity of the black hole scenario is what would hold if Penrose's 1969 cosmic censorship conjecture [46] were valid in some form. According to this vaguely worded conjecture, in the framework of a "realistic" theory of matter the singularities resulting (according to Penrose's earlier "preclassical" closed trapped surface theorem [33]) from gravitational collapse should generally be hidden within the horizon of a black hole with a regular exterior. However far from providing a satisfactory general proof of this conjecture, subsequent work on the question (of which there has not been as much as would be warranted) has tended to show that it can be valid only if interpreted in a rather restricted manner. Nevertheless, despite the construction of various more or less artificial counter-examples by Eardley, Smarr, Christodoulou and others [47] to at least the broader interpretations of this conjecture, it seems clear that there will remain an extensive range of "realistic" circumstances under which the formation of a regular black hole configuration is after all to be expected.

It remains a controversial question (and in any case one that is beyond the scope of this discussion of pure vacuum equilibrium states) just how broad a range of circumstances can lead to regular black hole formation, and whether or not "naked singularities" can sometimes be formed instead under "realistic" conditions. However that may be, all that is actually needed to establish the relevance of black hole for practical physical purposes (as a crucial test of Einstein's theory, and assuming the result is positive, as an indispensible branch of astrophysical theory) is the demonstration of effective stability with respect to small perturbations of at lease some example. This essential step was first achieved in a mathematically satisfactory manner for the special case of the original prototype black hole solution, namely the Schwarzschild solution, in a crucially important quasi-normal mode analysis [1] by Vishveshwara in 1970. Another important article [48] by Price provided a more detailed account of the rate at which the solution could be expected to tend towards the Schwarzschild form under realistic circumstances as seen from the point of view of an external observer. The work of Vishveshwara and Price put the physical relevance of the subject beyond reasonable doubt by demonstrating that this particular (spherical) example is not just stable in principle but that it will also be stable in the practical sense of tending to its stationary (in this particular case actually static) limit within a timescale that is reasonably short compared with other relevant processes: in the Schwarzschild case the relevant timescale for convergence, at a given order of magnitude of the radial distance from the hole, turns out just to be comparable with the corresponding light crossing timescale.

During the remainder of the "classical phase", important work [49, 50, 51] by Teukolsky, Press and others made substantial progress towards the confirmation that the Kerr solutions are *all* similarly stable so long as the specific angular momentum parameter $a = J/M$ is less than its maximum value, $a = M$. However the possibility that instability might set in at some intermediate value in the range $0 < a < M$ was not conclusively eliminated until much later on, in the "post-classical" era, when (following a deeper study of the problem by Kay and Wald [52]) the question was settled more conclusively by the publication of a powerful new method of analysis developed by Whiting [53].

While this work on the stability question was going on, one of the main activities characterising the "classical phase" of the subject was the systematic investigation (along lines pioneered [54, 55] by Christodoulou and Ruffini) of the general mechanical laws governing the behaviour of stationary and almost stationary black hole states. Work by a number of people including Hawking, Hartle, Bardeen, and myself [8, 56, 57, 58, 59] (and later, as far as the electromagnetic aspects [10, 60, 61, 62] are concerned, also Znajek and Damour) revealed a strong analogy with the thermodynamical behaviour of a viscous (and electrically resistive) fluid. (Following a boldly imaginative suggestion by Bekenstein [63], the suspicion that this analogy could be interpreted in terms of a deeper statistical mechanical reality was spectacularly confirmed [26] when Hawking laid the foundations of quantum black hole theory.)

It was the substantial theoretical framework built up in the way during the "classical" phase that decisively confirmed the crucial importance of the equilibrium state problem on which the present article is focussed. Returning to this more specialised topic, I would start by recalling that – as well as the provision of the first convincing demonstration that (contrary to what Israel [25] had at first been inclined to suspect) a black hole equilibrium state can indeed be stable – a noteworthy byproduct of Vishveshwara's epoch making paper [1] was its analysis of stationary as well as dynamical perturbations, which provided evidence favorable to my uniqueness conjecture [3, 4] in the form of a restricted "no hair" theorem to the effect that the only stationary pure vacuum generalisations obtainable from a Schwarzschild black hole by infinitesimal parameter variations are those of the Kerr family (in which relevant small parameter is the angular momentum J).

Encouraged by Vishveshwara's confirmation [1] of the importance of the problem, I undertook the first systematic attempt [7, 10] at verification of my uniqueness conjecture for the $a^2 \leq M^2$ Kerr solutions as stationary black hole states. For the sake of mathematical simplicity I restricted my attention to cases characterised by spherical topology and axial symmetry, conditions that could plausibly be guessed to be mathematically necessary in any case. I also ruled out consideration of conceivable cases in which closed timelike or null lines occur outside the black hole horizon (not just inside as in the Kerr case [4]), a condition that is evidently natural on physical grounds, but that is not at all obviously justifiable as a mathematically necessity.

Within this framework I was able in 1971 to make two decisive steps forward, at least for the generic case for which there is a non zero value of the decay parameter κ which (in accordance with the "zeroth" law of black hole thermodynamics [59]) must always be constant over the horizon in a stationary state. The first of these steps [7] was the reduction of the four dimensional vacuum black hole equilibrium problem to a two dimensional non-linear elliptic boundary problem, for which the relevant boundary conditions involve just two free parameters: the outer boundary conditions depend just on the mass M and the inner boundary conditions depend just on the horizon scale parameter c, which is proportional to the product of the decay parameter κ with the horizon area $\mathcal{A}$. The precise specification of this parameter c (originally denoted by the letter b, and commonly denoted in more recent literature by the alternative letter μ) is given by the definition $c = \kappa\mathcal{A}/4\pi$, and its value in the particular case of the Kerr black holes is given the formula $c = \sqrt{M^2 - a^2}$ with $a = J/M$.

The second decisive step obtained in 1971 was the demonstration [7] that the two dimensional boundary problem provided by the first step is subject to a "no hair" (i.e. no bifurcation) theorem to the effect that within a continuously differentiable family of solutions (such as the Kerr family) variation between neighbouring members is fully determined just by the corresponding variation of the pair of boundary value parameters, i.e. the solutions belong to disjoint 2-parameter families in which the individual members are fully specified just by the relevant values of M and c. The only known example of such a family was the Kerr solution, which of course includes the only spherical limit case, namely that of Schwarzschild. The theorem therefore implied that if, contrary to my conjecture, some other non-Kerr family of solutions existed after all, then it would have the strange property of being unable to be continuously varied to a non-rotating spherical limit. On the basis of experience with other equilibrium problems this strongly suggested that, even if other families did exist, they would be unstable and therefore physically irrelevant, unlike the Kerr solutions which, by Vishveshwara's work [1] were already known to be stable at least in the neighbourhood of the non-rotating limit.

Having drawn the conclusion from this plausible but debatable line of argument that for practical astrophysical purposes a pure vacuum black hole equilibrium could indeed be safely presumed to be described by a Kerr solution, I turned my attention to the problem of generalising this argument from the pure vacuum to the electrovac case. In the degenerate ($\kappa = 0$) case, it had been pointed out by Hartle and Hawking [42] that the Kerr-Newman [43] family did not provide the most general equilibrium solution, due to the existence of counter-examples provided by the Papapetrou-Majumdar solutions, but it remains plausible to conjecture that the most general non-degenerate solutions are indeed provided by the Kerr-Newman family (whose simple spherical limit is the Reissner-Nordstrom solution). The electrovac generalisation of the first step of my 1971 argument [7] turned out to be obtainable without much difficulty [10], the only difference being that the ensuing two-dimensional non-linear boundary problem now involved two extra parameters representing electric charge and magnetic monopole. As in the pure vacuum case, an essential trick was the use of a modified Ernst [64] transformation based on the *axial* Killing vector (instead of the usual time translation generator) so as to obtain a variational formulation for which – assuming *causality* – the action would be *positive definite*. The second step was more difficult: I did not succeed in constructing a suitable electrovac generalisation of the divergence identity – with a rather complicated but (like the action) *positive definite* right hand side – that had enabled me to establish the pure vacuum "no hair" theorem [7] for axisymmetric black holes in 1971, but an electrovac identity of the required – though even more complicated – form was finally obtained by Robinson [12] in 1974.

While this work on the electromagnetic generalisation was going on, a deeper investigation of the underlying assumptions was initiated by Hawking [8, 9, 11], who made very important progress towards confirmation of the supposition that the topology would be spherical, and that the geometry would be axisymmetric. The latter was achieved by what I call the "strong rigidity" theorem, which was originally advertised [11] as a demonstration that – assuming analyticity – the black hole equilibrium states would indeed have to be axisymmetric (and hence by my earlier "weak rigidity" theorem [56] *uniformly* rotating) except in the static case, which in the absence of external matter was already known – from the recent completion [65, 66] of the program initiated by Israel [6, 36] – to be not just axisymmetric but geometrically (not just topologically) spherical.

The claim to have adequately confirmed the property of axisymmetry [11] was however one of several exaggerations and overstatements that were too hastily put forward during that exciting "classical" period of breathlessly rapid progress. In reality, all that was mathematically established by the "strong rigidity" theorem was just that in the non axisymmetric case the equilibrium state would have to be non-rotating relatively to the asymptotic rest frame. The argument to the effect that this implied staticity depended on Hawking's generalisation [8] of the original Lichnerowicz [67] staticity theorem, which in turn assumed the absence of an "ergosphere" outside the (stationary non-rotating) horizon – a litigious supposition whose purported justification in the non-rotating case was based on heuristic considerations [11] that have since been recognised to be fundamentally misleading, due to the existence of counter-examples. A satisfactory demonstration that the non-rotating case must after all be static was not obtained until the comparatively recent development [20, 21, 22] (described in next section) of a new and much more effective approach that was initiated by Wald in the more serene "post classical" era.

A similarly overhasty announcement of my own during the hurry of the "classical phase" was the claim [10] to have obtained an electromagnetic generalisation of Hawking's Lichnerowicz type of staticity theorem using just the same litigious assumption (which fails anyway for the rotating case) of the absence of an ergosphere, i.e. strict positivity, $V > 0$, of the effective gravitational potential defined as the norm, $V = -k^\mu k_\mu$, of the stationarity generator. What was shown later by a more careful analysis [19] was that (after correction of a sign error in the original version [10]) an even stronger and more highly litigious inequality was in fact required – until it was made finally redundant by the more effective treatment recently developed by Wald and his associates [20, 21, 22]. (In dealing with the related "circularity" theorem, on which the treatment [7, 10] of the stationary case depends, I was more fortunate: my electromagnetic generalisation [70] of Papapetrou's pure vacuum prototype [71] has stood the test of time).

Among the other noteworthy overstatements from the hastily progressing "classical" period, a particularly relevant example is Wald's own premature claim [68] (based on what turns out to have been an essentially circular argument) to have gone beyond my 1971 "no hair" theorem [7] to get a more powerful uniqueness theorem of the kind that was not genuinely obtained until Robinson's 1975 generalisation from infinitesimal to finite differences of the divergence identity I had used. In achieving this ultimate *tour de force* Robinson [13] effectively strengthened the "no hair" theorem to a complete uniqueness theorem, thereby definitively excluding the – until then conceivable – existence of a presumably unstable non-Kerr branch of topologically spherical axisymmetric causally well behaved black hole solution.

Having already succeeded in generalising my original infinitesimal divergence identity [7] from the pure vacuum to the electrovac case [12], Robinson went on to try to find an analogous generalisation of his more powerful finite difference divergence identity [13] from the pure vacuum to the electrovac case. However this turned out to be too difficult, even for him, at least by the unsystematic, trial and error, search strategy that he and I had been using until then. As I guessed in a subsequent review [62], there was "a deep but essentially simple reason why the identities found so far should exist" and "the generalisation required to tie up the problem completely will not be constructed until *after* the discovery of such an explanation, which would presumably show one how to construct the required identities *directly*". It was only at a later stage, in the "post classical" period that, as Heusler [28] put it " this prediction was shown to be true" when, on the basis of a deeper understanding, such direct construction methods were indeed obtained by Mazur [17, 18] and, independently, using a different (less specialised) approach, by Bunting [15, 16].

4. The post-classical phase (since 1975)

Robinson's 1975 discovery of the finite-difference divergence identity [13] marked the end of what I call the "classical phase", whose focal event had been the 1972 Les Houches summer school [9, 10, 41] at which all the various aspects of black hole theory were assembled and treated together for the first and probably the last time. Since 1975 the subject has split into mutually non-interacting branches. On one hand there has been the new subject of quantum black hole theory: the discovery of the Hawking effect [26] aroused interest in the possible occurrence in the early universe of microscopic black holes for which such effects might be important, and this in turn lead to an interest into the conceivable effects (e.g. as potential contributors of black hole "hair") of various kinds of exotic (e.g. Yang Mill or dilatonic) fields that might have been relevant then. On the other hand, eschewing such rather wild speculations in favor of what more obviously exists in the real world, astrophysicists have been mainly interested in macroscopic black holes (of stellar mass and upwards) for which the only relevant long range interaction fields are still believed to be just gravitation and electromagnetism, but for which local effects of accreting plasma can create spectacular effects that are thought to be responsible for many observed phenomena ranging in scale and distance from quasars down to galactic X-ray sources such as Cygnus X-1.

After the development of these disconnected branches, work on the pure vacuum black hole equilibrium problem and its electrovac generalisation proceeded rather slowly. Starting with Bekenstein [69], most quantum black hole theorists were more concerned about generalising the problem to hypothetical fields of various new (e.g. dilatonic) types, whereas most astrophysical black hole theorists were concerned just with accreting matter that could be treated as a small perturbation on a pure vacuum background, whose equilibrium states they supposed to have been definitely established to consist just of the relevant, ($a \leq M$) subfamily. Only a handful of mathematically oriented theorists remained acutely aware that the definitive establishment of this naive supposition was still not complete. Another reason why progress in the theory of vacuum equilibrium states slowed down in the "post classical phase" was that the problems that had been solved in the "classical phase" had of course tended to be those that were the easiest.

In so far as the equilibrium problem is concerned, the most salient developments in the earlier "post classical" years were the completion referred to above by Mazur [17, 18] and Bunting [15, 16] of my work [7, 10] and Robinson's [12, 13] on the axisymmetric case, and the completion and streamlining [14, 72, 73] by Robinson, Simon, Bunting and Massood-ul-alam of the work [6, 36, 65, 66] initiated by Israel on the strictly static case.

Unlike the works just cited, which built upwards from the (not always entirely reliable) basis established in the classical period [7, 8, 10, 11, 65, 66], a more recent resurgence of activity [20, 21, 22, 23, 27, 74] – initiated by Wald and continued most recently by Chrusciel – has been more concerned with treating the shaky elements in the foundations of that underlying basis itself. This work has successfully closed an outstanding loophole in the previous line of argument by developing a new and more powerful kind of staticity theorem [20, 22] for "non-rotating" black holes: instead of the litigious assumption of a lower bound on V (which was needed in the now obsolete Hawking-Lichnerowicz approach) the new theorem depends on the justifiable [23] requirement that there exists a slicing by a maximal (space-like) hypersurface. It has thereby been possible to provide [21, 27, 74] a much more satisfactorily complete demonstration of what had been rather overconfidently asserted by Hawking and Ellis [11], namely that, subject to the assumptions of analyticity and of connectedness and non-degeneracy ($c \neq 0$) of the horizon, the black hole equilibrium state has to be axisymmetric or static.

5. What remains for the future?

As Chrusciel has emphasised [74], although many of the (declared and hidden) assumptions involved in the work during the "classical phase" have been disposed of, the more recent work on the black hole equilibrium problem is still subject to several important technical restrictions whose treatment remains as a challenge for the future.

One whose treatment should I think be given priority at this stage is the assumption of *analyticity* that has been invoked in all the work on the indispensible "strong rigidity" theorem that is needed to establish axisymmetry. It is to be remarked that if, as in my early work [7], axisymmetry is simply postulated at the outset, then analyticity will be demonstrable as an automatic consequence of the ellipticity of the differential system that is obtained as a result of the "circularity" property that is established by the generalised Papapetrou theorem [7, 70]. What I would guess is that it should be possible (and probably not more difficult than the other steps that have already been achieved) to prove the necessity of analyticity for a vacuum equilibrium state *without* assuming axisymmetry.

A more delicate question that remains to be settled is the possibility of equilibrium involving *several disconnected black holes.* As far as the pure vacuum problem is concerned, my conjecture is that such multi black hole solutions do not exist, but they have so far been rigorously excluded only in the strictly static case [73]. The axisymmetric case has recently been studied in some detail [75, 76] by Weinstein (who denotes the horizon scale parameter $c = \kappa\mathcal{A}/4\pi$ by the letter μ) but a definitive conclusion has not yet emerged. In the electrovac case the situation is certainly more complicated, since it is known [42] that there are counter-examples in which gravitational attraction is balanced by electrostatic repulsion. It is however to be noticed that the only counter-examples discovered so far, namely those of the Papapetrou-Majumdar family [44, 45] have horizons that are degenerate (in the sense of having a vanishing decay constant κ). It seems reasonable to conjecture that even in the electrovac case there are no non-degenerate multi black hole equilibrium states.

This last point leads on to a third major problem that still needs to be dealt with, namely the general treatment of the *degenerate* ($\kappa = 0$) case. The maximally rotating ($a = M$) Kerr solution is still the only known pure vacuum example, and I am still inclined to conjecture that it is unique, but the problem of proving this remains entirely unsolved. As far as the electrovac problem is concerned, the only known examples are those of the Kerr-Newman [43] and Papapetrou-Majumdar [44, 45] families. Recent progress by Heusler [78] has confirmed that the latter (whose equilibrium saturates a Bogomolny type mass limit [79]) are the only strictly static examples, but for the rotating degenerate case the problem remains wide open.

I wish to conclude by drawing attention to another deeper problem that, unlike the three referred to above, has been largely overlooked even by the experts in the field, but that seems to me just as interesting from a purely mathematical point of view, even if its physical relevance is less evident. This fourth problem, is that of solving the black hole equilibrium state problem without invoking the *causality axiom* on which nearly all the work described above depends (e.g. for obtaining the required positivity in the successive divergence identities [7, 12, 13, 15, 17, 18, 16] used in the axisymmetric case). As remarked in the introduction, all non static Kerr solutions contain closed timelike lines, though in the black hole subfamily with $J^2 \leq M^4$ they are entirely confined inside the horizon [4, 34]. Unlike analyticity, whose failure in shock type phenomena is physically familiar in many contexts, causality – meaning the absence of closed timelike lines – is a requirement that most physicists would be prepared to take for granted as an indispensible requirement for realism in any classical field model. However the example of sphalerons suggests that despite their unacceptability at a classical level, the mathematical existence of stationary black hole states with closed timelike lines outside the horizon might have physically relevant implications in quantum theory. The discovery of such exotic configurations would be a surprise to most of us, but would not contradict any theorem obtained so far. All that can be confidently asserted at this stage is that such configurations could not be static but would have to be of rotating type.

References

1. Vishveshwara, C.V. (1970), *Phys. Rev.* **D1**, 2870-79.
2. Carter, B. (1997), *Has the black hole equilibrium problem been solved?*, in *Proc. 8th Marcel Grossmann Meeting*, (Hebrew University, Jerusalem). [gr-qc/9712038]
3. Carter, B. (1967), *Stationary Axisymmetric Systems in General Relativity*, (Ph.D. Thesis, DAMTP, Cambridge).
4. Carter, B. (1968), *Phys. Rev.*, **174**, 1559-71.
5. Kerr, R.P. (1963), *Phys. Rev. Lett.*, **11**, 237-38.
6. Israel, W. (1967), *Phys.Rev.*, **164**, 1776-79.
7. Carter, B. (1971), *Phys. Rev. Letters*, **26**, 331-33.
8. Hawking, S.W. (1972), *Commun. Math. Phys.*, **25**, 152-56.
9. Hawking, S.W. (1973), in *Black Holes*, (Proc. 1972 Les Houches Summer School), ed. B. & C. DeWitt, 1-55 (Gordon and Breach, New York).
10. Carter, B (1973), in *Black Holes* (Proc. 1972 Les Houches Summer School), ed. B. & C. DeWitt, 125-210 (Gordon and Breach, New York).
11. Hawking, S.W., Ellis, G.F.R. (1973) *The Large Scale Structure of Space Time*, (Cambridge U.P., 1973).
12. Robinson, D.C. (1974), *Phys, Rev.*, **D10**, 458-60.
13. Robinson, D.C. (1975), *Phys, Rev. Lett.*, **34**, 905-06.
14. Robinson, D.C. (1977), *Gen. Rel. Grav.*, **8**, 695-98.
15. Bunting, G. (1983) *Proof of the Uniqueness Conjecture for Black Holes*, (Ph.D. Thesis, University of New England, Armadale N.S.W.).
16. Carter, B. (1985), *Commun. Math. Phys.*, **99**, 563-91.
17. Mazur, P.O. (1982), *J. Phys.*, **A15**, 3173-80.
18. Mazur, P.O. (1984), *Gen. Rel Grav.*, **16**, 211-15.
19. Carter, B. (1987), in *Gravitation in Astrophysics*, (NATO ASI **B156**, Cargèse, 1986), ed. B. Carter & J.B. Hartle, 63-122 (Plenum Press, New York).
20. Sudarsky,D., Wald, R.M. (1991), *Phys. Rev.*, **D46**, 1453-74.
21. Chrusciel, P.T., Wald, R.M. (1994), *Comm. Math. Phys.*, **163**, 561-604.

22. Sudarsky, D., Wald, R.M. (1993), *Phys. Rev.*, **D47**, 5209-13. [gr-qc/9305023]
23. Chrusciel, P.T., Wald, R.M. (1994), *Class. Quantum Grav.*, **11**, L147-52.
24. Thorne, K.S. (1994), *Black holes and time warps* (Norton, New York).
25. Israel, W. (1987), in *300 years of gravitation*, ed. S.W. Hawking, W. Israel, 199-276.
26. Hawking, S.W. (1975), *Comm. Math. Phys.*, **43**, 199-220.
27. Chrusciel, P.T. (1994), in *Differential Geometry and Mathematical Physics*, **170**, ed J. Beem, K.L. Dugal, 23-49 (American Math. Soc., Providence). [gr-qc/9402032]
28. Heusler, M. (1996), *Black hole uniqueness theorems* (Cambridge U.P.)
29. Schwarzschild, K. (1916), *Sitzber. Deut. Akad. Wiss. Berlin Kl. Math-Phys. Tech.*, 189-96.
30. Eisenstaedt (1982), J.*Arch. Hist. Exact Sci.*, **27**, 157-228.
31. Zel'dovich, Y.B., Novikov, I.D. (1967), *Relativistic Astrophysics* (Izdatel'stvo "Nauka", Moscow); English version ed. K.S Thorne, W.D. Arnett (University of Chicago Press, 1971).
32. Harrison, B.K., Thorne, K.S., Wakano, M., Wheeler, J.A (1965), *Gravitation theory and gravitational collapse*, (University of Chicago Press).
33. Penrose, R. (1965), *Phys. Rev. Letters*, **14**, 57-59.
34. Carter, B. (1978), *Gen. Rel. Grav.*, **11**, 437-50.
35. Vishveshwara, C.V. (1968), *J. Math. Phys.*, **9**, 1319-22.
36. Israel, W. (1968), *Commun. Math. Phys.*, **8**, 245-60.
37. Boyer, R.H., Price, T.G. (1965), *Proc. Camb. Phil. Soc.*, **61**, 531-34.
38. Boyer,R.H., Lindquist, R.W.(1967), *J. Math. Phys.*, **8**, 265-81.
39. Boyer, R.H. (1969), *Proc. Roy. Soc. Lond*, **A311**, 245-52.
40. Carter, B., *Phys. Rev.*, **141**, 1242-47.
41. Bardeen, J. (1973), in *Black Holes*, (Proc. 1972 Les Houches Summer School), ed. B. & C. DeWitt, 241-89 (Gordon and Breach, New York).
42. Hartle, J.B., Hawking, S.W. (1972), *Commun. Math. Phys.*, **26**, 87-101.
43. Newman, E.T., Couch, E., Chinnapared, K., Exton, A., Prakash, A., Torrence, R. (1965), *J. Math. Phys.*, **6** 918-19.
44. Papapetrou, A. (1945), *Proc. R. Irish Acad.*, **A51**, 191-204.
45. Majumdar, S.D., (1947), *Phys. Rev.*, **72**, 930-98.
46. Penrose, R. (1969), *Riv. Nuovo Cimento I*, **1**, 252-76.
47. Eardley, D. (1987), in *Gravitation in Astrophysics*, (NATO ASI **B156**, Cargèse, 1986), ed. B. Carter & J.B. Hartle, 229-35 (Plenum Press, New York).
48. Price, R.H. (1972), *Phys. Rev.*, **D5**, 2419-54.
49. Teukolsky, S.A. (1972), *Phys. Rev. Lett.*, **29**, 1114-18.
50. Teukolsky, S.A. (1973), *Astroph. J.*, **185**, 635.
51. Press, W.H., Teukolsky, S.A. (1963), *Astroph. J.*, **185**, 649.
52. Kay, B.S., Wald, R.M. (1987), *Class. Quantum Grav.*, **4**, 893-98.
53. Whiting, B. (1989), *J. Math. Phys.*, **30**, 1301-05.
54. Christodoulou, D. (1970), *Phys. Rev. Lett.*, **25**, 1596-97.
55. Christodoulou, D., Ruffini, R. (1971), *Phys. Rev.*, **D4**, 3552-55.
56. Carter, B. (1972), *Nature*, **238**, *Physical Science*, 71-72.
57. Hawking, S.W., Hartle, J.B. (1972), *Commun. Math. Phys.*, **27**, 283-90.
58. Hartle, J.B. (1973), *Phys. Rev.*, **D8**, 1010-24,
59. Bardeen, J., Carter, B., Hawking, S.W. (1973), *Comm. Math. Phys.*, **31**, 161-70.
60. Znajek, R.L. (1978), *Mon. Not. R. Astr. Soc.*, **182**, 639-46.
61. Damour, T., *Phys. Rev.*, **D18**, 3598-604.
62. Carter, B. (1979), in *General Relativity, an Einstein centenary survey*, ed. S.W. Hawking, W. Israel, 294-369 (Cambridge U.P.)
63. Bekenstein, J.D. (1973), *Phys. Rev.*, **D7**, 949-53.
64. Ernst, F.J. (1968), *Phys. Rev.*, **167**, 1175-78 and **168**, 1415-17.
65. Muller zum Hagen, H., Robinson, D.C., Seifert, H.J. (1973), *Gen. Rel. Grav.*, **4**, 53-78.

66. Muller zum Hagen, H., Robinson, D.C., Seifert, H.J. (1974), *Gen. Rel. Grav.*, **5**, 61-72.
67. Lichnerowicz, A. (1955), *Théories relativistes de la gravitation et de l'electro - magnétisme* (Masson, Paris).
68. Wald, R.M. (1971), *Phys. Rev. Lett.*, **26**, 1653-55.
69. Bekenstein, J.D. (1996), *preprint* (Hebrew University, Jerusalem). [gr-qc/9605059]
70. Carter, B. (1969), *J. Math. Phys.*, **10**, 1559-71.
71. Papapetrou, A. (1969), *Ann. Inst. H. Poincaré*, **4**, 83-85.
72. Simon, W. (1985), *Gen. Rel. Grav.*, **17**, 761-68.
73. Bunting, G.L., Massood-ul-Alam, A.K.M. (1987), *Gen. Rel. Grav.*, **19**, 147-54.
74. Chrusciel, P.T. (1996), *Helv. Phys. Acta*, **69**, 529-52.
75. Weinstein, G. (1990), *Comm. Pure. Appl. Math.*, **XLV** 1183-203.
76. Weinstein, G. (1994), *Trans. Amer. Math. Soc.*, **343**, 899-906.
77. Weinstein, G. (1996), *Comm. Part. Diff. Eq.*, in press. [gr-qc/9412036]
78. Heusler, M. (1996), *preprint* (University of Zurich). [gr-qc/9607001]
79. Gibbons, G.W., Hawking, S.W., Horowicz, G.T., Perry, M.J. (1983), *Comm. Math. Phys.*, **88**, 259-308.

2. STABILITY OF BLACK HOLES

Including reflections on the study of perturbations of black hole spacetimes

BERNARD F. WHITING
Department of Physics, University of Florida
Gainesville Fl 32611-8440, U.S.A.

1. Student's Guide

There is often something to be learned from a look back in history. From a current point of view, most would argue that our methods have become more sophisticated over the years. I would contend, and as you will surely see here, it is our problems which have become more sophisticated over the years, and our methods have changed so that we can continue to answer them. I doubt if this trend will alter substantially. Rather, I suggest that new challenges will keep rising to the fore. Let me give a recent, specific example[1].

Vishveshwara's original discussion of stability showed that there was no superficial case establishing instability, basically by dealing with single modes and by demonstrating the positivity of effective potentials. Recently it has been shown that an argument to establish *pointwise boundedness* required the use of some refined tools, resulting in a method which differs markedly in substance but not at all in essence from a relatively simple positive potential approach. In the intervening time there was a whole discussion of the relevance and methodology of gauge invariance, which really figured in neither of these two works. Now that at least one substantial question of stability is behind us, it turns out that the issue of gauge invariance is surfacing once more, and this time for real computations, being carried out with a view to being compared to real data in the not too distant future. But despite the extremely sophisticated methods employed in the stability analysis, in fact because of them, some rather difficult steps remain before we have a routine way of relating the degrees of freedom used in proofs with the degrees of freedom used in computations, let alone comparing computations with observation.

[1]For references, see later Sections.

B. R. Iyer and B. Bhawal (eds.), Black Holes, Gravitational Radiation and the Universe, 17–31.

So there is a clear challenge which is new. I would like to give here also an outlook which is novel - and hopefully fresh! It is often for their breadth that we have admired earlier contributers to our field. This is still a quality which, even today, we should not underestimate. Everyone should be aware that not all the tools used, nor all the influences felt, come entirely from within the domains of Physics or Mathematics. What you see, what you think, and how you do what you do, all matter. Frequently, it is unusual questions, out of the ordinary, and especially those which tantalize, which help maintain our interest long enough to lead to progress. In fact, I should like to stress that diverse influences play a significant rôle in providing the necessary inspiration. Hopefully, this knowledge will tempt others to resist that other temptation, to become too narrowly defined, and especially too soon. The cultivation of breadth is probably more important than the details of any particular piece of work.

2. Introduction

What problem does the question of black hole stability really address? The study of black holes has become so extensive that now the question of black hole stability no longer refers to one, single problem. In this article, attention will be confined to spacetimes arising as classical solutions to Einstein's field equations for General Relativity in four dimensions, usually with vacuum (or electromagnetic) source. This context is already diverse enough that, in the end, more than one type of problem will be seen to be addressed.

There is, by now, a traditional body of literature associated with the question of black hole stability. Reference to this literature will show that, generally, a perturbation problem was being discussed, and the question addressed was whether or not solutions to the perturbation problem could become unboundedly large. Solutions of the perturbation problem might take values in the entire background spacetime of the unperturbed geometry but, for black holes, interest was often confined simply to the domain of outer communication. In fact, this brand of perturbation problem was usually only well defined for perturbative initial data posed on some spacelike (or perhaps null) hypersurface in such a domain. A more recent problem, on black hole thermodynamics, which is also discussed here, arises in the context of Euclidean (not Lorentzian) boundary value data.

For almost all physical situations not involving gravitation, local stability of an individual system occurs when its energy is at a minimum. In the context of General Relativity, an immediate difficulty is encountered, since generalized notions of energy have been extraordinarily difficult to formulate. Thus, there is an obvious question of how to prove stability for a black hole spacetime. Perhaps far more instructive for the student is attention to the question of how to establish a situation in which stability could actually be tested. To begin with, the formulation of the perturbation problem refers to all ten of the metric components. These are governed by the ten linearized Einstein equations, subject to the four additional conditions corresponding to the freedom to make coordinate changes. In order to arrive at equations for isolated degrees of freedom for which the question of stability could be tested, several important steps were required.

The existence of background isometries was an essential ingredient in the solution for each different spacetime. The choice of a gauge in which to pose the formulation often proved very important, as did a realization of which quantities were (locally) gauge independent, since they naturally represented useful quantities in which to pursue an unambiguous investigation of stability. Finally, there was the crucial, and not always possible, step of decoupling the equations for these gauge independent quantities, to the point where single degrees of freedom could be examined in isolation, and perhaps tractably (through, say, a separation of variables being possible). Whereas gauge invariance is an essential aspect conceptually, decoupling (or separating) is so only from a practical perspective.

In this paper, attention will focus on those questions concerning black hole stability for which it has been possible to provide a reasonable answer. Generally, the conventions used will be compatible with those of Chandrasekhar [1], which are similar to those of many of the references cited. Emphasis will be on a clear statement of the problems encountered and their method of solution, and excessive technical detail will be avoided. For Schwarzschild, the linearized problem could be said to be well in hand. For Kerr, beyond mode stability the field is still open. For Kerr-Newman, any entirely satisfactory result has proved to be remarkably elusive.

3. The Schwarzschild Black Hole

The gravitational stability of the Schwarzschild black hole has been investigated throughout a period spanning some thirty years. In their seminal article of 1957, Regge and Wheeler built on a method of tensor decomposition used earlier in the study of electromagnetic waves [2]. They began by working directly with the metric perturbations, and after introducing their now famous gauge choice [3], they reduced the system of perturbation equations to a remarkably simple, single radial equation for one of the independent, gravitational degrees of freedom (now known as the axial mode, of 'odd' parity). Subsequently, Vishveshwara [4] made a number of significant breakthroughs. He extended their work with a more comprehensive analysis of the boundary conditions by reference to the globally acceptable

(Kruskal) coordinates, and included discussion of the non-propagating degrees of freedom (*i.e.*, $\ell = 0$ and $\ell = 1$ time independent perturbations). More importantly, making further use of their gauge choice, he was able to analyse the other independent degree of freedom (the 'even' parity, or polar mode), even though its dependence was not isolated into one equation for a single variable. That result was obtained shortly later through the work of Zerilli [5]. The net consequence of these labours was that the Schwarzschild black hole had been shown to be gravitationally stable against individual mode perturbations, and the independent degrees of freedom (of the gravitational field) were each governed by a single, separable equation, which has become instrumental in determining mode behavior.

The quantities which satisfied the Regge-Wheeler and Zerilli equations had been determined in a specific (the so called Regge-Wheeler) gauge. How should the quantities which satisfy those equations be determined when one is working in some other gauge? Using a gauge invariant Hamiltonian approach, Moncrief [6] was able to answer that question by finding which gauge invariant quantities the Regge-Wheeler and Zerilli functions were equivalent to in their chosen gauge. He thus lead the way to an entirely more systematic approach to the analysis of black hole perturbations, both in the discussion of stability and beyond. The question of gauge invariance for a general problem of spherical symmetry (including matter) was subsequently addressed by Gerlach and Sengupta [7].

Apart from the direct metric tensor approach championed by Regge and Wheeler, there exists an entirely different method of analysing spacetimes themselves, and also their perturbations, known as the Newman-Penrose spin coefficient formalism [8]. Because of their geometric symmetries, all the familiar black hole spacetimes fall within a certain restricted class known as Petrov type D, which is particularly amenable to a very compact and powerful refinement of the Newman-Penrose formalism put forward by Geroch, Held and Penrose [9], and of practical utility in the study of black hole perturbations[2].

Within the Newman-Penrose formalism, and because of the underlying spacetime symmetries, locally gauge (and tetrad) invariant components of the Weyl tensor can always be identified (as spacetime scalars) and used in an alternative perturbation analysis. Taking advantage of this, Bardeen and Press [11] obtained a pair of decoupled equations, different from those of Regge-Wheeler and Zerilli, also governing gauge invariant quantities (of opposite helicity). Like the equations of Regge-Wheeler and Zerilli, they were each a separable wave equation, with an effective potential determined by the underlying geometry. It turned out that these equations were not

[2]An exhaustive study of the decoupling and separation of variables for perturbations on type D spacetimes was begun by Kamran [10]

immediately suitable for a stability analysis, but their existence was later capitalized upon by Teukolsky [12] while investigating perturbations of the Kerr black hole.

There was, at this point, a surfeit of equations and, in fact, of gauge invariant quantities, since the gravitational field should have only two independent, gauge invariant, propagating degrees of freedom. Clearly, there must exist relations between the various solutions of the different equations. Even before the publication of the Bardeen-Press equation, Price [13] had studied the metric perturbations, and shown a relationship between an imaginary part of the Weyl tensor in the Newman-Penrose formalism and a solution to the Regge Wheeler equation [14]. But explicit forms for differential transformations relating solutions for the gauge invariant metric and Weyl tensor (Newman-Penrose) perturbations were given first by Chandrasekhar [15], who subsequently obtained, with Detweiler [16], a relation between solutions to the Regge-Wheeler and Zerilli equations.

Even with solutions to one or other of these equations, and with relations between solutions to the different equations, at least one other practical problem remains. Given some specific solution, what is the corresponding complete metric perturbation (up to gauge freedom) for the original spacetime? Although a number of authors [17, 18, 19] addressed this question, the most elegant, if somewhat formal solution to this problem was given by Wald [20]. Very recent work [21] has now found an explicit, concrete relation between initial data for metric perturbations and initial data for the Bardeen-Press equation in gauge invariant terms, and is currently being used to investigate a similar result for the perturbations of Kerr.

3.1. BEYOND MODE STABILITY

So far, discussion has been limited to the question of mode stability, which addresses behaviour exponentially in time. Energy integral arguments rule out exponential growth, and establish that the integral for a solution over a spacelike hypersurface is bounded by its value coming from the initial data. But study of the perturbations of the Schwarzschild spacetime did not rest with mode stability. There was seen a need to go beyond mode stability and to investigate pointwise boundedness for all future times. Efforts in this direction were begun by Wald [22]. and the most complete results to date were eventually obtained by Kay and Wald [23].

Wald [22] first proved pointwise boundedness of the individual terms in the spherical harmonic decomposition of gravitational perturbations with regular (*i.e.*, smooth, bounded), compact initial data (with support strictly outside the horizon). Using the spectral representation available for a self-adjoint operator, he proved stability against a global, power law (*i.e.*, $t^{1-\epsilon}$)

behaviour for finite linear sums of angular modes. Then, employing essentially a Sobelev inequality, he also eliminated the possibility of localized integrable singularities by proving pointwise boundedness of solutions for regular initial data, if radially compact away from the horizon. Dimock and Kay [24], in work on scalar fields, extended this result by removing the need for an angular decomposition.

In subsequent joint work, Kay and Wald [23], again working with scalar fields, removed the need for initial data to have support away from the horizon. They first showed an intermediate result: compact, bounded initial data, which simply vanished on the bifurcation 2-sphere, would remain well behaved, even where it crossed the future horizon. Then, by a cunning argument, they showed that, on some future spacelike hypersurface, the evolution of well behaved data which did not vanish initially on the bifurcation 2-sphere would, on and outside the future horizon, be the same as the evolution of certain, prescribed, good data which did vanish at the bifurcation 2-sphere on the initial hypersurface. The future evolution, on and outside the future horizon, would then be identical for the two data sets and would, therefore, be bounded everywhere on the exterior domain and its boundary. Through a further extension of their argument, they could then include geometries in which black holes formed by collapse. Collectively, these are currently the best results available for spherical symmetry, and they can be extended to higher spin perturbations without obstacle.

4. Charged Black Holes

It might be expected that the Kerr geometry, which is also a vacuum spacetime would be treated next, rather than the Reissner-Nordström black hole, which has an electromagnetic source. But the requisite method of treatment turns out to be quite different for Kerr, whereas that for Reissner-Nordström is similar in principle to what can be done for Schwarzschild, even if it has some new and physically interesting twist of its own. In fact, there is practically no astrophysical argument for the relevance of charged black holes, and studies [25] of quantum processes near black holes do little to change this, since the charge on a black hole will generally decay much more quickly than its mass. Nevertheless, particular physical interest stems from the fact that, in the coupled Einstein-Maxwell theory with non-zero background sources, incoming pure electromagnetic (or gravitational) radiation will, in general, excite outgoing radiation of *both* the electromagnetic and gravitational fields.

Following his earlier work for Schwarzschild, Moncrief also investigated the perturbations and stability of the Reissner-Nordström black hole [26], again from a Hamiltonian approach. He was able to establish mode stability, directly in terms of the metric and electromagnetic field perturbations. Shortly afterwards, Chitre [27] put forward an alternative description for the perturbations using the Newman-Penrose formalism. Although separation of variables and decoupling independent degrees of freedom was again possible for the charged, spherically symmetric black hole, what appears, so far, to be essential in this process is that the separation of variables must take place first, before the decoupling can proceed. Even then, the finally decoupled degrees of freedom each represent a mixture of the spin -1 and spin-2 fields initially used to set up the problem. This is the origin of the coupling between the electromagnetic and gravitational perturbations.

Whereas the methods of mode analysis were quite similar for Reissner-Nordström and for Schwarzschild, neither these methods, nor those used for the Kerr geometry have been found applicable to the charged, rotating Kerr-Newman black hole. The problem seems to be that the decoupling methods are so totally different in the charged, spherical and uncharged, axially symmetric cases as to be apparently incompatible in a spacetime which combines charge with axial symmetry. From our present perspective, lack of detailed knowledge about the angular eigenfunctions for the Kerr perturbations seems to be inhibitive, but Lee [28] has managed to analyse the coupled system in some detail, using the Newman-Penrose formalism. Charge introduces a non-linear dependence on the separation constant, as well as a modification of the singular point structure of the separated radial equation, so that the methods of the next section are not even applicable to the charged, non-rotating (spherical) black hole. However, there is some cause for hope yet, for although Moncrief did use the separation of variables in his Hamiltonian formulation, as for Lee, complete decoupling of the perturbed field equations was not necessary.

5. The Kerr Black Hole

Perhaps the most obvious question which first springs to mind is why it should be that the Kerr case is so different from that of Schwarzschild. The most immediate answer would be "symmetry", but the details of how that makes a difference is perhaps more illuminating. All three of the first works on the Schwarzschild problem [3, 4, 5] require explicit knowledge of the angular eigenfunctions on the sphere in order to use their tensor decomposition[3]. In the Kerr case, that essential spherical symmetry does not apply, and the symmetry which does remain seems to be of limited advantage.

[3]In fact, it was only within the last week that a report appeared [29] explicitly demonstrating the precise sense in which the decoupling of gauge invariant metric perturbations could be carried out in Schwarzschild without the prior separation of variables.

The long delay between the introduction of the Schwarzschild solution [30] and Kerr's discovery of an axisymmetric solution [31] to the vacuum Einstein equations is quite characteristic of the difficulties generally experienced in working with the Kerr geometry. Until the subsequent demonstration by Carter [32] that both the Hamilton-Jacobi equation for a free particle and the Klein-Gordon equation for a scalar field were separable, the Kerr geometry retained an air of relative intractability. However, Teukolsky's masterful extraction [33], using the Newman-Penrose formalism, of separable equations governing the gauge invariant perturbations for massless fields of spin half, one and two, all at once, suddenly gave rise to a general sense of euphoria. This was all the more heightened because of the degree of astrophysical interest which the Kerr geometry aroused: almost all astronomical objects rotate, some extremely rapidly. For the first time it become possible to examine vacuum spacetimes which were not just a small perturbation away from the Schwarzschild solution. There was also hope that the spacetime outside rotating astrophysical objects might similarly be more fully investigated.

It was not just the manner of obtaining equations which was different from the Schwarzschild case. For Kerr, there was no direct analogue of the metric perturbation treatment nor, correspondingly, the Regge Wheeler or Zerilli equations. However, as Vishveshwara had done for Schwarzschild, so by a detailed analysis [34], a precise description of the physically correct boundary conditions was obtained. These papers also contained the construction of the so called Teukolsky-Starobinsky transformations, relating solutions for the positive and negative helicity states. Wald [35] not only demonstrated equivalence between perturbations constructed from solutions of different helicities, but also gave a concise, though rather elusive, construction [20] for obtaining the complete perturbation corresponding to any one solution of Teukolsky's master equation.

The first suggestions that the Kerr black hole might also be stable came from the fact that numerical computations could not excite any instability. Furthermore, separability of the Teukolsky equations meant that mode stability could be examined in terms of the behaviour of modes for complex frequency. Explicit attempts to track complex modes which might cross the real axis to become unstable proved fruitless. In fact, it was shown [36] that instability could not occur *unless* the crossing of the real frequency axis took place in the superradiant regime[4], but no way was found to induce such a crossing. So, it was an accumulation of numerical results, combined with inconclusive analytic calculations, and knowledge of the Schwarzschild sta-

[4]Unexpectedly, for rotating black holes, there was a parameter regime in which integer spin perturbations could scatter with reflection amplitude greater than unity, which is known as the phenomenon of superradiance.

bility, which, for many years, lead to an awareness that the Kerr geometry might be stable, rather than a definitive proof.

Even for scalar fields on the Kerr geometry, a proof of mode stability is complicated by the existence of the ergoregion, inside which non-rotational timelike geodesic motion becomes impossible. Stability of axisymmetric scalar perturbations on Kerr can be shown by the methods used to demonstrate Schwarzschild stability. But for non-axisymmetric scalar perturbations the 'energy' integrand for the scalar field is not manifestly positive definite everywhere, and a new method is required. For non-scalar fields there is the added difficulty that the Teukolsky equations had complex-valued coefficients. Moreover, no transformations were known to functions analogous to the solutions of the Regge-Wheeler (or Zerilli) equations. Chandrasekhar and Detweiler [37] had obtained transformations which, in the Schwarzschild limit, were exactly the transformations from the Bardeen-Press to the Regge-Wheeler solutions, but in general they involved functions satisfying equations with a highly non-linear dependence on the unknown angular eigenvalue, which hindered further progress.

Two separate but related observations [38] were able to change this apparent stalemate[5]. The first came from an examination of the relation between the angular eigenfunctions for positive and negative helicity states, for each spin [34], and resulted in a demonstration that other similar transformations always existed. They each preserved the number and type of the singular points of the differential equations and never lead to any essential change in the angular eigenvalues, while generally resulting in new equations from which to determine the angular dependence of perturbations on the Kerr geometry. Up to an interchange under $\cos(\theta) \to \cos(\theta - \pi)$, there was effectively only one new independent equation introduced by these alternative transformations, with solution[5]

$$
{}^{m}T_{ls}(\theta) \propto \left[\sin\theta\left(\operatorname{sign}[s-m]\frac{\partial}{\partial\cos(\theta)} + a\omega + \frac{s + m\cos\theta}{\sin^2\theta}\right)\right]^{|s-m|} {}_{s}S^{m}{}_{l}(\theta),
$$

which has been found useful in the development of an analytic proof of mode stability for perturbations on the Kerr geometry.

The second observation concerned the construction of integral equations and, more generally, integral transformations, particularly with relation to radial solutions. Persides [40] had indicated that an integral equation (with a Laplace-like kernel) could be obtained for scalar perturbations on the Schwarzschild spacetime. Although other integral equations for black hole

[5]The remaining treatment in this section follows very closely that of Ref.[39].

[6]Boyer-Lindquist coordinates are used, with the notation adapted to that of Teukolsky for the angular eigenfunctions.

perturbations could be found by a generalization of his method [41], it has also been possible to demonstrate that an application of his original method exists, for any spin, in which integral transformations relate solutions of Teukolsky's equation to functions which again satisfy equations preserving the number and type of the singular points, and with no essential modification, in the new equations, to the linear occurrence of the constant of separation. For the Schwarzschild spacetime, these integral transformations related solutions of the Bardeen-Press equation to solutions of the Regge-Wheeler equation, just as Chandrasekhar [15] had been able to do with differential transformations. The general manner in which differential and integral transformations could lead to identical equations is not yet fully explored.

The form of the integral transformations, when written in terms of the original variables, is quite complicated and not very illuminating. However the general properties of the transformations can be easily described and are simple to understand. For the particular application of developing a proof of mode stability for perturbations of the Kerr black hole, the essential features, in addition to those stated above, are that i) the new radial functions are bounded in the whole radial range from the horizon to infinity and ii) the equation they satisfy can be written in terms of real-valued coefficients. The explicit construction involves a transformation from ${}_{-|s|}R_{lm}$ to some new radial function ${}_{m}K_{ls}$, and so depends on the explicit demonstration, due to Teukolsky and Starobinsky [34], that the positive and negative helicity modes can be independently transformed one into the other.

To complete a proof of mode stability for the Kerr black hole, (invertible) differential and integral transformations were chosen so that a new function could be constructed which satisfied a partial differential equation with a positive definite 'energy' integral. Mode stability for the Kerr black hole then became a familiar, solvable problem. However, because parameters in the transformations depended explicitly on the mode decomposition, application of the methods of Kay and Wald [22, 23, 24] would not provide a stronger result for the Kerr black hole without an additional result about mode completeness. Thus, scope for further, if difficult, work remains.

6. Black Hole Thermodynamics

To some it may seem rather surprising that any discussion pertaining to thermodynamics would appear in a work on classical, black hole solutions of the Einstein equations of General Relativity. However, black hole thermodynamics has played an extremely important rôle in theoretical physics over the last twenty five years, and not just in the domain of General Relativity. Much of that rôle depends, at least in part, on a Euclidean approach to equilibrium thermodynamics, and the investigation of Euclidean 4-geometries in connection with that approach. The Schwarzschild problem is by far the most studied, and will be the focus of attention here. While results do exist for other spherical symmetric systems, unsolved technical problems still remain, awaiting attention, in the treatment of rotating black holes.

Probably the most vexing question with respect to black hole thermodynamics is how to account for the entropy in some way which is reasonably well understood from a physical perspective. Confirming the value of the entropy in so many different ways has certainly not provided a definitive answer as to the nature of black hole entropy, nor has it identified exactly to which degrees of freedom that entropy must refer. Generally, any calculation of the entropy carries with it an implied assertion about the stability, of the particular system under study, with regard to thermal equilibrium (an aspect not yet addressed in Ref. [42]). Unlike the rest of this article, in the discussion here, attention will be confined just to those degrees of freedom which correspond to spherical symmetry, since these comprise the least set known to be sufficient for a calculation of the black hole entropy.

As is the rest of physics, the properly defined states of a thermal system are quantum states. A physical system in a state of thermal equilibrium is represented not by a pure state, but by a maximally mixed state, classically described as an ensemble. Feynman [43] showed clearly how, specifically in a system at fixed temperature, an effective Euclideanization of the underlying time dependence of a physical theory could lead to a concise formulation of the quantum description. In the case of General Relativity, the fields of the physical description actually determine the spacetime structure on which they live. Euclideanization of time in that structure also reflects back on the fields themselves, and influences what properties the question of stability will address.

As might be expected, the underlying quantum nature of the problem also influences what a determination of stability will require. For the sake of definiteness, a classical solution will be taken to be any Euclidean geometry for which the Euclideanized Einstein action is locally stationary. Then classical, but not necessarily thermodynamical, stability will depend on whether or not the Euclidean action is at a local minimum. Thermodynamical stability will depend on whether a functional integral for the partition function is predominantly determined by the value of the Euclidean action at a saddle point corresponding to a classically stable solution[7].

[7]The saddle point approximation in the neighbourhood of the classical solution will generally give lower order contribution unless the boundary data correspond to a system in the Planckian regime.

In the work where he first elaborated on the unstable nature of the canonical ensemble for a Schwarzschild black hole, Hawking [44] did not use Euclidean methods, but referred instead to the interaction of the black hole with a heat bath of thermal radiation. He found evidence that a black hole could not exist in equilibrium with an arbitrarily large box of radiation. Subsequently, following on the introduction of Euclidean methods, York [45] used a refined description of thermodynamic variables for gravitational systems, and the 'black hole in a box' has been an object of study ever since. It is now apparent that equilibrium problems in thermodynamics arise as boundary value problems in Euclidean General Relativity. Associated with any Euclidean solution will be the question of its classical stability as a solution of a given boundary value problem. Associated with any state of thermodynamic equilibrium will be the question of its stability as a statistical ensemble. And associated with any physical system will be the ultimate question of its stability in the quantum domain. Results are now known on all these aspects. In particular, going beyond the Euclidean approaches [46, 47], the problem of a quantum, spherically symmetric black hole in a finite box has now been treated in a Lorentzian Hamiltonian theory [48], and used as the basis for the calculation of a thermodynamic partition function.

In Ref. [46], account was taken of all smooth, spherically symmetric Euclidean geometries satisfying the given boundary conditions, but not necessarily being classical solutions. In the calculation of the partition function, the contribution of each geometry was determined by the value of its classical action, and many geometries could be shown to contribute with the same value. After the classical constraints were imposed, it was found that the partition function could be written as a single ordinary integral in a quantity determined by the size of the black hole horizon. More detailed calculations using both Euclidean path integration [47], and the analytically continued Lorentzian Hamiltonian quantum theory [48], confirm the broad features of this partition function and (most of) its implications[8,9]. In all cases, the expression for the partition function could be reduced to a single residual integral which allowed detailed comparison of the various forms obtained.

The most comprehensive result can be simply stated. For a black hole in a box of area $A = 4\pi r_o{}^2$ with its walls held at temperature $T = \beta^{-1}$,

[8]The main difference is an unusual, higher energy contribution to the density of states found in the first paper of Ref. [47], which allows for the possibility of a system being prepared in a state at negative temperature.

[9]In the Lorentzian Hamiltonian approach to evaluating the partition function, Euclidean considerations still appear unavoidable, for a choice is made, in determining the values of the analytically continued parameters, so that classical solutions to the reduced Hamiltonian theory with the prescribed boundary data are the Euclidean (or complex) Schwarzschild solutions which appear as saddle points in the Euclidean path integral.

the quantum partition function of the canonical ensemble is well defined for all data in which the size of the box is finite. But only under a certain condition, $32\pi r_o T < 27\hbar$, will the saddle point in the neighbourhood of a classical solution dominate the analytically continued partition function derived from the Lorentzian Hamiltonian, or alternatively, the Euclidean functional integral evaluation of the partition function. Outside this range, the partition function is still finite, but classical solutions no longer play a significant rôle in determining its behaviour, so in a precise physical sense the canonical ensemble would no longer be described as stable. Rather, it is non-classical geometries, containing miniscule black holes, which control the properties of the resulting partition function. Although Euclidean classical solutions may be found which satisfy the given boundary data, in the expanded range, $27\hbar > 32\pi r_o T \geq 12\sqrt{3}\hbar$, it would appear that they must be metastable, and undergo a phase transition to a non-classical geometry, possibly in a different topological sector, such as that of flat space occupied by hot quantum fields. The classical solutions for both the stable and metastable configurations occur at local minima of the classical action. For each such solution, there is also a classical solution occurring at a local maximum of the classical action, but this can never represent a stable configuration, and never controls the corresponding value of the partition function.

7. Epilogue

Except in the thermodynamic case, the proof of black hole stability has proceeded by the analysis of differential equations. Yet, very few equations of any kind appear in this article. In fact, by the time the relevant equations have been obtained, the battle has been essentially over. It is the analysis which has gone in beforehand to formulating the problem, and the associated conceptual difficulties, which have consumed the bulk of the effort expended in any proof of black hole stability. It is that which represents the work, and the contributions for which people are recognized. It is also there that future efforts must be directed in order to push the present results beyond where they currently rest. It is from there that the challenge goes out, especially to upcoming students, and interested researchers new to the field.

Acknowledgements

A review of this topic represents the product of more than twenty years of research in the field. Those who have contributed to its existence, either directly or indirectly - by encouragement, probing questions or stimulating discussion over the years, or by the influence they have had through the papers they have written and the seminars they have given - are too numerous to mention by name. But, before ending this article, that one person in particular should be especially thanked, in honour of whom this particular document has been brought into existence. May he, by that, see his influence flow forth to inspire anew further upcoming generations of students and researchers. Financial support for this undertaking was gratefully received from the National Science Foundation, through grant PHY-9408910.

References

1. S. Chandrasekhar, *The Mathematical Theory of Black Holes*, (OUP, Oxford, 1983).
2. J. A. Wheeler, "Geons", *Phys. Rev.*, **97**, 511 (1955).
3. T. Regge and J. A. Wheeler, *Phys. Rev.*, **108**, 1063 (1957).
4. C. V. Vishveshwara, *Phys. Rev. D.*, **10**, 2870 (1970).
5. F. J. Zerilli, *Phys. Rev. Lett.*, **24**, 737 (1970).
6. V. Moncrief, *Ann. Phys.*, **88**, 323 (1974).
7. U. H. Gerlach and U. K. Sengupta, *Phys. Rev. D.*, **19**, 2268 (1979).
8. E.T. Newman and R. Penrose, *J. Math. Phys.*, **3**, 566 (1962).
9. R. Geroch, A. Held and R. Penrose, *J. Math. Phys.*, **14**, 874 (1973).
10. N. Kamran, *Contributions to the Study of the Separation of Variables and Symmetry Operators for Relativistic Wave Equations on Curved Spacetime*, Ph. D. Thesis, University of Waterloo (1984).
11. J. M. Bardeen and W. H. Press, *J. Math. Phys.*, **14**, 7 (1973).
12. S. A. Teukolsky, *Phys. Rev. Lett.*, **29**, 1114 (1972).
13. Richard H. Price, *Phys. Rev. D.*, **5**, 2419 (1972).
14. Richard H. Price, *Phys. Rev. D.*, **5**, 2439 (1972).
15. S. Chandrasekhar, *Proc. Roy. Soc. Lond. A.*, **343**, 289 (1975).
16. S. Chandrasekhar and S. Detweiler, *Proc. R. Soc. Lond. A.*, **344**, 441 (1975).
17. J. M. Cohen and L. S. Kegeles, *Phys. Lett.*, **54A**, 5 (1975). See also, *Phys. Rev. D.*, **10**, 1070 (1974).
18. P. L. Chrzanowski, *Phys. Rev. D.*, **11**, 2042 (1975).
19. S. Chandrasekhar, in §79-96 of Ref. [1], gives a coherent account of his own earlier work.
20. R. M. Wald, *Phys. Rev. Lett.*, **41**, 203 (1978).
21. M. Campanelli and C. Lousto, "The Imposition of Initial Data in the Teukolsky Equation", University of Utah e-print e-print, gr-qc/9711008 (1997). See also M. Campanelli, W. Krivan and C. Lousto, "The Imposition of Initial Data in the Teukolsky Equation II: Numerical Comparison With the Zerilli-Moncrief Approach to Black Hole Perturbations", University of Utah e-print e-print, gr-qc/9801067 (1998).
22. R. M. Wald, *J. Math. Phys.*, **20**, 1056 (1979).
23. B. S. Kay and R. M. Wald, *Class. Quantum Grav.*, **4**, 893 (1987).
24. J. Dimock and B. S. Kay, *Ann. Phys.*, **175**, 366 (1987). See footnote 13 and Lemma 5.10.
25. W. T. Zaumen, *Nature*, **247**, 530 (1974). G. W. Gibbons, *Comm. Math. Phys.*, **44**, 245 (1975). B. Carter, *Phys. Rev. Lett.*, **33**, 558 (1974).
26. V. Moncrief, *Phys. Rev. D.*, **9**, 2707 (1974) and *Phys. Rev. D.*, **10**, 1057 (1974). See also, *Phys. Rev. D.*, **12**, 1526 (1975).
27. D. M. Chitre, *Phys. Rev. D.*, **13**, 2713 (1976).
28. C. H. Lee, *J. Math. Phys.*, **17**, 1226 (1976).
29. J. Jezierski, "Energy and Angular Momentum of the Weak Gravitational Waves on the Schwarzschild Background – Quasilocal Gauge-Invariant Formulation", Univer-

sity of Warsaw e-print, gr-qc/9801068 (1998).
30. K. Schwarzschild, Sitzber. Deut. Akad. Wiss. Berlin, *Kl. Math.-Phys. Tech.*, 189-196 (1916)
31. R. P. Kerr, *Phys. Rev. Lett.*, **11**, 237 (1963).
32. B. Carter, *Comm. Math. Phys.*, **10**, 280 (1968). See also, "Black Hole Equilibrium States" in *Black Holes*, eds C. DeWitt and B. S. DeWitt (Gordon and Breach, New York, 1973).
33. S. A. Teukolsky, *Astrophys. J.*, **185**, 635 (1973).
34. W. H. Press and S. A. Teukolsky, *Astrophys. J.*, **185**, 649 (1973); A. A. Starobinsky and S. M. Churilov, *Zh. Exp. i. Theoret. Fiz.*, **65**, 3 (1973), translated in Soviet *Phys. JETP*, **38**, 1 (1974).
35. R. M. Wald, *J. Math. Phys.*, **14**, 1453 (1973).
36. S. A. Teukolsky and W. H. Press, *Astrophys. J.*, **193**, 443 (1974).
37. S. Chandrasekhar and S. Detweiler, *Proc. Roy. Soc. A.*, **345**, 145 (1975), and *Proc. Roy. Soc. A.*, **350**, 165 (1976).
38. B. F. Whiting, *J. Math. Phys.*, **30**, 1301 (1989).
39. B. F. Whiting, *Proc. Fifth Marcel Grossman Meeting*, held in Perth, Western Australia, August 8-13, 1988, eds. D. G. Blair and M. J. Buckingham, 1207 (World Scientific, Singapore, 1989).
40. S. Persides, *J. Math. Phys.*, **14**, 1017 (1973).
41. E. L. Ince, *Ordinary Differential Equations*, Ch. VIII (Dover, New York, 1956).
42. R. M. Wald, *Phys. Rev. D.*, **48**, R3427 (1993). See also, V. Iyer and R. M. Wald, *Phys. Rev. D.*, **50**, 846 (1994), and *Phys. Rev. D.*, **52**, 4430 (1995).
43. R. P. Feynman and A. R. Hibbs, *Quantum Mechanics and Path Integrals*, Ch. 10 (McGraw-Hill, New York, 1965).
44. S. W. Hawking, *Phys. Rev. D.*, **13**, 191 (1976).
45. J. W. York, *Phys. Rev. D.*, **33**, 2092 (1986).
46. B. F. Whiting and J. W. York, *Phys. Rev. Lett.*, **61**, 1336 (1988).
47. J. Louko and B. F. Whiting, *Class. Quantum Grav.*, **9**, 457 (1992), and J. Melmed and B. F. Whiting, *Phys. Rev. D.*, **49**, 907 (1994).
48. J. Louko and B. F. Whiting, *Phys. Rev. D.*, **51**, 5583 (1995).

My personal journey along the black hole trail started in the sixties when I was a graduate student of Charles Misner at the University of Maryland. Also, that was when I first came to know about Vaidya and Raychaudhuri. Leepo Cheng was doing her master's thesis with Misner on the Vaidya metric. She was appalled by my ignorance when I told her that I did not know who Vaidya was. Later on, we were told Raychaudhuri was coming as a Visiting Professor. Again, my colleagues were suitably impressed by my ignorance when I confessed that I did not know who this Raychaudhuri was either. I went on not only to take a course on cosmology from him but also pass it with a little bit of honest cheating. Never did I dream that some day I would be delivering a lecture in honour of these two gentlemen.
Let me come back to black holes. That is not what they were called at that time. Schwarzschild singularity, which is a misnomer. Schwarzschild surface, which is better. The term 'Black Hole' was to be coined later on by John Wheeler. Perhaps, it was because of its intriguing name that so many people were enticed into working on the physics of the black hole. This is known as the Schickgruber Effect. Scholars have speculated on how human history might have been different if Mr. Schickgruber had not changed his name. But, he did change his name to Hitler. Misner proposed the following problem for my Ph.D. thesis. Take two of these entities that are now called black holes. Revolving around each other, they come close as energy is radiated away in the form of gravitational waves. They coalesce into an ellipsoidal 'Schwarzschild surface' still rotating and radiating. Study the whole process, computing all the characteristics of the emitted gravitational radiation. Fine, I said, thy will be done! At the time, I did not realise the magnitude of this problem. Had I pursued it, I might have entered the Guinness book of records as the oldest graduate student alive and that too without financial support. Anyway, this proposed problem required the understanding of two aspects of black holes: the geometrical structure of the black hole and the perturbations of its spacetime.

— C. V. VISHVESHWARA

3. SEPARABILITY OF WAVE EQUATIONS

E.G. KALNINS
Mathematics Department, University of Waikato
Hamilton, New Zealand

W. MILLER JR.
School of Mathematics, University of Minnesota
Minneapolis, Minn. 55455, U.S.A

AND

G.C. WILLIAMS
Mathematics Department, University of Waikato
Hamilton, New Zealand

1. Introduction

The method of separation of variables has proved a useful tool with which to address various aspects of the physics of black holes. In this article we will review the progress made using this method in solving various systems of equations in black hole space-time backgrounds. The equations of important physical interest that we shall discuss consist primarily of the linear perturbation equations for various fields. In particular: Maxwell's equations, Dirac's equation and the Weyl neutrino equation and the gravitational perturbation equations. We will also discuss generalised Hertz potentials, the Rarita-Schwinger equation and the properties of Teukolsky functions.

Separation of variables is a well known method of solution for problems in classical mechanics. Many of the fundamental (scalar) partial differential equations of mathematical physics in one unknown function can be reduced to ordinary differential equations using the method. The Hamilton-Jacobi, Helmholtz and Schrodinger equations are often amenable to such an approach. A comprehensive theory of separation of variables exists for these scalar wave equations, a review of which is given by Miller (1988). The development of a corresponding theory for the non-scalar equations of mathematical physics has however proved more elusive. A theory of separation of

B. R. Iyer and B. Bhawal (eds.), Black Holes, Gravitational Radiation and the Universe, 33–51.

variables has two principle aims. The first is to find separable solutions to a set of equations where such solutions exist. The second is to intrinsically characterise (i.e. in a frame independent way) the tetrad, coordinates and operators associated with the separation. A physical or geometrical interpretation of the separation constants, operators and frames may then be possible.

In the Kerr space-time the first separable solutions of significance were obtained by Carter (1968) for the Hamilton-Jacobi geodesic equations and Schrodinger's equation. The most significant results however were those of Teukolsky (1972) who obtained separable solutions for some components of the linearly perturbed electromagnetic and gravitational fields in the Kerr space-time. With such solutions available it became possible to consider electromagnetic and gravitational perturbations of the Kerr black hole. These ideas were extended to the Weyl neutrino field in the Kerr background independently by Teukolsky (1973) and Unruh (1973). A crucial development was the solution of the Dirac equation in the Kerr background via a separation of variables method. This was achieved by Chandrasekhar (1976) and later extended by Page (1976) to the Kerr-Newman background.

2. Black Hole Space-times

The Kerr-Newman solution is a Petrov type D electrovac solution. It appears to be the most general physically reasonable space-time for a stable isolated black hole. It is stationary, axisymmetric and asymptotically flat. (For a precise definition of these concepts see, for example, Heusler (1996)). The line element of the Kerr-Newman space-time in Boyer-Lindquist coordinates is

$$\begin{aligned} \mathrm{d}s^2 &= \rho'\bar{\rho}'\left(\frac{1}{\Delta}\mathrm{d}r^2 + \mathrm{d}\theta^2\right) + (r^2+a^2)\sin^2\theta\,\mathrm{d}\varphi^2 - \mathrm{d}t^2 \\ &\quad + \frac{2Mr}{\rho'\bar{\rho}'}(a\sin^2\theta\,\mathrm{d}\varphi - \mathrm{d}t)^2, \end{aligned} \tag{1}$$

where

$$\rho' = r + ia\cos\theta \quad \text{and} \quad \Delta = r^2 - 2Mr + a^2 + q^2. \tag{2}$$

This corresponds to a charged rotating black hole having mass M, angular momentum $J = aM$ and charge q. When $q = 0$ or $a = 0$ we recover the metrics of the Kerr or Reissner-Nordström space-times respectively. When both are zero we recover the Schwarzschild metric. With rotation absent ($a = 0$) the space-time has spherical symmetry.

In this article we will use the Newman-Penrose and abstract index formalisms as found in Penrose and Rindler (1984). We also use the Kinnersley null tetrad, the corresponding directional derivatives of which are

$$\begin{aligned} l^a\nabla_a &= D = \frac{1}{\Delta}((r^2+a^2)\partial_t + \Delta\partial_r + a\partial_\varphi) \\ m^a\nabla_a &= \delta = \frac{1}{\rho'\sqrt{2}}(ia\sin\theta\partial_t + \partial_\theta + i\csc\theta\partial_\varphi) \\ n^a\nabla_a &= \Delta = \frac{1}{2\rho'\bar{\rho}'}((r^2+a^2)\partial_t - \Delta\partial_r + a\partial_\varphi). \end{aligned} \tag{3}$$

This is precisely the same tetrad as used by Chandrasekhar (1983). The Kinnersley tetrad is a canonical one, the vectors l^a and n^a being aligned with the principal null directions of the space-time. Note that the symbol Δ is used in two different roles. Which it represents should be clear from context.

In the Kinnersley tetrad the non-zero spin coefficients are

$$\begin{aligned} &\rho = -\frac{1}{\bar{\rho}'}, \quad \pi = \frac{ia\sin\theta}{\sqrt{2}\bar{\rho}'^2}, \quad \tau = -\frac{ia\sin\theta}{\sqrt{2}\rho'^2}, \quad \mu = -\frac{\Delta}{2\rho'^2\bar{\rho}'}, \\ &\gamma = \mu + \frac{r-M}{2\rho'\bar{\rho}'}, \quad \beta = \frac{\cot\theta}{2\sqrt{2}\bar{\rho}'}, \quad \alpha = \pi - \bar{\beta}. \end{aligned} \tag{4}$$

The only non-zero components of the Weyl, Ricci and Maxwell spinors are

$$\Psi_2 = -\frac{M}{\bar{\rho}'^3} + \frac{q^2}{\rho'\bar{\rho}'^3}, \quad \Phi_{11} = \frac{q^2}{2\rho'^2\bar{\rho}'^2}, \quad \phi_1 = \frac{q}{2\bar{\rho}'^2}. \tag{5}$$

When considering the linear perturbation equations for a given field in the Kerr-Newman background one can make use of the stationarity and the axisymmetry to decompose each component of a perturbation into a superposition of modes of the form

$$f(r,\theta)e^{i\omega t + im\varphi} \tag{6}$$

where ω is real and m is a half-integer or integer depending on whether the perturbed field or wave equation under consideration is fermionic or bosonic respectively. When acting on quantities of the form (6) the directional derivatives (3) can be succinctly written in terms of the following radial and angular differential operators

$$\begin{aligned} \mathcal{D}_n &= \partial_r + i\frac{K}{\Delta} + 2n\frac{r-M}{\Delta}, \\ \mathcal{D}_n^\dagger &= \partial_r - i\frac{K}{\Delta} + 2n\frac{r-M}{\Delta}, \\ \mathcal{L}_n &= \partial_\theta + Q + n\cot\theta, \\ \mathcal{L}_n^\dagger &= \partial_\theta - Q + n\cot\theta, \end{aligned} \tag{7}$$

where

$$K = \omega(r^2 + a^2) + am \quad \text{and} \quad Q = \omega a \sin\theta + m \csc\theta. \tag{8}$$

When rotation is not present the Kerr-Newman space-time degenerates to one having spherical symmetry. Group theory then guarantees that any linearised perturbation equations can have all the angular dependence separated out (see for instance Gel'fand *et al.* (1963)). Solution of a given perturbation problem in the Schwarzschild or Reissner-Nordström space-times will therefore amount to solving for the remaining radial dependence of the various quantities. However, this is by no means a trivial problem. Solution of the combined electromagnetic and gravitational perturbation equations for the Reissner-Nordström space-time, for example, proceeds via an unexpected and not at all obvious decoupling of the radial equations (see Section 6).

The Kerr and Kerr-Newman space-times lack spherical symmetry. One cannot therefore expect, on the basis of an appeal to group theory, to effect a separation of variables for a given set of perturbation equations. Surprisingly most of the linearised perturbation equations in these two space-times, with one notable exception, continue to yield to separation of variables methods. Associated with each of the separable perturbation equations one typically finds a symmetry operator constructed from a Killing-Yano tensor.

A valence p Killing-Yano tensor is any totally anti-symmetric tensor $K_{a_1 a_2 \ldots a_p}$ which satisfies

$$\nabla_{(b} K_{a_1) a_2 \ldots a_p} = 0. \tag{9}$$

A valence p Killing spinor is a totally symmetric spinor $K_{A_1 A_2 \ldots A_p}$ satisfying

$$\nabla_{(BB'} K_{A_1 A_2 \ldots A_p)} = 0. \tag{10}$$

In the Kerr-Newman space-time there is only one valence two Killing-Yano tensor and one valence two Killing spinor. They are related by

$$K_{ab} = \frac{1}{2}(\epsilon_{A'B'} K_{AB} - \epsilon_{AB} \bar{K}_{A'B'}). \tag{11}$$

In addition the Killing spinor in the Kerr-Newman space-time also satisfies

$$\nabla_{BA'} K_A{}^B - \nabla_{AB'} \bar{K}_{A'}{}^{B'} = 0. \tag{12}$$

The vector

$$M_{AA'} = \frac{1}{3} \nabla_{BA'} K_A{}^B \tag{13}$$

is a Killing vector, that is, it satisfies

$$\nabla_{(b} M_{a)} = 0. \tag{14}$$

In the spin frame naturally associated with the Kinnersley tetrad the Killing spinor K_{AB} has components

$$K_{00} = 0, \quad K_{01} = -\bar{\rho}', \quad K_{11} = 0. \tag{15}$$

3. Maxwell's equations in the Kerr space-time

Maxwell's equations for a source free electromagnetic field are, in spinor form,

$$\nabla^{A}_{A'}\phi_{AB} = 0, \tag{16}$$

where ϕ_{AB} is a symmetric spinor. Expressed in the Newman-Penrose formalism Maxwell's equations read

$$\begin{aligned}(D - 2\rho)\phi_1 &= (\bar{\delta} + \pi - 2\alpha)\phi_0 - \kappa\phi_2,\\ (\delta - 2\tau)\phi_1 &= (\Delta + \mu - 2\gamma)\phi_0 - \sigma\phi_2,\\ (\bar{\delta} + 2\pi)\phi_1 &= (D - \rho + 2\epsilon)\phi_2 + \lambda\phi_0,\\ (\Delta + 2\mu)\phi_1 &= (\delta - \tau + 2\beta)\phi_2 + \nu\phi_0.\end{aligned} \tag{17}$$

If we write

$$\phi_k = (\rho'\sqrt{2})^{-k}\xi_k(r,\theta)e^{i\omega t + im\varphi} \tag{18}$$

then in the Kerr space-time and the Kinnersley tetrad Maxwell's equations become

$$\begin{aligned}\left(\mathcal{D}_0 + \frac{1}{\bar{\rho}'}\right)\xi_1 &= \left(\mathcal{L}_1 - \frac{ia\sin\theta}{\bar{\rho}'}\right)\xi_0,\\ \left(\mathcal{L}_0 + \frac{ia\sin\theta}{\bar{\rho}'}\right)\xi_1 &= \left(\mathcal{D}_0 - \frac{1}{\bar{\rho}'}\right)\xi_2,\\ \Delta\left(\mathcal{D}_0^\dagger + \frac{1}{\bar{\rho}'}\right)\xi_1 &= -\left(\mathcal{L}_1^\dagger - \frac{ia\sin\theta}{\bar{\rho}'}\right)\xi_2,\\ \left(\mathcal{L}_0^\dagger + \frac{ia\sin\theta}{\bar{\rho}'}\right)\xi_1 &= -\Delta\left(\mathcal{D}_1^\dagger - \frac{1}{\bar{\rho}'}\right)\xi_0.\end{aligned} \tag{19}$$

It is reasonably straightforward to obtain decoupled second order equations for ξ_0 and ξ_2. These second order equations admit separable solutions of the form

$$\begin{aligned}\xi_0 &= R_{+1}(r)S_{+1}(\theta),\\ \xi_2 &= R_{-1}(r)S_{-1}(\theta).\end{aligned} \tag{20}$$

Here $R_{\pm 1}(r)$ and $S_{\pm 1}(\theta)$ are Teukolsky functions (see Section 7). Having obtained solutions for ϕ_0 and ϕ_2 one can then obtain (Chandrasekhar, 1983)

$$\phi_1 = \frac{1}{\sqrt{2}\mathcal{C}_1\bar{\rho}'}\left[\mathcal{D}_0\mathcal{L}_1 - \frac{1}{\bar{\rho}'}(\mathcal{L}_1 + ia\sin\theta\mathcal{D}_0)\right]R_{-1}(r)S_{+1}(\theta)e^{i\omega t + im\varphi} \tag{21}$$

where

$$\mathcal{C}_1 = \sqrt{\lambda^2 - 4a\omega(a\omega + m)}. \tag{22}$$

The separation constant λ and the Starobinsky constant $\mathcal{C}_1$ have been shown by Kalnins *et al.* (1989a) to be characterised by symmetry operators

$$\Lambda^C_{(A}\phi_{B)C} = \frac{1}{2}\lambda\phi_{AB} \quad \text{and} \quad C^{AB}_{A'B'}\phi_{AB} = \mathcal{C}_1\bar{\phi}_{A'B'}, \tag{23}$$

where

$$\Lambda^C_A = \left(K_A{}^{A'EE'}\nabla_{EE'} - M_A{}^{A'}\right)\left(K^C{}_{A'}{}^{DD'}\nabla_{DD'} + 2M^C{}_{A'}\right) \tag{24}$$

and

$$C^{AB}_{A'B'} = \left(K^A{}_{(A'|}{}^{EE'}\nabla_{EE'} + M^A{}_{(A'|}\right)\left(K^B{}_{|B')}{}^{DD'}\nabla_{DD'} + 2M^B{}_{|B')}\right). \tag{25}$$

As can be seen the Killing-Yano tensor plays a crucial role in the above separation of Maxwell's equations.

Kalnins *et al.* (1986) have shown that the vectors defined by the null tetrad are in fact eigenvectors of the Killing-Yano tensor K_{ab} considered as a matrix. It should also be mentioned at this point that Torres del Castillo (1989c) has observed the relation between Killing spinors and the separability of Maxwell's equations and also extended the separability of Maxwell's equations to type D space-time backgrounds (Torres del Castillo, 1988a). Maxwell's equations can also be solved for the Robertson-Walker type space-times (Kalnins and Miller, 1991). Decoupled equations for Debye potentials of electromagnetic fields in space-times with local rotational symmetry have been obtained by Dhurandhar *et al.* (1980). The most general linear second order symmetry operators of Maxwell's equations in curved space-time have been found by Kalnins *et al.* (1992). Implicitly included in this study are the operators characterising λ and $\mathcal{C}_1$.

4. Dirac's equation, the Weyl Neutrino equation and the Rarita-Schwinger equation

Dirac's equation for a massive charged spin half particle in a curved space-time background with a background electromagnetic field has, in spinor notation, the form

$$\begin{aligned}(\nabla^{B'}_B - iq_eA^{B'}_B)\chi_{B'} &= -\frac{im_e}{\sqrt{2}}\phi_B, \\ (\nabla^B_{B'} - iq_eA^B_{B'})\phi_B &= \frac{im_e}{\sqrt{2}}\chi_{B'},\end{aligned} \tag{26}$$

where $A_{BB'}$ is the vector potential of the background electromagnetic field. Setting $m_e = 0,\ q_e = 0$ and $\chi_{A'} = \bar{\phi}_{A'}$ we recover two copies of the Weyl neutrino equation. In the Newman-Penrose formalism Dirac's equations have the form

$$
\begin{aligned}
(\bar{\delta} - iq_e A_{10'} + \pi - \alpha)\phi_0 - (D - iq_e A_{00'} + \epsilon - \rho)\phi_1 &= \frac{im_e}{\sqrt{2}}\chi_{0'}, \\
(\Delta - iq_e A_{11'} + \mu - \gamma)\phi_0 - (\delta - iq_e A_{01'} + \beta - \tau)\phi_1 &= \frac{im_e}{\sqrt{2}}\chi_{1'}i, \\
(D - iq_e A_{00'} + \bar{\epsilon} - \bar{\rho})\chi_{1'} - (\delta - iq_e A_{01'} + \bar{\pi} - \bar{\alpha})\chi_{0'} &= \frac{im_e}{\sqrt{2}}\phi_0, \qquad (27) \\
(\bar{\delta} - iq_e A_{10'} + \bar{\beta} - \bar{\tau})\chi_{1'} - (\Delta - iq_e A_{11'} + \bar{\mu} - \bar{\gamma})\chi_{0'} &= \frac{im_e}{\sqrt{2}}\phi_1.
\end{aligned}
$$

In contrast to the procedure carried out for Maxwell's equations where decoupling occured first followed by separation, the solution of Dirac's equation proceeds by postulating a separable ansatz for the solutions first and then subsequently decoupling the resulting equations for the functions of one variable. In the Kerr-Newman space-time the equations above have the solutions

$$
\begin{aligned}
\phi_0 &= R_{+\frac{1}{2}} S_{+\frac{1}{2}} e^{i\omega t + im\varphi}, & \phi_1 &= -\frac{1}{\bar{\rho}'} R_{-\frac{1}{2}} S_{-\frac{1}{2}} e^{i\omega t + im\varphi}, \\
\chi_{0'} &= R_{+\frac{1}{2}} S_{-\frac{1}{2}} e^{i\omega t + im\varphi}, & \chi_{1'} &= \frac{1}{\rho'} R_{-\frac{1}{2}} S_{+\frac{1}{2}} e^{i\omega t + im\varphi}. \qquad (28)
\end{aligned}
$$

where the functions $R_{\pm\frac{1}{2}}$ and $S_{\pm\frac{1}{2}}$ satisfy the equations

$$
\begin{aligned}
\left(\mathcal{D}_0 - \frac{iqq_e r}{\sqrt{2}\Delta}\right) R_{-\frac{1}{2}} &= (\lambda + im_e r) R_{+\frac{1}{2}}, \\
\left(\Delta \mathcal{D}^{\dagger}_{\frac{1}{2}} + \frac{iqq_e r}{\sqrt{2}}\right) R_{+\frac{1}{2}} &= (\lambda - im_e r) R_{-\frac{1}{2}}, \\
\mathcal{L}^{\dagger}_{\frac{1}{2}} S_{-\frac{1}{2}} &= (\lambda + am_e \cos\theta) S_{+\frac{1}{2}}, \qquad (29) \\
\mathcal{L}_{\frac{1}{2}} S_{+\frac{1}{2}} &= -(\lambda - am_e \cos\theta) S_{-\frac{1}{2}}.
\end{aligned}
$$

These equations are readily decoupled to give four second order equations for $R_{\pm\frac{1}{2}}$ and $S_{\pm\frac{1}{2}}$. The functions $R_{\pm\frac{1}{2}}$ and $S_{\pm\frac{1}{2}}$ are generalisations of Teukolsky functions. The separation parameter λ appearing above has been characterised by Carter and McLenaghan (1979) in terms of a non-trivial symmetry operator constructed from the Killing-Yano tensor $K_{AA'BB'}$. Their result expressed in spinor form is

$$
(K^{BB'}{}_{A}{}^{A'} \nabla_{BB'} + M_{A}{}^{A'})\chi_{A'} = -\frac{\lambda}{\sqrt{2}}\phi_A,
$$

$$(K^{BB'A}{}_{A'}\nabla_{BB'} - M^{A}{}_{A'})\phi_A = -\frac{\lambda}{\sqrt{2}}\chi_{A'}. \tag{30}$$

Physically this can be interpreted as a generalised angular momentum operator. As with Maxwell's equations the Killing-Yano tensor plays a central role in the separation of the Dirac equation. The most general first order formally self-adjoint operator that commutes with the Dirac operator has been found by McLenaghan and Spindel (1979). The general second order symmetry operator of the Weyl neutrino equation on curved space-time has been found by McLenaghan and Walker (1996).

Separation and solution of the Dirac equation in a class of Petrov type D space-times has been achieved by Güven (1977) and in a wide class of quite general space-times by Debever *et al.* (1984). Kamran and McLenaghan (1984a) have shown that the Weyl neutrino equation is separable in the class of solutions which are Petrov type D electrovac with non-singular aligned Maxwell field satisfying the generalised Goldberg-Sachs theorem and that the Dirac equation is separable in a sub-class of these solutions. They have also characterised the separable solution of the Dirac equation in an Einstein-Maxwell space-time (Kamran and McLenaghan, 1984b). Separation of the Dirac equation in space-times with local rotational symmetry has been studied by Iyer and Vishveshwara (1985), (1987) and Iyer and Kamran (1991).

The manner of separation that occurs with the Dirac equation in the Kerr space-time can be described using the idea of factorisable systems as formulated by Shapovalov and Ekle (1974a), (1974b) and Miller (1988). The notion of factorisable systems is not however enough for a comprehensive treatment of solutions of the Dirac equation. Fels and Kamran (1990) exhibit a space-time in which characterisation of the separation constants as the eigenfunctions of first order symmetry operators is not possible. Rather a complete characterisation requires the use of one second order symmetry operator. For a general discussion of mechanisms of separation for the Dirac equation in curved space-times see Rudiger (1984).

Explicit solutions of the Dirac equation have been studied extensively in Minkowski space-time. For a description of this work see Bagrov and Gitman (1990) and Shishkin and Villalba (1989). An analysis of the properties of the separable solutions appearing for the Dirac equation can be found in Cosgrove *et al.* (1983) and Kalnins and Miller (1992).

The separation and solution of the equations for various scalars associated with a Rarita-Schwinger field in a Petrov type D space-time was first achieved by Güven (1980). In proving the no-hair conjecture for the uncharged black holes of supergravity Güven showed that all solutions of the Rarita-Schwinger equation in the Kerr space-time that cannot be transformed away via a supersymmetry transformation are expressible in terms of the Teukolsky functions for $s = \pm\frac{3}{2}$. Kamran (1985) later extended the separability of the Rarita-Schwinger equation to all type D vacuum space-times (see also Torres del Castillo (1989a)). The Rarita-Schwinger equation can be written in spinor form as

$$\nabla^A_{A'}\theta_{ABB'} = 0, \tag{31}$$

where $\theta_{ABB'} = \theta_{(AB)B'}$. A related system of coupled equations can be constructed from the Rarita-Schwinger equation. Forming a new field ϕ_{ABC} from the Rarita-Schwinger field by

$$\phi_{ABC} = \nabla_{(AA'}\theta_{BC)}{}^{A'}. \tag{32}$$

it is not difficult to show that in a vacuum ϕ_{ABC} must satisfy

$$\nabla^A_{A'}\phi_{ABC} = \Psi_{BCKL}\theta^{KL}{}_{A'}. \tag{33}$$

Taken together the last two equations admit solutions for ϕ_{000} of the form

$$\phi_{000} = R_{-\frac{3}{2}}(r)S_{-\frac{3}{2}}(\theta)e^{i\omega t+im\varphi}, \tag{34}$$

where $R_{-\frac{3}{2}}$ and $S_{-\frac{3}{2}}$ are Teukolsky functions.

5. Gravitational perturbations of the Kerr space-time

Teukolsky (1972) found separable solutions for two of the components of the perturbed Weyl spinor in the Kerr space-time. Given a perturbation of the Kerr space-time the following relations hold true to first order in the perturbation. From the Bianchi identities and the Newman-Penrose equations we have

$$\begin{aligned} (\bar\delta - 4\alpha + \pi)\hat\Psi_0 - (D - 4\bar\rho)\hat\Psi_1 &= 3\hat\kappa\Psi_2, \\ (\Delta - 4\gamma + \mu)\hat\Psi_0 - (\delta - 4\tau - 2\beta)\hat\Psi_1 &= 3\hat\sigma\Psi_2, \\ (D - \rho - \bar\rho)\hat\sigma - (\delta - \tau + \bar\pi - \bar\alpha - 3\beta)\hat\kappa &= \hat\Psi_0. \end{aligned} \tag{35}$$

In these equations $\hat\Psi_0$, $\hat\Psi_1$, $\hat\sigma$ and $\hat\kappa$ are first order perturbations to the components of the Weyl spinor and the spin coefficients respectively. Making use of the identities that result from the commutation relations of the directional derivatives one can show that $\hat\Psi_0$ satisfies the decoupled equation

$$\begin{aligned} &[(D - 2\rho - \bar\rho)(\Delta - 4\gamma + \mu) - (\delta - 2\tau + \bar\pi - \bar\alpha - 3\beta)(\bar\delta - 4\alpha + \pi)]\hat\Psi_0 \\ &\quad = \Psi_2\hat\Psi_0. \end{aligned} \tag{36}$$

Writing $\hat{\Psi}_0 = \xi_0(r,\theta)e^{i\omega t+im\varphi}$ this equation reduces to

$$(\Delta\mathcal{D}_1\mathcal{D}_2^\dagger + \mathcal{L}_{-1}^\dagger\mathcal{L}_2 - 6i\omega\rho')\xi_0 \;=\; 0, \tag{37}$$

which has solution $\xi_0 = R_{+2}(r)S_{+2}(\theta)$. By considering the analogous set of equations for $\hat{\Psi}_4$ one obtains the following separable solution

$$\hat{\Psi}_4 = \frac{1}{4\bar{\rho}'^4}R_{-2}(r)S_{-2}(\theta)e^{i\omega t+im\varphi}. \tag{38}$$

The functions $R_{\pm 2}(r)$ and $S_{\pm 2}(\theta)$ are Teukolsky functions. The remaining components of the perturbed Weyl spinor are most easily found using the method of generalised Hertz potentials. The question as to the operator which characterises the separation constant and the interpretation of that operator remains unclear (see Kalnins *et al.* (1996)).

The use of generalised Hertz potentials to produce solutions to various equations in algebraically special space-times is due to Cohen and Kegeles (1975) and Kegeles and Cohen (1979). For an elegant explanation of the reasons behind the efficacy of this technique see Wald (1978). A generalised Hertz potential is a totally symmetric spinor $\bar{P}^{A_1'A_2'\ldots A_{2s}'}$ satisfying

$$\begin{aligned}\nabla^{B(A_1'}\left[\nabla_{BB'}\bar{P}^{A_2'\ldots A_{2s}')B'} - G_B{}^{A_2'\ldots A_{2s}')}\right]&\\ - (2s-1)(s-1)\bar{\Psi}_{B'C'}{}^{(A_1'A_2'}\bar{P}^{A_3'\ldots A_{2s}')B'C'} &= 0,\end{aligned} \tag{39}$$

where $G_B{}^{A_2'\ldots A_{2s}'}$ is an arbitrary gauge field. One can then construct a spin s field from the potential and gauge field as follows

$$\begin{aligned}\phi_{A_1A_2\ldots A_{2s}} \;=\; & \nabla_{(A_1A_1'}\nabla_{A_2A_2'}\cdots\nabla_{A_{2s-1}A_{2s-1}'}\\ & \left[\nabla_{A_{2s})A_{2s}'}\bar{P}^{A_1'A_2'\ldots A_{2s}'} - G_{A_{2s})}{}^{A_1'A_2'\ldots A_{2s-1}'}\right].\end{aligned} \tag{40}$$

In a flat space-time the resulting field $\phi_{A_1\ldots A_{2s}}$ will satisfy the massless spin s field equation

$$\nabla_{B'}^{B}\phi_{BA_2\ldots A_{2s}} = 0. \tag{41}$$

When the space-time is curved the field $\phi_{A_1\ldots A_{2s}}$ satisfies the massless spin s field equations for $s = \frac{1}{2}$ and $s = 1$ only. For $s = 2$ the method of generalised Hertz potentials can be used to construct a solution for the linear gravitational perturbations of the Kerr space-time.

Solving the equation for the generalised Hertz potential via separation of variables requires one make the right choice of gauge field. For the case of Maxwell's equations we have

$$\phi_{AB} = \nabla_{(AA'}\left[\nabla_{B)B'}\bar{P}^{A'B'} - G_{B)}^{A'}\right]. \tag{42}$$

The new field satisfies Maxwell's equations (16). In the Kinnersley tetrad and with the specific choice of gauge field $G_C{}^{B'} = -2\bar{U}_{CC'}\bar{P}^{C'B'}$ where $U_{CC'}$ has components

$$U_{00'} = \bar{\rho}, \quad U_{01'} = -\bar{\pi}, \quad U_{10'} = \bar{\tau}, \quad \text{and} \quad U_{11'} = -\bar{\mu}, \tag{43}$$

the Hertz potentials have solutions of the form

$$\bar{P}^{0'0'} = R_{-1}S_{+1}e^{i\omega t+im\varphi}, \quad \bar{P}^{0'1'} = 0, \quad \bar{P}^{1'1'} = 0 \tag{44}$$

or

$$\bar{P}^{1'1'} = \bar{\rho}'^2 R_{+1}S_{-1}e^{i\omega t+im\varphi}, \quad \bar{P}^{0'1'} = 0, \quad \bar{P}^{0'0'} = 0. \tag{45}$$

Both these forms of the Hertz potential lead to the same Maxwell spinors (by virtue of the Teukolsky-Starobinsky identities amongst the Teukolsky functions).

The interesting feature of the Hertz potential equation is that for the choice of gauge function

$$G_{BA'_2\ldots A'_{2s}} = -2sU_{BB'}\bar{P}^{B'}{}_{A'_2\ldots A'_{2s}} \tag{46}$$

solutions exist whose only non-zero components are of the form

$$\bar{P}^{0'\ldots0'} = R_{-s}S_{+s}e^{i\omega t+im\varphi} \quad \text{or} \quad \bar{P}^{1'\ldots1'} = \rho'^{2s}R_{+s}S_{-s}e^{i\omega t+im\varphi}, \tag{47}$$

where $R_{\pm s}$ and $S_{\pm s}$ are the spin s Teukolsky functions.

These potential functions for $s = 2$ have been used by Kegeles and Cohen (1979) to obtain components of the perturbed metric tensor corresponding to gravitational perturbations of a vacuum space-time. For $s = \frac{3}{2}$ they have been used by Torres del Castillo (1989b) in the context of Rarita-Schwinger fields. For gravitational perturbations the perturbed Einstein vacuum field equations $\delta R_{ab} = 0$ take the form

$$\nabla_k\nabla^k h_{ab} - 2\nabla_k\nabla_{(a}h_{b)}{}^k + \nabla_a\nabla_b h_k{}^k = 0, \tag{48}$$

where h_{ab} is the perturbation to the metric tensor. This can be thought of also as a spin 2 wave equation. In terms of the Hertz potential and gauge spinor h_{ab} is given by

$$\begin{aligned} h_{ABA'B'} = & \left[\nabla_{(AK'}\nabla_{B)L'}\bar{P}^{K'L'}{}_{A'B'} - \nabla_{(AK'}G^{K'}{}_{B)A'B'}\right] \\ & + \left[\nabla_{K(A'}\nabla_{LB')}P^{KL}{}_{AB} - \nabla_{K(A'}G^{K}{}_{B')AB}\right]. \end{aligned} \tag{49}$$

It should of course be understood that all that has been done for potentials with primed indices can also be repeated for potentials with unprimed indices. Hence the appearance of $P^{KL}{}_{AB}$ in (49) presents no problem.

The first order perturbation of the Weyl spinor is given by

$$\hat{\Psi}_{ABCD} = \nabla_{(AK'}\nabla_{BL'}h_{CD)}{}^{K'L'}. \tag{50}$$

Making one or other of the choices (47) results in explicit expressions for the perturbed metric in either the ingoing or outgoing radiation gauges ($h_{ab}l^b = 0$ or $h_{ab}n^b = 0$). An alternative method of solution of the gravitational perturbation problem has been given by Chandrasekhar (1983). In his solution the expressions for the perturbed components of the metric are more complex.

6. Combined electromagnetic and gravitational perturbations of charged black holes

The Kerr-Newman space-time has a background electromagnetic field. A first order perturbation in this field gives rise to first order perturbations in the energy-momentum tensor resulting in first order gravitational perturbations. Consequently in the Kerr-Newman and Reissner-Nordström space-time linear electromagnetic or gravitational perturbations cannot be considered separately. This makes the solution of the corresponding perturbation equations more difficult. In fact the method of separation of variables appears to fail in the Kerr-Newman case. In what follows we will distinguish the value of a quantity in the perturbed space-time from its value in the background space-time by placing a circumflex above it.

Central to the solution of the Kerr-Newman perturbation problem are the following equations. From the Newman-Penrose equations we have, to first order in the perturbation,

$$(\Delta + \mu + \bar{\mu} + 3\gamma - \bar{\gamma})\hat{\lambda} - (\bar{\delta} + \pi + 3\alpha + \beta - \tau)\hat{\nu} = -\hat{\Psi}_4. \tag{51}$$

From the Bianchi identities we have, to first order

$$\begin{aligned}(\bar{\delta} + 2\alpha + 4\pi)\hat{\Psi}_3 - (D + 4\epsilon - \rho)\hat{\Psi}_4 + (\bar{\delta} + 2\alpha - 2\bar{\tau})\hat{\Phi}_{21}& \\ = \hat{\lambda}(3\Psi_2 + 2\Phi_{11}),& \qquad (52)\\ (\Delta + 2\gamma + 4\mu)\hat{\Psi}_3 - (\delta + 4\beta - \tau)\Psi_4 - (\Delta + 2\gamma + 2\bar{\mu})\hat{\Phi}_{21}& \\ = \hat{\nu}(3\Psi_2 - 2\Phi_{11}).& \qquad (53)\end{aligned}$$

The fourth perturbation equation arises by considering what is essentially one of the Ricci identities, namely

$$\begin{aligned}[\hat{\bar{\delta}} + 2\hat{\pi}, \hat{\Delta} + 2\hat{\mu}]\hat{\phi}_1 \;=\; & (\hat{\bar{\tau}} - \hat{\bar{\beta}} - \hat{\alpha})\hat{\Delta}\hat{\phi}_1 + (\hat{\bar{\mu}} - \hat{\bar{\gamma}} + \hat{\gamma})\hat{\bar{\delta}}\hat{\phi}_1 \\ & - \hat{\nu}\hat{D}\hat{\phi}_1 + \hat{\lambda}\hat{\delta}\hat{\phi}_1 - 2\hat{\phi}_1\hat{\Delta}\hat{\pi} + 2\hat{\phi}_1\hat{\bar{\delta}}\hat{\mu}. \qquad (54)\end{aligned}$$

A similar set of four equations exist between the quantities $\hat{\Psi}_0$, $\hat{\Psi}_1$, $\hat{\Phi}_{01}$, $\hat{\kappa}$ and $\hat{\sigma}$. Using Maxwell's equations to substitute for the derivatives of $\hat{\phi}_1$ in equation (54), applying two of the Newman-Penrose equations, then swapping the order of differentiation in each of the two second order differential operators occuring on the left hand side of the resulting equation, one can obtain, to first order

$$\begin{aligned} &(\delta + 3\beta - \bar{\alpha} + \bar{\pi})[(\bar{\delta} + 2\alpha + 3\pi)\hat{\phi}_2 - 2\hat{\lambda}\phi_1] \\ -\, &(D + 3\epsilon + \bar{\epsilon} - \bar{\rho})[(\Delta + 2\gamma + 3\mu)\hat{\phi}_2 - 2\hat{\nu}\phi_1] = 2(2\hat{\Psi}_3\phi_1 - 3\Psi_2\hat{\phi}_2). \quad (55) \end{aligned}$$

By making the following change of variables

$$\begin{aligned} u_3 &= \tfrac{1}{4}(3\Psi_2\hat{\phi}_2 - 2\hat{\Psi}_3\phi_1), & u_4 &= -\hat{\Psi}_4\phi_1, \\ v_0 &= (\bar{\delta} + 2\alpha + 3\pi)\hat{\phi}_2 - 2\hat{\lambda}\phi_1, & v_1 &= (\Delta + 2\gamma + 3\mu)\hat{\phi}_2 - 2\hat{\nu}\phi_1, \quad (56) \end{aligned}$$

equations (51) - (54) can be cast in a simpler and more symmetric form:

$$\begin{aligned} (\delta + 3\beta - \bar{\alpha} + \bar{\pi})v_0 - (D + 3\epsilon + \bar{\epsilon} - \bar{\rho})v_1 &= -8u_3, & (57) \\ (\bar{\delta} + 3\pi + 3\alpha + \bar{\beta} - \bar{\tau})v_1 - (\Delta + 3\mu + 3\gamma + \bar{\mu} - \bar{\gamma})v_0 &= 2u_4, & (58) \\ -2(\bar{\delta} + 2\alpha + 6\pi)u_3 + (D + 4\epsilon - 3\rho)u_4 &= -\frac{1}{2}v_0\Omega_+, & (59) \\ -2(\Delta + 2\gamma + 6\mu)u_3 + (\delta + 4\beta - 3\tau)u_4 &= -\frac{1}{2}v_1\Omega_-, & (60) \end{aligned}$$

where $\Omega_\pm = 3\Psi_2 \pm 2\Phi_{11}$. The quantities u_3, u_4, v_0 and v_1 are also invariant under a change in the tetrad of first order of smallness (see Wald (1979) and Torres del Castillo (1987)).

The four equations (57) - (60) do not appear to admit separable solutions. The presence of terms in the Ricci spinor on the right hand side of equations (59) and (60) prevents an obvious decoupling of the equations as has occured in the case of the Kerr metric gravitational perturbations. Torres del Castillo (1988c) has shown that the Hertz potential methods used in the Kerr case can be extended to the problem of the Kerr-Newman perturbation. However instead of one potential two are now required. The equations satisfied by these potentials are closely related to equations (57) - (60). This observation implies the existence of a symmetry operator for the complex potential equations (Kalnins and Williams, 1997a). Consequently if we have just one solution to the equations we can generate more by application of the symmetry operator. Indeed the equations (57) - (60) can be reformulated in terms of integral equations using the Green's function of the Teukolsky operator for spins 1 and 2. The resulting formulation then provides a solution in principle by repeated iteration of the resulting integral operator (Kalnins and Williams, 1997a).

Chandrasekhar in his computations with these equations sought a decoupling analogous to that of the Kerr case. He was unsuccessful but computer algebra computations have shown that this is indeed possible (Kalnins and Williams, 1997b) resulting in partial differential equations of eighth order. However the condition appears to be without utility.

In the Reissner-Nordström space-time, solution of equations (57) - (60) is possible. By writing each of the four functions u_3, u_4, v_0 and v_1 as a product of a radial function and an angular function and by making use of the spherical symmetry of the Reissner-Nordström space-time one obtains a solution for these four functions of the form

$$\begin{aligned} v_0 &= -iu^l_{m2}(\theta)V_0(r), \qquad & v_1 &= u^l_{m1}(\theta)V_1(r), \\ u_3 &= u^l_{m1}(\theta)U_3(r), \qquad & u_4 &= -iu^l_{m2}(\theta)U_4(r), \end{aligned} \tag{61}$$

where the functions $u^l_{mn}(\theta)$ are generalised spherical functions (Gel'fand *et al.*, 1963). The functions U_3, U_4, V_0 and V_1 satisfy four radial equations which can be written in matrix form as

$$\begin{bmatrix} (\mathcal{D}_0 - 3r^{-1}) & \mu & 0 & -2r^{-1} \\ \mu & \Delta(\mathcal{D}^\dagger_{-1} + 3r^{-1}) & -r^{-1} & 0 \\ 0 & 8q^2r^{-1} - 6M & (\mathcal{D}_0 - 3r^{-1}) & -\mu \\ 4q^2r^{-1} - 6M & 0 & -\mu & \Delta(\mathcal{D}^\dagger_{-1} + 3r^{-1}) \end{bmatrix} \times \begin{bmatrix} V_1 \\ V_0 \\ U_4 \\ U_3 \end{bmatrix} = \begin{bmatrix} 0 \\ 0 \\ 0 \\ 0 \end{bmatrix}. \tag{62}$$

where $\mu = \sqrt{(l+2)(l-1)}$. It remains to decouple these equations. Following Chandrasekhar (1979) we make the change of variables

$$\begin{bmatrix} V_1 \\ V_0 \\ U_4 \\ U_3 \end{bmatrix} = \begin{bmatrix} \mu & 0 & -\mu & 0 \\ 0 & -\mu & 0 & \mu \\ -q_1 & 0 & q_0 & 0 \\ 0 & q_0 & 0 & -q_1 \end{bmatrix} \begin{bmatrix} F_0 \\ G_0 \\ F_1 \\ G_1 \end{bmatrix} \tag{63}$$

in equation (62) and then multiply on the left by

$$\frac{1}{\mu(b_0 - b_1)} \begin{bmatrix} b_0 & 0 & \mu & 0 \\ 0 & b_1 & 0 & \mu \\ b_1 & 0 & \mu & 0 \\ 0 & b_0 & 0 & \mu \end{bmatrix}, \tag{64}$$

where b_0 and b_1 are defined by the relations

$$b_0 + b_1 = 6M \quad \text{and} \quad b_0 b_1 = -4\mu^2 q^2. \tag{65}$$

The four equations decouple into the following two pairs (here $i = 0,\ 1$)

$$\begin{aligned} \left(\mathcal{D}_0 - \frac{3}{r}\right)F_i - \left(\mu + \frac{2b_i}{\mu r}\right)G_i &= 0, \\ \Delta\left(\mathcal{D}^\dagger_{-1} + \frac{3}{r}\right)G_i - \left(\mu + \frac{b_{1-i}}{\mu r}\right)F_i &= 0. \end{aligned} \tag{66}$$

If we make the substitutions

$$F_i = r^3 Y_i \quad \text{and} \quad G_i = \frac{\Delta}{r^3}X_i \tag{67}$$

then these two pairs of equations become,

$$\begin{aligned} r^3\mathcal{D}_0 Y_i - \frac{\Delta}{r^3}\left(\mu + \frac{2b_i}{\mu r}\right)X_i &= 0, \\ \frac{\Delta^2}{r^3}\mathcal{D}^\dagger_0 X_i - r^3\left(\mu + \frac{b_{1-i}}{\mu r}\right)Y_i &= 0. \end{aligned} \tag{68}$$

Eliminating X_i from the second of equations (68) results in the second order ordinary differential equation

$$\Delta\left(\mathcal{D}^\dagger_{-1} + \frac{6}{r}\right)\left[\left(\mu + \frac{2b_i}{\mu r}\right)^{-1}\mathcal{D}_0 Y_i\right] - \left(\mu + \frac{b_{1-i}}{\mu r}\right)Y_i = 0. \tag{69}$$

7. Teukolsky Functions and Teukolsky-Starobinsky Identities

The spin s Teukolsky functions $R_{\pm s}(r)$ and $S_{\pm s}(\theta)$ satisfy the second order differential equations (here $\Delta = r^2 - 2Mr + a^2$)

$$\begin{aligned} (\Delta\mathcal{D}^\dagger_{1-s}\mathcal{D}_0 + 2(2s-1)i\omega r)R_{-s} &= \lambda R_{-s}, \\ (\Delta\mathcal{D}_1\mathcal{D}^\dagger_s - 2(2s-1)i\omega r)R_{+s} &= \lambda R_{+s}, \\ (\mathcal{L}^\dagger_{1-s}\mathcal{L}_s + 2(2s-1)\omega a\cos\theta)S_{+s} &= -\lambda S_{+s}, \\ (\mathcal{L}_{1-s}\mathcal{L}^\dagger_s - 2(2s-1)\omega a\cos\theta)S_{-s} &= -\lambda S_{-s}. \end{aligned} \tag{70}$$

The Teukolsky functions $R_{\pm s}$ and $S_{\pm s}$ have some interesting properties. In fact the equations satisfied by the $S_{\pm s}$ functions are a form of confluent Heun equation. Fackerell and Crossman (1977) and Leaver (1986) have studied in detail expansion theorems and the spectrum of the parameter λ and its various asymptotic forms.

Kalnins *et al.* (1989b) have shown the Teukolsky-Starobinsky identities hold for $s = \frac{1}{2},\ 1,\ \frac{3}{2},\ 2,\ \frac{5}{2},\ \ldots$. The identities are

$$\Delta^s \mathcal{D}_0^{2s}(\Delta \mathcal{D}_{1-s}^\dagger \mathcal{D}_0 + 2(2s-1)i\omega r) = (\Delta \mathcal{D}_{1-s} \mathcal{D}_0^\dagger - 2(2s-1)i\omega r)\Delta^s \mathcal{D}_0^{2s}, \tag{71}$$

$$\mathcal{L}_{1-s}\mathcal{L}_{2-s}\ldots\mathcal{L}_{s-1}\mathcal{L}_s(\mathcal{L}_{1-s}^\dagger \mathcal{L}_s + 2(2s-1)\omega a\cos\theta) = (\mathcal{L}_{1-s}\mathcal{L}_s^\dagger - 2(2s-1)\omega a\cos\theta)\mathcal{L}_{1-s}\mathcal{L}_{2-s}\ldots\mathcal{L}_{s-1}\mathcal{L}_s. \tag{72}$$

Two very similar identities are obtained by taking the complex conjugate of the first identity ($\mathcal{D} \longleftrightarrow \mathcal{D}^\dagger$) and by letting $\theta \to \pi - \theta$ in the second identity ($\mathcal{L} \longleftrightarrow -\mathcal{L}^\dagger$). They read

$$\Delta^s \mathcal{D}^{\dagger 2s}_0(\Delta \mathcal{D}_{1-s} \mathcal{D}_0^\dagger - 2(2s-1)i\omega r) = (\Delta \mathcal{D}_{1-s}^\dagger \mathcal{D}_0 + 2(2s-1)i\omega r)\Delta^s \mathcal{D}^{\dagger 2s}_0, \tag{73}$$

$$\mathcal{L}_{1-s}^\dagger\mathcal{L}_{2-s}^\dagger\ldots\mathcal{L}_{s-1}^\dagger\mathcal{L}_s^\dagger(\mathcal{L}_{1-s}\mathcal{L}_s^\dagger - 2(2s-1)\omega a\cos\theta) = (\mathcal{L}_{1-s}^\dagger \mathcal{L}_s + 2(2s-1)\omega a\cos\theta)\mathcal{L}_{1-s}^\dagger\mathcal{L}_{2-s}^\dagger\cdots\ldots\mathcal{L}_{s-1}^\dagger\mathcal{L}_s^\dagger. \tag{74}$$

Note that for any function $f(r)$

$$\mathcal{D}_n(\Delta^s f(r)) = \Delta^s \mathcal{D}_{n+s} f(r). \tag{75}$$

Note also that R_{-s} and $\Delta^s R_{+s}$ satisfy complex conjugate equations. Similarly, $S_{+s}(\pi - \theta)$ satisfies the same equation as $S_{-s}(\theta)$.

A direct consequence of the Teukolsky-Starobinsky identities is the following: If one acts on the function R_{-s} with both sides of (71) then one finds that $\Delta^s \mathcal{D}_0^{2s} R_{-s}$ is a solution of the Teukolsky equation for $\Delta^s R_{+s}$. Similarly, by acting on the function $\Delta^s R_{+s}$ with both sides of (73) one finds that $\Delta^s \mathcal{D}^{\dagger 2s}_0 \Delta^s R_{+s}$ is a solution of the Teukolsky equation for R_{-s}. By suitably choosing the relative normalisation of the functions R_{-s} and R_{+s}, we can write

$$\begin{aligned} \Delta^s \mathcal{D}_0^{2s} R_{-s} &= \mathcal{C}_s \Delta^s R_{+s}, \\ \Delta^s \mathcal{D}^{\dagger 2s}_0 \Delta^s R_{+s} &= \bar{\mathcal{C}}_s R_{-s} \end{aligned} \tag{76}$$

where $\mathcal{C}_s$ is a complex constant. A similar treatment of the angular identities results in the relations

$$\begin{aligned} \mathcal{L}_{1-s}\mathcal{L}_{2-s}\ldots\mathcal{L}_{s-1}\mathcal{L}_s S_{+s} &= \mathcal{B}_s S_{-s}, \\ \mathcal{L}_{1-s}^\dagger\mathcal{L}_{2-s}^\dagger\ldots\mathcal{L}_{s-1}^\dagger\mathcal{L}_s^\dagger S_{-s} &= (-1)^{2s}\mathcal{B}_s S_{+s}. \end{aligned} \tag{77}$$

where $\mathcal{B}_s$ is a real constant. The constants $\mathcal{C}_s$ and $\mathcal{B}_s$ are referred to as Starobinsky constants.

By use of the mapping $r \longleftrightarrow \xi = ia\cos\theta$ one can make the above radial and angular identities (76) and (77) almost formally identical. By such means one can show that

$$\mathcal{B}_s^2 = (-1)^{2s}|\mathcal{C}_s|^2\,|_{M=0} \tag{78}$$

Computation of $|\mathcal{C}_s|^2$ can be achieved by examining the identity

$$\Delta^s \mathcal{D}^{\dagger 2s}_0 \Delta^s \mathcal{D}_0^{2s} R_{-s} = |\mathcal{C}_s|^2 \tag{79}$$

and replacing the occurences of second order and higher derivatives of R_{-s} with first and zeroth order terms by means of the Teukolsky equation that R_{-s} must satisfy. Such a procedure is very protracted for all but the first few small values of s. Chandrasekhar (1990) has shown how to determine the general expression for the magnitude of the Teukolsky-Starobinsky constant $|\mathcal{C}_s|^2$ in the form of a determinant resulting from considerations following from the notion of algebraically special solutions. Torres del Castillo (1988b) has also investigated aspects of the applicability of Teukolsky-Starobinsky identities in type D vacuum backgrounds with cosmological constant. The magnitude of the Starobinsky constants for spin $\frac{1}{2}$, 1, $\frac{3}{2}$ and 2 are

$$|\mathcal{C}_{\frac{1}{2}}|^2 = \lambda, \tag{80}$$

$$|\mathcal{C}_1|^2 = \lambda^2 - 4\omega^2\tilde{\alpha}^2, \tag{81}$$

$$|\mathcal{C}_{\frac{3}{2}}|^2 = \lambda^2(\lambda+1) - 16\omega^2(\lambda\tilde{\alpha}^2 - a^2), \tag{82}$$

$$\begin{aligned}|\mathcal{C}_2|^2 = {} & \lambda^2(\lambda+2)^2 - 8\omega^2\tilde{\alpha}^2\lambda(5\lambda+6) + 96\omega^2a^2\lambda, \\ & + 144\omega^4\tilde{\alpha}^4 + 144\omega^2M^2,\end{aligned} \tag{83}$$

where

$$\tilde{\alpha}^2 = a^2 + \frac{am}{\omega}. \tag{84}$$

We also note here that it is possible to invent a relativistically invariant system of equations which has the general spin s Teukolsky functions appearing in their solution. In fact the coupled system of equations

$$\begin{aligned}&\nabla^{KA'}\phi_{KA_2\ldots A_{2s}} = (2s-1)(s-1)\Psi^{KL}{}_{(A_2A_3}\theta^{A'}{}_{A_4\ldots A_{2s})KL}, \\ &I\nabla_{(A_1K'}\theta^{K'}{}_{A_2\ldots A_{2s})} - \frac{1}{6}(2s-3)\theta^{K'}{}_{(A_2\ldots A_{2s}}\nabla_{A_1)K'}I = I\phi_{A_1A_2\ldots A_{2s}},\end{aligned} \tag{85}$$

where $I = \Psi_{ABCD}\Psi^{ABCD}$, typically has solutions of the form

$$\begin{aligned}\phi_{0\ldots0} &= R_{+s}S_{+s}e^{i\omega t+im\varphi}, \\ \theta^{0'}_{0\ldots0} &= -\frac{1}{\sqrt{2}\rho'^2(2s-1)(s-1)\Psi_2}\Delta\left(\mathcal{D}^\dagger_s - \frac{2s-1}{\bar{\rho}'}\right)\phi_{0\ldots0}, \\ \theta^{1'}_{0\ldots0} &= -\frac{1}{\sqrt{2}\bar{\rho}'(2s-1)(s-1)\Psi_2}\left(\mathcal{L}^\dagger_s - \frac{(2s-1)ia\sin\theta}{\bar{\rho}'}\right)\phi_{0\ldots0},\end{aligned} \tag{86}$$

with similar expressions existing for components having indices consisting of all 1's.

8. Comments

We have seen that the solution of systems of differential equations that arise in General Relativity via some sort of separation of variables contains elegant examples but as yet no theoretical basis. More work at a mathematical level needs to be done to improve this theoretical situation as well as to solve the problems specific to General Relativity itself e.g. the combined perturbation problem for the Kerr-Newman black hole. A theory of separation of variables for the non-scalar wave equations of mathematical physics needs to explain the separability observed in these important physical examples.

References

Bagrov, V.G. and Gitman, D.M. (1990) *Exact solutions of relativistic equations*, Kluwer, Dortrecht.

Carter, B. (1968) *Comm. Math. Phys.*, **10**, 280.

Carter, B. and McLenaghan, R.G. (1979) *Phys. Rev.*, **D19**, 1093.

Chandrasekhar, S. (1976) *Proc. R. Soc. Lond.*, **A349**, 571.

Chandrasekhar, S. (1979) *Proc. R. Soc. Lond.*, **A365**, 453.

Chandrasekhar, S. (1983) *The Mathematical Theory of Black Holes*, Oxford University Press.

Chandrasekhar, S. (1990) *Proc. R. Soc. Lond.*, **A430**, 433.

Cohen, J.M. and Kegeles, L.S. (1975) *Phys. Rev.*, **D10**, 1070.

Dhurandhar, S.V., Vishveshwara, C.V. and Cohen, J.M. (1980) *Phys. Rev.*, **D21**, 2794.

Cosgrove, C.M., Fackerell, E.D. and Suffern K.G. (1983) *J. Math. Phys.*, **24**, 1350.

Debever, R., Kamran, N. and McLenaghan, R.G. (1984) *J. Math. Phys.*, **25**, 1955.

Fackerell, E.D. and Crossmann, R.G. (1977) *J. Math. Phys.*, **18**, 1849.

Fels, M. and Kamran, N. (1990) *Proc. R. Soc. Lond.*, **A428**, 229.

Gel'fand, I.M., Minlos, R.A. and Shapiro, Z.Ya. (1963) *Representations of the Rotation and Lorentz Groups and their applications*, Pergamon Press.

Güven, R. (1977) *Proc. R. Soc. Lond.*, **A356**, 465.

Güven, R. (1980) *Phys. Rev.*, **D22**, 2327.

Heusler, M. (1996) *Black Hole Uniqueness Theorems*, Cambridge University Press.

Iyer, B.R. and Vishveshwara, C.V. (1985) J. Math. Phys., **26**, 1034.

Iyer, B.R. and Vishveshwara, C.V. (1987) J. Math. Phys., **28**, 1377.

Iyer, B.R. and Kamran, N. (1991) *J. Math. Phys.*, **32**, 2497.

Kalnins, E.G., Miller Jr., W. and Williams, G.C. (1986) *J. Math. Phys.*, **27**, 1893.

Kalnins, E.G., Miller Jr., W. and Williams, G.C. (1989a) *J. Math. Phys.*, **30**, 2630.

Kalnins, E.G., Miller Jr., W and Williams, G.C. (1989b) *J. Math. Phys.*, **30**, 2925.

Kalnins, E.G. and Miller Jr., W. (1991) *J. Math. Phys.*, **32**, 698.

Kalnins, E.G., McLenaghan, R.G. and Williams, G.C. (1992a) *Proc. R. Soc. Lond.*, **A439**, 103.

Kalnins, E.G. and Miller Jr., W. (1992b) *J. Math. Phys.*, **33**, 286.

Kalnins, E.G., Williams, G.C. and Miller Jr., W. (1996) *Proc. R. Soc. Lond.*, **A452**, 997.

Kalnins, E.G. and Williams, G.C. (1997a) Approachs to the Kerr-Newman perturbation problem: Symmetry operators and Green's functions, *(preprint)*.

Kalnins, E.G. and Williams, G.C. (1997b) Integrability conditions for the coupled electromagnetic-gravitational perturbation equations of the Kerr-Newman space-time, *(preprint)*.
Kamran, N. and McLenaghan, R.G. (1984b) *Bull. Cl. Sci. Acad. R. Belg.*, **70**, 596.
Kamran, N. and McLenaghan, R.G. (1984a) *J. Math. Phys.*, **25**, 1019.
Kamran, N. and McLenaghan, R.G., (1984c) *Phys. Rev.*, **D30**, 357.
(1985) *J. Math. Phys.*, **26**, 1740.
Kegeles, L.S. and Cohen, J.M. (1979) *Phys. Rev.*, **D19**, 1641.
Leaver, E.W. (1986) *J. Math. Phys.*, **27**, 1238.
McLenaghan, R.G. and Spindel, Ph. (1979) *Phys. Rev.*, **D20**, 409.
McLenaghan, R.G. and Walker, D.M. (1996) Symmetry Operators for the Weyl Neutrino Equations on Curved Space-Time, *(preprint)*.
Miller Jr., W. (1988) Mechanisms for variable separation in partial differential equations and their relationship to group theory. In *Symmetries and Non-linear Phenomena*, eds. D. Levi and P. Winternitz, World Scientific (1988) 188.
Page, D.N. (1976) *Phys. Rev.*, **D14**, 1509.
Penrose, R. and Rindler, W. (1984) *Spinors and space-time Vol. 1: Two-spinor calculus and relativistic fields*, Cambridge University Press.
Rudiger, R. (1984) *J. Math. Phys.*, **25**, 649.
Shapovalov, V.N. and Ekle, G.G. (1974a) *Izv. vyssh. ucheb. zaved. Fiz.*, **2**, 83.
Shapovalov, V.N. and Ekle, G.G. (1974b) *Izv. vyssh. ucheb. zaved. Fiz.*, **2**, 88.
Shishkin, G.V. and Villalba, V.V. (1989) *J. Math. Phys.*, **30**, 2132.
Teukolsky, S.A. (1972) *Phys. Rev. Lett.*, **29**, 1114.
Teukolsky, S.A. (1973) *Astrophys. J.*, **185**, 635.
Torres del Castillo, G.F. (1987) *Class. Quantum. Grav.*, **4**, 1011.
Torres del Castillo, G.F. (1988a) *J. Math. Phys.*, **29**, 971.
Torres del Castillo, G.F. (1988b) *J. Math. Phys.*, **29**, 2078.
Torres del Castillo, G.F. (1988c) *Class. Quantum. Grav.*, **5**, 649.
Torres del Castillo, G.F. (1989a) *J. Math. Phys.*, **30**, 446.
Torres del Castillo, G.F. (1989b) *J. Math. Phys.*, **30**, 1323.
Torres del Castillo, G.F. (1989c) *J. Math. Phys.*, **30**, 2614.
Unruh, W. (1973) *Phys. Rev. Lett.*, **31**, 1265.
Wald, R.M. (1978) *Phys. Rev. Lett.*, **41**, 203.
Wald, R.M. (1979) *Proc. R. Soc. Lond.*, **A369**, 67.

"By reason of its faster and faster infall the surface of the imploding star moves away from the distant observer more and more rapidly. The light is shifted to the red. It becomes dimmer millisecond by millisecond, and in less than a second is too dark to see...The star, like the Cheshire cat, fades from view. One leaves behind only its grin, the other, only its gravitational attraction. Gravitational attraction, yes; light, no. No more than light do any particles emerge. Moreover, light and particles incident from outside ... and going down the black hole only add to its mass and increase its gravitational attraction."

— JOHN WHEELER (1967)

Black Hole was Wheeler's new name. Within months it was adopted enthusiastically by relativity physicists, astrophysicists, and the general public, in East as well as West – with one exception: In France, where the phrase *trou noir* (black hole) has obscene connotations, there was resistance for several years.

— KIP THORNE

At this stage of exploration, the black hole physicists realised that almost all of their results had been anticipated in the latter half of the last century by an Oxford mathematics don who wrote under the name of Lewis Carroll. Alice's fall is symbolic of gravitational collapse. The white rabbit's watch cannot maintain regular time as the rabbit flits in and out of the strong gravitational field. Alice opens up like a telescope because of tidal distortion. The Cheshire cat's grin represents the gradual disappearance of a collapsing star as viewed by an outside observer. As the red queen says, an observer has to do all the running he can in order to stay at the same place, as the stationary limit is reached.

— C. V. VISHVESHWARA

4. ENERGY-CONSERVATION LAWS FOR PERTURBED STARS AND BLACK HOLES

V. FERRARI
Dipartimento di Fisica "G. Marconi",
Università di Roma "La Sapienza" and Sezione INFN ROMA1,
p.le A. Moro 5, I-00185, Roma, Italy.

1. Introduction

When a gravitational wave impinges on a star or a black hole, there is no exact way to evaluate which fraction of wave will be absorbed and which reflected. For such scattering processes, an energy conservation law cannot be derived in the framework of the exact theory of General Relativity. However, due to the small amplitude that gravitational waves are expected to possess, the effect they induce can be considered as a small perturbation of the background geometry. Under this assumption a scattering theory can be constructed and conservation laws governing the generation and propagation of the energy flux can be found, but with relevant differences between stars and black holes.

It is well known that after the separation of variables, accomplished by expanding all perturbed tensors in suitably defined angular harmonics, the perturbations of black holes are described by radial Schrödinger-like equations, with potential barriers generated by the spacetime curvature [1]-[5]. The underlying scattering problem is one by a one-dimensional potential barrier, extending from $r_* = -\infty$ (the black hole horizon) to $r_* = +\infty$, where r_* is the *tortoise* radial coordinate usually introduced to study these problems. In this case, an energy conservation law can be constructed in terms of Wronskians of independent solutions of the wave equations, the constancy of which implies that given a wave of unitary amplitude incident on the potential barrier, the sum of the squared amplitudes of the transmitted and reflected waves is still equal to unity [6]. If the black hole is non-rotating, an alternative approach allows to construct, directly from the linearized Einstein-Maxwell equations, a vector $\vec{\mathbf{E}}$ which satisfies an

B. R. Iyer and B. Bhawal (eds.), Black Holes, Gravitational Radiation and the Universe, 53–68.

ordinary divergence-free equation, and therefore represents a *conserved current* [7]. It can be shown that this current coincides with that obtained by taking the second variations of the time component of the Einstein pseudotensor for the gravitational field, added to a suitably defined Noether operator for the Maxwell field if the black hole is charged [8, 9]. Obviously, they both reduce to the energy flux found in terms of Wronskians, after the separation of variables is accomplished. The relations between these different approaches will be discussed in sections 2, 3 and 6. Finally, since the Einstein-Maxwell equations arise from a Lagrangian variational principle, the conserved current can also be derived by a suitably defined symplectic current constructed from the background field and a pair of linearized solutions off of this background [10].

A gravitational wave incident on a star, no matter how compact the star is, interacts with the fluid present in the interior, and the consequent exchange of energy changes the nature of the scattering problem with respect to the black hole case. We shall restrict the following analysis to non-rotating stars. In this case, after separating the variables [1], the linearized Einstein equations and the equations of hydrodynamics split into two distinct sets, the *axial* and the *polar*. The axial perturbations induce only a stationary rotation in the fluid, and they are described by a radial Schrödinger-like equation with a potential barrier which depends on how the energy density and the pressure are distributed in the unperturbed star. The underlying scattering problem is now one by a central potential barrier and, as for black holes, the energy flux can still be constructed in terms of Wronskians of independent solutions. However, due to the particular boundary conditions the solution must satisfy, this flux can be explicitly evaluated only by applying the Regge theory of potential scattering, essentially in its original form [11]. This will be shown in section 4.

For the polar perturbations, which do excite pulsations in the fluid, the conserved current can be constructed either by direct manipulation of the linearized Einstein + hydrodynamical equations [7], or by the second variations of the Einstein pseudotensor and of a suitably defined Noether operator for the fluid [12]. The explicit evaluation of this conserved current requires the generalization of the Regge theory of complex angular momentum [11] to the system of coupled differential equations describing the polar perturbations, as will be shown in section 5.

[1] The separation of variables in the perturbed equations of a non-rotating star or black hole can be accomplished by expanding the perturbed metric tensor and the stress-energy tensor of the source in tensorial spherical harmonics. These harmonics have a different behaviour under the angular transformation $\theta \to \pi - \theta$, $\varphi \to \pi + \varphi$, and consequently the separated equations split in two sets, the *polar* or *even*, belonging to the parity $(-1)^{\ell}$, and the *axial* or *odd*, belonging to the parity $(-1)^{(\ell+1)}$.

2. The derivation of a conserved current from the field equations

The axisymmetric perturbations of a static, spherically symmetric spacetime with prevailing sources (a Maxwell field or a perfect fluid) are governed by flux integrals that can be derived *ab initio* directly from the linearized equations [7]. I will briefly illustrate how such flux integrals can be obtained. We start writing the metric of the unperturbed spacetime in the following form

$$ds^2 = e^{2\nu}dt^2 - e^{2\mu_2}dr^2 - e^{2\mu_3}d\theta^2 - e^{2\psi}d\varphi^2, \tag{1}$$

where the metric functions depend only on r and ϑ. The axisymmetric perturbations of the spacetime (1) can be described by the line-element [2]

$$ds^2 = e^{2\nu}dt^2 - e^{2\psi}(d\varphi - q_2dr^2 - q_3d\theta - \omega dt)^2 - e^{2\mu_2}dr^2 - e^{2\mu_3}d\theta^2. \tag{2}$$

As a consequence of a generic perturbation, the metric functions undergo small changes with respect to their unperturbed values

$$\begin{aligned} \nu &\longrightarrow \nu + \delta\nu, \qquad & \mu_2 &\longrightarrow \mu_2 + \delta\mu_2, \\ \psi &\longrightarrow \psi + \delta\psi, \qquad & \mu_3 &\longrightarrow \mu_3 + \delta\mu_3, \\ \omega &\longrightarrow \delta\omega, \qquad & q_2 &\longrightarrow \delta q_2 \quad , \qquad q_3 \longrightarrow \delta q_3. \end{aligned} \tag{3}$$

In addition to the metric perturbations, if the source of the background geometry is a Maxwell field, the perturbations of the electromagnetic tensor have to be considered

$$F_{\mu\nu} \longrightarrow F_{\mu\nu} + \delta F_{\mu\nu}. \tag{4}$$

If the prevailing source is a perfect fluid (a star), as a consequence of a perturbation each element of fluid moves away from its equilibrium position and consequently the energy density and the pressure change by an infinitesimal amount

$$\epsilon \longrightarrow \epsilon + \delta\epsilon, \qquad\qquad p \longrightarrow p + \delta p. \tag{5}$$

The perturbations will be assumed to take place adiabatically. We shall impose, with no loss of generality, that all perturbed functions have a common time dependence given by $e^{i\sigma t}$.

The approach used in ref.[7] to find the conserved current is the following.

i) We write the perturbed Einstein-Maxwell equations or the Einstein and

[2] The number of unknown functions in eq. (2) is seven, one more than needed. However, this extra degree of freedom is removed when the boundary conditions of the problem are specified.

the hydrodynamical equations respectively in the case of charged solutions or in the case of stars. These equations are written in terms of the perturbed variables, say δX_i. ii) We multiply each equation by an assigned δX_i^*, where the $*$ indicates complex conjugation and σ is assumed to be real, and subtract the complex conjugate equation. iii) The resulting equations are combined and rearranged, and finally put in the form of two divergence-free equations, one for the *polar*, one for the *axial* perturbations

$$\frac{\partial}{\partial x^\alpha} E^\alpha = 0, \qquad \alpha = (x^2 = r, x^3 = \cos\theta). \tag{6}$$

For the *polar* equations, which couple the perturbations of the diagonal part of the metric $[\delta\nu, \delta\psi, \delta\mu_2, \delta\mu_3]$ with the perturbations of the source, the vector $\vec{\mathbf{E}}$ is

Einstein-Maxwell spacetime

$$\begin{aligned} E_2^{pol} &= e^{\alpha_2}\Big\{\Big[\delta\mu_3, \delta\mu_3^*\Big]_2 + \Big[\delta\psi, \delta\psi^*\Big]_2 \\ &- \Big(\delta\nu_{,2}\delta(\psi+\mu_3)^* - c.c\Big) + \Big(\delta\mu_2\delta(\psi+\mu_3)^*_{,2} - c.c\Big) \\ &+ 4F_{03}e^{-\psi+\mu_2}\Big(Y\delta\psi^* - Y^*\delta\psi\Big) + 2e^{-2\psi}\Big[Y, Y^*\Big]_2\Big\} \\ E_3^{pol} &= e^{\alpha_3}\Big\{\Big[\delta\mu_2, \delta\mu_2^*\Big]_3 + \Big[\delta\psi, \delta\psi^*\Big]_3 \\ &- \Big(\delta\nu_{,3}\delta(\psi+\mu_2)^* - c.c\Big) + \Big(\delta\mu_3\delta(\psi+\mu_2)^*_{,3} - c.c\Big) \\ &- 4F_{02}e^{-\psi+\mu_3}\Big(Y\delta\psi^* - Y^*\delta\psi\Big)2e^{-2\psi}\Big[Y, Y^*\Big]_3\Big\}, \end{aligned} \tag{7}$$

where $Y = \dfrac{e^{\psi+\nu}}{i\sigma}\delta F_{23}$, and

$$\begin{aligned} \alpha 2 &= \psi + \nu + \mu_3 - \mu_2 \\ \alpha 3 &= \psi + \nu + \mu_2 - \mu_3. \end{aligned} \tag{8}$$

Spacetimes with a fluid source

$$\begin{aligned} E_2^{pol} &= e^{\alpha_2}\Big\{\Big[\delta\mu_3, \delta\mu_3^*\Big]_2 + \Big[\delta\psi, \delta\psi^*\Big]_2 - \Big(\delta\nu_{,2}\delta(\psi+\mu_3)^* - c.c\Big) \\ &+ \Big(\delta\mu_2\delta(\psi+\mu_3)^*_{,2} - c.c\Big) \\ &+ \Big(2[(\epsilon+p)\delta(\psi+\mu_3-\mu_2)^* - \delta p^*]e^{\nu+\mu_2}\xi_2 - c.c\Big)\Big\}, \\ E_3^{pol} &= e^{\alpha_3}\Big\{\Big[\delta\mu_2, \delta\mu_2^*\Big]_3 + \Big[\delta\psi, \delta\psi^*\Big]_3 - \Big(\delta\nu_{,3}\delta(\psi+\mu_2)^* - c.c\Big) \end{aligned}$$

$$+ \left(\delta\mu_3\delta(\psi+\mu_2)^*_{,3} - c.c\right)$$
$$+ \left(2[(\epsilon+p)\delta(\psi+\mu_2-\mu_3)^* - \delta p^*]e^{\nu+\mu_3}\xi_3 - c.c\right)\Big\}, \qquad (9)$$

where [3] ξ_2 and ξ_3 are the components of the lagrangian displacement of the perturbed elements of fluid. The terms $\left[\delta X_i, \delta X_i^*\right]_2$ and $\left[\delta X_i, \delta X_i^*\right]_3$ are the Wronskians of two complex conjugate solutions:

$$\left[\delta X_i, \delta X_i^*\right]_k = \delta X_{i,k}\delta X_i^* - \delta X_{i,k}^*\delta X i. \qquad (10)$$

When the source terms are zero, Eqs.(7) and (9) coincide, and refer to the polar perturbations of a Schwarzschild black hole.

The axial perturbations involve the off-diagonal part of the metric plus source terms. By applying the same procedure as in the polar case, one finds that the vector $\vec{\mathbf{E}}$ in the Einstein-Maxwell case is

$$\begin{aligned} E_2^{ax} &= e^{-3\psi+\nu-\mu_2+\mu_3}\left[X, X^*\right]_2 + 4e^{-\psi+\nu-\mu_2+\mu_3}\left[Y, Y^*\right]_2 \\ &- 4F_{03}e^{-2\psi+\nu+\mu_3}\left(YX^* - Y^*X\right) \\ E_3^{ax} &= e^{-3\psi+\nu-\mu_3+\mu_2}\left[X, X^*\right]_3 + 4e^{-\psi+\nu-\mu_3+\mu_2}\left[Y, Y^*\right]_3 \\ &+ 4F_{02}e^{-2\psi+\nu+\mu_2}\left(YX^* - Y^*X\right) \end{aligned} \qquad (11)$$

where now

$$\begin{aligned} Y &= e^{\psi+\nu}F_{01} \\ X &= e^{3\psi+\nu-\mu_2-\mu_3}\left(\delta q_{2,3} - \delta q_{3,2}\right). \end{aligned} \qquad (12)$$

In ref. [7] the axial conserved current was derived only for the Einstein-Maxwell case. However, as we shall show later, a conserved flux can also be derived in the fluid case through a different procedure.

The meaning of eq. (6) is clarified by the application of Gauss's theorem: given any two closed contours, say C_1 and C_2, taken one inside the other in the (x^2, x^3)-plane

$$\int_{C_1}(E_2 dx^3 - E_3 dx^2) = \int_{C_2}(E_2 dx^3 - E_3 dx^2), \qquad (13)$$

[3]It should be noted that in the expression (9) for the vector $\vec{\mathbf{E}}$ the terms containing the hydrodynamical variables have opposite sign with respect to those of ref. [7] Eqs.133-134. This is due to the fact that the convention which is used in this paper is

$$G_{ab} = +2T_{ab},$$

while in ref [7] the source term had opposite sign.

provided **E** is non-singular inside the area included between C_1 and C_2. If the closed contour is a circle of radius r eq. (6) becomes

$$< E_2 >= \int_0^\pi r^2 E_2 \sin\theta d\theta = \text{const}, \tag{14}$$

which expresses the conservation of the flux of the vector **E** across a spherical surface of radius r surrounding the central source.

Having established that the axisymmetric perturbations of a static spherically symmetric spacetime, with a Maxwell field or a perfect fluid as source, are governed by flux integrals, it remains to be ascertain what is the physical meaning of the vector $\vec{\mathbf{E}}$ and whether we are entitled to say that it does indeed represent an *energy flux*. We shall answer this question first by re-analysing the role of the vector $\vec{\mathbf{E}}$ in the framework of the theory of scattering applied to the perturbations of a non-rotating black hole.

3. The conserved current and the scattering theory

Let us consider the case when the conserved current $\vec{\mathbf{E}}$ refers to the polar perturbations of a Schwarzschild black hole, i.e all source terms in the above equations (7) or (9) vanish. Let us consider the polar case first. In order to explicitly evaluate the conserved flux, we need to expand the metric tensor in tensorial spherical harmonics. If the gauge is chosen as in eq. (2), the polar metric functions have the following angular dependence

$$\begin{aligned} \delta\nu &= N_\ell(r)P_\ell(\cos\theta), \qquad & \delta\mu_2 &= L_\ell(r)P_\ell(\cos\theta), \\ \delta\mu_3 &= T_\ell(r)P_\ell + V_\ell(r)P_{\ell,\theta,\theta}, \qquad & \delta\psi &= T_\ell(r)P_\ell + V_\ell(r)P_{\ell,\theta}\cot\theta, \end{aligned} \tag{15}$$

By this substitution, the polar equations can be separated and reduced to the Zerilli equation

$$\begin{aligned} &\frac{d^2 Z_\ell^{pol}}{dr_*^2} + [\sigma^2 - V_\ell^{pol}(r)]Z_\ell^{pol} = 0, \\ &V_\ell^{pol}(r) = \frac{2(r-2M)}{r^4(nr+3M)^2}[n^2(n+1)r^3 + 3Mn^2r^2 + 9M^2nr + 9M^3], \end{aligned} \tag{16}$$

where

$$n = \frac{1}{2}(\ell+1)(\ell-2), \qquad \frac{dr}{dr_*} = 1 - \frac{2M}{r}, \tag{17}$$

and the radial function $Z_\ell^{pol}(r)$ is

$$Z_\ell^{pol}(r) = \frac{r}{nr+3M}\left(3MV_\ell(r) - rL_\ell(r)\right). \tag{18}$$

By making use of Eqs.(15), of the background equations and of the relations among the perturbed quantities given in chapter 4 of ref. [6], it is possible to show that the conserved flux reduces to

$$< E_2^{pol} >= \int_0^\pi r^2 E_2^{pol} \sin\theta d\theta = \frac{4\ell(\ell+1)}{2\ell+1}\left[Z_\ell^{pol}, Z_\ell^{pol*}\right]_{r_*} = \text{const.} \qquad (19)$$

A similar derivation can be done for the axial perturbations. In that case the perturbed equations reduce to the Regge-Wheeler equation

$$\frac{d^2 Z_\ell^{ax}}{dr_*^2} + \left[\sigma^2 - V_\ell^{ax}(r)\right] Z_\ell^{ax} = 0,$$
$$V_\ell^{ax}(r) = \frac{1}{r^3}\left(1 - \frac{2M}{r}\right)[\ell(\ell+1)r - 6M], \qquad (20)$$

written for the radial function (cf. eq. 12)

$$Z_\ell^{ax}(r) = X(r)/r, \qquad (21)$$

and it is easy to show that in this case

$$\begin{aligned} < E_2^{ax} > &= \int_0^\pi r^2 E_2^{ax} \sin\theta d\theta \\ &= \frac{9}{(\ell+\frac{1}{2})(\ell+2)(\ell+1)(\ell-1)}\left[Z_\ell^{ax}, Z_\ell^{ax*}\right]_{r_*} = \text{const.} \end{aligned} \qquad (22)$$

Thus, for a Schwarzschild black hole the axial and polar conserved currents are the Wronskians of two complex conjugate solutions of the wave equations. We shall now show that the constancy of these Wronskians expresses indeed an energy conservation law. First we note that, due to the short-range character of the potentials of the Zerilli and of the Regge-Wheeler equations, the asymptotic behaviour of the solution at $r_* = \pm\infty$ is, in general, a superposition of outgoing and incoming waves $Z \sim e^{\pm i\sigma r_*}$. Consider two solutions of the wave equations, say Z_1 and Z_2, satisfying respectively the following boundary conditions

$$\begin{aligned} r_* &\to +\infty \qquad Z_1 \to e^{-i\sigma r_*}, \qquad \text{pure outgoing wave} \\ r_* &\to -\infty \qquad Z_2 \to e^{+i\sigma r_*} \qquad \text{pure incoming wave.} \end{aligned} \qquad (23)$$

(Z_1, Z_1^*) and (Z_2, Z_2^*), are pairs of independent solutions, since their Wronskians are different from zero. Therefore, we can write Z_1 as a linear combination of (Z_2, Z_2^*) and vice-versa:

$$\begin{aligned} Z_1 &= A(\sigma)Z_2 + B(\sigma)Z_2^*, \\ Z_2 &= C(\sigma)Z_1 + D(\sigma)Z_1^*. \end{aligned} \qquad (24)$$

We now divide Z_2 by $D(\sigma)$ and define

$$Z_R = \frac{Z_2}{D(\sigma)} = R_1(\sigma) Z_1 + Z_1^*, \tag{25}$$

where $R_1(\sigma) = \frac{C(\sigma)}{D(\sigma)}$, and similarly

$$Z_L = \frac{Z_1}{B(\sigma)} = R_2(\sigma) Z_2 + Z_2^*, \tag{26}$$

where $R_2(\sigma) = \frac{A(\sigma)}{B(\sigma)}$. The solutions Z_R and Z_L satisfy the following boundary conditions

$$Z_R \to \begin{cases} T_1(\sigma) e^{+i\sigma r_*} & r_* \to -\infty \\ e^{+i\sigma r_*} + R_1(\sigma) e^{-i\sigma r_*} & r_* \to +\infty \end{cases} \tag{27}$$

$$Z_L \to \begin{cases} e^{-i\sigma r_*} + R_2(\sigma) e^{+i\sigma r_*} & r_* \to -\infty \\ T_2(\sigma) e^{-i\sigma r_*} & r_* \to +\infty \end{cases}$$

where $T_1(\sigma) = \frac{1}{D(\sigma)}$, and $T_2(\sigma) = \frac{1}{B(\sigma)}$. Thus, Z_R represents a wave of unitary amplitude incident on the potential barrier from $+\infty$ which gives rise to a reflected wave of amplitude $R_1(\sigma)$ and to a transmitted wave of amplitude $T_1(\sigma)$, whereas Z_L is a unitary wave incident from $-\infty$ which is partially reflected $(R_2(\sigma))$ and partially transmitted $(T_2(\sigma))$. By computing the Wronskian of the two solutions at $\pm\infty$, it is easy to verify that

$$\begin{aligned} [Z_L, Z_R]_{r_*} &= -2i\sigma T_2(\sigma) & r_* \to +\infty \\ [Z_L, Z_R]_{r_*} &= -2i\sigma T_1(\sigma) & r_* \to -\infty, \end{aligned} \tag{28}$$

and, since the Wronskian is constant,

$$T_1(\sigma) = T_2(\sigma) = T(\sigma). \tag{29}$$

Moreover

$$\begin{aligned} [Z_L, Z_L^*]_{r_*} &= 2i\sigma(|R_2(\sigma)|^2 - 1) & r_* \to -\infty \\ [Z_L, Z_L^*]_{r_*} &= -2i\sigma|T_2(\sigma)|^2 & r_* \to +\infty, \end{aligned} \tag{30}$$

and consequently

$$|R_2|^2 + |T_2|^2 = 1. \tag{31}$$

By a similar procedure applied to Z_R we easily find

$$|R_1|^2 + |T_1|^2 = 1. \tag{32}$$

This means that R_1 and R_2 can differ only by a phase factor and that

$$|R|^2 + |T|^2 = 1 \tag{33}$$

holds in general. This equation, which derives from the constancy of the Wronskians, establishes the symmetry and the unitarity of the S-matrix, and expresses the conservation of energy because it says that if a wave of unitary amplitude is incident on one side of the potential barrier, it splits into a reflected and a transmitted wave such that the sum of the square of their amplitudes is still one. The relation between the conserved flux integrals given in Eqs.(19) and (22), and the energy conservation law (33) is thus established.

For a Reissner-Nordstrom black hole it can similarly be shown that the conserved polar flux can be written as

$$< E_2^{pol} >= \frac{4\ell(\ell+1)}{2\ell+1}\left[Z_{i\ell}^{pol}, Z_{i\ell}^{pol*}\right]_{r_*} = \text{const}, \qquad i = 1,2 \tag{34}$$

where $Z_{1\ell}^{pol}$ and $Z_{2\ell}^{pol}$ are a combination of the two kind of radiation, electromagnetic and gravitational, that prevail and similarly for $< E_2^{ax} >$ (cf. ref. [7]). The functions $(Z_{1\ell}^{pol}, Z_{2\ell}^{pol})$ and $(Z_{1\ell}^{ax}, Z_{2\ell}^{ax})$ satisfy two pairs of decoupled Shroedinger-like wave equations with one-dimensional potential barriers (a complete account on the perturbations of the Reissner-Nordstrom spacetime can be found in ref. [6], ch. 5). Thus, for charged black holes $< E_2 >$ is the total conserved energy flux. By evaluating the contribution of the electromagnetic field (given by the terms in Y in Eqs.7 and 11) and that of the gravitational field, from this conserved flux it is possible to establish how waves of different type interact and mutually exchange energy.

4. The energy flux for the axial perturbations of stars

The interpretation and the evaluation of the conserved current in terms of Wronskians of complex conjugate solutions of the wave equations governing the perturbations of black holes, raises the question whether a similar approach can be used for stars. In the following I shall show that the "Wronskian approach" still works, though with some relevant modification, *in the case of the axial perturbations of stars.*

After separating the variables by the standard expansion in tensorial spherical harmonics, it is easy to show that the radial behaviour of the axial perturbations is described by a function $Z_\ell^{ax}(r)$, defined as in eq. (21), which satisfies the following Schrödinger-like equation

$$\frac{d^2 Z_\ell^{ax}}{dr_*^2} + [\sigma^2 - V_\ell^{ax}(r)]Z_\ell^{ax} = 0, \tag{35}$$

where $r_* = \int_0^r e^{-\nu+\mu_2} dr$, and

$$V_\ell^{ax}(r) = \frac{e^{2\nu}}{r^3}[l(l+1)r + r^3(\epsilon - p) - 6m(r)], \qquad \nu_{,r} = -\frac{p_{,r}}{\epsilon + p}. \tag{36}$$

Thus, the potential barrier depends on how the energy-density and the pressure are distributed inside the star in the equilibrium configuration. Outside the star ϵ and p vanish and V_ℓ^{ax} reduces to the Regge-Wheeler potential barrier given in eq. (20). If σ is real, the solution of eq. (35) which is free of singularities at the origin has the expansion

$$Z_\ell^{ax} \sim r^{l+1} + \frac{1}{2(2l+3)}\{(l+2)[\frac{1}{3}(2l-1)\epsilon_0 - p_0] - e^{2\nu(r=0)}\sigma^2\}r^{l+3} + \dots, \tag{37}$$

where ϵ_0 and p_0 are the values of the energy-density and of the pressure at the center of the star.

Since the axial perturbations do not excite the fluid motion, except for a stationary rotation, an incident axial wave experiences a potential scattering as it does in the case of a Schwarzschild black hole, but with a basic difference: the Schrödinger-like equation for a Schwarzschild black hole describes a problem of scattering by a one-dimensional potential barrier, whereas in the case of a star it describes the scattering by a central potential (see figure 1).

In the expressions of the conserved current (19), (22) and (34) derived for black holes, the Wronskians were constructed from complex conjugate solutions of the wave equations, under the assumption that σ is real. Indeed, the wave equations governing the axial and polar perturbations of black holes allow such solutions. In seeking a solution of the equation (35) for a star, we need to impose the regularity condition (37) at the origin. If σ is real, this is a real condition, and consequently Z_ℓ^{ax} is real, $\left[Z_\ell^{ax}, Z_\ell^{ax*}\right]_{r_*} = 0$, and $< E_2^{ax} >= 0$, which means that no net flux of radiation is associated with the real solutions of the axial equation.

However, the physical requirement that the oscillations of a star must be damped by the emission of gravitational waves implies that the solution at radial infinity behaves as a pure outgoing wave. This is a complex condition that, together with the imposed regularity at the origin, identifies the *quasi-normal modes* of vibration of the star. It should be reminded that

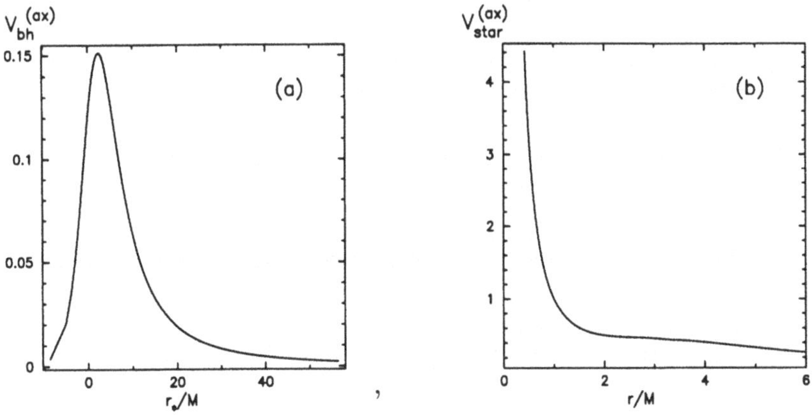

Figure 1. *The potential of the Regge-Wheeler equation versus the tortoise coordinate r_*/M is plotted in (a). In (b) the potential barrier of the axial equation for a perturbed star versus r/M is plotted for a polytropic star with polytropic index n=1.5. Both potentials are evaluated for $\ell = 2$.*

the complex eigenfrequencies of the quasi-normal modes (*quasi-stationary states* in Quantum Mechanics) are the poles of the scattering amplitude (cf. ref. [6] and ref. [13]). Let us assume that σ is complex, and that, consequently, the solution of eq. (35) Z_ℓ^{ax} is complex. We shall try to construct the Wronskians needed to evaluate the energy flux by using this assumption. We multiply the wave equation by Z^*, and subtract the complex conjugate equation. The result is (to hereafter we shall omit the indices '*ax*' and 'ℓ')

$$\frac{d}{dr_*}\Big[Z, Z^*\Big]_{r_*} = -4i\sigma_0\sigma_I |Z|^2, \qquad \to \qquad \Big[Z, Z^*\Big]_{r_*} = -4i\sigma_0\sigma_I \int_0^r |Z|^2 dr_*. \tag{38}$$

Since the solution must behave as a pure outgoing wave at infinity, i.e. $Z \sim e^{-i(\sigma_0 + i\sigma_I)}$, it follows that

$$|Z|^2 \sim e^{2i\sigma_I r_*},$$

and the integral in eq. (38) explodes. This approach does not work.

We shall now show how the Regge theory of potential scattering [14], applied essentially in its original form, allows to circumnvent the above difficulty. The Regge theory is applicable to problems of potential scattering when the wave equation is separable, and the wave function can be written in terms of a radial function and a Legendre polynomial $P_l(\cos\theta)$. The wave equation can be written in the following form

$$\frac{d^2 Z}{dr^2} + \left[\sigma^2 - \frac{\ell(\ell+\ell)}{r^2} - U(r)\right] Z = 0, \tag{39}$$

where the "centrifugal" part of the potential barrier has been explicitly separated from $U(r)$, which is a short range, central potential. We shall impose that the function Z is an analytic function of the frequency σ and of the angular momentum ℓ, both assumed to be complex.

In ref. [14] it has been shown that, to any given pole of the scattering amplitude, i.e. to any resonance $(\sigma_0 + i\sigma_i; \ell_0)$, corresponding to a fixed integral value of the angular momentum ℓ_0, there exists a pole in the complex ℓ-plane, $(\sigma_0; \ell_0 + i\ell_i)$, belonging to the same quasi-stationary state, i.e. to the same σ_0. Let us assume that we know the location of a pole $(\sigma_0 + i\sigma_i)$ in the complex σ-plane, and the location of the corresponding $(\ell_0 + i\ell_i)$ in the complex ℓ-plane. If $|\ell_i| << \ell_0$, which is equivalent to say that the "angular life " of the mode is small, eq. (39) becomes

$$\frac{d^2Z}{dr^2} + \left[\sigma^2 - \frac{\ell_0(\ell_0+1)}{r^2} - U(r)\right] Z = i\ell_i \frac{(2\ell_0+1)}{r^2} Z + \mathcal{O}(\ell_i^2). \qquad (40)$$

Multiplying equation (40) by Z^* and subtracting from the resulting equation its complex conjugate with respect to ℓ, we find that

$$\frac{d}{dr}\left[Z, Z^*\right]_r = 2i\ell_i \frac{(2\ell_0+1)}{r^2}|Z|^2. \qquad (41)$$

It is easy to show that if $|\ell_i| << \ell_0$ the solution of eq. (39) can be written as

$$Z(r; \sigma_0, \ell_0 + i\ell_i) \simeq Z_R(r; \sigma_0, \ell_0) + i\ell_i \left[\frac{\partial}{\partial l} Z_R(r; \sigma_0, \ell_0)\right], \qquad (42)$$

and can be derived by the only knowledge of the real solution $Z_R(r; \sigma_0, \ell_0)$. Since Z_I is of order ℓ_i (cfr. eq.(42)), $|Z|^2 = |Z_R|^2 + \mathcal{O}(l_i^2)$, and from equation (41) it follows that

$$\left[Z, Z^*\right]_r = 2i\ell_i(2\ell_0+1)\int_0^r \frac{dr}{r^2}|Z_R|^2. \qquad (43)$$

The integral on the right-hand side converges for $r \to \infty$ and it is positive definite. In Quantum Mechanics the non-constancy of the Wronskian shown in eq. (43) is interpreted as the emission of a new particle in the field volume. (see for example Landau & Lifshitz 1977, page 588). It should be stressed that the knowledge of the pole (ℓ_0, ℓ_i) is essential to evaluate eq. (43).

Let us return to the axial perturbations of stars. In this case the radial wave equation is obtained by expanding the wave function in Gegenbauer polynomials $C_{l+2}^{-\frac{3}{2}}$. This may affect the amplitude of the residue associated with each pole in the scattering amplitude, but none of the conclusion we have reached so far would change. If we write the axial equation in a form analogous to eq. (39):

$$\frac{d^2 Z}{dr_*^2} + \left[\sigma^2 - \frac{e^{2\nu}}{r^2}\ell(\ell+1) - U(r)\right] Z = 0, \tag{44}$$

where

$$U(r) = e^{2\nu}\left[(\epsilon - p) - \frac{6M}{r^3}\right], \tag{45}$$

and assume that $\sigma = \sigma_0$ is the real part of the frequency of a quasi-normal mode previously determined, and $\ell = \ell_0 + i\ell_i$ is the corresponding pole in the complex l-plane, equation (44) can be written in a form equivalent to eq. (40)

$$\frac{d^2 Z}{dr_*^2} + \left[\sigma_0^2 - \frac{e^{2\nu}}{r^2}\ell_0(\ell_0+1) - U(r)\right] Z = i\ell_i(2\ell_0+1)\frac{e^{2\nu}}{r^2}Z, \tag{46}$$

where $Z = Z(r_*, \sigma_0, \ell_0 + i\ell_i)$. Multiplying by Z^* and subtracting from the complex conjugate equation we finally find

$$[Z, Z^*]_{r_*} = 2i\ell_i(2\ell_0+1)\int_0^{r_*} \frac{e^{2\nu}}{r^2}|Z_R|^2 dr_*. \tag{47}$$

In analogy with the interpretation of equation (43) the right-hand side of eq. (47) can be interpreted as a measure of the gravitational energy which crosses a sphere of radius r_*.

5. The energy flux for the polar perturbations of stars

The polar perturbations of stars are the relativistic generalization of the newtonian tidal perturbations, and, as in newtonian theory, they couple the fluid motion with the perturbed gravitational field [4]. Outside the star the equations reduce to the Zerilli wave equation, and the interior and exterior solutions must be matched continuously across the surface.

The evaluation of the polar current (9) derived in section 2 requires an extension of the Regge theory of potential scattering beyond its standard domain of applicability. In fact, the polar equations reduce to a wave equation only outside the star. In the interior fifth order set (reducible to

[4]In ref. [15] it has been shown that it is possible to derive a set of equations that describe exclusively the perturbations of the gravitational field, and then find the fluid variables by solving a few algebraic equations involving the gravitational quantities and their first derivatives. This approach shows that the information on the dynamical evolution of the fluid and its modes is also encoded in the gravitational radiation which is emitted by the perturbed star.

fourth) of equations must be solved, and no simple analogue of the theorem that assures the constancy of the Wronskian of any two linearly independent solutions of a wave equation exists for such systems. Furthermore, the scattering problem is not a simple scattering by a central potential: a polar gravitational wave incident on a star excites the fluid motion, the oscillations of which are responsible for the polar *quasi-normal modes.* Nevertheless, the Regge theory can be generalized to solve the present problem [11]. Let us first write the conserved current (14) by substituting the separated variables (15) in Eq.(9) and by integrating over the angle

$$
\begin{aligned}
< E_2 >= &-\frac{4(n+1)}{(2l+1)} \int_0^r e^{\nu+\mu_2}[(N+L)X^* - (N+L)^* X] dr \\
&+ e^{\nu+\mu_2} \left\{ (n+1) r [(N+X)F^* - (N+X)^* F] + r^3 (\Pi F^* - \Pi^* F) \right\} \\
&+ r^2 e^{\nu-\mu_2} \left\{ \frac{1}{2} r \nu_{,r} (UF^* - U^* F) + \frac{1}{2(\epsilon+p)} (\Pi U^* - \Pi^* U) \right\}, \qquad (48)
\end{aligned}
$$

where we have defined $X = nV$, $F = (n-1)V - T$, and U and Π are, respectively, the radial part of the r-component of the lagrangian displacement $\vec{\xi}$ of a perturbed element of fluid, and the radial part of the perturbed pressure

$$
\begin{aligned}
&2(\epsilon+p) e^{\nu+\mu_2} \xi_2(r,\theta) = U(r) P_\ell \\
&\delta p = \Pi(r) P_\ell(\cos\theta). \qquad (49)
\end{aligned}
$$

It can be shown that outside the star, where all source terms are zero, eq. (48) reduces to eq. (19).

Equation (48) has been formally derived from the equations describing the polar perturbations, by considering a solution $(\delta\psi, \delta\mu_2, \delta\mu_3, \delta\nu, \xi_2, \delta p)$, and the complex conjugate solution of the same equations, for *real* σ *and real* l as shown in section 2. But as for the axial perturbations, the boundary conditions of the problem (cf. [15] §6) do not allow a physically meaningful complex conjugate solution for real σs and real ℓs, and eq.(48) cannot be used as it stands. However, we can assume that the polar perturbations also can be described by functions that are analytic in the complex ℓ-plane (cf. eq. (42))

$$
\begin{aligned}
N &= N_R(r;\sigma_0,\ell_0) + i\ell_i \Big(\frac{\partial}{\partial \ell} N_R(r;\sigma_0,\ell) \Big), \\
X &= X_R(r;\sigma_0,\ell_0) + i\ell_i \Big(\frac{\partial}{\partial \ell} X_R(r;\sigma_0,\ell) \Big), \\
&\;\;.
\end{aligned}
\qquad (50)
$$

where $N_R(r;\sigma_0,\ell)$, $X_R(r;\sigma_0,\ell)$, etc. are solutions of the perturbed equations corresponding to real $\sigma = \sigma_0$, and real ℓ. It can be shown [11] that this extension is indeed possible, and that in correspondence to the poles in the complex σ-plane (σ_0, σ_i), which identify the polar *quasi-normal modes*, there exist poles (ℓ_0, ℓ_i) belonging to the same quasi-normal mode. Under these premises, the analytic extension of $< E_2^{pol} >$ in the complex ℓ-plane can also be performed, and the right-hand side of eq.(48) can be evaluated in terms of the extended solution (50) to give the total flux of polar gravitational energy through the star. In the next section the interpretation of $< E_2 >$ as an energy flux will be further motivated.

6. The conserved current and the Einstein pseudotensor

It is known that if $T^{\mu\nu}$ is the stress-energy tensor of any kind of source, the covariant divergence-free equations it satisfies by virtue of the Bianchi identities

$$T^{\mu\nu}{}_{;\nu} = 0 \tag{51}$$

fail to provide a local conservation law because these do not include the contribution of the gravitational field. One then introduces the energy-momentum pseudotensor of the gravitational field $t^{\mu\nu}$, such that the sum of the two satisfies an ordinary divergence-free equation

$$\frac{\partial}{\partial x^\nu}\left[\sqrt{-g}\,(T^{\mu\nu} + t^{\mu\nu})\right] = 0, \tag{52}$$

from which a local conservation law can be inferred. Since for the vacuum $T^{\mu\nu} = 0$, eq. (52) gives

$$\frac{\partial}{\partial x^\nu}\left[\sqrt{-g} t^{\mu\nu}\right] = 0, \tag{53}$$

and for this reason Einstein called $t^{\mu\nu}$ the energy component of the gravitational field. One may argue that, if $< E_2 >$ is an energy flux, it should be possible to re-derive its expression by taking the second variation of the time-component of the divergence-free equation satisfied by $\sqrt{-g}(t^{\mu\nu} + T^{\mu\nu})$.

The problem is that the pseudotensor $t^{\mu\nu}$ is defined up to a divergenceless term, as eq. (52) clearly shows, and we need to decide which $t^{\mu\nu}$ to choose. A possible definition of $t^{\mu\nu}$ is that given by Landau and Lifschitz [16], which has the advantage of being a symmetric pseudotensor, but unfortunately its second variation does not reproduce the vector $\vec{\mathbf{E}}$, both for the Einstein-Maxwell spacetimes and in the fluid case.

The reason for this failure was clarified in refs. [8] and [9]. The only pseudotensor from which our conserved current can be obtained is the Einstein pseudotensor [5], because its second variation retains the divergence-free property, provided only the equations governing the static spacetime and its linear perturbations are satisfied. This property derives from the fact that the Einstein pseudotensor is a Noether operator for the gravitational field. The Landau-Lifshitz pseudotensor fails to reproduce the conserved current because it does not satisfy the foregoing requirements. In addition, the contribution of the source should be introduced not by adding the second variation of the stress-energy tensor of the source $T^{\mu\nu}$, as one may naively have thought, but through a suitably defined Noether operator, the form of which has been derived by Raphael Sorkin for an electromagnetic field [8], and by Vivek Iyer for a fluid source [12]. Though these operators do not coincide with $T^{\mu\nu}$, they give the same conserved quantities.

Having established that our flux integral can be obtained from the time component of the divergence free equations satisfied by the Einstein pseudotensor added to a Noether operator for the source, we are entitled to say that $< E_2 >$ does represent the total energy flow, developed as a consequence of an axisymmetric perturbation of a static, spherically symmetric spacetime.

References

1. Regge T., Wheeler J.A. (1957) *Phys. Rev.*108, 1063
2. Zerilli F.J. (1970) *Phys. Rev.*, **D2**, 2141
3. V. Moncrief (1974) *Phys. Rev.*, **D9**, 2707; (1974) *Phys. Rev.*, **D10**, 1507; (1975) *Phys. Rev.*, **D12**, 1526.
4. F.J. Zerilli (1974) *Phys. Rev.*, **D9**, 860
5. Teukolsky S.(1972) *Phys. Rev. Lett.*, **29**, 1114; (1973) *Ap. J.*, **185**, 635.
6. Chandrasekhar S. (1984) *The mathematical theory of black holes* Oxford: Clarendon Press
7. Chandrasekhar S., Ferrari V. (1990) *Proc. R. Soc. Lond.*, **A428**, 325
8. Sorkin R. (1991) *Proc. R. Soc. Lond.*, **A435**, 635
9. Chandrasekhar S., Ferrari V. (1991) *Proc. R. Soc. Lond.*, **A435**, 645
10. Burnett G. , Wald R. (1990) *Proc. R. Soc. Lond.*, **A430**, 57
11. Chandrasekhar S., Ferrari V. (1992) *Proc. R. Soc. Lond.*, **A437**, 133
12. Iyer V. (1997) *Phys. Rev. D*, **D55**, 3411
13. Chandrasekhar S., Ferrari V., Winston R.(1991) *Proc. R. Soc. Lond.*, **A434**, 635
14. Alfaro, V., Regge, T. (1963) *Potential scattering*, Amsterdam: North Holland Press.
15. Chandrasekhar S., Ferrari V. (1991) *Proc. R. Soc. Lond.*, **A432**, 247
16. Landau, L.D., Lifshitz, E.M.(1977) *Quantum mechanics: non-relativistic theory*, London: Pergamon Press.

[5]It should be mentioned that the first variation of the energy-momentum pseudotensors identically vanishes.

5. GRAVITATIONAL COLLAPSE AND COSMIC CENSORSHIP

ROBERT M. WALD
Enrico Fermi Institute and Department of Physics, University of Chicago, 5640 S. Ellis Avenue, Chicago, Illinois 60637-1433, USA.

1. Introduction

It has long been known that under a wide variety of circumstances, solutions to Einstein's equation with physically reasonable matter must develop singularities [1]. In particular, if a sufficiently large amount of mass is contained in a sufficiently small region, trapped surfaces must form [2] or future light cone reconvergence should occur [3], in which case gravitational collapse to a singularity must result. One of the key outstanding issues in classical general relativity is the determination of the nature of the singularities that result from gravitational collapse.

A key aspect of this issue is whether the singularities produced by gravitational collapse must always be hidden in a black hole–so that no "naked singularities", visible to a distant observer, can ever occur. The conjecture that, generically, the singularities of gravitational collapse are contained in black holes is known as the *weak cosmic censorship conjecture* [4]. (A much more precise statement of this conjecture will be given in Sec. 2 below.) A closely related conjecture, known as the *strong cosmic censorship conjecture* [5], asserts that, generically, timelike singularities never occur, so that even an observer who falls into a black hole will never "see" the singularity. This paper will focus exclusively on the weak cosmic censorship conjecture.

In the past seven years, there have been two articles in the New York Times indicating the demise of weak cosmic censorship [6]. The main theme of this paper is that such reports of the death of cosmic censorship are greatly exaggerated: Although very little progress has been made towards a general proof of the weak cosmic censorship conjecture, it remains in

B. R. Iyer and B. Bhawal (eds.), Black Holes, Gravitational Radiation and the Universe, 69–85.

good health, and, indeed, is probably healthier today than at any time in the past.

2. Formulation of the weak cosmic censorship conjecture

The issue at the heart of the weak cosmic censorship conjecture can be expressed in graphic terms by posing the following question: Could a mad scientist–with arbitrarily large but finite resources–destroy the universe? We know that, in principle, such a mad scientist could produce a spacetime singularity by gathering together a sufficiently large amount of mass into a sufficiently small region. The essential content of the weak cosmic censorship conjecture is the assertion that, if he were to do so, neither the singularity he would produce nor any of its effects could ever propagate in such a way as to reach a distant observer. Of course, even if weak cosmic censorship fails, the universe might still be protected from destruction by mad scientists, since even if naked singularities were produced, they might always be of a benign character, with well defined rules–presumably arising from quantum gravity–governing dynamical evolution in their presence. However, if weak cosmic censorship fails, then in a literal sense, we would come face-to-face with the laws of quantum gravity whenever gravitational collapse to a naked singularity occurs in distant regions of our universe.

To formulate the statement of the weak cosmic censorship conjecture more precisely, we first need to make precise the notion of the "finiteness" of the resources available to our mad scientist. This notion is well modelled by restricting consideration to spacetimes containing an asymptotically flat initial data surface, i.e., a hypersurface Σ on which the induced spatial metric, h_{ab} becomes Euclidean at asymptotically large distances from some compact ("central") region, and the extrinsic curvature, K_{ab}, of Σ goes to zero at a suitable rate at infinity. If matter fields are present, additional asymptotic fall-off conditions for the matter fields also would be required. (The precise asymptotic conditions on this initial data that would be most suitable to impose probably would best be left open until further progress is made in investigations of cosmic censorship, and will not be considered here.) In particular, it should be noted that the restriction to asymptotically flat initial data ensures that our mad scientist initially has only a finite total amount of energy at his disposal, but it does not place any direct restrictions on the initial conditions he might set up in the "central region" of the spacetime.

The basic idea of the weak cosmic censorship conjecture is that, starting from these initial conditions, any sufficiently distant observer will neither encounter any singularities nor any effects arising from–i.e., propagating out of–singularities. To make this idea more precise, we need a suitable notion of a spacetime being asymptotically flat "at large distances and at late times". Such a notion is provided by the requirement that the spacetime be asymptotically flat at future null infinity. The standard definition of asymptotic flatness at future null infinity requires that one be able to conformally embed the spacetime in a suitable way into a spacetime with a boundary, $\mathcal{I}^+$, which, roughly speaking, provides endpoints for the null geodesics which propagate to asymptotically large distances. The precise details of this definition are not crucial to our discussion and can be found in standard references (see, e.g., [1, 7]). The precise smoothness requirements most suitable to impose at $\mathcal{I}^+$ undoubtedly will depend on the precise choice of asymptotic conditions on the initial data (see above), and will not be considered here.

For a spacetime, (M, g_{ab}), which is asymptotically flat at future null infinity, the *black hole* region, $\mathcal{B}$, of the spacetime is defined by

$$\mathcal{B} = M - I^-(\mathcal{I}^+) , \tag{1}$$

where the chronological past, I^-, is taken in the ("unphysical") conformally completed spacetime. The *event horizon*, $\mathcal{H}$, of the black hole is defined to be the boundary of $\mathcal{B}$ in M

$$\mathcal{H} = \partial \mathcal{B}. \tag{2}$$

We are now in a position to give a relatively precise formulation of weak cosmic censorship:

Weak cosmic censorship conjecture: Let Σ be a 3-manifold which, topologically, is the connected sum of $\mathcal{R}^3$ and a compact manifold. Let (h_{ab}, K_{ab}, ψ) be nonsingular, asymptotically flat initial data on Σ for a solution to Einstein's equation with *suitable* matter (where ψ denotes the appropriate initial data for the matter). Then, *generically*, the maximal Cauchy evolution of this data is a spacetime, (M, g_{ab}) which is asymptotically flat at future null infinity, with complete $\mathcal{I}^+$.

In this formulation of the conjecture, the asymptotic flatness of (M, g_{ab}) (with complete $\mathcal{I}^+$) ensures that sufficiently distant observers can live out their lives in their entirety, free from the effects of any catastrophic events occurring in the central region of the spacetime. Furthermore, the fact that these observers lie in the domain of dependence of Σ implies that they also are free from any non-deterministic effects that might occur if singularities are produced. If singularities are produced, they cannot be seen from $\mathcal{I}^+$.

The above conjecture remains somewhat imprecise on account of the two words written in italics. In order for the matter to be "suitable", it clearly is necessary that the coupled Einstein-matter field equations have a well posed initial value formulation. It undoubtedly also should be required that the matter stress-energy tensor satisfy suitable energy conditions, such as the dominant energy condition. In addition, it would seem reasonable to require that the matter be such that, in any fixed, globally hyperbolic, background spacetime (such as Minkowski spacetime), one always obtains globally nonsingular solutions of the matter field equations starting from regular initial data; otherwise, any "naked singularities" produced in the dynamical evolution of the Einstein-matter equations may have nothing to do with gravitational collapse. Note that this latter condition would rule out fluids as "suitable matter" (in particular, on account of "shell crossing" singularities and shocks), although fluid examples remain quite valuable as simple matter models for testing behavior related to cosmic censorship (see, e.g., [8]). It is not clear whether any further restrictions should be imposed–or, indeed, if some of the above restrictions should be weakened somewhat. In any case, the "suitable" matter fields certainly should include the Maxwell field and the Klein-Gordon scalar field.

The "generic" condition was inserted in the above conjecture because it would not be fatal to the physical content of the conjecture if examples exist where dynamical evolution produces naked singularities, provided that the initial data required for these examples is so special that it would be physically impossible to achieve. A way to express the idea that no generic violations of weak cosmic censorship occur would be to require that all initial data giving rise to violations of the behavior specified in the conjecture is confined to a "set of measure zero" and/or a "set whose closure has empty interior". Unfortunately, it is far from clear precisely what measure or topology should be imposed on the space of initial data. Undoubtedly, it will be necesary to develop a much deeper insight into the dynamics implied by Einstein's equation before a natural choice of measure or topology will emerge, and I feel that the precise definition of "generic" would best be left open until that point.

Does the weak cosmic censorship conjecture hold? To answer this question, we would need to know a great deal about the global properties of solutions to Einstein's equation. Global existence of solutions with nearly flat initial data has been proven [9]. Thus, weak cosmic censorship holds for nearly flat data–where the nonlinear effects of general relativity are too weak to produce any singularities at all. However, mathematical techniques have not progressed to a stage where a direct attempt at a general proof of the weak cosmic censorship conjecture would be feasible. Thus, the evidence both for and against the validity of weak cosmic censorship has been largely of an "anecdotal" or "circumstantial" nature, or has been confined to some very restricted cases (such as spherical symmetry). In the remaining sections of this paper, I shall briefly summarize much of this evidence.

3. Some evidence in favor of cosmic censorship

In this section, I shall describe some analyses in support of cosmic censorship. Most of the basic ideas described here are at least 20 years old, but some notable recent progress has occurred.

3.1. STABILITY OF BLACK HOLES

If weak cosmic censorship fails, then gravitational collapse can (generically) result in a naked singularity rather than a black hole. If so, then it is quite possible that the formation of a black hole would be a non-generic outcome of collapse. In that case, one might expect to see evidence of this in linear perturbation theory of a background spacetime containing a black hole. Specifically, one might expect an initial, smooth perturbation to grow without bound on the black hole horizon, signalling the conversion of the black hole into a naked singularity. (In linear perturbation theory, a blow-up of the perturbation on the horizon could occur only at asymptotically late times, but it could occur at a finite time in the nonlinear theory.) Thus, the study of linear perturbations of a black hole background provides an excellent testing ground for weak cosmic censorship. A demonstration of the linear instability of black holes would effectively disprove weak cosmic censorship, whereas a demonstration of their linear stability would provide some notable evidence in support of it.

The first analysis of the linear stability of the Schwarzschild black hole was given in 1970 by C.V. Vishveshwara [10], who established its stability to axial (i.e., "odd parity") perturbations. Shortly thereafter, a convincing demonstration of the stability of the Schwarzschild black hole–together with detailed information about the decay properties of the perturbations–was given by Price [11]. More recently, a completely rigorous proof of the boundedness of perturbations at asymptotically late times has been given [12].

The analysis of the stability of a Kerr black hole is much less tractable than the Schwarzschild case. Nevertheless, Whiting [13] succeeded in proving that no unstable modes exist. Thus, it appears that weak cosmic censorship has passed the crucial test of stability of black holes to general, linear perturbations.

Another test of stability can be performed for extremal charged Kerr black holes, i.e., charged Kerr black holes whose mass, M, angular momentum, J, and charge, Q, satisfy

$$M^2 = Q^2 + (J/M)^2 \, . \tag{3}$$

No stationary black hole solutions exist when the right side of Eq. (3) exceeds the left side. Thus, if one can get an extremal black hole to "swallow" an object whose charge and/or angular momentum is sufficiently large compared with its mass, there would be no black hole final state available for the system to settle down to. Presumably, a naked singularity then would result. However, an analysis of test particle motion in an extremal charged Kerr background indicates that it is not possible to get a black hole to swallow too much charge or angular momentum [14]. Interestingly, a similar anaylsis of test particle motion for certain extremal "black hole" solutions of the Einstein-Maxwell equations with a positive cosmological constant–where the spacetime is asymptotically DeSitter rather than asymptotically flat–indicates that "over-charging" can occur in this case [15]. However, these "black holes" are only rough analogs of black holes in asymptotically flat spacetimes; in particular, in these solutions there are naked singularities as seen from the DeSitter analog of $\mathcal{I}^+$. Even if these "black holes" could be "destroyed" by perturbations, this would not contradict the formulation of weak cosmic censorship given above for asymptotically flat spacetimes–though it would provide evidence against some formulations of strong cosmic censorship.

3.2. FAILED COUNTER-EXAMPLES

A class of possible counter-examples to weak cosmic censorship involving collapsing shells of null dust was proposed by Penrose [16] and generalized by Gibbons [17] over 25 years ago. A similar class of possible counter-examples also can be given for time symmetric initial data. In the analysis of these classes of possible counterexamples, a key role is played by the following two results from the theory of black holes in general relativity (see, e.g., [1, 7]):

- If weak cosmic censorship holds, then every trapped surface must be entirely contained within a black hole. (Here, a *trapped surface*, S, is a compact, 2-dimensional surface having the property that the convergence of both the outgoing and ingoing null geodesics normal to S is everywhere positive.)
- If weak cosmic censorship holds and if matter satisfies the null energy condition (i.e., if $T_{ab}k^a k^b \geq 0$ for all null k^a), then the area of the event horizon of a black hole cannot decrease with time.

Now consider a convex shell of null dust with flat interior, which collapses from infinity down to an infinite density singularity. (Note that such null dust presumably would not qualify as "suitable matter" in our formulation of the weak cosmic censorship conjecture, but if the inequality given below could be violated for null dust, it presumbly also could be violated for "suitable matter".) In this example, there are two free functions of two variables which may be specified arbitrarily, characterizing the initial "shape" and initial mass density of the shell. Except in the case of a spherical shell with constant mass density, the solution exterior to the shell is not known. Nevertheless, enough information about the solution can be deduced to make an interesting test of cosmic censorhip. Specifically, the Bondi mass, M, infinitesimally outside of the shell at past null infinity can be computed, and it seems reasonable to expect that there will exist exterior solutions with total ADM mass equal to M (or, at least, arbitrarily close to M). Furthermore, by integrating the Raychaudhuri equation for outgoing null geodesic congruences across the delta-function mass distribution on the null shell one can determine the presence of trapped surfaces lying infinitesimally outside of the shell.

Suppose, now, that a trapped surface, S, lies infinitesimally outside of the shell, and let $A(S)$ denote its area. As mentioned above, if weak cosmic censorship holds, then S must lie within a black hole. Let A_0 denote the area of the 2-surface obtained by intersecting the event horizon, $\mathcal{H}$, of this black hole with the null shell. Then, assuming the validity of weak cosmic censorship, the following string of inequalities should hold:

$$A(S) \leq A_0 \leq 16\pi M_{\text{bh}}^2 \leq 16\pi M^2 . \tag{4}$$

Here, the first inequality follows from the fact that S lies within the black hole and the shell is infalling. The second inequality follows from the area nondecrease theorem (see above) together with the fact that the maximum possible area of a stationary black hole of mass M_{bh} is achieved by the Schwarschild value, $16\pi M_{\text{bh}}^2$. (Here it is assumed that the black hole settles down to a stationary final state.) The final inequality expresses the fact that the mass of the final black hole cannot exceed the total ADM mass of the spacetime.

Remarkably, the inequality $A(S) \leq 16\pi M^2$ implied by Eq. (4) involves only quantities which can be computed without knowing the solution exterior to the shell, and thus can be readily checked. Failure of this inequality to hold in any example would be nearly fatal to cosmic censorship, as only a few small loopholes would remain–such as the possible "unsuitablity" of the null dust matter, the possibly "non-generic" nature of the example, and the (very remote) possiblity that the black hole does not become asymptotically stationary.

By making use of the fact that the interior of the shell is flat, the issue of whether the above inequality holds can be reduced to the issue of whether a certain "isoperimetric inequality" holds for (topological) spheres of non-negative mean curvature embedded in Euclidean 3-space. Until very recently, the inequality $A(S) \leq 16\pi M^2$ had been proven only in some special cases. However, recent results of Trudinger [18] on strengthened isoperimetric inequalities has enabled a general proof to be given that $A(S) \leq 16\pi M^2$ in all cases; see [19] for further discussion.

A similar argument–showing that an analog of (4) must hold if weak cosmic censorship is valid–can be given for the case of time symmetric initial data (i.e., $K_{ab} = 0$) on a spacelike hypersurface Σ. A *minimal surface*, S, on Σ is a compact (without boundary) 2-surface on which $p \equiv h^{ab} p_{ab} = 0$ everywhere, where p_{ab} is the extrinsic curvature of S in Σ. In the case where $K_{ab} = 0$, both sets of null geodesics orthogonal to a minimal surface S will have vanishing expansion, so S will be marginally trapped. It then follows that S must lie within a black hole. Now, let S_{out} be the outermost minimal surface on Σ. Since the intersection of the black hole horizon, $\mathcal{H}$, with Σ must lie outside of (or coincide with) S_{out}, and since S_{out} is the outermost minimal surface, it follows that the area, A_0, of $\mathcal{H} \cap \Sigma$ cannot be smaller than $A(S_{\text{out}})$. Hence, in analogy with Eq. (4), we obtain

$$A(S_{\text{out}}) \leq A_0 \leq 16\pi M^2 , \tag{5}$$

where M denotes the ADM mass of the spacetime. Again, both $A(S_{\text{out}})$ and M can be calculated directly from the initial data given on Σ, without the necessity of evolving the data off Σ. In essence, Eq. (5) shows that if cosmic censorship is valid, then a strengthened version of the positive mass theorem must hold for time symmetric initial data whenever minimal surfaces are present.

An argument for the validity of Eq. (5) was given in [20], but it relied on the assumed existence of a particular foliation of Σ by 2-surfaces, as first proposed by Geroch [21] in an argument for the positivity of total mass. However, very recently, a proof of the existence of the required foliation has been given [22], thus establishing that Eq. (5) does, indeed, hold for general, time symmetric initial data.

Of course, the proof of the inequalities (4) and (5) for the very restricted classes of spacetimes to which they apply is a far cry from even the beginnings of a general proof of weak cosmic censorship. Nevertheless, it is very hard to understand *why* these highly nontrivial inequalities should hold unless weak cosmic censorship can be thought of as providing the underlying physical reason behind them. Thus, while it is not clear how much "objective evidence" in favor of cosmic censorship is provided by the failure of these counter-examples, their failure has given many researchers considerable confidence in the validity of weak cosmic censorship.

4. The hoop conjecture

Although the results of the previous section provide some suggestive evidence in favor of cosmic censorship, it is not surprising that not all researchers have been convinced by this evidence, and some serious doubts about the general validity of weak cosmic censorship have been expressed. Many of these doubts have centered upon an idea known as the *hoop conjecture*, which has been formulated as follows (see [23] or box 32.3 of [24]):

Hoop conjecture: Black holes with horizons form when and only when a mass M gets compacted into a region whose circumference in EVERY direction is $\mathcal{C} \leq 4\pi M$.

(Here, one envisions "passing a hoop" of circumference $4\pi M$ around the matter in every direction to test this criterion.) The basic idea intended to be expressed by this conjecture is that gravitational collapse in all three spatial dimensions must occur in order for a black hole to form. If collapse occurs in fewer dimensions (i.e., to a 2-dimensional "pancake" configuration or a 1-dimensional "spindle" configuration), then the hoop conjecture is normally interpreted as asserting that a naked singularity should result.

Clearly, the above statement of the hoop conjecture in not intended to be mathematically precise, and there are some obvious mathematical difficulties with its formulation. Probably the most serious of these difficulties arises from the fact that the "only when" portion of the conjecture cannot be expected hold unless "gravitational energy" is included in the "mass M", since there should be no difficulty forming a black hole out of a sufficient concentration of gravitational radiation. However, there is no local notion of "gravitational energy" in general relativity. In addition, there is no obvious notion of the "circumference" a world tube in spacetime (such as the world tube containing the "mass M" of the conjecture), since arbitrarily near (in spacetime) to any given 2-dimensional surface exterior to a world tube are 2-surfaces (approximated by suitably chosen broken null surfaces) with arbitraily small circumference in every direction. In other words, one can pass an arbitrarily small hoop around *any* concentration of mass by making segments of the hoop move in a suitable, relativistic manner as one moves the hoop around the mass. In order for the conjecture to have meaning, it is necessary to specify the choice of spacelike slicing on which the circumference is to be measured, but it is not obvious how this should be done.

I do not know how to overcome the difficulties in the formulation of the "only when" portion of the hoop conjecture. In the theorem of Schoen and Yau [2]–establishing the existence of trapped (or anti-trapped) surfaces in certain situations when a sufficiently large amount of matter is compacted into a sufficiently small region–the difficulties in defining the "circumference" are avoided by using the internal geometry on a spacelike slice to measure the "size" of the region. (This measure of the size also can be made arbitrarily small by appropriate choices of slicing, but the matter energy-momentum density on the slice is then necessarily affected in a corresponding manner.) It should be noted that the theorem of Schoen and Yau does not actually establish a version of the "when" half of the hoop conjecture unless one *assumes* weak cosmic censorship, since the presence of a trapped surface necessarily implies the existence of a black hole only under that assumption (see the beginning of subsection 3.2).

The main motivation for the "when" half of the hoop conjecture appears to have arisen from the study of the collapse of matter with cylindrical symmetry (i.e., rotational symmetry about an axis together with translational symmetry along that axis). It has long been known that cylindrically symmetric fluids can collapse to singularities, but no trapped surfaces ever form [23, 25], and the singularities are "visible" from infinity. These examples do not directly provide counter-examples to weak cosmic censorship because the cylindrically symmetric spacetimes are not asymptotically flat in the required sense: The matter distribution and curvature extend to infinity along the axis, and even in the directions perpendicular to the axis, the metric approaches flatness too slowly. However, these examples might seem to suggest that weak cosmic censorship could be violated in the collapse of a very long but finite "spindle" of matter.

However, although cylindrically symmetric fluids can collapse to a singularity, it does not appear that such singularities occur generically or that they occur at all with "suitable matter". Specifically, for a dust cylindrical shell, an arbitrarily small amount of rotation causes a "bounce" [26]. Furthermore, it has been shown that in the vacuum and Einstein-Maxwell cases, no singularities whatsoever occur in cylindrical collapse [27]. Thus, even if it turns out that the collapse of a finite "spindle" of fluid matter can produce naked singularities similar to those which occur in the exactly cylindrical case, it seems unlikely that spindle collapse will produce naked singularities that satisfy either the "suitable matter" or "generic" provisions of the weak cosmic censorship conjecture.

Nevertheless, Shapiro and Teukolsky [28] performed numerical calculations of the collapse of highly prolate gas spheroids and found behavior which they interpreted as both supporting the hoop conjecture and providing likely counter-examples to the weak cosmic censor conjecture. In their calculations, they evolved a number of spacetimes describing collapsing gas spheriods using maximal time slicing. They found that when the spheroid was highly prolate, a singularity formed just exterior to the ends of the spheroid–at which point, of course, they could no longer continue their numerical evolution. They then searched for trapped surfaces on the maximal hypersurfaces and found that none were present. They interpreted this as indicating that the singularity might be naked.

As already indicated above, even if the singularity they found were shown to be naked, it would be far from clear that the "suitable matter" and "generic" provisions of the weak cosmic censor conjecture could be satisfied. However, it also should be emphasized that the absence of a trapped surface lying on their maximal slices in the portion of the spacetime that they constructed does not provide much evidence that the singularity they found is naked. Indeed, even in Schwarzschild spacetime, it is possible to choose a (highly non-spherically-symmetric) time slice which comes arbitrarily close to the singularity inside the black hole, and yet has no trapped surfaces contained within its past [29]. A good test of whether something like this might be occurring in the Shapiro-Teukolsky examples is provided by the Penrose-Gibbons spacetimes described above in subsection 3.2. One can choose the shell of null dust to be highly prolate and one can arrange the collapse of the shell so that–in the flat hyperplane slicing of the flat interior of the shell (the analog of maximal slicing in the Shapiro-Teukolsky examples)–the singularity occurs at the ends first. One may then analyze whether trapped surfaces occur just outside the null shell and, if so, where they are located. Examples can be given where trapped surfaces exist, but no trapped surface is wholly contained to the past of the "last hyperplane" prior to encountering the singularity [30]. These examples suggest that if a significantly larger portion of the Shapiro-Teukolsky spacetimes were constructed (which could be done by using different choices of time slicing), then trapped surfaces enclosing the singularity might be found. Thus, at present, there do not appear to be strong reasons to believe that the singularities found by Shapiro and Teukolsky are actually naked.

In summary, if weak cosmic censorhip holds, then the Schoen and Yau theorem [2] can be viewed as giving a precise statement and proof of a version of the "when" portion of the hoop conjecture–although, as pointed out to me by N. O'Murchadha, the applicability of this theorem is extremely restrictive. However, formidable difficulties would have to be overcome even to give a precise formulation of the "only when" portion of the hoop conjecture, since one would need a notion of the gravitational contribution to the "mass M". Furthermore, even if some version of the "only when" portion of the conjecture turns out to be valid, there need not be any conflict with weak cosmic censorship, since it could well be the case that if suitable matter fails to be sufficiently compacted in all three spatial directions, then, generically, no singularity forms. In any case, I am not aware of any results related to the hoop conjecture which cast a serious doubt on the validity of weak cosmic censorship.

5. The spherically symmetric Einstein-Klein-Gordon System

As previously emphasized, a general analysis of the validity of cosmic censorship would appear to require a much greater mastery of the global properties of solutions to Einstein's equation than is presently achievable. Therefore, it is natural to focus attention on more tractable special cases. The assumption of spherical symmetry greatly simplifies the analysis of gravitational collapse, and, largely for that reason, has been widely studied. It should be kept in mind that there is no guarantee that phenomena found in spherically symmetric gravitational collapse are representative of general phenomena. In particular, a phenomenon which is generic under the restriction to spherical symmetry need not be generic when that restriction is removed. Nevertheless, it is instructive to explore the phenomena that occur in spherical gravitational collapse, and to determine to what degree cosmic censorship holds in that case.

By Birkhoff's theorem the gravitational field itself has no dynamical degrees of freedom in the spherical case, so it is essential to have matter degrees of freedom present. Spherical collapse has been most widely studied with fluid matter, particularly "dust", i.e, a perfect fluid with $P = 0$. Numerous examples have been found where naked singularities occur (see, e.g., [31]-[33]). However, most of these examples appear to be non-generic in character, and all of them appear to rely on properties of fluid matter that also would allow one to produce singularities during evolution in flat spacetime. In particular, the "shell crossing" and "shell focusing" naked singularities found in the collapse of dust matter appear to depend crucially on the ability to "aim" the dust so as to produce infinite density before the self-gravitation of the dust becomes large. Thus, in order to obtain more insight into the validity of cosmic censorship in the spherically symmetric case, it would appear to be necessary to study examples with a more "suitable" form of matter.

A suitable form of matter which provides an excellent testing ground for cosmic censorship is provided by a massless Klein-Gordon scalar field, ϕ. The complete system of equations for the Einstein-Klein-Gordon system are

$$\nabla^a \nabla_a \phi = 0 \, , \tag{6}$$

$$G_{ab} = 8\pi[\nabla_a \phi \nabla_b \phi - \frac{1}{2} g_{ab} \nabla_c \phi \nabla_c \phi] \, . \tag{7}$$

When restricted to spherical symmetry, Eqs.(6) and (7) simplify greatly from the general case, but, as will be indicated further below, they still provide a rich dynamics.

In a series of papers, Christodoulou [33]–[37] has given a remarkably complete analysis of the singularities that can arise in spherically symmetric solutions to the Einstein-Klein-Gordon equations. Christodoulou considered evolution from initial data posed on a future null cone, C_0^+, with vertex on the world line Γ corresponding to the center of spherical symmetry, $r = 0$. (Here r denotes the usual Schwarzschild radial coordinate defined by $4\pi r^2 = A$, where A denotes the area of the orbit the rotation group.) The initial data on C_0^+ can be characterized by the function

$$\alpha \equiv \frac{d}{dr}(r\phi) , \tag{8}$$

which may be freely specified on C_0^+ subject to asymptotic conditions and boundary conditions at $r = 0$. In Ref.[35] Christodoulou showed that unique solutions of bounded variation (defined precisely in Ref.[35]) exist provided that the initial data is such that the function α is of bounded variation on C_0^+.

In Ref.[36] Christodoulou investigated the global behavior spherically symmetric Einstein-Klein-Gordon solutions which possess the following additional conformal symmetry: There exists a one-parameter group of diffeomorphisms (parametrized by λ) under which, for some constant k,

$$g_{ab} \to \lambda^2 g_{ab}, \quad r \to \lambda r, \quad \phi \to \phi - k \ln \lambda \, . \tag{9}$$

Such solutions will not be asymptotically flat, but the initial data for these solutions can be suitably "truncated" so as to yield asymptotically flat data, and consequently they are relevant for testing cosmic censorship. By analyzing these solutions, Christodoulou proved [36] that *there exist choices of asymptotically flat initial data which evolve to solutions with a naked singularity.* In these solutions, a singularity first forms at the origin and then propagates out to infinity along a (singular) future null cone, reaching $\mathcal{I}^+$ at a finite retarded time. Thus, $\mathcal{I}^+$ is incomplete in the maximally evolved spacetime. The null cone singularity is of a rather mild type, in that the curvature remains bounded as one approaches the cone away from the vertex, although derivatives of the curvature blow up on the cone. Christodoulou also proved that there exist choices of asymptotically flat initial data which evolve to what he referred to as "collapsed cone singularities". The solutions with collapsed cone singularities can be thought of as describing black holes of vanishing mass which possess a singular event horizon. In these solutions, $\mathcal{I}^+$ is complete–so the conditions of our formulation of the weak cosmic censorship conjecture are satisfied–but the singularity is not really hidden in a black hole and from infinity one can "see" events which are arbitrarily close to the singularity.[1]

[1]It should be noted that all of Christodoulou's examples of naked singularities arise from initial data on C_0^+ of a low differentiability class (but, of course, with α of bounded variation, so that the initial value formulation is well posed). His examples with collapsed cone singularities can have initial data of arbitrarily high (but finite) differentiability.

Thus, the analysis of [36] established for the first time that naked singularities can arise by the gravitational collapse of "suitable matter".[2] A great deal of insight into the circumstances under which naked singularities are produced was provided by Choptuik's [39] numerical investigations of the behavior of Einstein-Klein-Gordon solutions which are "just on the verge" of collapsing to a black hole. Choptuik considered various one-parameter families of initial data with the property that for small values of the parameter, the incoming scalar waves are weak and disperse back to infinity, whereas for large values of the parameter, the incoming scalar waves are strong and collapse to a Schwarzschild black hole. Choptuik then tuned the parameter to the "borderline value" where collapse first occurs, and used mesh refinement techniques to study the properties of the "borderline solution" near $r = 0$ in detail. Remarkably, he found that for all of the one-parameter families he considered, the borderline solution always asymptotically approached a particular "universal solution". Furthermore, this universal solution was found to possess a discrete self-similarity, i.e., it admits a diffeomorphism (as opposed to a one-parameter group of diffeomorphisms) satisfying Eq. (9) with $k = 0$. Neither the universality of the borderline collapse behavior nor the discrete self-similarity of the universal solution had been anticipated prior to Choptuik's analysis. Most importantly for the issue of cosmic censorship, the numerical investigations by Choptuik (confirmed by others [40]) indicated that the borderline solutions possess naked singularities of a nature similar to Christodoulou's examples.

Similar discrete or continuous self-similarity has been found to occur in the borderline solutions for a number of other systems (see, in particular, [41]). Presumably, these borderline solutions also possess naked singularities (although I am not aware of demonstrations of this). However, in some systems where there exist unstable, stationary, nonsingular solutions–in particular, in the Einstein-Yang-Mills system–some borderline solutions approach one of these stationary, nonsingular solutions rather than a self-similar solution [42].

The fact that naked singularities in the spherically symmetric Einstein-Klein-Gordon system were encountered in Choptuik's numerical calculations only for the borderline solutions suggests that the occurrence of naked singularities is non-generic. Clearly, no definitive conclusions in this regard can be drawn from numerical studies. However, an analytic demonstration of the non-generic character of naked singularities has recently been given

[2]The existence of a naked singularity had previously been claimed for a particular scale invariant solution of the Einstein-Klein-Gordon equations (i.e., a solution satisfying Eq. (9) with $k = 0$), which has been studied by a number of authors [38] (see also [35]). However, the analysis given in [36] shows that this solution (or, more precisely, suitable asymptotically flat "truncations" of this solution) actually corresponds to a collapsed cone singularity rather than a naked singularity.

by Christodoulou [37]. In order to state this result, it is useful to classify solutions arising from the maximal evolution of asymptotically flat initial data on C_0^+ as follows:

case (i): No singularities at all occur at any finite advanced time and $\mathcal{I}^+$ is future complete.[3] This case necessarily arises when the initial data is sufficiently "small" [35].
case (ii): A singularity forms at a finite advanced time, but it is entirely contained within a black hole, as in the "standard picture" of gravitational collapse. In particular, $\mathcal{I}^+$ is complete, and in order to reach the singularity, an observer must pass through a (non-singular) event horizon.
case (iii): Neither case (i) nor case (ii) holds.

Note that under this classification, case (iii) includes all solutions with naked singularities and collapsed cone singularities, as well as any solutions with other, as yet undiscovered, pathologies that might be viewed as contrary to the spirit of the weak cosmic censorship conjecture.

Christodoulou proved the following [37]: *Consider any initial data–characterized by the function α_0–which evolves to a spacetime in category (iii) above. Then there exists a continuous function f such that for any real number $c \neq 0$, the initial data characterized by $\alpha = \alpha_0 + cf$ evolves to a spacetime in category (ii).*[4] In other words, by an arbitrarily small perturbation of the initial data, a spacetime containing a naked singularity (or a collapsed cone singularity or other pathology) can be converted to a black hole. Thus, within the class of initial data of bounded variation, solutions with naked singularities are non-generic in the above, precise sense.

Although restricted to the case of the spherically symmetric Einstein-Klein-Gordon equations, the above result provides the first true cosmic censorship theorem for a nontrivial system.

6. Conclusions

Although the question of whether weak cosmic censorship holds remains very far from being settled, there appears to be growing evidence in support of its validity. This evidence consists primarily of the stability of black holes (see subsection 3.1), the proof of the failure of certain classes of counter-

[3]In this case, the spacetime also must be future timelike and null geodesically complete, with the possible exception of the "central geodesic" Γ at $r = 0$, whose completeness was not explored by Christodoulou.

[4]The borderline solutions studied by Choptuik presumably comprise a surface, $\mathcal{S}$, of co-dimension 1 in the space of initial data. In order to have the above property at $\alpha_0 \in \mathcal{S}$, Christodoulou's one parameter family must be tangent to $\mathcal{S}$, and $\mathcal{S}$ must be convex (as viewed from the "black hole side") in the direction defined by this family.

examples (see subsection 3.2), and the proof of a cosmic censorship theorem for the spherically symmetric Einstein-Klein-Gordon system (see Sec. 5).

Acknowledgements

I wish to thank Demetrios Christodoulou for reading the manuscript. This research was supported in part by NSF grant PHY 95-14726 to the University of Chicago.

References

1. S.W. Hawking and G.F.R. Ellis, *The Large Scale Structure of Space-Time*, Cambridge University Press (Cambridge, 1973).
2. R. Schoen and S.-T. Yau, Commun. Math. Phys., **90**, 575 (1983).
3. R. Penrose, in *Black Holes and Relativistic Stars*, ed. R.M. Wald, University of Chicago Press (in press).
4. R. Penrose, Revistas del Nuovo Cimento, **1**, 252 (1969).
5. R. Penrose, in *General Relativity, an Einstein Centennary Survey*, ed. S.W. Hawking and W. Israel, Cambridge University Press (Cambridge, 1979)
6. *Computer defies Einstein's theory*, New York Times, March 10, 1991, section 1, P.21; *A Bet on a Cosmic Scale, and a Concession, Sort of*, New York Times, February 12, 1997, section 1, P.1.
7. R.M. Wald, *General Relativity*, University of Chicago Press (Chicago, 1984).
8. P.S. Joshi and I.H. Dwivedi, Commun. Math. Phys., **146**, 333 (1992)
9. D. Christodoulou and S. Klainerman, *The Global Nonlinear Stability of Minkowski Space*, Princeton University Press (Princeton, 1993).
10. C.V. Vishveshwara, Phys. Rev., **D1**, 2870 (1970).
11. R. Price, Phys. Rev., **D5**, 2419 and 2439 (1972).
12. B.S. Kay and R.M. Wald, Class. Quant. Grav., **4**, 893 (1987).
13. B.F. Whiting, J. Math. Phys., **30**, 1301 (1989).
14. R.M. Wald, Ann. Phys., **82**, 548 (1974); I. Semiz Class. Quant. Grav., **7**, 353 (1990).
15. D.R. Brill, G.T. Horowitz, D. Kastor, and J. Traschen, Phys. Rev., **D49**, 840 (1994).
16. R. Penrose, Ann. N.Y. Acad. Sci., **224**, 125 (1973).
17. G. Gibbons, Commun. Math. Phys., **27**, 87 (1972).
18. N.S. Trudinger, Ann. Inst. Henri Poincare, **11**, 411 (1994).
19. G. Gibbons, *Collapsing shells and the isoperimetric inequality for black holes*, hep-th/9701049.
20. P.S. Jang and R.M. Wald, J. Math. Phys., **18**, 41 (1977).
21. R. Geroch, Ann. N.Y. Acad. Sci., **224**, 108 (1973).
22. G. Huisken and T. Ilmanen, *Proof of the Penrose Inequality*, to appear.
23. K.S. Thorne, in *Magic Without Magic: John Archibald Wheeler*, ed. J. Klauder, W.H. Freeman (San Francisco, 1972).
24. C.W. Misner, K.S. Thorne, and J.A. Wheeler, *Gravitation*, W.H. Freeman (San Francisco, 1973).
25. P.T. Chrusciel, Ann. Phys., **202**, 100 (1990).
26. T.A. Apostolatos and K.S. Thorne, Phys. Rev., **D46**, 2435 (1992).
27. B. Berger, P.T. Chrusciel, and V. Moncrief, Ann. Phys., **237**, 322 (1995).
28. S. Shapiro and S.A. Teukolsky, Phys. Rev., Lett **66**, 994 (1991).
29. V. Iyer and R.M. Wald, Phys. Rev., **D44**, 3719 (1991).
30. K.P. Tod, Class. Quant. Grav., **9**, 1581 (1992); K.P. Tod, unpublished; M. Pelath, unpublished.
31. P. Yodzis, H.-J. Seifert, and H. Muller zum Hagen, Commun. Math. Phys., **34**, 135 (1973).

32. D.M. Eardley and L. Smarr, Phys. Rev., **D19**, 2239 (1979); D. Christodoulou, Commun. Math. Phys., **93**, 171 (1984); P.S. Joshi and I.H. Dwivedi, Phys. Rev., **D47**, 5357 (1993).
33. D. Christodoulou, Commun. Math. Phys., **109**, 613 (1987).
34. D. Christodoulou, Commun. Pure & Applied Math., **XLIV**, 339 (1991).
35. D. Christodoulou, Commun. Pure & Applied Math., **XLVI**, 1131 (1993).
36. D. Christodoulou, Ann. Math., **140**, 607 (1994).
37. D. Christodoulou, Ann. Math., (in press).
38. T. Maithreyan, Ph.D. thesis, Boston University, 1984 (unpublished); M.D. Roberts, Gen. Rel. and Grav. **21**, 907 (1989); P.R. Brady, Class. Quant. Grav. **11**, 1255 (1994); Y. Oshiro, K. Nakamura, and A. Tomimatsu, Prog. Theor. Phys., **91**, 1265 (1994).
39. M.W. Choptuik, Phys. Rev. Lett., **70**, 9 (1993).
40. R.S. Hamade and J.M. Stewart, Class. Quant. Grav., **13**, 497 (1996).
41. A.M. Abrahams and C.R. Evans, Phys. Rev. Lett., **70**, 2980 (1993); C.R. Evans and J.S. Coleman, Phys. Rev. Lett., **72**, 1782 (1994); E.W. Hirschmann and D.M. Eardley, Phys. Rev., **D51**, 4198 (1995).
42. M.W. Choptuik, T. Chmaj, and P. Bizon, Phys. Rev. Lett., **77**, 424 (1996).

Thus the black hole is characterised entirely by mass, angular momentum and charge and no other parameter. This uniqueness theorem is often referred to colloquially as the 'no-hair theorem'. The assertion that the black hole has no hair is probably the ultimate application of Ockham's razor!

— C. V. VISHVESHWARA

The singularity, naked or clothed, is the biggest problem of all. It belongs today to the realm of conjecture and speculation, since physical laws, including general relativity, break down when confronted with the infinite space-time curvature. Will quantum effects come to the rescue, stop the collapse at the last moment and remedy the situation? No one knows. In this respect we might as well echo Faust and say, " We do not need what we know, and we do not know what we need". But one thing is certain. The black hole is by no means merely the silent finale to the symphony of the stars, nor is it the dark crystal ball in which the fate of general relativity is revealed. The unequivocal proof of the black hole's existence and a proper perspective of its role in the cosmic order both lie in the domain of the future.

— C. V. VISHVESHWARA

6. DISTURBING THE BLACK HOLE

JACOB D. BEKENSTEIN
Racah Institute of Physics, Hebrew University of Jerusalem, Givat Ram, Jerusalem 91904, ISRAEL

Foreword

It is a pleasure to dedicate this contribution to C.V. Vishveshwara on his sixtieth birthday. It is accompanied by warm wishes for a long and healthy life.

As a graduate student, I first met Vishu at the GR-6 conference in Copenhagen back in 1971. I was by then familiar with Vishveshwara's theorem [1] that the infinite redshift surface of a static black hole is always the horizon. At that time black hole physics was just getting started, and such neat relations between black hole features were rare. Vishu's theorem was a welcome hard fact in the middle of much folklore, and helped clarify in mind what black holes were about. At the conference I had a long talk with him, and I vividly remember being impressed by the range of research problems he had going simultaneously.

Vishu is also connected with another special event in my life. In 1974, just five or six days after taking up my first faculty job, at Ben–Gurion University in Israel, I went to Tel–Aviv to the GR-7 meeting. Vishu was there. I recall going with him and Yavuz Nutku on a long rambling walk through the streets of Tel–Aviv. We talked about many problems, and also speculated about why all the apartment buildings in Tel–Aviv have empty spaces for ground floors. Many things are better understood in black hole physics today since those days when Vishu's theorem was one of the few pegs to hang one's thoughts on, but I am afraid I still do not understand why those ground floors are left empty.

1. Introduction

Does the event horizon area of a black hole always grow under external perturbations? Hawking's area theorem [2] would suggest an affirmative

B. R. Iyer and B. Bhawal (eds.), Black Holes, Gravitational Radiation and the Universe, 87–101.

answer whenever classical fields obeying the weak energy condition are involved. Nevertheless, one can categorize a variety of situations in which an external perturbation transmitted through common fields is slowly applied and relaxed, and does *not* lead to area increase. This classical "adiabatic invariance" of horizon area, not yet a theorem but a collection of examples, is obviously consistent with the entropy character of black hole area [3, 4] because in classical thermodynamics entropy is invariant under slow changes of an insulated system in thermodynamic equilibrium. It is also an important motivation in an approach to black hole quantization [5, 6, 7, 8] which has received increasing attention in the last couple of years [9]. Keeping the ultimate application of the adiabatic property to black hole quantization in the back of the mind will help in grasping the significance of the medley of examples garnered here. In collecting these I have had in mind generating interest in turning the observation into a precise theorem.

Consider a small patch of event horizon area δA; it is formed by null generators whose tangents are $l^\alpha = dx^\alpha/d\lambda$, where λ is an affine parameter along the generators. By definition of the convergence ρ of the generators, δA changes at a rate

$$d\delta A/d\lambda = -2\rho\delta A. \tag{1}$$

Now ρ itself changes at a rate given by the optical analogue of the Raychaudhuri equation (with Einstein's equations already incorporated; I use units such that $G = c = 1$) [10, 11]

$$d\rho/d\lambda = \rho^2 + |\sigma|^2 + 4\pi T_{\alpha\beta}\, l^\alpha l^\beta, \tag{2}$$

where σ is the shear of the generators and $T_{\alpha\beta}$ the energy momentum tensor. The shear evolves according to

$$d\sigma/d\lambda = 2\rho\sigma + (3\epsilon - \bar{\epsilon})\sigma + C_{\alpha\beta\gamma\delta}\, l^\alpha m^\beta l^\gamma m^\delta, \tag{3}$$

where $C_{\alpha\beta\gamma\delta}$ is the Weyl conformal tensor, m^α one of the Newman–Penrose tetrad legs which lies in the horizon, and ϵ a pure imaginary parameter.

Many types of classical matter obey the weak energy condition

$$T_{\alpha\beta}\, l^\alpha l^\beta \geq 0. \tag{4}$$

Whenever this is true, ρ can – according to Eq.(2) – only *grow* or remain unchanged along the generators. Now were ρ to become positive at any event along a generator of our horizon patch, then by Eq.(2) it would remain positive henceforth, and indeed grow bigger. The joint solution of Eqs.(1)-(2) shows that δA would shrink to nought in a finite span of λ [2, 12]. This result of vanishing with its implied extinction of generators would constitute a singularity on the horizon. But it is an axiom of the subject

[2] that event horizon generators cannot end in the future. The only way out is to accept that $\rho \leq 0$ everywhere along the generators, which by Eq.(1) signifies that the horizon patch's area can never decrease. This is the essence of Hawking's area theorem.

It will be noticed that to keep the horizon area constant requires $\rho = 0$ which by Eqs.(2)-(3) implies that both $C_{\alpha\beta\gamma\delta}\, l^\alpha m^\beta l^\gamma m^\delta$ and $T_{\alpha\beta}\, l^\alpha l^\beta$ vanish at the horizon. Vanishing of $C_{\alpha\beta\gamma\delta}\, l^\alpha m^\beta l^\gamma m^\delta$ requires that the geometry be quasi-stationary to prevent gravitational waves, which are quantified by $C_{\alpha\beta\gamma\delta}$, from impinging on the horizon. Thus with a quasi-stationary geometry, preservation of the horizon's area requires

$$T_{\alpha\beta}\, l^\alpha l^\beta = 0 \quad \text{on the horizon.} \tag{5}$$

Contrary to the folklore which considers increase of horizon area to be an almost compulsory consequence of changes in the black hole, I shall here exhibit a variety of situations in which the conditions that keep horizon area unchanged occur naturally. The rule that seems to emerge is that quasi-stationary changes of the black hole occasioned by an external influence will leave the horizon area unchanged. This means that an "adiabatic theorem" for black holes must exist.

2. Black Hole Disturbed by Scalar Charges

Consider a Schwarzschild black hole with exterior metric

$$ds^2 = -(1 - 2M/r)dt^2 + (1 - 2M/r)^{-1}dr^2 + r^2(d\theta^2 + d\varphi^2). \tag{6}$$

Suppose sources of a *minimally coupled* scalar field Φ are brought up slowly from infinity to a finite distance from the hole, and then withdrawn equally slowly. Does this changing influence increase the horizon's area? Given that the changes are quasi-static, the question is just whether condition (5) is satisfied for all time. As we shall point out in Sec. 3, a potential barrier at $r \sim 3M$ screens out quasi-static scalar perturbations. For this reason our analysis becomes more than academic only when the sources actually enter the region $2M < r < 3M$.

If the scalar's sources are weak, one may regard Φ as a quantity of first order, and proceed by perturbation theory. The scalar's energy–momentum tensor,

$$T_\alpha{}^\beta = \nabla_\alpha \Phi \nabla^\beta \Phi - \frac{1}{2}\,\delta_\alpha^\beta\, \nabla_\gamma \Phi\, \nabla^\gamma \Phi, \tag{7}$$

will be of second order of smallness. I shall suppose the same is true of the energy–momentum tensor of the sources themselves. Thus to first order the metric (6) is unchanged. Neglecting for the moment time derivatives, the scalar equation outside the scalar's sources can be written in the form

$$\partial/\partial r\left[(r^2-2Mr)\partial\Phi/\partial r\right]-\hat{L}^2\,\Phi=0, \tag{8}$$

where $\hat{L}^2$ is the usual squared angular momentum operator (but without the $\hbar^2$ factor). This equation suggests looking for a solution of the form

$$\Phi=\Re\sum_{\ell=0}^{\infty}\sum_{m=-\ell}^{\ell}f_{\ell m}(r)\,Y_{\ell m}(\theta,\varphi), \tag{9}$$

where the $Y_{\ell m}$ are the familiar spherical harmonic (complex) functions. Since the $Y_{\ell m}$ form a complete set in angular space, any function $\Phi(r,\theta,\varphi)$ can be so expressed. And since $\hat{L}^2Y_{\ell m}=\ell(\ell+1)Y_{\ell m}$, the radial and angular variables separate, and one finds for $f_{\ell m}$ the equation

$$d/dr\left[(r^2-2Mr)df_{\ell m}/dr\right]-\ell(\ell+1)f_\ell=0. \tag{10}$$

Since the index m does not figure in this equation, I write just plain $f_\ell(r)$; one may obviously pick $f_\ell(r)$ to be real.

Let us change from variable r to $x\equiv r/M-1$ and define $F_\ell(x)\equiv f_\ell(r)$, so that Eq.(10) becomes

$$d/dx\left[(1-x^2)F_\ell\right]+\ell(\ell+1)F_\ell=0. \tag{11}$$

This is the Legendre equation of order ℓ. Its solutions, regular at the singular point $x=1$ of the equation are the well known Legendre polynomials $P_\ell(x)$. Independent solutions are furnished by the Legendre associated functions Q_ℓ which have the general form [13]

$$Q_\ell(x)=\frac{1}{2}\ln\left[\frac{x+1}{x-1}\right]P_\ell(x)+\text{polynomial of order }(\ell-1)\text{ in }x. \tag{12}$$

The associated solutions are thus singular at the horizon $x=1$, and must not be included in Φ, as the following argument makes clear.

We obviously require that the event horizon remains regular under the scalar's perturbation; otherwise the black hole is destroyed and our discussion is over before it began. A minimal requirement for regularity is that physical invariants like $\Upsilon_1\equiv T_\alpha{}^\alpha$, $\Upsilon_2\equiv T_\alpha{}^\beta T_\beta{}^\alpha$, $\Upsilon_3\equiv T_\alpha{}^\beta T_\beta{}^\gamma T_\gamma{}^\alpha$, etc., be bounded, for divergence of any of them would surely induce curvature singularities via the Einstein equations. By Eq.(7) the invariant Υ_k is always proportional to $(\Phi_{,\alpha}\,\Phi^{,\alpha})^k$. Now to lowest order of smallness the metric components that enter into Υ_k are just the Schwarzschild ones. In particular, in our static case $\Phi,_\alpha \Phi,^\alpha = (1-2M/r)\Phi,_r^2 + \cdots$. One must thus require

$$(1-2M/r)^{1/2}\, \Phi,_r \text{ bounded at horizon.} \tag{13}$$

Since this condition must hold for every θ and φ, it follows from the independence of the various $Y_{\ell m}$ in Eq.(9) that

$$\forall \ell : \quad (1-2M/r)^{1/2}\, df_\ell/dr \text{ bounded.} \tag{14}$$

But according to Eq.(12) the Q_ℓ are too singular to satisfy this equation, *i.e.*, $\sqrt{x-1}\,Q'_\ell(x) \to \infty$ as $x \to 1$. Thus one must discard the Q_ℓ from the set of radial solutions relevant in the region between the innermost source and the horizon.

Thus in this inner region of the black hole exterior we have, to first order in perturbation theory, the exact solution

$$\Phi = \Re \left(\sum_{\ell=0}^{\infty} \sum_{m=-\ell}^{\ell} C_{\ell m}\, P_\ell(r)\, Y_{\ell m}(\theta, \varphi) \right), \tag{15}$$

where the (complex) coefficients $C_{\ell m}$ permit us to match the solution to every distribution of sources by the usual methods. As those sources are moved around slowly, the $C_{\ell m}$ will change slowly (we shall investigate the question of changes at finite speed in Sec. 3). Now according to Eq.(7),

$$T_r{}^r - T_t{}^t = (1-2M/r)\, \Phi,_r^2\,. \tag{16}$$

Because Φ is completely regular down to the horizon, this shows that

$$\lim_{r \to 2M} (T_r{}^r - T_t{}^r) = 0. \tag{17}$$

We now explain the significance of this general result [14, 15] for our specific problem.

Any 3D–hypersurface of the form $\{\forall t, r = \text{const.}\}$ has a tangent $\tau^\alpha = \delta_t{}^\alpha$ with norm $-(1-2M/r)$ as well as the normal $\eta_\alpha = \partial_\alpha(r - \text{const.}) = \delta_\alpha{}^r$ with norm $(1-2M/r)$. The vector $N^\alpha \equiv \tau^\alpha + (1-2M/r)\eta^\alpha$ is obviously null, and as $r \to 2M$, both its covariant and contravariant forms remain well defined. Indeed, as $r \to 2M$ the other two vectors become null; in contrast with them, N^α remains well behaved at the horizon, so that it must there be proportional (with finite non-vanishing proportionality constant) to l^α, the tangent to the horizon generator. This can be verified by remarking that N^α, just as l^α, is future pointed ($N^t > 0$) as well as outgoing ($N^r > 0$). Now at any point $r \geq 2M$

$$T_{\alpha\beta} N^\alpha N^\beta = T_t{}^t\, N_t\, N^t + T_r{}^r\, N_r\, N^r = (T_r{}^r - T_t{}^r)\, N_r\, N^r. \tag{18}$$

But $N_r N^r \to 0$ at the horizon, so that in view of Eq.(17), $T_{\alpha\beta} l^\alpha l^\beta = 0$.

Since Eq.(5) is satisfied, the horizon's area does not change under the action of the scalar field. This conclusion is obviously conditional on the changes of the scalar's sources taking place sufficiently slowly, for otherwise Eq.(8) would be invalid, and the components of $T_\mu{}^\nu$ I employed would be affected. Our result supports the notion that adiabatic perturbations of a Schwarzschild black hole do not change the area, so that horizon area is an adiabatic invariant.

Notice that $T_{\alpha\beta} l^\alpha l^\beta$ vanishes because it is a product of *two* vanishing factors. This suggests that the area invariance result will remain valid under small "perturbations" to our scenario. Let us investigate the issue of time dependence of the scalar field; it is important because I am contemplating moving the sources of the scalar field so that it is never perfectly static as assumed heretofore.

3. The Time-Dependent Problem

Let us retain in the scalar equation the time derivatives:

$$-\frac{r^4}{(r^2 - 2Mr)}\frac{\partial^2 \Phi}{\partial t^2} + \frac{\partial}{\partial r}\left[(r^2 - 2M)\frac{\partial \Phi}{\partial r}\right] - \hat{L}^2\,\Phi = 0. \tag{19}$$

In analogy with Eq.(15) we now look for a solution of the form

$$\Phi = \Re \int_0^\infty d\omega \sum_{\ell=0}^{\infty} \sum_{m=-\ell}^{\ell} C_{\ell m}(\omega)\, f_\ell(\omega, r)\, Y_{\ell m}(\theta, \varphi) e^{-\imath\omega t}. \tag{20}$$

In terms of Wheeler's "tortoise" coordinate $r^* \equiv r + 2M \ln(r/2M - 1)$, for which the horizon resides at $r^* = -\infty$, the equation satisfied by the new radial function $H_{\ell\omega}(r^*) \equiv r f_\ell(\omega, r)$ is [12]

$$-\frac{d^2 H_{\ell\omega}}{dr^{*2}} + \left(1 - \frac{2M}{r}\right)\left(\frac{2M}{r^3} + \frac{\ell(\ell+1)}{r^2}\right) H_{\ell\omega} = \omega^2 H_{\ell\omega}. \tag{21}$$

The analogy between Eq.(21) and the Schrödinger eigenvalue equation permits the following analysis [12] of the effects of distant scalar sources on the black hole horizon. Waves with "energy" ω^2 on their way in from a distant source run into a positive potential, the product of the two parentheses in Eq.(21). The potential's peak is situated at $r \approx 3M$ for all ℓ. Its height is $0.0264M^{-2}$ for $\ell = 0$, $0.0993M^{-2}$ for $\ell = 1$ and $\approx 0.038\,\ell(\ell+1)M^{-2}$ for $\ell \geq 2$. Therefore, waves with any ℓ and $\omega < 0.163M^{-1}$ coming from sources at $r \gg 3M$ have to tunnel through the potential barrier to get near the horizon. As a consequence, the wave amplitudes that penetrate to the horizon are small fractions of the initial amplitudes, most of the waves being reflected back. In fact, the tunnelling coefficient vanishes in the limit $\omega \to 0$ [12]. This means that adiabatic perturbations by distant sources (which surely means they only contain Fourier components with $\omega \ll M^{-1}$) perturb the horizon very weakly (this is just an inverse of Price's theorem [12] that a totally collapsed star's asymptotic geometry preserves no memories of the star's shape). Thus one would not expect significant growth of horizon area from scalar perturbations originating in distant sources.

What if the scalar's sources are moved into the region $2M < r < 3M$ inside the barrier? They will now be able to perturb the horizon; do they change its area? To check I look for the solutions of Eq.(21) in the region near the horizon where the potential is small compared to ω^2; according to the theory of linear second order differential equations, they are of the form

$$H_{\ell\omega}(r^*) = \exp(\pm\imath\omega r^*) \times [1 + \theta(1 - 2m/r)]. \tag{22}$$

The Matzner boundary condition [16] that the physical solution be an ingoing wave as appropriate to the absorbing character of the horizon selects the sign in the exponent as negative. Hence the typical term in Φ is

$$\frac{1 + \theta(1 - 2m/r)}{r} P_\ell(\cos\theta)\,\cos\psi; \qquad \psi \equiv \omega(r^* + t) - m\varphi. \tag{23}$$

Is the perturbation well behaved at the horizon? In particular, are all the invariants Υ_k of Sec. 2 (or equivalently $\Phi,_\alpha \Phi,^\alpha$) bounded there? Let us first look at a Φ composed of a single mode like that in Eq.(23). An explicit calculation on the Schwarzschild background using $dr^*/dr = (1 - 2M/r)^{-1}$ gives, after a miraculous cancellation of terms divergent at the horizon (pointed out by A. Mayo),

$$\Phi,_\alpha \Phi,^\alpha \propto \frac{m^2 P_\ell{}^2 \sin^2\psi}{r^4 \sin^2\theta} + \left(\frac{dP_\ell}{d\theta}\right)^2 \frac{\cos^2\psi}{r^4} + \frac{\omega \sin(2\psi)}{r^3} P_\ell^2 + \cdots, \tag{24}$$

where " $\cdots$ " here and henceforth denote terms that vanish as $r \to 2M$. This expression is bounded at the horizon. Now suppose Φ is the sum of two modes like (23). Let us label the various parameters with subscripts "1" and "2". Then a calculation gives $\Phi,_\alpha \Phi,^\alpha$ as consisting of three groups of terms, two of them of form (24) with subscripts 1 and 2, respectively, and a third of the form

$$\frac{m_1 m_2 P_{\ell_1} P_{\ell_2} \sin\psi_1 \sin\psi_2}{r^4 \sin^2\theta} + \left(\frac{dP_{\ell_1}}{d\theta}\right)\left(\frac{dP_{\ell_2}}{d\theta}\right) \frac{\cos\psi_1 \cos\psi_2}{r^4} +$$
$$+ \frac{\omega_1 \sin\psi_1 \cos\psi_2 + \omega_2 \sin\psi_2 \cos\psi_1}{r^3} P_{\ell_1} P_{\ell_2} + \cdots \tag{25}$$

This is also bounded. By induction any Φ of form (20) will give a bounded $\Phi_{,\alpha}\,\Phi_{,}{}^{\alpha}$. Thus all the Υ_k are bounded at $r = 2M$, and a generic scalar perturbation does not disturb the horizon unduly.

The extent by which the *shape* of the horizon is perturbed must be linear in the magnitude of the invariant Υ_1 (Einstein's equations have $T_{\alpha\beta}$ as source, not $T_\alpha{}^\gamma T_\gamma{}^\beta$). It is then clear from both our results that this perturbation is of order $\theta(\omega^0)$ generically, and of $\theta(\omega)$ in the monopole case. I now show that the change in the horizon area is of $\theta(\omega^2)$, so that for small ω the area is (relatively) invariant.

Because now there is time variation and $T_t{}^r \neq 0$, Eq.(18) has to be generalized to the form

$$T_{\alpha\beta}N^\alpha N^\beta = (T_r{}^r - T_t{}^r)\,N_r\,N^r + 2T_t{}^r N_r N^t. \tag{26}$$

From N^α's definition in Sec. 2 we have $N_r N^r = 1 - 2M/r$ and $N_r N^t = 1$. And from Eq.(7) it is clear that $T_r{}^r - T_t{}^t = \Phi_{,r}\,\Phi_{,}{}^r - \Phi_{,t}\,\Phi_{,}{}^t$ while $T_t{}^r = \Phi_{,t}\,\Phi_{,}{}^r$. Thus

$$T_{\alpha\beta}N^\alpha N^\beta = [\Phi_{,t} + (1-2M/r)\Phi_{,r}\,]^2. \tag{27}$$

If one now substitutes a Φ made up of a single mode like in Eq.(23), one concludes that

$$T_{\alpha\beta}l^\alpha l^\beta \propto \frac{\omega^2 P_\ell^2 \sin^2\psi}{r^2} + \cdots. \tag{28}$$

A quick way to this result is to recognize that $l^\alpha \propto \tau^\alpha \equiv (\partial/\partial t)^\alpha$ because the horizon generators must lie along the only Killing vector of the problem. In view of Eq.(7) and the null character of l^α,

$$T_{\alpha\beta}l^\alpha l^\beta \propto (\Phi_{,\alpha}\,\tau^\alpha)^2 = (\partial\Phi/\partial t)^2, \tag{29}$$

which reproduces Eq.(28). And if one substitutes the generic Φ, the proportionality to the square of frequency will obviously remain.

Thus, when scalar field sources are moved inside the barrier, they perturb the geometry, and the horizon's shape in particular, by an amount which does not, in general, vanish as the perturbations change very slowly. By contrast, the rate of change of the horizon area vanishes as the square of the typical Fourier frequency of the perturbation. In this sense the horizon area is an adiabatic invariant.

4. Generalization to Non-minimally Coupled Field

To what extent is our result generic? For example, does it depend on the nature of the interaction? One can probe this question by replacing our minimally coupled scalar field by one coupled non-minimally to curvature. The scalar equation is replaced by [17]

$$(\nabla_\alpha\nabla^\alpha - \xi R)\Phi = 0, \tag{30}$$

where R denotes the Ricci scalar and $\xi \neq 0$ is a real constant measuring the extent of non-minimal coupling; $\xi = 1/6$ corresponds to a conformally invariant scalar equation. The corresponding energy–momentum tensor is [18]

$$T_\alpha{}^\beta = \nabla_\alpha\Phi\nabla^\beta\Phi - \frac{1}{2}\nabla_\alpha\Phi\,\nabla^\alpha\Phi\,\delta_\alpha^\beta - \xi(\nabla_\alpha\nabla^\beta\Phi^2 - \delta_\alpha^\beta\nabla_\mu\nabla^\gamma\Phi^2 - G_\alpha{}^\beta\Phi^2), \tag{31}$$

where $G_\alpha{}^\beta$ denotes the Einstein tensor. Evaluating it with help of the Einstein equations, one obtains, in the region outside the scalar's sources where they do not contribute to $T_\alpha{}^\beta$,

$$T_\alpha{}^\beta = \frac{\nabla_\alpha\Phi\nabla^\beta\Phi - \frac{1}{2}\delta_\alpha^\beta\,\nabla_\gamma\Phi\,\nabla^\gamma\Phi - \xi(\nabla_\alpha\nabla^\beta\Phi^2 - \delta_\alpha^\beta\nabla_\gamma\nabla^\gamma\Phi^2)}{1 - 8\pi G\xi\Phi^2}. \tag{32}$$

The following analysis is carried out neglecting the time variation as in Sec. 2. If again one regards Φ as of first order of smallness, then it is clear that $T_\alpha{}^\beta$ is of second order and consequently the lowest correction to the Schwarzschild metric, Eq.(6), is of second order. Likewise, R in Eq.(30) is of second order. Therefore, in the static case one may again get Φ to first order by solving Eq.(8), and Φ is given by the series in Eq.(9) with radial functions which are superpositions of Legendre polynomials P_ℓ and Legendre functions Q_ℓ.

To select out the physical combinations of these, I again require - the argument is exactly like in Sec. 2 - that every *diagonal* component of $T_\alpha{}^\beta$ be bounded. In particular one has from Eq.(32) that

$$T_t^t - T_\varphi^\varphi = \frac{2\xi(1-3M/r)\Phi\,\Phi_{,r}}{r(1-8\pi G\xi\Phi^2)} \quad \text{bounded at horizon.} \tag{33}$$

It is clear from this that Φ itself must be bounded at $r = 2M$, for if it were to diverge there, then $|T_t^t - T_\varphi^\varphi| \sim |d\ln\Phi/dr| \to \infty$. I thus discard all the Q_ℓ functions from Φ, thereby returning to the form (15): Φ is regular at the horizon.

Now

$$T_t{}^t - T_r{}^r = \left(1 - \frac{2M}{r}\right)\frac{(2\xi-1)\Phi_{,r}^2 + 2\xi\,\Phi\,\Phi_{,rr}}{1 - 8\pi G\xi\Phi^2}. \tag{34}$$

From this it is clear that $T_t{}^t - T_r{}^r \to 0$ at $r = 2M$ unless $8\pi G\xi\Phi^2 \to 1$ there. This last possibility must be excluded for it would be equivalent to having $\ln(1 - 8\pi G\xi\Phi^2) \to -\infty$, whereby $d\ln(1 - 8\pi G\xi\Phi^2)/dr$ would necessarily diverge at $r = 2M$. But this derivative occurs as a factor in $T_t^t - T_\varphi^\varphi$ and its divergence would not be compensated for; thus Eq.(33) would be contradicted. Since $T_t{}^t - T_r{}^r \to 0$ on the horizon the arguments at the end of Sec. 2 can be repeated to show that the area of the black hole is unchanged by the scalar perturbation. Although we shall not go here into the time dependent problem, it seems that coupling the field non-minimally makes little difference regarding the adiabatic invariance of the horizon.

A. Mayo [19] has worked out the effect of static electromagnetic perturbations on the horizon and shown that they also leave its area unchanged.

5. Waves Impinging on a Kerr Black Hole

In the above two examples when the perturbation is removed, the black hole exterior must return to the exact Schwarzschild form because a static spherical black hole with no electric charge cannot retain scalar hair [20]. Further, the black hole returns to its original mass since, for the Schwarzschild case, the horizon area and the mass are related one–to–one, and the area has not changed. The question thus arises: Does adiabatic invariance of the area continue to hold when the black hole's perturbation is accompanied by a net change in some of the Wheeler black hole parameters, mass, charge and angular momentum? We now show that the answer is affirmative for adiabatic scalar perturbations of the Kerr black hole.

In the Kerr case the meaning of "adiabatic" needs to be refined. It is well known that a static ($\omega = 0$) but non-axisymmetric perturbation of a Kerr black hole, such as would be caused by field sources held in its vicinity at rest with respect to infinity, necessarily causes an increase in horizon area [21]. However, static perturbations in this sense are not adiabatic from the local point of view. Because of the dragging of inertial frames [12], any non-axisymmetric static field is perceived by momentarily radially stationary local inertial observers as endowed with temporal variation as these observers are necessarily dragged through the field's spatial inhomgeneity. At the horizon the dragging frequency is the hole's rotational frequency Ω, and a field component with azimuthal "quantum" number m is seen to vary with temporal frequency $m\Omega$ which need not be small. Evidently, "adiabatic" here must mean that *according to momentarily radially stationary local inertial observers*, the perturbation has only low frequency Fourier components.

To see these concepts in practice, consider a Kerr black hole of mass M and angular momentum J. I shall not need the metric; all that is important here is that the horizon area is

$$A = 4\pi \left[\left(M + \sqrt{M^2 - (J/M)^2} \right)^2 + (J/M)^2 \right] . \tag{35}$$

Let sources distributed well away from the hole radiate upon it a *weak* scalar wave of the form

$$\Phi = f(r)\, Y_{\ell m}(\theta, \varphi)\, e^{-\imath\omega t}. \tag{36}$$

In the spirit of perturbation theory I shall neglect the gravitational waves so produced. The black hole geometry will eventually be changed by interaction with this wave, but since the latter is taken to be weak, I shall assume that the change amounts to a transition from one Kerr geometry to another with slightly different M and J. In the final analysis such assumption is justified by the stability of the Kerr geometry. Since the geometry thus remains axisymmetric and stationary after the change, the wave preserves its form (36) over all time. Long ago Starobinskii [22] showed that for small $(\omega - m\Omega)$ the absorption coefficient for a wave like (36) has the form

$$\Gamma = K_{\omega\ell m}(M, J) \cdot (\omega - \Omega\, m)\,, \tag{37}$$

where

$$\Omega \equiv \frac{J/M}{r_{\mathcal{H}}{}^2 + (J/M)^2}; \quad \text{with} \quad r_{\mathcal{H}} \equiv M + \sqrt{M^2 - (J/M)^2} \tag{38}$$

is the rotational angular frequency of the hole, while $K_{\omega\ell m}(M, J)$ is a positive coefficient. If one chooses $\omega = m\Omega$ (static perturbation in the eyes of the local inertial observers), the wave is perfectly reflected and no change of the black hole parameters ensues.

By choosing $(\omega - \Omega\, m)$ slightly positive, one arranges for a small fraction of the wave to get absorbed. Now imagine surrounding the system by a large spherical mirror. The part of the wave reflected off the black hole gets reflected perfectly back towards it by the mirror. The wave thus bounces back and forth between the two and a sizable fraction of its energy and angular momentum will eventually get absorbed by the hole. Similarly, if one makes $(\omega - m\Omega)$ slightly negative, the wave gets amplified upon reflection off the hole (Zel'dovich–Misner superradiance [23]) and the repeated reflections make it stronger than originally. Since both processes envisioned take place over many cycles of reflection (a long interval of t time), the consequent (substantial) changes of M and J occur *adiabatically* according to distant observers. This is in addition to the adiabatic nature of the perturbing wave as seen by local observers.

A simple calculation shows that small changes of the horizon area are related to those of M and J by [24, 3, 25]

$$\Delta A = (\Delta M - \Omega \Delta J)\,/\Theta_K, \tag{39}$$

where

$$\Theta_K \equiv \frac{1}{2} A^{-1} \sqrt{M^2 - (J/M)^2} \tag{40}$$

is the surface tension [24] or surface gravity [25] of the black hole. The overall changes ΔM and ΔJ must stand in the ratio ω/m. This can be worked out from the energy–momentum tensor, but is immediately clear if one thinks of the wave as composed of quanta, each with energy $\hbar\omega$ and angular momentum $\hbar m$, and using conservation energy and angular momentum. Since $\omega \approx \Omega m$, it is seen from Eq.(39) that if the black hole is not extremal ($J \neq M^2$ and so $\Theta_K \neq 0$), $\Delta A \approx 0$ to the accuracy of the former equality. Therefore, Kerr horizon area is invariant during adiabatic changes of the mass and angular momentum as judged globally. These changes come about from field perturbations which are adiabatic from the perspective of local inertial observers, as stated earlier. It is under these two conditions that the horizon area is an adiabatic invariant.

The last conclusion is, however, inapplicable to the extremal Kerr black hole ($J = M^2$). In this case $\Theta_K = 0$ so one cannot use Eq.(39) to calculate the change in area, but must work directly with Eq.(35). From Eq.(38) one learns that $\Omega = 1/2M$ so that $\Delta J = \Omega^{-1}\Delta M = 2M\Delta M$. Replacing $M \to M + \Delta M$ and $J \to J + 2M\Delta M$ in Eq.(35), and subtracting the original expression gives

$$\Delta A = 8\pi(2+\sqrt{2})M\Delta M + \theta((\Delta M)^2). \tag{41}$$

This is not a small quantity; a generic addition of mass ΔM will give a ΔA of the same order. Thus horizon area of an extremal Kerr hole is not an adiabatic invariant. Sec. 6 gives one more example of the departure of extremal black horizon area from the adiabatic invariance.

For non-extremal black holes the conclusion that horizon area is an adiabatic invariant extends to perturbations by electromagnetic waves. One only has to replace the $Y_{\ell m}(\theta,\varphi)$ in the wave by an electric or magnetic type *vector* spherical harmonic to describe the electromagnetic modes. The conclusion is the same.

6. Particle Absorption by Reissner–Nordström Black Hole

Thus far I have illustrated the adiabatic invariant character of black hole horizon area under the field–black hole interaction. The present example focuses rather on point particle–black hole interaction. It is none other than the Christodoulou reversible process [26] for a Reissner–Nordström black hole.

Consider a Reissner–Nordström black hole of mass M and positive charge Q. The exterior metric is

$$ds^2 = -\chi\, dt^2 + \chi^{-1}\, dr^2 + r^2(d\theta^2 + d\varphi^2) \tag{42}$$

with

$$\chi \equiv 1 - 2M/r + Q^2/r^2. \tag{43}$$

At infinity one shoots in radially a classical point particle of mass m and positive charge ε with total relativistic energy adjusted to the value

$$E = \varepsilon Q/r_{\mathcal{H}}. \tag{44}$$

Here $r_{\mathcal{H}}$ is the r coordinate of the infinite red shift surface ($\chi = 0$), which by Vishveshwara's theorem [1] coincides with the event horizon:

$$r_{\mathcal{H}} = M + \sqrt{M^2 - Q^2}. \tag{45}$$

In Newtonian terms this particle should marginally reach the horizon where its potential energy just exhausts the total energy. The relativistic equation of motion leads to the same conclusion.

Let us consider the action for the radial motion,

$$S = \int \left[-m\sqrt{\chi\,(dt/d\tau)^2 - (dr/d\tau)^2/\chi} - \varepsilon A_t\; dt/d\tau \right] d\tau, \tag{46}$$

where τ, the proper time, acts as a path parameter, and $A_t = Q/r$ is the only non-trivial component of the electromagnetic 4–potential. The stationary character of the background metric and field means that there exists a conserved quantity, namely

$$E = -\delta S/\delta(dt/d\tau) = \frac{m\,\chi}{\sqrt{\chi\,(dt/d\tau)^2 - (dr/d\tau)^2/\chi}}\,\frac{dt}{d\tau} + \frac{\varepsilon\, Q}{r}. \tag{47}$$

We also know that the norm of the velocity is conserved. This together with the definition of proper time gives $\sqrt{\chi\,(dt/d\tau)^2 - (dr/d\tau)^2/\chi} = 1$. Substituting $dt/d\tau$ from here in Eq.(47) gives

$$E = m\sqrt{\chi + (dr/d\tau)^2} + \frac{\varepsilon\, Q}{r}. \tag{48}$$

It is easy to see that this is precisely the total energy of the particle, for at large distances from the hole, $E \approx m + mv^2/2 - m\,M/r + \varepsilon\, Q/r$ (sum of rest, kinetic, gravitational and electrostatic potential energies). Setting $E = \varepsilon\, Q/r_{\mathcal{H}}$ shows that the radial motion has a turning point ($dr/d\tau = 0$) precisely at the horizon $[\chi(r_{\mathcal{H}}) = 0]$.

Because the particle's motion has a turning point at the horizon, it gets accreted by it. The area of the horizon is originally

$$A = 4\pi {r_{\mathcal{H}}}^2 = 4\pi\left(M + \sqrt{M^2 - Q^2}\right)^2, \tag{49}$$

and the (small) change inflicted upon it by the absorption of the particle is

$$\Delta A = (\Delta M - Q\,\Delta Q/r_{\mathcal{H}})/\Theta_{RN} \tag{50}$$

with

$$\Theta_{RN} \equiv \frac{1}{2} A^{-1} \sqrt{M^2 - Q^2} \tag{51}$$

being the surface gravity analogous to the previous Θ_K. Thus if the black hole is not extremal so that $\Theta_{RN} \neq 0$, $\Delta A = 0$ because $\Delta M = E = \varepsilon Q / r_{\mathcal{H}}$ while $\Delta Q = \varepsilon$. Therefore, the horizon area is invariant under the accretion of the particle from a turning point.

To a momentarily radially stationary local inertial observer, the particle in question hardly moves radially as it is accreted. Thus its assimilation is adiabatic. By contrast, if E were larger than in (44), the particle would not try to turn around at the horizon, and the local observer would see it moving radially at finite speed and being assimilated quickly. And the horizon's area would increase upon its accretion, as is easy to check from the previous argument. Thus invariance of the area goes hand in hand with adiabatic changes at the horizon as judged by local observers at the horizon. Needless to say, the changes in M and Q also occur very slowly as judged by distant observers. This is the double sense in which the area is an adiabatic invariant.

The above conclusions fail for the extremal Reissner–Nordström black hole. When $Q = M$, $\sqrt{M^2 - Q^2}$ in Eq.(49) is unchanged to $\theta(\varepsilon^2)$ during the absorption, so that $\Delta A = 8\pi M E$. This is not a small change, so the horizon's area is not an adiabatic invariant. It is clear, as already noted earlier, that extremal black holes behave differently from generic black holes in this as in other phenomena.

7. Conclusions

The examples collated here suggest the existence of a theorem which would state that, classically, under suitably adiabatic changes (two such are characterized here) of a black hole in equilibrium, the area of its event horizon does not change. Aside from harmonizing with the understanding that horizon area represents entropy, this theorem would provide a formal motivation for quantizing the black hole in the spirit of the "old quantum theory" or Bohr–Sommerfeld quantization. The implications of such quantization have already been considered [5, 6].

I thank Avraham Mayo for criticism. This contribution is based on research supported by a grant from the Israel Science Foundation.

References

1. Vishveshwara, C. V. (1968) *J. Math. Phys.*, **9**, 1319.
2. Hawking, S. W. (1971) *Phys. Rev. Letters*, **26**, 1344 .
3. Bekenstein, J. D. (1973) *Phys. Rev. D*,**7**, 2333.
4. Hawking, S. W. (1975) *Commun. Math. Phys.*, **43**, 212.

5. Bekenstein, J. D. (1974) *Lett. Nuovo Cimento*, **11**, 467.
6. Mukhanov, V. F. (1986) *JETP Letters*, **44**, 63; Mukhanov, V. F. (1990) in *Complexity, Entropy and the Physics of Information: SFI Studies in the Sciences of Complexity*, vol. III, Zurek, W. H., ed., Addison–Wesley, New York.
7. Bekenstein, J. D. and Mukhanov, V. F. (1995) *Phys. Lett. B*, **360**, 7.
8. Bekenstein, J. D. (1996) in *XVII Brazilian National Meeting on Particles and Fields*, da Silva, A. J. et. al, eds., Brazilian Physical Society; Bekenstein, J. D. (1998) in *Proceedings of the VIII Marcel Grossmann Meeting on General Relativity*, Piran, T. and Ruffini, R., eds., World Scientific, Singapore.
9. For a full list of references see the second of Refs. 8; the published ones include Kogan, I. (1986) *JETP Letters*, **44**, 267; Danielsson, U. H. and Schiffer, M. (1993) *Phys. Rev. D*, **48**, 4779; Maggiore, M. (1994) *Nucl. Phys. B*, **429**, 205; Peleg, Y. (1995) *Phys. Lett. B*, **356**, 462; Lousto, C. O. (1995) *Phys. Rev. D*, **51**, 1733; Kastrup, H. (1996) *Phys. Lett. B*, **385** 75; Barvinskii, A. and Kunstatter, G. (1996) *Phys. Lett. B*, **329** 231; Louko, J. and Mäkelä, J. (1996) *Phys. Rev. D*, **54**, 4982; Mäkelä, J. (1997) *Phys. Lett. B*, **390**, 115; Berezin, V. (1997) *Phys. Rev. D*, **55**, 2139; Brotz, C. and Kiefer, K. (1997) *Phys. Rev. D*, **55**, 2186.
10. Newman, E. and Penrose, R. (1962) *J. Math. Phys.*, **3**, 566.
11. Pirani, F. A. E. (1965) in *Lectures on General Relativity: Brandeis Summer Institute in Theoretical Physics*, Vol. I, Deser, S. and Ford, K. W., eds., Prentice–Hall, Englewood Cliffs, NJ.
12. Misner, C. W. , Thorne, K. S. and Wheeler, J. A. (1973) *Gravitation*, Freeman, San Francisco.
13. Mathews, J. and Walker, R. L. (1970) *Mathematical Methods of Physics*, Second Edition, Benjamin, Menlo Park, California.
14. Achucarro, A., Gregory, R. and Kuijken, K. (1995) *Phys. Rev. D*, **52**, 5729; Núñez, D., Quevedo, H. and Sudarsky, D. (1996) *Phys. Rev. Letters*, **76**, 571.
15. Mayo, A. E. and Bekenstein, J. D. (1996) *Phys. Rev. D*, **54**, 5059.
16. Matzner, R. A. (1968) *Phys. Rev. D*, **9**, 163.
17. This is a slight generalization of the conformally invariant field equation introduced in Penrose, R. (1965) *Proc. Roy. Soc. London A*, **284**, 159.
18. Parker, L. (1973) *Phys. Rev. D*, **7**, 976 gives the $\xi = 1/6$ case; for the general case see, for example, Ref. 15.
19. Mayo, A. E., publication in preparation.
20. The no scalar hair theorems for the minimally coupled massless or massive field are given in Bekenstein, J. D. (1972) *Phys. Rev. Letters*, **28**, 452; Bekenstein, J. D. (1973) *Phys. Rev. D*, **5**, 1239 and 2403. Generalizations to non-minimal coupling (to curvature) have been given by Saa, A. (1996) *J. Math. Phys.*, **37**, 2346 and Ref.15. For a general review of no hair theorems see Heusler, M. (1996) *Black Hole Uniqueness Theorems*, Cambridge University Press, Cambridge.
21. Hawking, S. W. and Hartle, J. B (1972) *Commun. Math. Phys.*, **27**, 283.
22. Starobinskii, A. A. (1973) *Sov. Phys. JETP*, **37**, 28.
23. Zel'dovich, Ya. B. (1971) *JETP Letters*, **14**, 180; Misner, C. W., unpublished.
24. Bekenstein, J. D. (1972), Princeton University Dissertation (unpublished).
25. Bardeen, J., Carter, B. and Hawking, S. W. (1973) *Commun. Math. Phys.*, **31**, 181.
26. Christodoulou, D. (1970), *Phys. Rev. Letters*, **25**, 1596; Christodoulou, D. and and Ruffini, R. (1971) *Phys. Rev. D*, **4**, 3552.

In the year 1756 Siraj-ud-daula, the Nawab of Bengal tried to create a black hole by squeezing a large amount of matter, in the readily available human form, into a rather small volume. The experiment was a failure and the political consequences were disastrous. But the name black hole came into existence. Only in the last few years have we come to realize that the Nawab's method of manufacturing black holes is feasible after all, but the compressions required are of cosmological proportions. These are the primordial black holes which may have nothing to do with politics.

About two hundred years later, the term black hole was introduced into astrophysics. Since then it has come to represent many things in popular imagination: symbol of Dantesque doom, panacea for astrophysical ills, inexhaustible source of energy, free ticket to time travel, gateway to other universes that may not exist, explanation for the missing mass in the universe, the missing solar neutrinos, the missing link – in general anything that is missing. Fortunately, behind all this folklore and fantasy, fallacy and fiction, stands the black hole of the sane scientist, which is a product of Einstein's general theory of relativity and a manifestation of high space-time curvature.

— C. V. VISHVESHWARA

7. NOTES ON BLACK HOLE FLUCTUATIONS AND BACK-REACTION

B. L. HU
Department of Physics, University of Maryland, College Park, Maryland 20742, USA.
ALPAN RAVAL
Department of Physics, University of Wisconsin-Milwaukee, Milwaukee, Wisconsin 53201, USA
AND
SUKANYA SINHA
Raman Research Institute, Bangalore, India

1. Classical and Semiclassical Back-reaction of Metric Perturbations

The idea of viewing a black hole (particle detector) interacting with a quantum field as a dissipative system, and the Hawking [1]- Unruh [2] radiation as a manifestation of a fluctuction-dissipation relation was first proposed by Candelas and Sciama [3, 4]. Even though, as we will soon see, the fluctuations in the thought of these earlier authors are not the correct ones and the relations proposed were not really addressing the back-reaction of quantum fields in a classical black hole spacetime, the idea remains attractive. Indeed one of us (BLH) found it so attractive that he launched a systematic investigation into the statistical mechanical properties of particle/spacetime and quantum field interactions. This involved the introduction of statistical mechanical ideas such as quantum open sytems [5] and field-theoretical methods such as the influence functional [6] and Schwinger-Keldysh formalisms [7] for the establishment of a quantum statistical field theory in curved spacetime (for a review, see [8, 9]). It was found that the back-reaction of quantum fields (through processes like particle creation) on a classical background spacetime can be described by an Einstein-Langevin equation [10], which is a generalization of the semiclassical Einstein equa-

B. R. Iyer and B. Bhawal (eds.), Black Holes, Gravitational Radiation and the Universe, 103–120.

tion to include stochastic sources due to created particles. Indeed it was also found from first principles that the back-reaction can be encapsulated in the form of a Fluctuation-Dissipation Relation (FDR) [12, 13], which takes into account the mutual influence of the quantum field and the background spacetime (or detector, in the case of Unruh radiation). We expect it to hold also for black hole systems, both in the familiar static condition where black hole thermodynamics based on the Bekenstein-Hawking relation was constructed, and for dynamical collapse problems. This is the major theme in our current program of research.

Note that it is nontrivial that such a relation exists at least on two counts. First, that the dynamics of spacetime interacting with a quantum field can indeed be treated like a classical particle on a trajectory with stochastic components as depicted in quantum Brownian motion [16, 17]; and second, although in statistical thermodynamics the FDR is usually derived for near-equilibrium conditions (via linear response theory), these semiclassical gravity processes can serve as an illustration that such a relation can indeed exist for nonequilibrium processes. This was conjectured by one of us in 1989 [8] and demonstrated in succeeding works [10, 11, 12, 14]. We will follow this line of thought to pursue black hole back-reaction problems. In this section we will describe in general terms the classical and semiclassical back-reaction of metric perturbations. In Sec. 2 we describe stochastic semiclassical gravity in a cosmological setting, focusing on the derivation of an Einstein-Langevin equation. In Sec. 3 we describe metric fluctuations and back-reaction in semiclassical gravity, and the FDR for a black hole in equilibrium with its thermal radiation. We comment on the inadequacy in Mottola's [18] derivation of a FDR for back-reaction of static black holes. In Sec. 4 we discuss dynamical black holes, identifying the Bardeen metric as a fruitful avenue for semiclassical back-reaction analysis. We also comment on the nature of Candelas and Sciama's [3] FDR and its shortcoming for the description of back-reaction. We then summarize Bekenstein's theory of black hole fluctuations and note the difference from our approach. In these notes we are merely preparing the ground for our investigation by sorting out the issues and identifying the inadequacies of previous approaches. Our findings will appear in later publications elsewhere.

For brevity we shall use schematic expressions to discuss the ideas here while relegating the details to research papers in progress. Let us start with the classical Einstein equation for a spacetime with metric $g_{\alpha\beta}$

$$G_{\mu\nu}(g_{\alpha\beta}) = \kappa T^{\rm cm}_{\mu\nu} \tag{1.1}$$

where $G_{\mu\nu}$ is the Einstein tensor, $\kappa = 8\pi G_N$, G_N being the Newton constant, and $T^{\rm cm}_{\mu\nu}$ is the energy momentum tensor of some classical matter (cm). Consider perturbations of the metric tensor at the classical level, i.e.,

$$g_{\alpha\beta} = g^{(0)}_{\alpha\beta} + \epsilon h^{(1)}_{\alpha\beta} + \epsilon^2 h^{(2)}_{\alpha\beta} + \ldots \tag{1.2}$$

where the superscript n in parentheses on $h^{(n)}_{\alpha\beta}$ indicates the order of the perturbation. Most studies of gravitational waves and instability are carried out for linear perturbations $n = 1$.

The linear perturbations (in harmonic gauge) satisfy the linearized Einstein equations in the form [19]

$$\Box h_{\alpha\beta} = 2\kappa \delta T^{\rm cm}_{\alpha\beta} \tag{1.3}$$

The problem of gravitational perturbations in a cosmological Friedmann-Lemaitre-Robertson-Walker (FLRW) spacetime in relation to galaxy formation was first treated in detail by Lifshitz [20]. Perturbations in a Schwarzschild black hole spacetime was treated by Regge and Wheeler, Vishveshwara, Zerilli et al [21] and in a Kerr black hole by Teukolsky, Chandrasekhar and others [22].

For quantum fields in a classical background spacetime with metric $g^{(0)}_{\mu\nu}$, the wave equation for, say, a massive (m) minimally coupled scalar field Φ is

$$\Box\Phi + m^2\Phi = 0 \tag{1.4}$$

The first order metric perturbations $h^{(1)}_{\alpha\beta}$ in a vacuum ($T_{\alpha\beta} = 0$) obey an equation similar in form to the above with $m = 0$ (the Lifshitz equation [20]) and can thus be identified as two components (because of the 2 polarizations) of a massless minimally coupled scalar field.

Now let us consider the back-reaction of gravitational perturbations or quantum fields in classical and semiclassical gravity. At the classical level, assuming a vacuum background, the linear perturbations $h^{(1)}_{\mu\nu}$ would contribute a source to the $O(\epsilon^0)$ equation in the form [19] (Eq. 35.70)

$$G_{\mu\nu}(g^{(0)}) = \kappa \langle T^{\rm gw}_{\mu\nu} \rangle_I \tag{1.5}$$

where

$$T^{\rm gw}_{\mu\nu} \equiv \frac{1}{32\pi}[\bar{h}^{(1)}_{\alpha\beta|\mu}\bar{h}^{(1)}_{\alpha\beta|\nu}], \tag{1.6}$$

is the (inhomogeneous) energy momentum tensor of the gravitational waves (gw) described (in a transverse-traceless gauge) by $\bar{h}^{(1)}_{\alpha\beta} \equiv h^{(1)}_{\alpha\beta} - \frac{1}{2}h^{(1)}g^{(0)}_{\alpha\beta}$. Here the $\langle\,\rangle_I$ around $T^{\rm gw}_{\mu\nu}$ denotes the Isaacson average over the inhomogeneous sources (taken over some intermediate wavelength range larger than the natural wavelength of the waves but smaller than the curvature radius of the background spacetime). This is an example of back-reaction at the classical level. (For related work see [23].)

At the semiclassical level, the back-reaction comes from particles created in the quantum field (qf) on the background spacetime [24, 25]. Schematically the semiclassical Einstein equation takes the form

$$G_{\mu\nu}(g^{(0)}) = \kappa \langle T^{\rm qf}_{\mu\nu} \rangle_V \tag{1.7}$$

where

$$T^{\rm qf}_{\mu\nu} \equiv \Phi_{,\mu}\, \Phi_{,\nu} + \frac{1}{2} m^2 \Phi^2 g_{\mu\nu} \tag{1.8}$$

is the energy momentum tensor of, say, a massive minimally coupled scalar field [1]. Here $\langle \rangle_V$ denotes expectation value taken with respect to some vacuum state with symmetry commensurate with that of the background spacetime. Studies of semiclassical Einstein equation have been carried out in the last two decades by many authors for cosmological [28] and black hole spacetimes [29].

This is the point where our story begins. In the last decade we have been able to move one step ahead in the semiclassical back-reaction problem, extending the above semiclassical framework to a stochastic semiclassical theory, where noise and fluctuations from particle creation are accounted for and included in. Spacetime dynamics is now governed by a stochastic generalization of the semiclassical Einstein equation known as the Einstein-Langevin equation — the conventional theory of semiclassical gravity with sources given by the vacuum expectation value of the energy momentum tensor being a mean field approximation of this new theory. We now describe the essence of this new theory, using gravitational perturbations, again in a schematic form. In this context we can see the distinction between metric perturbations and metric fluctuations on the one hand and the proper meaning of metric fluctuations on the other.

2. Stochastic Semiclassical Gravity: Einstein-Langevin Equation

For concreteness let us consider the background spacetime to be a spatially flat FLRW universe with metric $\tilde{g}^{RW}_{\mu\nu}$ plus small perturbations $\tilde{h}_{\mu\nu}$,

$$\tilde{g}_{\mu\nu}(x) = \tilde{g}^{RW}_{\mu\nu} + \tilde{h}_{\mu\nu} \equiv e^{2\alpha(\eta)} g_{\mu\nu}. \tag{2.1}$$

It is conformally related [with conformal factor $\exp(2\alpha(\eta))$] to the Minkowski metric $\eta_{\mu\nu}$ and its perturbations $h_{\mu\nu}(x)$:

$$g_{\mu\nu} = \eta_{\mu\nu} + h_{\mu\nu}(x). \tag{2.2}$$

[1] By virtue of what we discussed above, the gravitons – quantized linear perturbations of the background metric – obey an equation similar in form to that of a massless minimally coupled quantum scalar field. For graviton production in cosmological spacetimes see [26, 27].

Here η is the conformal time related to the cosmic time t by $dt = \exp[\alpha(\eta)]d\eta$. The perturbations $\tilde{h}_{\mu\nu}$ can be either anisotropic, as in a Bianchi type (Type I case is treated by Hu and Sinha [12]), or inhomogeneous (treated by Campos and Verdaguer [14]). Here we follow these works.

The classical action for a free massless conformally coupled real scalar field $\Phi(x)$ is given by

$$S_f[\tilde{g}_{\mu\nu}, \Phi] = -\frac{1}{2}\int d^n x\sqrt{-\tilde{g}}\left[\tilde{g}^{\mu\nu}\partial_\mu\Phi\partial_\nu\Phi + \xi(n)\tilde{R}\Phi^2\right], \tag{2.3}$$

where $\xi(n) = (n-2)/[4(n-1)]$, and $\tilde{R}$ is the Ricci scalar for the metric $\tilde{g}_{\mu\nu}$. Define a conformally related field $\phi(x) \equiv \exp[(n/2-1)\alpha(\eta)]\Phi(x)$, the action S_f (after one integration by parts)

$$S_f[g_{\mu\nu}, \phi] = -\frac{1}{2}\int d^n x\sqrt{-g}\left[g^{\mu\nu}\partial_\mu\phi\partial_\nu\phi + \xi(n)R\phi^2\right] \tag{2.4}$$

takes the form of an action for a free massless conformally coupled real scalar field $\phi(x)$ in a spacetime with metric $g_{\mu\nu}$, *i.e.* a nearly flat spacetime. As the physical field $\Phi(x)$ is related to the field $\phi(x)$ by a power of the conformal factor, a positive frequency mode of the field $\phi(x)$ in flat spacetime will correspond to a positive frequency mode in the conformally related space. One can thus establish a quantum field theory in the conformally related space by use of the conformal vacuum (see [25]). Quantum effects such as particle creation and trace anomalies arise from the breaking of conformal flatness of the spacetime produced by the perturbations $h_{\mu\nu}(x)$.

The stochastic semiclassical Einstein equation differs from the semiclassical Einstein equation (SCE) by a) the presence of a stochastic term measuring the fluctuations of quantum sources (arising from the difference of particles created in neighbouring histories, see, [11]) and b) a dissipation term in the dynamics of spacetime. Thus it endows the form of a Einstein-Langevin equation [10]. Two points are noteworthy: a) The fluctuations and dissipation (kernels) obey a fluctuation-dissipation relation, which embodies the back-reaction effects of quantum fields on classical spacetime, b) The stochastic source term engenders metric fluctuations. The semiclassical Einstein equation depicts a mean field theory which one can retrieve from the Einstein-Langevin equation by taking a statistical average with respect to the noise distribution.

The stochastic semiclassical Einstein equation, or Einstein-Langevin equation takes on the form

$$\begin{aligned}
\tilde{G}^{\mu\nu}(x) &= \kappa\left(T_{\rm c}^{\mu\nu} + T_{\rm qs}^{\mu\nu}\right), \\
T_{\rm qs}^{\mu\nu} &\equiv \langle T^{\mu\nu}\rangle_{\rm q} + T_{\rm s}^{\mu\nu}, \\
T_{\rm s}^{\mu\nu} &\equiv 2e^{-6\alpha}F^{\mu\nu}[\xi].
\end{aligned} \tag{2.5}$$

Here, $T_c^{\mu\nu}$ is due to classical matter or fields, $\langle T^{\mu\nu}\rangle_q$ is the vacuum expectation value of the stress tensor of the quantum field, and $T_{\rm qs}^{\mu\nu}$ is a new stochastic term. Up to first order in $h_{\mu\nu}$ they are given by

$$\langle T^{\mu\nu}_{(0)}\rangle_q = \lambda\left[\tilde{H}^{\mu\nu}_{(0)} - \frac{1}{6}\tilde{B}^{\mu\nu}_{(0)}\right],$$

$$\langle T^{\mu\nu}_{(1)}\rangle_q = \lambda\Bigg[\ (\tilde{H}^{\mu\nu}_{(1)} - 2\tilde{R}^{(0)}_{\alpha\beta}\tilde{C}^{\mu\alpha\nu\beta}_{(1)}) - \frac{1}{6}\tilde{B}^{\mu\nu}_{(1)}$$
$$+\ 3e^{-6\alpha}\left(-4(C^{\mu\alpha\nu\beta}_{(1)}\alpha)_{,\alpha\beta} + \int d^4y A^{\mu\nu}_{(1)}(y)\mathrm{K}(x-y;\bar{\mu})\right)\ \Bigg], \quad (2.6)$$

where the constant $\lambda = 1/(2880\pi^2)$ characterizes one-loop quantum correction terms (which include the trace anomaly and particle creation processes [30]) and $\bar{\mu}$ is a renormalization parameter. Here $C_{\mu\alpha\nu\beta}$ is the Weyl curvature tensor and the tensors $B^{\mu\nu}(x)$, $A^{\mu\nu}(x)$ and $H^{\mu\nu}(x)$ are given by (see, e.g., [14, 30, 31] and earlier references)

$$\begin{aligned}
B^{\mu\nu}(x) &\equiv \frac{1}{2}g^{\mu\nu}R^2 - 2RR^{\mu\nu} + 2R^{;\mu\nu} - 2g^{\mu\nu}\Box_g R,\\
A^{\mu\nu}(x) &\equiv \frac{1}{2}g^{\mu\nu}C_{\alpha\beta\rho\sigma}C^{\alpha\beta\rho\sigma} - 2R^{\mu\alpha\beta\rho}R^{\nu}{}_{\alpha\beta\rho} + 4R^{\mu\alpha}R_{\alpha}{}^{\nu}\\
&\quad -\frac{2}{3}RR^{\mu\nu} - 2\Box_g R^{\mu\nu} + \frac{2}{3}R^{;\mu\nu} + \frac{1}{3}g^{\mu\nu}\Box_g R,\\
H^{\mu\nu}(x) &\equiv -R^{\mu\alpha}R_{\alpha}{}^{\nu} + \frac{2}{3}RR^{\mu\nu} + \frac{1}{2}g^{\mu\nu}R_{\alpha\beta}R^{\alpha\beta} - \frac{1}{4}g^{\mu\nu}R^2. \quad (2.7)
\end{aligned}$$

All terms mentioned so far in the semiclassical Einstein equation, including the dissipative kernel K, are familiar from back-reaction calculations done in the seventies and eighties (see. e.g., [28]). The new result in the nineties is in the appearance of a stochastic source [12, 14], the tensor $F^{\mu\nu}(x)$

$$F^{\mu\nu}(x) = -2\partial_\alpha\partial_\beta\xi^{\mu\alpha\nu\beta}(x), \quad (2.8)$$

which is symmetric and traceless, *i.e.* $F^{\mu\nu}(x) = F^{\nu\mu}(x)$ and $F^{\mu}_{\mu}(x) = 0$ (implying that there is no stochastic correction to the trace anomaly). It accounts for the noise associated with fluctuations of the quantum field. For spacetimes with linear metric perturbations as considered here, the stochastic correction to the stress tensor has vanishing divergence (to first order in the metric perturbations).

The stochastic source given by the noise tensor $\xi_{\mu\nu\alpha\beta}(x)$ (which for this problem has the symmetries of the Weyl tensor) is characterized completely by the two point correlation function which is the noise kernel $N(x-y)$ (the probability distribution for the noise is Gaussian) [12, 14]

$$\begin{aligned}
\langle\xi_{\mu\nu\alpha\beta}(x)\rangle_\xi &= 0,\\
\langle\xi_{\mu\nu\alpha\beta}(x)\xi_{\rho\sigma\lambda\theta}(y)\rangle_\xi &= T_{\mu\nu\alpha\beta\rho\sigma\lambda\theta}\mathrm{N}(x-y), \quad (2.9)
\end{aligned}$$

where $T_{\mu\nu\alpha\beta\rho\sigma\lambda\theta}$ is the product of four metric tensors (in such a combination that the right-hand side of the equation satisfy the Weyl symmetries of the two stochastic fields on the left-hand side). Its explicit form is given in [14]

If we now take the mean value of equation (2.5) with respect to the stochastic source ξ we find, as a consequence of the noise correlation, the relation,

$$\langle T_{\text{eff}}^{\mu\nu} \rangle_\xi = \langle T^{\mu\nu} \rangle_q \tag{2.10}$$

and we recover the semiclassical Einstein equations.

The stochastic term $2F^{\mu\nu}$ will produce a stochastic contribution $h_{\mu\nu}^{\text{st}}$ to the spacetime inhomogeneity, *i.e.* $h_{\mu\nu} = h_{\mu\nu}^{\text{c}} + h_{\mu\nu}^{\text{st}}$, which we call metric fluctuations. Considering a flat background spacetime for simplicity (setting $\alpha = 0$), one obtains (by adopting the harmonic gauge condition $(h_{\mu\nu}^{\text{st}} - \frac{1}{2}\eta_{\mu\nu}h^{\text{st}})^{,\nu} = 0$) a linear equation for the metric fluctuations $h_{\mu\nu}^{\text{st}}$

$$\begin{aligned} \Box h_{\mu\nu}^{\text{st}} &= 2\kappa S_{\mu\nu}^{\text{st}}, \\ S_{\text{st}}^{\mu\nu} &= 2F^{\mu\nu} = -4\partial_\alpha\partial_\beta\xi^{\mu\alpha\nu\beta}, \end{aligned} \tag{2.11}$$

The solution of these equations and the computation of the noise correlations have been given by Campos and Verdaguer in their beautiful paper, from where the above schematic description is adapted and where further details can be found.

We believe this new framework is fruitful for investigation into metric fluctuations and back-reaction effects. We and others [12, 14, 32] have applied it to study quantum effects in cosmological spacetimes. Work on black hole spacetimes is just beginning. Let us first review what has been done before, what we regard as deficient and describe the setup of this problem in our approach.

3. Metric fluctuations and Back-reaction in Semiclassical Gravity

It is perhaps useful to begin by emphasizing the difference in meaning of 'metric fluctuations' used in our approach and that used by others. In the glossary of almost all other authors metric fluctuations have been used in a test field context, referring to the two-point function of gravitational perturbations $h_{\mu\nu}$ in the classical sense or the expectation value of graviton two-point function $\langle h_{\mu\nu}(x)h_{\rho\sigma}(y)\rangle$ in a semiclassical sense – semiclassical in that the background remains classical even though the perturbations are quantized. It is in a test field context because one considers gravitational perturbations and their two-point functions from a fixed background geometry . This is a useful concept, but says nothing about back-reaction. It is useful as a measure of the fluctuations in the gravitational field at particular regions of spacetime. Ford has explored this aspect in great detail. For example, the recent work of Ford and Svaiter [33] shows that black hole horizon fluctuations are much smaller than Planck dimensions for black holes whose mass exceeds the Planck mass. For these black holes they induced that the semiclassical derivation of the Hawking radiance should remain valid, and that contrary to some recent claims [34, 35], there is no drastic effect near the horizon arising from metric fluctuations. However, for back-reaction considerations, where the background spactime metric varies in accordance with the behavior of quantum fields present, the graviton 2-point function calculated with respect to a fixed background is not the relevant quantity to consider.

In contrast, metric fluctuations $h^{\rm st}_{\mu\nu}$ (see Eq.(2.11)) in our work [10, 11, 12] and that of Campos and Verdaguer [14] are defined for semiclassical gravity in a manifestly back-reaction context. They are classical stochastic quantities arising from the flucutations in the quantum fields present and are important only at the Planck scale[2]. We see that they are derived from the noise kernel, which involves graviton 4-point functions. It is this quantity which enters into the fluctutation-dissipation relation – not the usual graviton 2 point function – which encaptures the semiclassical back-reaction. This is an important conceptual point which has not been duly recognized.

3.1. FLUCTUATION-DISSIPATION RELATION DESCRIPTION OF SEMICLASSICAL BACK-REACTION

With this understanding let us now expound the typical form of fluctuation-dissipation relation (FDR) in quantum field theory, starting with the paradigm of quantum Brownian motion (see, for example, [6]). Consider a quantum mechanical detector or atom coupled linearly to a quantized, otherwise free field which is initially in some quantum state, pure or mixed (typically taken to be a thermal state). After the coupling is switched on, the atom will "relax" to equilibrium over a time scale which depends on the coupling constant. When the atom reaches equilibrium, the equilibrium fluctuations of its observables depend on the quantum state of the field (for example, if they are thermal fluctuations, the associated temperature will be the temperature of the field). There are therefore two relevant processes: dissipation in the atom as it approaches equilibrium, and fluctuations at equilibrium. These two processes are generally related by a fluctuation-dissipation relation.

[2]The two point function of gravitons are not stochastic variables and so in a stricter sense they should not be called metric 'fluctuations'. To avoid confusion we may at times call our quantities $h^{\rm st}_{\mu\nu}$, induced metric fluctuations.

For quantum fields, let us consider the model of a simplified atom or detector with internal coordinate Q, coupled to a quantized scalar field ϕ via a bilinear interaction with coupling constant e: $L_I(t) = eQ(t)\phi(x(t))$, t being the atom's proper time and $x(t)$ denoting its parametrized trajectory. It can be shown [10, 15] that the semiclassical dynamics of Q is given by stochastic equations of the form

$$\frac{d}{dt}\frac{\partial L}{\partial \dot{Q}} - \frac{\partial L}{\partial Q} + 2\int^t \gamma(t,s)\dot{Q}(s) = \xi(t), \tag{3.1}$$

where $\xi(t)$ is a stochastic force arising out of quantum or thermal fluctuations of the field. Its correlator is defined as $\langle \xi(t)\xi(t')\rangle = \hbar\nu(t,t')$.

The functions γ and ν can be written in terms of two-point functions of the field bath surrounding the atom. Thus,

$$\begin{aligned} \mu(t,t') &= \frac{d}{d(t-t')}\gamma(t-t') = \frac{e^2}{2}G(x(t),x(t')) = -i\frac{e^2}{2}\langle[\phi(x(t),\phi(x(t'))]\rangle \\ \nu(t,t') &= \frac{e^2}{2}G^{(1)}(x(t),x(t')) = \frac{e^2}{2}\langle\{\phi(x(t),\phi(x(t'))\}\rangle, \end{aligned} \tag{3.2}$$

where G and $G^{(1)}$ are respectively the Schwinger (commutator) and Hadamard (anticommutator) functions of the free field ϕ evaluated at two points on the atom's trajectory. Both functions are evaluated in whatever quantum state the field is initially in, not necessarily a vacuum state. Because of the way μ and ν enter the equations of motion, they are referred to as *dissipation* and *noise* kernels, respectively. It should be noted that these two kernels, although independent of Q, ultimately determine the rate of energy dissipation of the internal coordinate Q and its quantum/thermal fluctuations.

The statement of the FDR for such cases is that ν and γ are related by a linear non-local relation of the form

$$\nu(t-t') = \int d(s-s')K(t-t',s-s')\gamma(s-s'). \tag{3.3}$$

For thermal states of the field, and in $3+1$ dimensions, it can be shown that

$$K(t,s) = \int_0^\infty \frac{d\omega}{\pi}\omega\coth\left(\frac{1}{2}\beta\hbar\omega\right)\cos(\omega(t-s)), \tag{3.4}$$

where β is the inverse temperature. Equations (3.3) and (3.4) constitute the general form of a FDR. With these ideas in mind, we will now discuss the two forms of FDR which have appeared in the literature in the context of static (following) and dynamic (next section) black holes.

3.2. FLUCTUATION AND BACK-REACTION IN STATIC BLACK HOLES

Back-reaction in this context usually refers to seeking a consistent solution of the semiclassical Einstein equation for the geometry of a black hole in equilibrium with its Hawking radiation (enclosed in a box to ensure relative stability). Much effort in the last 15 years has been devoted to finding a regularized energy-momentum for the back-reaction calculation. See Ref.[36] for recent status and earlier references. One important early work on back-reaction is by York [37], while the most thorough is carried out by Hiscock, Anderson et al [38].

Since the quantum field in such problems is assummed to be in a Hartle-Hawking state, concepts and techniques from thermal field theory are useful. Hartle and Hawking [39], Gibbons and Perry [40] used the periodicity condition of the Green function on the Euclidean section to give a simple derivation of the Hawking temperature for a Schwarzschild black hole. In the same vein, Mottola [18] showed that in some generalized Hartle-Hawking states a FDR exists between the expectation values of the commutator and anti-commutator of the energy-momentum tensor This FDR is similar to the standard thermal form common in linear response theory:

$$S_{abcd}(x,x') = \int_{-\infty}^{\infty} \frac{d\omega}{2\pi} e^{-i\omega(t-t')} \coth\left(\frac{1}{2}\beta\omega\right) \tilde{D}_{abcd}(\mathbf{x},\mathbf{x}';\omega), \tag{3.5}$$

where S and D are the anticommutator and commutator functions of the energy-momentum tensor, respectively, and $\tilde{D}$ is the temporal Fourier transform of D. That is,

$$\begin{aligned} S_{abcd}(x,x') &= \langle\{\hat{T}_{ab}(x),\, \hat{T}_{cd}(x')\}\rangle_\beta \\ D_{abcd}(x,x') &= \langle[\hat{T}_{ab}(x),\, \hat{T}_{cd}(x')]\rangle_\beta . \end{aligned} \tag{3.6}$$

He also identifies the two-point function D as a dissipation kernel by relating it to the time rate of change of the energy density when the metric is slightly perturbed. Thus, Eq.(3.5) represents a bonafide FDR relating the fluctuations of a certain quantity (say, energy density) to the time rate of change of the very same quantity.

However, this type of FDR has rather restricted significance as it is based on the assumption of a specific background spacetime (static in this case) and state (thermal) of the matter field(s). It is not adequate for the description of back-reaction where the spacetime and the state of matter are determined in a self-consistent manner by their dynamics and mutual influence. One should look for a FDR for a parametric family of metrics (belonging to a general class) and a more general state of the quantum matter (in particular, for Boulware and Unruh states). We expect the derivation of such an FDR will be far more complicated than the simple case where the

Green functions are periodic in imaginary time, and where one can simply take the results of linear response theory almost verbatim.

Even in this simple case, it is worthwhile to note that there is a small departure from standard linear response theory for quantum systems. This arises from the observation that the dissipation kernel in usual linear response analyses is given by a two-point commutator function of the underlying quantum field, which is independent of the quantum state for free field theory. In this case, we are still restricted to free fields in a curved background. However, since the dissipation now depends on a two-point function of the stress-tensor, it is a four-point function of the field, with appropriate derivatives and coincidence limits. This function is, in general, state-dependent. We have seen examples from related cosmological back-reaction problems where it is possible to explicitly relate the dissipation to particle creation in the field, which is definitely a state-dependent process. For the black-hole case, this would imply a quantum-state-dependent damping of semiclassical perturbations. To obtain a causal FDR for states more general than the Hartle-Hawking state, one needs to use the in-in (or Schwinger-Keldysh) formalism applied to a class of quasistatic metrics (generalization of York [37]) and proceed in a way leading to the noise kernel similar to that illustrated in Sec. 2. This research is presently pursued by Campos, Hu, Phillips and Raval [41, 42].

4. Fluctuations and Back-reaction in Dynamical Black Holes

4.1. QUASISTATIONARY APPROXIMATION AND BARDEEN METRIC

Back-reaction for dynamical (collapsing) black holes are much more difficult to treat than static ones, and there are fewer viable attempts. For situations with black hole masses much greater than the Planck mass, one important work which captures the overall features of dynamical back-reaction is that by Bardeen [43], who, using a generalization of a classical model geometry (Vaidya metric for stars with outgoing perfect fluid), argued that the mass of a radiating black hole decreases at a rate given by its luminosity, as expected from energy considerations. That is,

$$\frac{dM}{dt} = -L. \tag{4.1}$$

In particular, for a black hole emitting Hawking radiation, the luminosity goes as $L = \alpha\hbar M^{-2}$, M being the black hole mass, and α some constant. Far from the horizon ($r > \mathcal{O}(6M)$), Bardeen's geometry takes the form

$$ds^2 = -\left(1 - \frac{2m(u)}{r}\right) du^2 - 2du\, dr + r^2 d\Omega^2, \tag{4.2}$$

with

$$m(u) = \int^{u} du\, L(u), \tag{4.3}$$

$L(u)$ being the Hawking luminosity, and $m(u)$ the Bondi mass. With this Bardeen argued that the semiclassical picture of black hole evaporation should hold until the black hole reaches Planck size.

More recently Parentani and Piran [44], using a spherically symmetric geometry and a simplified scalar field model which neglects the potential barrier, carried out a numerical integration of the coupled quantized scalar field and semiclassical Einstein equations, and showed that the solution of the semiclassical theory in this model is the geometry described by Bardeen. Using the same model, Massar [45] recently showed that the emitted particles continue to have a thermal distribution with a time-dependent Hawking temperature $(8\pi M(t))^{-1}$. We refer readers to the latter work for details.

With this as a backdrop, the goal of our current program is to
1) derive a fluctuation-dissipation relation embodying the back-reaction for this quasistationary regime;
2) apply the stochastic field-theoretic formalism to the near-Planckian regime and derive an Einstein-Langevin equation for the dynamical metric including the effects of induced metric fluctuations.

4.2. FLUCTUATION-DISSIPATION RELATION OF CANDELAS AND SCIAMA

On the first issue, historically Candelas and Sciama [3] first proposed a fluctuation-dissipation relation for the depiction of dynamic black hole evolutions. But as we will point out, their relation does not include back-reaction in full and is not a FDR in the correct statistical mechanical sense.

As a starting point they considered the *classical* relation, due to Hartle and Hawking [46], between energy flux transmitted across the horizon of a perturbed black hole and the shear[3]:

$$\frac{d^2E}{dtd\Omega} = \frac{M^2}{\pi} \mid \sigma(2M) \mid^2, \tag{4.4}$$

where $\sigma(2M)$ is the perturbed shear of the null congruence which generates the future horizon.

In turn, the dissipation of horizon area with respect to the advanced null coordinate is related to the energy flux across the horizon, and the

[3]The analysis of Candelas and Sciama holds for a Kerr black hole. However, we simplify this to the Schwarzschild case in the interest of clarity; no qualitative features are lost in this simplification.

above equation becomes (see, for example [47])

$$\frac{dA}{dv} = 4M \int_H |\sigma|^2 \, dA, \tag{4.5}$$

the integral being performed over the horizon.

The classical formula above immediately suggests a fluctuation dissipation description: the dissipation in area is related linearly to the squared absolute value of the shear amplitude. This description is even more relevant when the gravitational perturbations are quantized. Then the integrand of the right-hand-side of Eq.(4.5) is $\langle \sigma^* \sigma \rangle$, the expectation value being taken with respect to an appropriate quantum state. Candelas and Sciama choose this state to be the Unruh vacuum, arguing that it is the vacuum which approximates best a flux of radiation from the hole at large radii.

The details of this expectation value are available in Ref.[3]. We simply note here that, with the substitution of this quantity in (4.5), the left-hand-side of that equation now represents the dissipation in area due to the Hawking flux of gravitational radiation, and the right-hand-side comes from pure quantum fluctuations of gravitons (as opposed to semiclassical fluctuations of gravitational perturbations, which are induced by the presence of quantum matter). It is tempting to regard this as a quantum FDR characteristic of the Hawking process, as do the authors of Ref.[3].

However, Eq.(4.5) is not yet an FDR in a truly statistical mechanical sense because it does not relate dissipation of a certain quantity (in this case, horizon area) to the fluctuations of *the same quantity*. To do so would require one to compute the two point function of the area, which, in turn, is a four-point function of the graviton field, and intimately related to a two-point function of the stress tensor. The stress tensor is the true "generalized force" acting on the spacetime via the equations of motion, and the dissipation in the metric must eventually be related to the fluctuations of this generalized force for the relation to qualify as an FDR. The calculation of 4-point functions of the metric perturbations h_{ab} in black hole spacetimes is currently being pursued by Hu, Phillips and Raval [42], and the derivation of such an FDR for quantum matter-black hole spacetime systems by Hu, Raval and Sinha [48].

4.3. BEKENSTEIN'S THEORY OF BLACK HOLE FLUCTUATIONS

The importance attributed to the correlation of mass function was a central point in Bekenstein's theory of black hole fluctuations [49]. Because it bears some similarity in conceptual emphasis to our approach we'd like to refresh the reader's memory of this work. We will also comment on the basic difference from our approach.

Bekenstein considers the mass fluctuations (and fluctuations of other parameters) of an *isolated* black hole due to the fluctuations in the radiation emitted by the hole, and considers the question of when such fluctuations can be large. For simplicity, only mass fluctuations are considered. The basic assumption is that the black hole mass $M(t)$ is a stochastic function with some probability distribution due to the stochastic emission of field quanta, and furthermore, that energy is conserved during the stochastic emission of quanta. As we shall see, this latter assumption leads to some startling predictions. With these assumptions, one may express the black hole mass at some time $t + dt$ in terms of the mass at an earlier time t by the equation

$$M(t+dt) = M(t) - m(dt), \tag{4.6}$$

where $m(dt)$ is the energy taken away by radiation in time dt. Averaging the above equation yields

$$\begin{aligned} \frac{d}{dt}\langle M \rangle &= -\frac{m(dt)}{dt} = -\langle L \rangle \\ &= -\alpha\hbar\langle M^{-2} \rangle, \end{aligned} \tag{4.7}$$

where Eq. (4.1) with the Hawking luminosity is used here. Furthermore, squaring Eq. (4.6) before taking the average yields,

$$\frac{d}{dt}\langle M^2 \rangle = 2\alpha\hbar\langle M^{-1} \rangle + \beta\hbar^2\langle M^{-3} \rangle, \tag{4.8}$$

the second term in the above expression being obtained by Bekenstein from an approximate expression for the energy distribution of quanta emitted in time dt.

Defining $\sigma_M = \langle M^2 \rangle - \langle M \rangle^2$, and using the moment expansions

$$\begin{aligned} \langle M^{-1} \rangle &= \frac{\langle M^2 \rangle}{\langle M \rangle^3} + \mathcal{O}(\langle M^3 \rangle), \\ \langle M^{-2} \rangle &= -\frac{2}{\langle M \rangle^2} + \frac{3\langle M^2 \rangle}{\langle M \rangle^4} + \mathcal{O}(\langle M^3 \rangle), \\ \langle M^{-3} \rangle &= -\frac{5}{\langle M \rangle^3} + \frac{6\langle M^2 \rangle}{\langle M \rangle^5} + \mathcal{O}(\langle M^3 \rangle), \end{aligned} \tag{4.9}$$

Eqs. (4.8) and (4.7) together imply

$$\frac{d}{dt}\sigma_M = \frac{\hbar}{\langle M \rangle^3}\left(-4\alpha\sigma_M + \beta\hbar(1 + 6\sigma_M)\right). \tag{4.10}$$

The above differential equation for σ_M possesses the approximate solution

$$\sigma_M \sim \chi\hbar\left(\frac{M_0^4}{\langle M \rangle^4} - 1\right), \tag{4.11}$$

χ being a constant related to α and β, and M_0^4 an integration constant to be interpreted as the mass when it is sharply resolved (i.e. when $\sigma_M = 0$). As pointed out by Bekenstein, one of the many consequences of this solution is that the fluctuations σ_M can grow as large as $\langle M \rangle^2$ for $\langle M \rangle = M_c \sim \hbar^{1/6} M_0^{2/3}$. According to this picture, therefore, depending on the initial mass, mass fluctuations can grow large far before the Planck scale. Once the critical mass M_c is reached, the dynamics of the black hole differs drastically from the standard semiclassical picture, because the equation for $\langle M \rangle$, Eq. (4.7), is itself driven by the fluctuations in mass.

A crucial assumption for the validity of such a scenario for black hole evaporation is, of course, the stochastic energy balance equation (4.6) and the related Eq. (4.7). It is possible that different assumptions for the stochastic dynamics lead to drastically different conclusions about the late stages of the evaporation process. The prediction of Bekenstein's theory is at variance with Bardeen's in which the semiclassical picture of black hole evaporation remains valid until the hole reaches Planck size. Our stochastic semiclassical gravity theory would also support Bardeen's scenario as the fundamental stochastic (Einstein-Langevin) equation for the black hole mass is expected to take the form

$$\frac{dM}{dt} = -\frac{\alpha\hbar}{\langle M \rangle^2} + \xi(t), \tag{4.12}$$

where $\xi(t)$ is a stochastic term with vanishing average value. If such an equation were to hold, the average mass would be independent of the fluctuations, while the fluctuations would be slaved to the dynamics of the average mass, where back-reaction will be incorporated in a self-consistent manner. As we showed in Sec. 2 this type of behavior of the mean field (semiclassical metric) and fluctuations is also found to occur in the treatment of back-reaction in cosmological spacetimes.

We hope to find an Einstein-Langevin equation of the form (4.12) for the description of black hole evaporation (in contrast to Bekenstein's equations of the form (4.7)). The Einstein-Langevin equation when averaged over the probability distribution for the noise ξ should yield the semiclassical Einstein equation as a mean field theory.

5. Prospects

Here we have discussed some representative work related to ours on semiclassical black hole fluctuations and back-reaction (other noteworthy proposals A include that of 'tHooft [50], and work on 2D dilaton gravity [51]). We have also sketched our approach, and marked out some important points of departure. This include 1) metric fluctuations and their role in back-reaction – our definition of (induced) metric fluctuations is in terms of graviton 4 point functions; 2) the true statistical mechanical meaning of the fluctuation-dissipation relation and its embodiment of the back-reaction effects. Our formulation testifies to the existence of a stochastic regime in semiclassical gravity where the dynamics of spacetime is governed by an Einstein-Langevin Equation which incoporates metric fluctuations induced by quantum field processes. We wish to explore the physics in this new stochastic regime, including possible phase transition characteristics near the Planck scale and its connection with low energy string theory predictions.

For the implementation of this program we are currently engaged in
a) setting up the CTP effective action for the quasistatic and dynamic cases,
b) computing the fluctuations of the energy momentum tensor for fields both in the Hartle-Hawking state and the Unruh state, and
c) exploring the interior solution of a collapsing black hole as this bears closer resemblance to a cosmological problem (Kantowski Sachs universe) [38, 52].

This program will take a few years to fruition and we hope to report on some results in Vishu's 65th birthday celebration.

Acknowledgement

This work is supported in part by NSF grant PHY94- 21849 to the University of Maryland and PHY95-07740 to the University of Wisconsin-Milwaukee. Part of this work was reported by BLH at the Second Symposium on Quantum Gravity in the Southern Cone held in January 1998 in Bariloche, Argentina.

References

1. S. W. Hawking, Nature **248**, 30 (1974); Com. Math. Phys. **43**, 199 (1975).
2. W. G. Unruh, Phys. Rev. **D14**, 3251 (1976).
3. P. Candelas and D. W. Sciama, Phys. Rev. Lett. **38**, 1372 (1977).
4. D. W. Sciama, in *Relativity, Quanta and Cosmology – Centenario di Einstein* ed. F. DeFinis (Editrici Giunti Barbera Universitaria, Firenze, Itay, 1979). D. Sciama, P. Candelas and D. Deutsch, Adv. Phys. 30, 327 (1981). D. J. Raine and D. W. Sciama, Class. Quantum Grav. 14, A325 (1997).
5. See, e.g., E. B. Davies, *The Quantum Theory of Open Systems* (Academic Press, London, 1976); K. Lindenberg and B. J. West, *The Nonequilibrium Statistical Mechanics of Open and Closed Systems* (VCH Press, New York, 1990); U. Weiss, *Quantum Dissipative Systems* (World Scientific, Singapore, 1993).
6. R. Feynman and F. Vernon, Ann. Phys. (NY) **24**, 118 (1963). R. Feynman and A. Hibbs, *Quantum Mechanics and Path Integrals*, (McGraw - Hill, New York, 1965). A. O. Caldeira and A. J. Leggett, Physica **121A**, 587 (1983). H. Grabert, P. Schramm and G. L. Ingold, Phys. Rep. **168**, 115 (1988). B. L. Hu, J. P. Paz and Y. Zhang, Phys. Rev. **D45**, 2843 (1992); **D47**, 1576 (1993).
7. J. Schwinger, J. Math. Phys. **2** (1961) 407; P. M. Bakshi and K. T. Mahanthappa,

J. Math. Phys. 4, 1 (1963), 4, 12 (1963). L. V. Keldysh, Zh. Eksp. Teor. Fiz. **47**, 1515 (1964) [Engl. trans. Sov. Phys. JEPT **20**, 1018 (1965)]; G. Zhou, Z. Su, B. Hao and L. Yu, Phys. Rep. **118**, 1 (1985); Z. Su, L. Y. Chen, X. Yu and K. Chou, Phys. Rev. **B37**, 9810 (1988); B. S. DeWitt, in *Quantum Concepts in Space and Time* ed. R. Penrose and C. J. Isham (Claredon Press, Oxford, 1986); R. D. Jordan, Phys. Rev. **D33** , 44 (1986). E. Calzetta and B. L. Hu, Phys. Rev. **D35**, 495 (1987).

8. B. L. Hu, Physica A158, 399 (1979).
9. B. L. Hu, *Quantum Statistical Fields in Gravitation and Cosmology* in *Proc. Third International Workshop on Thermal Field Theory and Applications*, eds. R. Kobes and G. Kunstatter (World Scientific, Singapore, 1994) gr-qc/9403061.
10. B. L. Hu and A. Matacz, Phys. Rev. **D51**, 1577 (1995).
11. E. Calzetta and B. L. Hu, Phys. Rev. D49, 6636 (1994).
12. B. L. Hu and S. Sinha, Phys. Rev. **D51**, 1587 (1995).
13. A. Raval, B. L. Hu and J. Anglin, Phys. Rev. **D 53**, 7003 (1996).
14. A. Campos and E. Verdaguer, Phys. Rev. **D53**, 1927 (1996).
15. A. Raval, B. L. Hu and D. Koks, Phys. Rev. **D 55**, 4795 (1997).
16. B. L. Hu, J. P. Paz and Y. Zhang, Phys. Rev. **D45**, 2843 (1992); **D 47**, 1576 (1993).
17. M. Gell- Mann and J. B. Hartle, Phys. Rev. **D47**, 3345 (1993).
18. E. Mottola, Phys. Rev. **D33**, 2136 (1986).
19. C. W. Misner, K. S. Thorne, J. A. Wheeler, *Gravitation* (W. H. Freeman and Company, New York, 1970).
20. E. M. Lifshitz, Zh. Eksp. Teor. Phys. **16**, 587 (1946) [J. Phys. USSR **10**, 116 (1946)].
21. T. Regge and J. A. Wheeler, Phys. Rev. 108, 1063 (1957); C. V. Vishveshwara, Phys. Rev. D1, 2870 (1970); F. J. Zerrili, Phys. Rev. D2, 2141 (1970).
22. S. Teukolsky, Phys. Rev. Lett. 29, 1114 (1972) S. Chandrasekhar, *The Mathematical Theory of Black Holes* (Oxford University Press, Oxford, 1992).
23. G. E. Tauber, Tensor 1, (1970).
24. L. Parker, Phys. Rev. **183**, 1057 (1969); R. U. Sexl and H. K. Ubantke, Phys. Rev. **179**, 1247 (1969); Ya. B. Zeldovich, Zh. Eksp. Teor. Fiz. Pis'ma Red. **12**, 443 (1970) [JETP Lett.**32**, 307 (1970)]; B. L. Hu, Phys. Rev. D **9**, 3263 (1974).
25. N. D. Birrell and P. C. W. Davies, *Quantum Fields in Curved Space* (Cambridge University Press, Cambridge, 1982).
26. L. Grishchuk, Zh. Eksp. Teor. Fiz. 67, 825 (1974) [Sov. Phys.- JETP 40, 409 (1975)].
27. L. Ford and L. Parker, Phys. Rev. **D16**, 1601 (1977).
28. Ya. Zel'dovich and A. Starobinsky, Zh. Eksp. Teor. Fiz **61**, 2161 (1971) [Sov. Phys.- JETP **34**, 1159 (1971)]; JETP Lett. (1977) B. L. Hu and L. Parker, Phys. Lett. 63A, 217 (1977); Phys. Rev. **D17**, 933 (1978); F. V. Fischetti, J. B. Hartle and B. L. Hu, Phys. Rev. **D20**, 1757 (1979); J. B. Hartle and B. L. Hu, Phys. Rev. **D20**, 1772 (1979); **21**, 2756 (1980); A. A. Starobinsky, Phys. lett. 91B, 99 (1980); B. L. Hu, Phys. Lett. 103B, 331 (1981); P. A. Anderson, Phys. Rev. D28, 271 (1983); D29, 615 (1984) E. Calzetta and B. L. Hu, *Phys. Rev.* **D35**, 495 (1987).
29. J. M. Bardeen, Phys. Rev. Lett. 46, 382 (1981); P. Hajicek and W. Israel, Phys. Lett. 80A, 9 (1980); J. W. York, Jr., Phys. Rev. D31, 775 (1986).
30. B. S. De Witt, Phys. Rep. **C19**, 295 (1975).
31. S. A. Fulling and L. Parker, Ann. Phys. 87, 176 (1974).
32. E. Calzetta, A. Campos, and E. Verdaguer, Phys. Rev. D56, 2163 (1997).
33. L. H. Ford and N. F. Svaiter, Phys. Rev. D56, 2226 (1997).
34. A. Casher et al, *Black Hole Fluctuations* hep-th/9606016.
35. R. D. Sorkin, *How Rinkled is the Surface of the Black Hole?* gr-qc/9701056.
36. B. P. Jenson, J. G. McLaughlin and A. C. Ottewill, Phys. Rev. D51, 5676 (1995).
37. J. W. York, Jr., Phys. Rev. D28, 2929 (1983); D31, 775 (1985); D33, 2092 (1986).
38. P. R. Anderson, W. A. Hiscock and D. A. Samuel, Phys. Rev. Lett. 70, 1739 (1993); Phys. Rev. D51, 4337 (1995); W. A. Hiscock, S. L. Larson and P. A. Anderson, Phys. Rev. D56, 3571 (1997).
39. J. B. Hartle and S. W. Hawking, Phys. Rev. D13, 2188 (1976).

40. G. Gibbons and M. J. Perry, Proc. Roy. Soc. Lon. A358, 467 (1978).
41. A. Campos, B. L. Hu and A. Raval, *Fluctuation-Dissipation Relation for a Quantum Black Hole in Quasi-Equilibrium with its Hawking Radiation* (in preparation).
42. B. L. Hu, N. Phillips and A. Raval, *Fluctuations of the Energy Momentum Tensor of a Quantum Field in a Black Hole Spacetime* (in preparation).
43. J. M. Bardeen, Phys. Rev. Lett. 46, 382 (1981).
44. R. Parentani and T. Piran, Phys. Rev. Lett. **73**, 2805 (1994).
45. S. Massar, Phys. Rev. **D 52**, 5857 (1995).
46. J. B. Hartle and S. W. Hawking, Commun. Math. Phys. **27**, 283 (1973).
47. S. W. Hawking, in *Black Holes*, edited by B. S. DeWitt and C. DeWitt (Gordon and Breach, New York, 1973).
48. B. L. Hu, A. Raval and S. Sinha, *Back-reaction of a Radiating Quantum Black Hole and Fluctuation-Dissipation Relation* (in preparation).
49. J. D. Bekenstein, in *Quantum Theory of Gravity*, edited by S. M. Christensen (Adam Hilger, Bristol, 1984).
50. G. 'tHooft, Nucl. Phys. B256, 727 (1985), B335, 138 (1990); C. R. Stephens, G. 'tHooft and B. F. Whiting, Class. Quant. Grav. 11, 621 (1994).
51. S. Bose, J. Louko, L. Parker, and Y. Peleg, Phys. Rev. D **53**, 5708 (1996).
52. B. L. Hu, J. Louko, N. Phillips and J. Simon, in preparation.

8. BLACK HOLES IN HIGHER CURVATURE GRAVITY

R.C. MYERS
Department of Physics, McGill University,
Montréal, Québec, Canada H3A 2T8.

1. Introduction

The idea that the Einstein action should be modified by the addition of interactions involving higher powers of the Riemann curvature tensor has a long history stretching back to the early days of general relativity. Such higher curvature theories originally appeared in proposals by Weyl and Eddington for a geometric unification of electromagnetism and gravity [1]. Much later, interest arose in higher derivative theories of gravity because they provided renormalizable quantum field theories [2, 3]. Unfortunately, the new massive spin-two excitations, which tame the ultraviolet divergences in such theories, result in the instability of the classical theory [4] and the loss of unitarity in the quantum theory [3, 5]. While higher curvature theories have thus proven inadequate as the foundation of quantum gravity, they still have a role to play within the modern paradigm of effective field theories [6].

Irrespective of the fundamental nature of quantum gravity, there should be a low energy effective action which describes the dynamics of a "background metric field" for sufficiently weak curvatures and sufficiently long distances. On general grounds, this effective gravity action will consist of the usual Einstein action plus a series of covariant, higher-dimension interactions, *i.e.*, higher curvature terms, and also higher derivative terms involving the "low-energy" matter fields. The appearance of such interactions can be seen, for example, in the renormalization of quantum field theory in curved space-time [7], or in the construction of low-energy effective actions for string theory [8, 9]. In this context, the higher curvature interactions simply produce benign perturbative corrections to Einstein gravity coupled to conventional matter fields [10, 11].

Naive dimensional analysis would suggest that the coefficients of the

B. R. Iyer and B. Bhawal (eds.), Black Holes, Gravitational Radiation and the Universe, 121–136.

higher dimension terms in such an effective Lagrangian should be dimensionless numbers of order unity times the appropriate power of the Planck length. Thus one might worry that all of the effects of the higher curvature terms would be the same order as those of quantum fluctuations, and so there would seem to be little point in studying modifications of "classical" black holes from higher dimension terms. One motivation for studying a classical higher curvature theory is that it is, of course, possible that the coefficients of some higher dimension terms are larger than what would be expected from simple dimensional analysis. Moreover, it is interesting to explore black holes in generalized gravity theories in order to discover which properties are peculiar to Einstein gravity, and which are robust features of all generally covariant theories of gravity. Further even within this framework, one may still discover clues as to the ultimate nature of the underlying theory.

With these introductory remarks, we will go on to discuss some of the recent investigations of black holes in higher curvature theories of gravity. The remainder of this chapter is organized as follows: In section 2, we briefly review various black hole solutions which appear in the literature. In section 3, we focus on black hole thermodynamics [12] in the context of higher curvature gravity. Finally in section 4, we provide a brief discussion and indicate some of the open questions.

At this point, let us add that many of the recent candidates for a theory of quantum gravity, especially those which attempt to unify gravity with other interactions, are theories for which the space-time dimension is greater than four. Thus much of the following discussion refers to black holes in higher dimensional space-times [13]. While this idea may seem unusual and/or unappealing to some readers, it is certainly one familiar to our esteemed colleague, C.V. Vishveshwara [14, 15], to whom this volume is dedicated. Throughout, we also employ the conventions of ref. [16] in any formulae.

2. Black Hole Solutions

When faced with the task of finding solutions of higher curvature gravity (or any higher derivative theory), one must realize that it is incorrect to assume that adding higher derivative correction terms with "small" coefficients will *only* produce small modifications of the solutions of the unperturbed theory [11]. Any higher curvature theory will, in fact, contain whole classes of new solutions unavailable to the classical Einstein theory. In particular, the set of maximally symmetric vacuum solutions will now include some number of (anti-)deSitter vacua, as well as flat space [17, 18]. A common feature of these new solutions is that they are not analytic in the coefficients of the higher derivative interactions, and hence in the context of an effective field theory, they should be regarded as unphysical. Systematic procedures have been developed to exclude these spurious solutions [11, 19]. However, in the context of effective field theory where the domain of validity of the equations of motion is expected to be limited, it suffices to treat the higher curvature contributions perturbatively.

For example, let us consider the field equations: $R_{ab} = \alpha\, H_{ab}$ where R_{ab} is the Ricci tensor, H_{ab} is some higher derivative contribution, and α is the (dimensionful) coefficient of the higher curvature term in the action. Now making the obvious expansion of the solution, $g_{ab} = g^0_{ab} + \alpha\, g^1_{ab} + \ldots$, one must solve

$$R_{ab}(g^0) = 0\,, \tag{1}$$

$$\left(\Delta[g^0]\, g^1\right)_{ab} = H_{ab}(g^0)\,, \tag{2}$$

where $\Delta[g^0]$ is the second order operator obtained from linearizing the Ricci tensor about the metric g^0_{ab}. Solving Eq.(1) amounts to selecting a solution of the unperturbed Einstein theory. Solving Eq.(2) is solving for a linearized fluctuation in that background with some source term. As well as imposing appropriate boundary conditions on g^1_{ab} (*e.g.*, that it preserves asymptotic flatness and regularity of the horizon), one would (usually) choose it to preserve the symmetries of the original solution (*e.g.*, spherical symmetry, stationarity). While the above discussion was phrased in terms of purely gravitational solutions, it would be straightforward to include matter fields, as one might in considering charged black holes. Further this approach extends in an obvious way to developing the perturbation expansion to higher orders, which may involve including additional higher order interactions.

In this expansion, the coupling constant α has the dimension of length to some (positive) power n, *e.g.*, $n = 2$ if H_{ab} arises from a curvature-squared interaction. The true dimensionless expansion parameter in the above analysis is then α/L^n where L is the local curvature scale. Hence this expansion will always break down in the black hole interior near the singularity, but it will be reliable in the exterior region of a large black hole where the curvatures are small. Thus within this context, one can expect that for large black holes the asymptotic regions feel only minor corrections due to the higher curvature terms. Near the singularity the higher curvature contributions become strong, but this implies that one has left the realm in which the effective action is to be trusted. It would seem that at this point one must come to grips with the full underlying fundamental theory. However, one might attempt to make models of high curvature behavior [20, 21] to guide our intuition.

Within this framework, it is interesting to consider the propagation of metric disturbances in the black hole background. One would organize the disturbance with the same α expansion as above: $g_{ab} = (g^B + h)_{ab} = (g^0 + h^0)_{ab} + \alpha(g^1 + h^1)_{ab}$ where $g^B_{ab} = g^0_{ab} + \alpha\, g^1_{ab}$ is the background metric satisfying Eqs.(1) and (2), above. The disturbance then satisfies

$$\left(\Delta[g^0]\,h^0\right)_{ab} = 0\,, \tag{3}$$

$$\left(\Delta[g^0]\,h^1\right)_{ab} = J_{ab}(g^0, h^0)\,, \tag{4}$$

where $J_{ab}(g^0, h^0)$ is the linearization of $H_{ab}(g^0 + h^0)$. Note that within this scheme, the propagation of metric disturbances and hence the causal structure are completely determined by the original background metric g^0_{ab}. Hence we are guaranteed that the black hole really remains a black hole, and the event horizon remains an event horizon.

At some level, the conclusions of the previous paragraph may actually seem somewhat surprising. For instance, one would conclude that rather than following null geodesics in the perturbed background $g^B_{ab} = g^0_{ab} + \alpha\, g^1_{ab}$, "high frequency" gravity waves still follow null geodesics of the original metric g^0_{ab}. In either case, one would expect these conclusions to apply in an approximation: $L^n >> \lambda^n >> \alpha$ where λ is the wavelength of the disturbance. The first inequality justifies a geometric optics approximation [16], whereas the second is required for the reliability of the effective action. Now in examining the propagation of wavefronts in curved space, one expects curvature corrections to appear at the order λ/L, whereas the modifications to the metric due to the higher curvature interactions are of the order α/L^n. Hence given the above inequalities, one has $\lambda/L >> (\lambda/L)^n >> \alpha/L^n$, and so the discrepancy between using either g^B_{ab} or g^0_{ab} to define null geodesics is certainly a subleading correction in determining the propagation of gravitational disturbances. However, using g^0_{ab} seems more consistent in the application of perturbation theory [22, 23].

Eqs.(3) and (4) are also relevant in addressing the important question of the (linearized) stability of the horizon [26] - a topic to which Vishu has made seminal contributions [27]. Given that the original Einstein theory was stable, one knows that there are no runaway solutions to Eq.(3). Eq.(4) simply extends the latter equation by the addition of a regular source term. Hence it is also clear that h^1_{ab} has no runaway solutions by the application of the original stability analysis. Thus one can immediately deduce that the black hole solution remains stable within this perturbative framework.

A perturbative approach has been applied in examining modifications of the four-dimensional Schwarzschild black hole within the context of renormalized Einstein gravity [28, 29]. In this case, curvature-squared interactions do not produce any modifications for solutions of the four-dimensional vacuum Einstein equations. Hence this analysis considered Einstein gravity perturbed by the addition of terms involving three Riemann curvatures.

The main observation resulting from this analysis [28] was that the relations between the black hole's mass and its thermodynamic parameters [12] are modified. In particular, the black hole entropy was no longer proportional to the area of the horizon.

Various perturbative analyses have also been made to study black holes in string theory [30]–[33]. These include considering the effects of curvature-squared terms on spherically symmetric black holes in arbitrary dimensions [31], and on four-dimensional black holes with angular momentum [32] or with charge [33].[1] In certain supersymmetric string theories, the leading higher curvature interaction can be shown to contain four curvatures [35]. The effect of these terms on spherically symmetric black holes in arbitrary dimensions has also been studied [30]. Apart from modifications to the usual thermodynamic properties of the black holes, one of the remarkable observations here was the fact that the higher curvature terms induce various new forms of long-range scalar field hair on the black holes. However, this new hair may be regarded as secondary [36], in that it is completely determined by the black hole's primary hair, *e.g.*, the mass and charge. In other words, there are no new constants of integration, and hence these solutions do not violate the spirit of the no-hair theorems [37] Essentially, the new hair arises because the scalar fields have non-minimal couplings to the higher curvature terms.

Motivated originally by string theory, a great deal of attention has been focussed on Lovelock gravity [38]. The latter is defined by a Lagrangian which is the sum of dimensionally extended Euler densities. In four dimensions, all of the higher curvature terms are total derivatives, and hence the theory reduces to Einstein gravity. However, in higher dimensions, the new interactions do make nontrivial contributions. A distinguishing feature of these Lagrangians is that the resulting equations of motion contain no more than second derivatives in time [38]. As quantum theories then, they are free of ghosts when expanding about flat space [39], and so they evade the problem of unitarity loss — however, they remain nonrenormalizable. Exact spherically symmetric solutions were first found for the curvature-squared or Gauss-Bonnet theory [18]. These results were quickly extended to arbitrary Lovelock theories [40]–[42], as well as charged black holes [43]. These solutions displayed a rich structure of multiple horizons, and unusual thermodynamic properties [41]. For example, certain (uncharged) solutions could be found with vanishing Hawking temperature [44]. The topic of Lovelock black holes is, in fact, another area upon which Vishu's research has touched [15].

[1]Note that in string theory, charged black holes typically differ from the Reissner-Nordstrom geometry as a result of nonminimal couplings between the matter fields [34].

While in four dimensions, the Lovelock action yields only Einstein gravity, one can also consider the Kaluza-Klein compactification [45] of a higher dimensional Lovelock theory down to four dimensions. The resulting theory consists of Einstein gravity coupled to various scalar and vector fields with nonminimal higher-derivative interactions [46]. In this case, there are four-dimensional black holes which carry secondary scalar hair [47].

More recently researchers [48, 49], interested in whether the secondary scalar hair found in ref. [31]–[33] survived beyond perturbation theory, investigated exact solutions of a (four-dimensional) dilatonic Gauss-Bonnet theory. In the latter, the curvature-squared interaction is modified by the addition of a nonminimal scalar coupling. The full equations of this theory are difficult enough that they could not be solved analytically. However, analytic arguments and numerical evidence indicates that the modified black holes do carry secondary scalar hair [48]. These results were also extended to black holes carrying charge [49].

Some work [50, 51] has also been done on black holes in theories where the Lagrangian density takes the form $\sqrt{-g}f(R)$ where $f(R)$ is a polynomial in the Ricci scalar. These models are amenable to analysis because they can be mapped to a theory of Einstein gravity and a minimally coupled scalar with an unusual potential [50, 52]. Note that if other matter fields are included then the latter develop unusual nonminimal couplings in the Einstein-scalar theory. In the case $f = R + a_2 R^2$, a uniqueness theorem was proven for certain classes of matter fields in four dimensions [50]. Provided[2] that $a_2 > 0$, no new hair can arise and so the only black hole solutions are identical to those of Einstein gravity. Ref. [51] extended this work to spherically symmetric solutions for general polynomial actions in arbitrary dimensions. One finds that, with the same restriction on the quadratic term and irrespective of the remaining terms in the polynomial f, the only asymptotically flat black holes are still the Schwarzschild solutions.

The latter investigations of the Lovelock, dilatonic Gauss-Bonnet and polynomial-in-R theories all go beyond the perturbative approach originally described and consider exact solutions of the full higher derivative equations of motion. Certainly amongst the exact solutions, one will find some which lend themselves to a Taylor expansion in the higher curvature coupling constants. These solutions could then be considered the result of carrying out the perturbation expansion to infinite order. To be of interest in the effective field theory framework though, one would have to know that there are no additional higher curvature interactions at higher orders. In general, this seems an unlikely scenario, and in string theory, certainly one that does

[2]Note that this condition is also required for the stability of the theory [53, 54].

not apply [55][3] Although the physical motivation may not be strong, one can still set out to study these theories and their solutions as a mathematical problem in its own right. In this respect, an advantageous feature, which the three theories discussed here seem to share in common, is the absence of negative energy ghosts [39, 53, 54, 57] – at least with certain restrictions on the coupling constants.

From this point of view, one should re-address the important question of the stability of the horizon for black hole solutions in these theories. Of course, the linearized stability for the full higher derivative equations is an extremely difficult problem[4], however, some limited results have been achieved. Ref. [54] shows that the four-dimensional Schwarzschild black hole is stable in a general fourth order gravity theory. There also some limited results indicating stability of spherically symmetric four-dimensional black holes in the dilatonic Gauss-Bonnet theory [59].

A more fundamental question which should also be considered is the actual causal structure of the "black hole" solutions. It would seem that for the three theories of interest here, the Lovelock, dilatonic Gauss-Bonnet and polynomial-in-R gravities, that the full equations are (or can be mapped to) second-order hyperbolic systems. However, the characteristic surfaces of these equations need not coincide with null cones in the background space-time, opening the possibility that gravitational disturbances could propagate "faster than light." Note that such acausal behavior would result immediately if the theory in question had negative energy ghosts [60], which, although a feature of generic higher curvature theories, is not the case for these three theories. The possibility that gravitational disturbances could escape a "black hole" would have profound consequences for these theories. In the case of the Lovelock theories, investigations have been made of the modified characteristics [61], and in this case, a preliminary study [62] of radial wavefronts in static black holes suggests that disturbances can not escape the horizon. The same result would necessarily seem to apply for the polynomial-in-R theories, which are essentially mapped to Einstein gravity (with a conformally related metric) coupled to a massive scalar field [52]. We stress, however, that we view the motivation in studying these theories as more mathematical than physical.

3. Black Hole Thermodynamics

Black hole thermodynamics [12] is certainly one of the most remarkable features of the area of physics to which this volume is devoted. It produces

[3]In certain cases, however, one can use supersymmetry to argue that special solutions are exact to all orders despite the higher curvature corrections to the action [56].

[4]Note that going to higher dimensions introduces complications by itself [58].

a confluence of ideas from thermodynamics, quantum field theory and general relativity. Much of the interest in black hole thermodynamics comes from the hope that it will provide some insight into the nature of quantum gravity. While originally developed in the context of Einstein gravity, it is easy to see that much of the framework should extend to higher curvature theories, as well. Hawking's celebrated result [63] that a black hole emits thermal radiation with a temperature proportional to its surface gravity, κ:

$$k_B T = \frac{\hbar\kappa}{2\pi c} \tag{5}$$

is a prediction of quantum field theory in space-time containing a horizon [7], independent of the details of the dynamics of the gravity theory. Alternatively, this result will follow from the evaluation of the of the black hole partition function using the Euclidean path integral method [64]. Applying the latter approach in higher curvature gravity, it is also clear that one can derive a version of the First Law of black hole mechanics

$$\frac{\kappa}{2\pi c}\delta\mathcal{S} = c^2\delta M - \Omega^{(\alpha)}\delta J_{(\alpha)} , \tag{6}$$

where M, $J_{(\alpha)}$ and $\Omega^{(\alpha)}$ are the black hole mass, canonical angular momentum, and the angular velocity of the horizon [65]. Given the identification of the black hole temperature with the surface gravity in Eq.(5), the black hole entropy is naturally identified as $S = (k_B/\hbar)\mathcal{S}$. In the context of Einstein gravity, this gives the famous Bekenstein-Hawking entropy [63, 67]:

$$S_{BH} = \frac{k_B c^3}{\hbar G}\frac{A_H}{4} . \tag{7}$$

where A_H is the area of the event horizon [66].

Many of the early investigations [28, 31, 30, 41, 43, 44] which examined particular black hole solutions in higher curvature theories noted that Eq.(7) no longer applied - see ref. [68] for a review. Important conceptual progress was made in ref. [69], where it was realized that $\mathcal{S}$ should take the form of a geometric expression evaluated at the event horizon. This result was explicitly demonstrated there for the special case of Lovelock gravity, by extending a Hamiltonian derivation of the First Law [70]. Various other techniques were then developed [71]–[76] to expand this result to other higher curvature theories. In particular though, Wald [71] developed an elegant new derivation of the First Law which applies for any diffeomorphism invariant theory - see below. His derivation makes clear that in Eq.(6), $\mathcal{S}$ may always be expressed as a *local* geometric density integrated over a space-like cross-section of the horizon.

3.1. BLACK HOLE ENTROPY AS NOETHER CHARGE

Here, we will provide a brief introduction to Wald's derivation of the First Law. The interested reader is referred to Refs. [71, 73, 74, 77] for a complete description. In the following, we also adopt the standard convention of setting $\hbar = c = k_B = 1$.

An essential element of Wald's approach is the Noether current associated with diffeomorphisms [78]. Let L be a Lagrangian built out of some set of dynamical fields, including the metric, collectively denoted as ψ. Under a general field variation $\delta\psi$, the Lagrangian varies as

$$\delta(\sqrt{-g}L) = \sqrt{-g}E\cdot\delta\psi + \sqrt{-g}\,\nabla_a\theta^a(\delta\psi) \quad , \tag{8}$$

where "$\cdot$" denotes a summation over the dynamical fields including contractions of tensor indices. Then the equations of motion are $E = 0$. With symmetry variations for which $\delta(\sqrt{-g}L) = 0$, θ^a is the Noether current which is conserved when the equations of motion are satisfied — *i.e.*, $\nabla_a\theta^a(\delta\psi) = 0$ when $E = 0$. For diffeomorphisms, where the field variations are given by the Lie derivative $\delta\psi = \mathcal{L}_\xi\psi$, the variation of a covariant Lagrangian is a total derivative, $\delta(\sqrt{-g}L) = \mathcal{L}_\xi(\sqrt{-g}L) = \sqrt{-g}\nabla_a(\xi^a\,L)$. Thus one constructs an improved Noether current,

$$J^a = \theta^a(\mathcal{L}_\xi\psi) - \xi^a L \quad ,$$

which satisfies $\nabla_a J^a = 0$ when $E = 0$.

A fact [79], which may not be well-appreciated, is that for *any* local symmetry, there exists a globally-defined Noether potential Q^{ab}, satisfying $J^a = \nabla_b Q^{ab}$ where $Q^{ab} = -Q^{ba}$. The Noether potential Q^{ab} is a local function of the dynamical fields and a linear function of the symmetry parameter (*i.e.*, ξ^a in the present case). Of course, this equation for J^a is valid up to terms which vanish when the equations of motion are satisfied. Given this expression for J^a, it follows that the Noether charge contained in a spatial volume Σ can be expressed as a boundary integral $\oint_{\partial\Sigma} d^{D-2}x\,\sqrt{h}\,\epsilon_{ab}Q^{ab}$, where h_{ab} and ϵ_{ab} are the induced metric and binormal form on the boundary $\partial\Sigma$ respectively.

Another key concept that enters in Wald's construction is that of a Killing horizon. Given a Killing vector field which generates an invariance for a particular solution — *i.e.*, $\mathcal{L}_\xi\psi = 0$ for all fields — a Killing horizon is a null hypersurface whose null generators are orbits of the Killing vector. If the horizon generators are geodesically complete to the past (and if the surface gravity is nonvanishing), then the Killing horizon contains a spacelike cross-section B, the *bifurcation surface*, on which the Killing field χ^a vanishes [80]. It can be shown that the event horizon of any black hole, which is static or is stationary with a certain t–$\phi^{(\alpha)}$ orthogonality condition, must be a Killing horizon [81].[5] A stronger result holds in general relativity, where it can be proven that the event horizon for any stationary black hole is Killing [83].

The key to Wald's derivation of the First Law is the identity

$$\delta H = \delta \int_\Sigma dV_a J^a - \int_\Sigma dV_a \nabla_b(\xi^a\theta^b - \xi^b\theta^a), \tag{9}$$

where H is the Hamiltonian generating evolution along the vector field ξ^a, and Σ is a spatial hypersurface with volume element dV_a. This identity is satisfied for arbitrary variations of the fields away from any background solution. If the variation is to another solution, then one can replace J^a by $\nabla_b Q^{ab}$, so the variation of the Hamiltonian is given by surface integrals over the boundary $\partial\Sigma$. Further, if ξ^a is a Killing vector of the background solution, then $\delta H = 0$, and in this case, one obtains an identity relating the various surface integrals over $\partial\Sigma$.

Suppose that the background solution is chosen to be a stationary black hole with horizon-generating Killing field $\chi^a\partial_a = \partial_t + \Omega^{(\alpha)}\partial_{\phi(\alpha)}$, and the hypersurface Σ is chosen to extend from asymptotic infinity down to the bifurcation surface where χ^a vanishes. The surface integrals at infinity then yield precisely the mass and angular momentum variations, $\delta M - \Omega^{(\alpha)}\delta J_{(\alpha)}$, appearing in Eq.(6), while the surface integral at the bifurcation surface reduces to $\delta \oint_B d^{D-2}x\,\sqrt{h}\,\epsilon_{ab}Q^{ab}(\chi)$. Finally, it can be shown that the latter surface integral always has the form $(\kappa/2\pi)\delta\mathcal{S}$, where κ is the surface gravity of the background black hole, and $\mathcal{S} = 2\pi \oint_B d^{D-2}x\,\sqrt{h}\,\epsilon_{ab}Q^{ab}(\tilde\chi)$, with $\tilde\chi^a$, the Killing vector scaled to have unit surface gravity.

By construction Q^{ab} involves the Killing field $\tilde\chi^a$ and its derivatives. However, this dependence can be eliminated as follows [71]: Using Killing vector identities, Q^{ab} becomes a function of only $\tilde\chi^a$ and the first derivative, $\nabla_a\tilde\chi_b$. At the bifurcation surface, though, $\tilde\chi^a$ vanishes and $\nabla_a\tilde\chi_b = \epsilon_{ab}$, where ϵ_{ab} is the binormal to the bifurcation surface. Thus, eliminating the term linear in $\tilde\chi^a$ and replacing $\nabla_a\tilde\chi_b$ by ϵ_{ab} yields a completely geometric functional of the metric and the matter fields, which may be denoted $\tilde Q^{ab}$. One can show that the resulting expression,

$$\mathcal{S} = 2\pi \oint d^{D-2}x\,\sqrt{h}\,\epsilon_{ab}\,\tilde Q^{ab}, \tag{10}$$

yields the correct value for $\mathcal{S}$ when evaluated not only at the bifurcation surface, but in fact on an arbitrary cross-section of the Killing horizon [73, 74].

[5]Note that the Zeroth Law, *i.e.*, the constancy of of the surface gravity over a stationary event horizon, follows if the latter is also a Killing horizon [81, 82]. This is significant since the Zeroth Law is actually an essential ingredient to the entire framework of black hole thermodynamics.

Using Wald's technique, the formula for black hole entropy has been found for a general Lagrangian of the following form:

$$L = L(g_{ab}, R_{abcd}, \nabla_e R_{abcd}, \nabla_{(e_1}\nabla_{e_2)} R_{abcd}, \ldots; \psi, \nabla_a \psi, \nabla_{(a_1}\nabla_{a_2)}\psi, \ldots) \ ,$$

involving the Riemann tensor and symmetric derivatives of R_{abcd} (and the matter fields, denoted by ψ) up to some finite order n. $\mathcal{S}$ may then be written [72]–[74]

$$\mathcal{S} = -2\pi \oint d^2x\sqrt{h} \sum_{m=0}^{n} (-)^m \nabla_{(e_1} \ldots \nabla_{e_m)} Z^{e_1 \cdots e_m : abcd} \epsilon_{ab}\, \epsilon_{cd} \tag{11}$$

where the Z-tensors are defined by

$$Z^{e_1 \cdots e_m : abcd} \equiv \frac{\delta L}{\delta \nabla_{(e_1} \ldots \nabla_{e_m)} R_{abcd}} \ . \tag{12}$$

As a more explicit example, consider a polynomial-in-R action

$$I = \frac{1}{16\pi G} \int d^4x \sqrt{-g} \left(R + a_2 R^2 + a_3 R^3\right) \tag{13}$$

for which one finds the simple result

$$\mathcal{S} = \frac{1}{4G} \oint d^{D-2}x \sqrt{h} \left(1 + 2a_2 R + 3a_3 R^2\right) \ . \tag{14}$$

Similarly in the Gauss-Bonnet theory with action

$$I = \frac{1}{16\pi G} \int d^4x \sqrt{-g} \left(R + \alpha(R_{abcd}R^{abcd} - 4R_{ab}R^{ab} + R^2)\right) \tag{15}$$

one finds from Eq.(11) that

$$\mathcal{S} = \frac{1}{4G} \oint d^{D-2}x \sqrt{h} \left(1 + 2\alpha \tilde{R}(h)\right) \ . \tag{16}$$

where $\tilde{R}(h)$ is the Ricci scalar calculated for the induced metric h_{ab}. In both Eqs.(14) and (16), the first term yields the expected contribution for Einstein gravity, namely $A/(4G)$. Thus just as the Einstein term in the action is corrected by higher-curvature terms, the Einstein contribution to the black hole entropy receives higher-curvature corrections.

4. Epilogue

While higher curvature theories are typically pathological when considered as fundamental, they may still be studied within the framework of effective field theory [6] where they produce minor corrections to Einstein gravity. Still perturbative investigations [28]–[33] of black holes in this context have revealed modifications to black hole thermodynamics and the generation of new scalar hair in these theories. It may be of interest to the perturbative framework on a more formal footing, *e.g.*, addressing questions such as whether or not solutions to Eq.(2) always exist which leave the event horizon a regular surface.

The absence of ghosts in the Lovelock, dilatonic Gauss-Bonnet and polynomial-in-R theories also seems an interesting question to investigate more fully. This feature seems to make these theories an interesting mathematical framework in which to study exact black hole solutions of the full higher derivative equations. It may be of interest then to find more general stationary solutions, *i.e.*, rotating black hole solutions. The stability of the event horizon, however, remains an important question to be addressed for even the solutions with spherical symmetry.

Ultimately it was the studies of various black holes solutions that led to the beautiful generalization of black hole thermodynamics for higher curvature theories. Wald's new derivation of the First Law [71] demonstrates the black hole entropy is always determined by a local geometric expression evaluated at the event horizon, and provides a general and explicit formula (11) for $\mathcal{S}$. It may be that these results may be used to provide an even more refined test of our understanding of black hole entropy in superstring theory, given the recent dramatic progress in that area [9]. One interesting question related to Wald's derivation is determining the minimal requirements for a stationary event horizon to be a Killing horizon.

While Eq.(11) provides an elegant expression for $\mathcal{S}$, it should be noted that amongst the details overlooked in section 3.1 was the fact that a number of ambiguities arise in the construction of $\tilde{Q}^{ab}$ [73, 74]. Hence Eq.(11) should be understood as the result of making certain (natural) choices in the calculation. None of these ambiguities have any effect when $\mathcal{S}$ is evaluated on a stationary horizon, but they might become significant for non-stationary horizons. As an example, it should be noted that for the Gauss-Bonnet theory, Eqs.(11) and (16) are strictly not the same because the latter expression relies on the additional information that the extrinsic curvature of the bifurcation surface vanishes [69]. Hence the two expressions should not be expected to coincide on a non-stationary horizon.

An important guide to resolving these ambiguities should be the Second Law. Certainly if $\mathcal{S}$ is to play the role of an entropy, it should also satisfy the Second Law of black hole thermodynamics as a black hole evolves, *i.e.*, there should be a classical increase theorem for any dynamical processes. Within Einstein gravity, the Second Law is established by Hawking's area theorem [83]. So far only limited results have been produced for higher curvature theories [84] – see also comments in ref. [77]. One can show for quasi-stationary processes that the Second Law is in fact a direct consequence of the First Law (and a local positive energy condition for the matter fields), independent of the details of the gravitational dynamics. For polynomial-in-R theories, one can establish the Second Law with certain restrictions on the coupling constants (and again a positive energy condition on the matter sector). One proof of the latter involves studying the properties of the null rays along the event horizon with an extension of the Raychaudhuri equation. A valuable extension of these results [84] would be to establish the Second Law within a perturbative framework. Of course, another important question is to determine the validity of the generalized Second Law [85] for evaporating black holes in the higher curvature theories.

A final question which I pose here is related to the conformal anomaly [86]. In quantum field theory [7], the conformal anomaly may be characterized as higher derivative modifications of the renormalized gravitational equations of motion which do not result from the variation of a local action. It appears that these terms will modify the expression for the black hole entropy [87] but they can not in general be addressed with Wald's construction – see, however, [88]. It would be interesting to improve on the latter to systematically include the effects of these contributions, and to determine whether $\mathcal{S}$ or its variation in the First Law still retains its local geometric character. The investigation of a two-dimensional toy model would seem to indicate that the answer to the last question is negative [88].

I would like to gratefully acknowledge useful discussions with T. Jacobson. I am also grateful for the hospitality of the Institute for Theoretical Physics at the University of California, Santa Barbara, where this manuscript was completed. This work was supported by NSERC of Canada, and at the ITP, UCSB by NSF Grant PHY94–07194.

References

1. H. Weyl, *Ann. Phys.* (Leipzig) **59** (1919) 101; *Phys. Zeit.* **22** (1921) 473; A. Eddington, *The Mathematical Theory of Relativity*, 2nd ed. (Cambridge University Press, 1924).
2. R. Utiyama and B.S. DeWitt, *J. Math. Phys.* **3** (1962) 608.
3. K.S. Stelle, *Phys. Rev.* **D16** (1977) 953; *Gen. Rel. and Grav.* **9** (1978) 353.
4. D.G. Boulware, S. Deser and K.S. Stelle, *Phys. Lett.* **168B** (1986) 336; in *Quantum Field Theory and Quantum Statistics*, eds. I.A. Batalin, C.J. Isham and C.A. Vilkovisky (Hilger, Bristol, 1987).
5. See for example: D.G. Boulware, A. Strominger and E.T. Tomboulis in: *Quantum Theory of Gravity*, ed. S. Christensen (Hilger, Bristol, 1984).
6. See for example: S. Weinberg, *The Quantum Theory of Fields*, Vols. 1 and 2, (Cambridge University Press, 1995).
7. See for example: N.D. Birrell and P.C.W. Davies, *Quantum fields in curved space*, (Cambridge University Press, 1982).
8. See for example: M.B. Green, J.H. Schwarz and E. Witten, *Superstring Theory*,

(Cambridge University Press, 1987); D. Lust and S. Theisen, *Lectures on String Theory*, (Springer, Berlin, 1989); J. Polchinski, *String Theory* (Cambridge University Press, 1998).
9. See the contribution of S. Wadia in this volume.
10. See for example: J.F. Donoghue, "Introduction to the Effective Field Theory Description of Gravity," gr-qc/9512024.
11. J.Z. Simon, *Phys. Rev.* **D41** (1990) 3720; **D43** (1991) 3308; L. Parker and J.Z. Simon, *Phys. Rev.* **D47** (1993) 1339 [gr-qc/9211002].
12. See the contribution of J.D. Bekenstein in this volume.
13. R.C. Myers and M.J. Perry, *Ann. Phys.* **172** (1986) 304.
14. B.R. Iyer and C.V. Vishveshwara, *Pramana* **32** (1989) 749; *CQG* **5** (1988) 961;
15. B.R. Iyer, S. Iyer and C.V. Vishveshwara, *CQG* **6** (1989) 1627.
16. R.M. Wald, *General Relativity* (University of Chicago Press, Chicago, 1984).
17. A.A. Starobinsky, *Phys. Lett.* **91B** (1980) 99; M.S. Madsen, *Nucl. Phys.* **B323** (1989) 242.
18. D.G. Boulware and S. Deser, *Phys. Rev. Lett.* **55** (1985) 2656.
19. X. Jaén, J. Llosa and A. Molina, *Phys. Rev.* **D34** (1986) 2302; See also: D.A. Eliezer and R.P. Woodard, *Nucl. Phys.* **B325** (1989) 389.
20. V.P. Frolov, M.A. Markov and V.F. Mukhanov, *Phys. Lett.* **B216** (1989) 272.
21. D. Morgan, *Phys. Rev.* **D43** (1991) 3144; M. Trodden, V.F. Mukhanov and R.H. Brandenberger, *Phys. Lett.* **B316** (1993) 483 [hep-th/9305111].
22. For static black holes, one can develop an "improved" expansion [30, 31] of the metric with: $g_{tt} = g^0_{tt}(1+\alpha h_{tt})$, $g_{rr} = g^0_{rr}(1+\alpha h_{rr})$, *etcetera*. Within this approach, it should be obvious that the event horizon of the original metric really does remain a horizon in the perturbed background, irrespective of the previous discussion.
23. We would advocate that the same reasoning should be applied in the recent studies which suggest that as a result of nonminimal couplings to curvature in an effective action, photon propagation in curved space backgrounds can be "superluminal" [24]. One should not ascribe any physical significance to the results derived there without a thorough investigation of the expected curvature corrections to the propagation of wave packets [25].
24. See for example: R.D. Daniels and G.M. Shore, *Nucl. Phys.* **B425** (1994) 634 [hep-th/9310114]; *Phys. Lett.* **B367** (1996) 75 [gr-qc/9508048]; G.M. Shore, *Nucl. Phys.* **B460** (1996) 379 [gr-qc/9504041]; R. Lafrance and R.C. Myers, *Phys. Rev.* **D51** (1995) 2584 [hep-th/9411018].
25. A.D. Dolgov and I.B. Khriplovich, *Sov. Phys. JETP* **58** (1983) 671.
26. See the contributions of B.F. Whiting and R.M. Wald in this volume.
27. C.V. Vishveshwara, *Phys. Rev.* **D1** (1970) 2870; L.A. Edelstein and C.V. Vishveshwara, "Differential Equations for Perturbations on the Schwarzschild Metric," *Phys. Rev.* **D1** (1970) 3514.
28. M. Lu and M.B. Wise, *Phys. Rev.* **D47** (1993) 3095 [gr-qc/9301021].
29. A. Dobado and A. López, *Phys. Lett.* **B316** (1993) 250; Erratum, **B321** (1994) 435 [hep-ph/9309221].
30. R.C. Myers, *Nucl. Phys.* **B289** (1987) 701.
31. C.G. Callan, R.C. Myers and M.J. Perry, *Nucl. Phys.* **B311** (1988) 673.
32. B. Campbell, M. Duncan, N. Kaloper and K.A. Olive, *Phys. Lett.* **B251** (1990) 34; B. Campbell, N. Kaloper and K.A. Olive, *Phys. Lett.* **B263** (1991) 364; **B285** (1992) 199; P. Kanti and K. Tamvakis, *Phys. Rev.* **D52** (1995) 3506 [hep-th/9504031].
33. S. Mignemi and N.R. Stewart, *Phys. Rev.* **D47** (1993) 5259 [hep-th/9212146]; S. Mignemi, *Phys. Rev.* **D51** (1995) 934 [hep-th/9303102]; M. Natsuume, *Phys. Rev.* **D50** (1994) 3949 [hep-th/9406079].
34. See for example: G.T. Horowitz in: *String theory and quantum gravity '92*, Trieste 1992 Proceedings [hep-th/9210119]; M. Cvetic in: *International Conference on Supersymmetries in Physics (SUSY 96)*, eds. R.N. Mohapatra and A. Rasin, (North-Holland, Amsterdam, 1997) [hep-th/9701152].

35. D.J. Gross and E. Witten, *Nucl. Phys.* **B277** (1986) 1; M.T. Grisaru and D. Zanon, *Phys. Lett.* **B177** (1986) 347.
36. S. Coleman, J. Preskill and F. Wilczek, *Nucl. Phys.* **B378** (1992) 175 [hep-th/9201059].
37. See for example: M. Heusler, *Helv. Phys. Acta* **69** (1996) 501 [gr-qc/9610019]; *Black Hole Uniqueness Theorems*, (Cambridge University Press, 1996); and references therein; See contribution of B. Carter in this volume.
38. D. Lovelock, *J. Math. Phys.* **12** (1971) 498; **13** (1972) 874.
39. B. Zweibach, *Phys. Lett.* **156B** (1985) 315; B. Zumino, *Phys. Rep.* **137** (1986) 109.
40. J.T. Wheeler, *Nucl. Phys.* **B268** (1986) 737; **B273** (1986) 732.
41. R.C. Myers and J.Z. Simon, *Phys. Rev.* **D38** (1988) 2434; B. Whitt, *Phys. Rev.* **D38** (1988) 3000.
42. Solutions within a particular class of these theories were also recently considered: M. Bãnados, C. Teitelboim and J. Zanelli, *Phys. Rev.* **D49** (1994) 975 [gr-qc/9307033].
43. D.L. Wiltshire, *Phys. Lett.* **B169** (1986) 36; *Phys. Rev.* **D38** (1988) 2445.
44. R.C. Myers and J.Z. Simon, *Gen. Rel. Grav.* **21** (1989) 761.
45. T. Appelquist, A. Chodos and P.G.O. Freund, *Modern Kaluza-Klein Theories* (Addison-Wesley, 1987); A. Salam and J. Strathdee, *Ann. Phys.* **141** (1982) 316.
46. F. Müller-Hoissen, *Phys. Lett.* **201B** (1988) 325; *CQG* **5** (1988) L35.
47. E. Poisson, *CQG* **8** (1991) 639.
48. P. Kanti, N.E. Mavromatos, J. Rizos, K. Tamvakis and E. Winstanley, *Phys. Rev.* **D54** (1996) 5049 [hep-th/9511071]; S.O. Alexeyev and M.V. Pomazanov, *Phys. Rev.* **D55** (1997) 2110 [hep-th/9605106]; "Singular Regions in Black Hole Solutions in Higher Order Curvature Gravity," gr-qc/9706066; S.O. Alexeyev, *Grav. Cosmol.* **3** (1997) 161 [gr-qc/9704031].
49. T. Torii, H. Yajima and K. Maeda, *Phys. Rev.* **D55** (1997) 739 [gr-qc/9606034]; P. Kanti and K. Tamvakis, *Phys. Lett.* **B392** (1997) 30 [hep-th/9609003].
50. B. Whitt, *Phys. Lett.* **B145** (1984) 176.
51. S. Mignemi and D.L. Wiltshire, *Phys. Rev.* **D46** (1992) 1475 [hep-th/9202031].
52. P.W. Higgs, *Nuov. Cim.* **11** (1959) 816; G. Magnano, M. Ferraris and M. Francaviglia, *Gen. Rel. Grav.* **19** (1987) 465; *CQG* **7** (1990) 557; K. Maeda, *Phys. Rev.* **D39** (1989) 3159; A. Jakubiec and J. Kijowski, *Phys. Rev.* **D37** (1988) 1406; J.D. Barrow and S. Cotsakis, *Phys. Lett.* **B214** (1988) 515; G. Magnano and L.M. Sokolowski, *Phys. Rev.* **D50** (1994) 5039 [gr-qc/9312008]; D. Wands, *CQG* **11** (1994) 269, [gr-qc/9307034].
53. A. Strominger, *Phys. Rev.* **D30** (1984) 2257.
54. B. Whitt, *Phys. Rev.* **D32** (1985) 379.
55. See for example: D.J. Gross and J.H. Sloan, *Nucl. Phys.* **B291** (1987) 41.
56. See for example: G.T. Horowitz and A.A. Tseytlin, *Phys. Rev. Lett.* **73** (1994) 3351 [hep-th/9408040]; E. Bergshoeff, R. Kallosh and Tomas Ortin, *Phys. Rev.* **D50** (1994) 5188 [hep-th/9406009]; R. Kallosh and T. Ortin, *Phys. Rev.* **D50** (1994) 7123 [hep-th/9409060]; A.A. Tseytlin in: *Advances in Astrofundamental Physics*, eds. N. Sanchez and A. Zichichi (World Scientific, 1995) [hep-th/9410008].
57. S. Deser and Z. Yang, *CQG* **6** (1989) L83.
58. R. Gregory and R. Laflamme, *Phys. Rev.* **D51** (1995) 305 [hep-th/9410050]; *Nucl. Phys.* **B428** (1994) 399 [hep-th/9404071]; *Phys. Rev. Lett.* **70** (1993) 2837 [hep-th/9301052]; E. Witten, *Nucl. Phys.* **B195** (1982) 481.
59. P. Kanti, N.E. Mavromatos, J. Rizos, K. Tamvakis and E. Winstanley, "Dilatonic Black Holes in Higher Curvature String Gravity II: Linear Stability," hep-th/9703192.
60. S. Coleman in: *Subnuclear Phenomena*, ed. A. Zichichi (Academic Press, New York, 1970).
61. Y. Choquet-Bruhat, *J. Math. Phys.* **29** (1988) 1891; C. Aragone in: *SILARG 6 Proceedings*, ed. M. Novello (World Scientific, 1988); G.W. Gibbons and P.J. Ruback, *Phys. Lett.* **B171** (1986) 390.

62. T. Jacobson and R.C. Myers, unpublished.
63. S.W. Hawking, *Commun. Math. Phys.* **43** (1975) 199.
64. G.W. Gibbons and S.W. Hawking, *Phys. Rev.* **D15** (1977) 2752; S.W. Hawking in *General Relativity: An Einstein Centenary Survey*, eds. S.W. Hawking and W. Israel (Cambridge University Press, Cambridge, 1979).
65. In higher dimensions, one has the possibility of commuting rotations in totally orthogonal planes. Hence the black hole is characterized by the set of angular momenta $J_{(\alpha)}$ in the maximal set of orthogonal planes. See ref. [13].
66. Note that in D dimensions, a space-like cross-section of the horizon is a $(D{-}2)$-dimensional space, and so the surface area A_H refers to the volume of such a slice.
67. J.D. Bekenstein, *Lett. Nuov. Cim.* **4** (1972) 737; *Phys. Rev.* **D7** (1973) 2333; **D9** (1974) 3292.
68. M. Visser, *Phys. Rev.* **D48** (1993) 583 [hep-th/9303029].
69. T. Jacobson and R.C. Myers, *Phys. Rev. Lett.* **70** (1993) 3684 [hep-th/9305016].
70. D. Sudarsky and R.M. Wald, *Phys. Rev.* **D46** (1992) 1453; R.M. Wald in: *Directions in General Relativity*, eds. B.L. Hu, M. Ryan and C.V. Vishveshwara (Cambridge University Press, 1993) [gr-qc/9305022]; For a closely related method, see also: J.D. Brown, E.A. Martinez and J.W. York, *Phys. Rev. Lett.* **66** (1991) 2281; J.D. Brown and J.W. York, *Phys. Rev.* **D47** (1993) 1407; 1420.
71. R.M. Wald, *Phys. Rev.* **D48** (1993) 3427 [gr-qc/9307038].
72. M. Visser, *Phys. Rev.* **D48** (1993), 5697 [hep-th/9307194].
73. V. Iyer and R.M. Wald, *Phys. Rev.* **D50** (1994) 846, [gr-qc/9403028].
74. T. Jacobson, G. Kang and R.C. Myers, *Phys. Rev.* **D49** (1994) 6587 [gr-qc/9312023].
75. J. Louko, J.Z. Simon and S.N. Winters-Hilt, *Phys. Rev.* **D55** (1997) 3525 [gr-qc/9610071].
76. M. Banados, C. Teitelboim and J. Zanelli, *Phys. Rev. Lett.* **72** (1994) 957 [gr-qc/9309026].
77. R.M. Wald, "Black Holes and Thermodynamics," [gr-qc/9702022] to appear in: *Black Holes and Relativistic Stars*, ed. R.M. Wald.
78. J. Lee and R.M. Wald, *J. Math. Phys.* **31** (1990) 725.
79. R.M. Wald, *J. Math. Phys.* **31** (1990) 2378; See also: I.M. Anderson in *Mathematical Aspects of Classical Field Theory*, eds. M. Gotay, J. Marsden and V. Moncrief, *Cont. Math.* **132** (1992) 51.
80. I. Rácz and R.M. Wald, *CQG* **9** (1992) 2643.
81. B. Carter, *Phys. Rev. Lett.* **26** (1971) 331; in: *Black Holes*, ed. C. DeWitt and B.S. DeWitt (Gordon and Breach, New York, 1973.
82. I. Rácz and R.M. Wald, *CQG* **13** (1996) 539 [gr-qc/9507055].
83. S.W. Hawking, *Comm. Math. Phys.* **25** (1972) 152; S.W. Hawking and G.R.F. Ellis, *The Large Scale Structure of Spacetime* (Cambridge University Press, Cambridge, 1973).
84. T. Jacobson, G. Kang and R.C. Myers, *Phys. Rev.* **D52** (1995) 3518 [gr-qc/9503020].
85. W. Zurek, *Phys. Rev. Lett.* **49** (1982) 1683; W.H. Zurek and K.S. Thorne, *Phys. Rev. Lett.* **54** (1985) 2171; K.S. Thorne, W.H. Zurek and R.H. Price in: *Black Holes: The Membrane Paradigm*, eds. K.S. Thorne, R.H. Price and D.A. MacDonald (Yale University Press, New Haven, 1986); W.G. Unruh and R.M. Wald, *Phys. Rev.* **D25** (1982) 942; *Phys. Rev.* **D27** (1983) 2271; correction: M.J. Radzikowski and W.G. Unruh, *Phys. Rev.* **D37** (1988) 3059; R.M. Wald in: *Highlights in Gravitation and Cosmology*, eds. B.R. Iyer *et al* (Cambridge University Press, Cambridge, 1988); V.P. Frolov and D.N. Page, *Phys. Rev. Lett.*, **71** (1993) 3902 [gr-qc/9302017].
86. M.J. Duff, *CQG* **11** (1994) 1387 [hep-th/9308075].
87. D.V. Fursaev, *Phys. Rev.* **D51** (1995) 5352 [hep-th/9412161]; S.N. Solodukhin, "Entropy of Schwarzschild Black Hole and String-Black Hole Correspondence," hep-th/9701106.
88. R.C. Myers, *Phys. Rev.* **D50** (1994) 6412 [hep-th/9405162]; J.D. Hayward, *Phys. Rev.* **D52** (1995) 2239 [gr-qc/9412065].

9. MICRO-STRUCTURE OF BLACK HOLES AND STRING THEORY

SPENTA R. WADIA
Theoretical Physics Group,
Tata Institute of Fundamental Research,
Homi Bhabha Rd., Mumbai-400 005, India.
AND
Jawaharlal Nehru Centre for Advanced Scientific Research,
Jakkur P.O., Bangalore 560064, India.

1. Introduction

Black holes are a subject of much interest in various areas of astro-physics and physics. There is little doubt among astronomers about black hole candidates in binary systems and there is also evidence that the matter of galaxies is belched in 'active galactic nuclei' which are black holes with a million times the solar mass. Why do black holes occur? The standard answer is gravitational collapse and the existence of black hole solutions in general relativity. The basic characteristic of a black hole within the framework of general relativity is that it has a curvature singularity where all the matter is supposed to have fallen in and that this singularity is 'surrounded' by an event horizon. This is a null surface that divides the space-time into two distinct regions and plays a fundamental role in the description of the properties of a black hole. The linear stability analysis for the Schwarzschild black hole is a significant contribution of Vishveshwara [1].

The general theory of relativity has been very successful as a classical theory, but a consistent quantization is still lacking. Our present understanding indicates that this problem does not have a solution within the framework of local quantum field theory. One may argue that the consistency problem of quantum gravity arises only at the Planck length $l_p^2 = \frac{\hbar G}{c^3}$ (G is Newton's constant and c is the speed of light) and hence one may expect general relativity to be a consistent description for low energy phe-

B. R. Iyer and B. Bhawal (eds.), Black Holes, Gravitational Radiation and the Universe, 137–147.

nomena involving macroscopic objects, like black holes. However this turns out not to be true. When one turns on quantum effects black holes behave like black bodies which absorb and emit radiation with a characteristic temperature determined by the macroscopic parameters of the black hole. Hence it is inevitable that they can radiate and evaporate. Complete black hole evaporation means that the final state of whatever went into a black hole is thermal radiation. Hawking's calculation [2] implies that this radiation is exactly thermal and this is in conflict with the unitary evolution of pure states in quantum mechanics. This fact is called the 'information paradox'. Besides the temperature, the thermodynamic state of the black hole is also characterized by its 'entropy', given by the Bekenstein-Hawking formula

$$S = \frac{A}{4l_p^2}, \tag{1}$$

where A refers to the area of the horizon of the black hole. In cgs units $l_p = 10^{-33}$ cms. This is one of the most remarkable formulas of physics because it involves the Planck length at low energies. The formula is of geometrical origin and if one equates it to Boltzmann's formula for the entropy of a macroscopic system then it is natural to ask the question about its microscopic origin.

While graviton loops, propagating in a background Minkowski spacetime, render quantized gravity inconsistent at Planck distances, the existence of black holes leads to the information paradox on a macroscopic scale, and raises fundamental questions about the true degrees of freedom of quantum gravity. It is well known that string theory resolves the problem of ultra-violet divergences. The basic physical reason is that the strings cannot come closer than the string length l_s, which is related to the Planck length (in four dimensions) as $l_p = gl_s$, where g is the string coupling. In this article we will present recent mounting evidence that string theory has the ingredient to show that the Bekenstein-Hawking formula is a consequence of Boltzman's law [3, 4, 5, 6, 7]

$$S = \ln \Omega\,. \tag{2}$$

The number of black hole states Ω at given values of macroscopic parameters are derived from the 'underlying' microscopic degrees of freedom of string theory. The thermodynamics of near extremal black holes can also be derived from the known properties of the highly degenerate black hole wave function using the standard principles of quantum statistical mechanics [8, 9, 10, 11, 12, 13, 14, 15, 16]. The issue of directly accessing the black hole micro states by 'external experiments' can also be addressed with success. It seems that unitarity in quantum mechanics is not violated at least in the examples amenable to rigorous study. What seems to undergo a revision is our ideas of space-time at the microscopic level. The success of string theory in explaining and accounting for the various features of black holes should be considered as providing evidence for the "String Paradigm".

2. Black Hole Thermodynamics

Before we discuss the recent progress on these questions it is worth recalling the main points leading to black hole radiance within the framework of General Relativity. For simplicity of presentation we discuss the Schwarzschild black hole characterized by its mass M. One considers the propagation of matter in the background of a black hole. For large M, the metric in the vicinity of the horizon, in the Kruskal coordinates is (upto a scaling) that of flat Minkowski space-time. These coordinates are related to the Schwarzschild coordinates by the well known Rindler transform. Now we know how to do quantum field theory in Minkowski coordinates. The Fock vacuum here is what is usually called the Minkowski vacuum and would be the vacuum of an observer who is falling freely into a black hole. Under the Rindler transform this description gets radically transformed: the Minkowski vacuum appears as a density matrix to the observer who does quantum field theory in Schwarzschild (Rindler) coordinates. This density matrix has the associated Hawking temperature $T = \frac{\hbar c^3}{8\pi MG}$ of the black hole and the spectrum is that of a black body. The appearance of the density matrix stems from the fact that the Hilbert space of the Rindler observer is a direct sum of Hilbert spaces corresponding to observables 'inside' and 'outside' the horizon. Since the observables inside the horizon are not accessible to the Rindler observer, one has to perform a trace over them. As yet one cannot say that there is information loss because the thermal spectrum arises from an average that the Rindler observer performs. The problem arises when the black hole reaches the end point of evaporation, i.e. when most of its mass has been radiated away and the end point is a purely thermal state. In such a situation one cannot even in principle recover the correlations present in the imploding pure state that created the black hole. This is the information paradox: a pure quantum mechanical initial state ends in a thermal (mixed) state.

We note that in the case of the Schwarzschild black hole the area of the horizon in Eq.(1) is given by $A = 16\pi M^2$. This plus ththe formula for the Hawking temperature leads to the first law of black hole thermodynamics.

Hawking asserts that the black hole thermodynamics and information loss is intrinsic to general relativity because of the focusing properties of the gravitational field. It is this contention that has been questioned most vehemently by t'Hooft [17] and Susskind [3] and many other physicists who believe that black hole thermodynamics must have a statistical basis in a unitary quantum theory, just like the standard thermodynamics of a macroscopic system.

As we have already mentioned in this article we will review some evidence in favor of the latter view point.

3. The 5-dimensional Black Hole of IIb String Theory

We will base our discussion on a black hole which can be analyzed using present techniques. This black hole occurs as a 5-dimensional solution of the supergravity theory which describes the low energy limit of type IIb string theory. We shall denote by (x_1, x_2, x_3, x_4) the 4 non-compact dimensions of space and by $(x_5, x_6, x_7, x_8, x_9)$ the compact dimensions which form a 5-torus. The 5th circle has a radius R and the 4-torus has a volume V. Here we choose $R \gg V^{1/4} \sim l_s$. The theory has various massless fields, but for our purpose we consider only the metric, the dilaton, and the anti-symmetric 3-form which gives rise to 'electric' and 'magnetic' charges Q_1 and Q_5 respectively in 4 space-dimensions. The (t, x_5) component of the metric gives rise to a Kaluza-Klein charge N. The dilaton field is a finite constant at the horizon.

The black hole metric in 5-dimensions is given by

$$ds_5^2 = -f(r)^{-2/3}\left(1-\frac{r_0^2}{r^2}\right)dt^2 + f(r)^{1/3}\left[\left(1-\frac{r_0^2}{r^2}\right)^{-1}dr^2 + r^2 d\Omega_3^2\right] \quad (3)$$

where

$$f = \left(1+\frac{r_0^2\sinh^2\alpha}{r^2}\right)\left(1+\frac{r_0^2\sinh^2\gamma}{r^2}\right)\left(1+\frac{r_0^2\sinh^2\sigma}{r^2}\right), \quad (4)$$

and the boost angles α, β, γ are parameters that characterize the black-hole. Note that in 5 space-time dimensions, the above represents a black hole with the inner horizon at $r = 0$, and the outer horizon at $r = r_0$. Note that the location of the inner and outer horizons is a coordinate dependent statement and it is possible to choose coordinates so that the outer horizon is located at $r = A^{1/3}$.

The charges Q_1, Q_5, N are given by the formulae

$$Q_1 = \frac{V r_0^2}{2g}\sinh 2\alpha, \quad (5)$$

$$Q_5 = \frac{r_0^2}{2g}\sinh 2\gamma, \quad (6)$$

$$N = \frac{R^2 V r_0^2}{2g^2} \sinh 2\sigma \ . \tag{7}$$

In order to have a meaningful semi-classical limit where the string coupling $g \to 0$, the charges Q_1, Q_5 and N have to be chosen large holding gQ_1, gQ_5 and g^2N fixed. This also implies that the area of the horizon is large compared to the string scale l_s as it should be for a black hole. Let us now present some relevant formulae of black hole thermodynamics for the above black hole. The energy, entropy and the temperature are given by

$$E = \frac{RVr_0^2}{2g^2} (\cosh 2\alpha + \cosh 2\gamma + \cosh 2\sigma), \tag{8}$$

$$S = \frac{2\pi RVr_0^3}{g^2} \cosh \alpha \cosh \gamma \cosh \sigma, \tag{9}$$

$$T^{-1} = 2\pi r_0 \cosh \alpha \cosh \gamma \cosh \sigma. \tag{10}$$

We note that in the limit $\sigma \to \infty$ and $r_0 \to 0$, holding $r_0 \exp \sigma$ fixed, we have a Bogomolny-Prasad-Sommerfeld (BPS) solution which corresponds to an extremal black hole with $T = 0$ but a large non-zero entropy $S = \sqrt{Q_1 Q_5 N}$. We are interested in the case with a slight deviation from extremality, when the temperature is tiny but non-zero and the specific heat at fixed values of the charges is positive. In this limit we can present a sound understanding of black hole thermodynamics in terms of string theory. This circumstance is in contrast to the situation of the Schwarzschild black hole which has negative specific heat so that the black hole becomes hotter as it radiates. However inspite of the difference the near extremal black hole also presents the information paradox as it radiates towards its ground state.

Our strategy is to give a microscopic model of the above black hole. The situation is very much like giving a microscopic basis to the soliton of the chiral model of pions. We of course know that the soliton solution of the weakly coupled chiral model is the baryon of $SU(N)$ QCD and it is composed of N quarks. In the chiral model the baryon number appears as a topological invariant in terms of the chiral field while in QCD the baryon number is a counting problem. One simply assigns a baryon charge $\frac{1}{N}$ to the quarks. If the black hole entropy is indeed given by Boltzmann's law then it is natural to ask about the string theory analogues of the quarks of QCD.

In the next section we will introduce D-branes which are special soliton solutions of string theory. The main hypothesis is that these solitons are the constituents of the black holes. Just as the quarks of QCD interact via gluons, the D-brane interactions are mediated via open strings which end on them. It is worth emphasizing that these degrees of freedom are not evident in the standard formulation of general relativity in much the same way that in the chiral model of pions it is hard to see any trace of the quark building blocks.

4. String Theory Solitons: D-branes

D-branes are like the heavy quarks of string theory [18] except that unlike the quarks of QCD they appear as soliton solutions of string theory. The mass of a D-brane is $M \sim \frac{1}{l_s g}$, where g is the string coupling. For small g they are indeed very heavy objects. Their gravitational field is given by $MG \sim g$ because Newton's constant scales as $G \sim g^2$. D-branes also carry one unit of Ramond charge. It should be noted that perturbative excitations of string theory have zero Ramond charge. In that sense Ramond charge is a soliton charge. In theories with more than one supersymmetry these solitons are stable because they saturate the so called BPS bound, which essentially says that their mass, in appropriate units is equal to their Ramond charge.

Another feature of D-branes is that these solitons are not necessarily point like. Their world volume can be anywhere between zero to nine dimensional. D-branes are generalized domain walls and their precise characterization is in terms of Dirichlet boundary conditions imposed on the directions of the open strings which end on them. This also explains why they are called D-branes. The other important feature is that a brane configuration is invariant under only half of the supersymmetries of the bulk theory.

5. Effective Gauge Theory

For a review of the initial material in this section see Ref.[19]. The type IIb string theory in which Eq.(3) is a solution has D_1–branes and D_5–branes as solitons. The D_1–branes are wrapped around the 5th circle and the D_5–branes are wrapped around the 5-torus in the directions 5, 6, 7, 8 and 9. Note that the 5th direction is common to both the branes. There are Q_1 D_1–branes and Q_5 D_5–branes. There are interactions between these branes which are mediated by open strings that end on the branes. It is one of the most remarkable facts, that properties of branes which are near each other (or are sitting on top of each other) are described by the massless modes of the open string theory, namely gauge theories. Let us enumerate the relevant massless fields for the above configuration of branes. Clearly we have 3 types of open massless strings depending upon which brane they end on: 1) (1,1) strings with both ends on a D_1 brane. There are $4Q_1^2$ massless components which arrange themselves as 4 $Q_1 \times Q_5$ matrices. The

4 refers to the directions (6,7,8,9) which lie in the 5-branes. We have set the strings along the non-compact space-directions to zero. Since these branes are indistinguishable we can rotate the matrix by an element of the unitary group $U(Q_1)$. 2) (5,5) strings with both ends on a D_5 brane. These massless strings arrange themselves once again into 4 $Q_5 \times Q_5$ matrices with $U(Q_5)$ as the symmetry group. 3) The third set of massless strings (1,5) and (5,1) have one end on a D_1 brane and another on a D_5 brane. They form $4Q_1Q_5$ massless modes with a $U(Q_1) \times U(Q_5)$ symmetry group. It turns out that the interacting system of branes is a $U(Q_1) \times U(Q_5)$ non-abelian gauge theory in the 2-dimensions (x_5,t) with N=(4,4) supersymmetry. The restriction to 2-dimensions roughly speaking comes about because x_5 is the only direction common to the 2 sets of branes. Why N=(4,4) supersymmetry? This is because half the 32 supersymmetries of the bulk IIb theory are broken by the set of 5-branes and the other half by the set of 1-branes. This leaves us with 8 supersymmetries whose charges are 4-left movers and 4-right movers in the 2-dimensional field theory.

The analysis of the low lying modes of the above gauge theory is quite involved and it was only recently completed [16]. We shall not discuss it here in any detail, except summarize the result. The modes of the gauge theory which have long wavelengths from the view point of the bulk theory are described by a N=(4,4) super-conformal field theory (SCFT) with central charge c=6. The bosonic co-ordinates of the SCFT Y^m form a vector representation of $SO(4)_I$, which is the tangent space group of the internal directions (6,7,8,9). Their 4-fermionic partners are also spinors of $SO(4)_E$, which is the rotational symmetry of the 4 external space-dimensions. The bosonic coordinates are scalars under $SO(4)_E$ and the fermionic coordinates are scalars under $SO(4)_I$. An important feature of this SCFT is that its fields are fractionally moded w.r.t. the original 5th circle of radius R and the minimum momentum is given by $\frac{1}{RQ_1Q_5}$ which is vanishingly small in the limit when the charges are very large. This is consistent with the fact that the absorption cross section at zero wave number is non-zero and equals the area of the horizon of the black hole. One of the most important aspects of this story is that as long as we are dealing with low energy phenomena the above SCFT is independent of the strength of the coupling constant of the gauge and hence the open string theory. This is one of the miracles of supersymmetric field theory with N=(4,4) supersymmetry.

Let us now discuss the wave function of the extremal black hole. The states can be easily constructed within the sub-space of the total Fock space of the SCFT. This sub-space is defined by the requirement that the black hole state is either extremal (BPS) or atmost near extremal. As we had noted before the extremal black hole which is also a BPS state is characterized by the Kaluza-Klein momentum $P = \frac{N}{R}$. If we identify this momentum with $L_0 - \bar{L_0}$ which generates rotations along x_5 in the SCFT then we get

$$P = \frac{NQ_1Q_5}{RQ_1Q_5} . \tag{11}$$

The BPS condition implies that $\bar{L_0} = H - P = 0$, and hence we get the level of the SCFT to be $L_0 = NQ_1Q_5$. This means that the ground state of the black hole is characterized by all the degenerate states of the SCFT which are at this level. There is a standard formula for the number of states at a certain level n in a SCFT of central charge c, provided n is very large:

$$\Omega \sim \exp 2\pi\sqrt{cn/6} . \tag{12}$$

Substituting $n = NQ_1Q_5$ and $c = 6$ in Eq.(12), reproduces the Bekenstein-Hawking formula, viz. $S = 2\pi\sqrt{Q_1Q_5N}$.

In order to describe a near extremal black hole we relax the BPS condition $\bar{L_0} = 0$, and choose the conditions

$$L_0 = Q_1Q_5(N + n_l), \tag{13}$$

$$\bar{L_0} = Q_1Q_5n_l, \tag{14}$$

$n_l \ll NQ_1Q_5$. The above condition ensures that the deviation from the BPS state is small and the condition $P = \frac{N}{R}$ is maintained. The entropy of the above configuration turns out to be

$$S = 2\pi\sqrt{Q_1Q_5}(\sqrt{N + n_l} + \sqrt{n_l}) . \tag{15}$$

This formula also exactly matches the entropy deduced from the near extremal black hole solution (3).

6. Black Hole Thermodynamics

Now that we have a precise understanding of the micro states of a near extremal black hole, we can address the important question of Hawking radiation. In order to discuss that we need to know two things. Firstly, we have to know whether the states of the SCFT can be treated as a thermal ensemble. In our discussion we will simply assume this. This is not really a serious assumption because we are dealing with a system with very large degrees of freedom. Secondly we need to know how the modes of the closed string couple to the SCFT. Since the latter has fractional moding on a circle of radius R, at low energies, only the zero modes of the closed field with respect to the 5th dimension have a non-zero coupling. Further, this coupling is located at the origin of the 4-dimensional space, $x_i = 0$, $i = 1, 2, 3, 4$. The next question is about the derivation of the effective coupling. Here we appeal to symmetry principles, (like in the discussion of the effective pion nucleon coupling in QCD) which in our problem are the global $SO(4)_I$ and $SO(4)_E$ symmetries. As an example one can write down the coupling of the closed string states which correspond to deformations of the internal torus. These are scalar particles in 5-dimensions denoted by the field h_{mn},

$$T_{\text{eff}} \int dt \int_0^{2\pi R\widetilde{Q}_1\widetilde{Q}_5} d\widetilde{\sigma}(\delta_{mn} + h_{mn}(t))\partial_\alpha \widetilde{Y}^m \partial^\alpha \widetilde{Y}^n . \tag{16}$$

$T_{\text{eff}} = \frac{1}{\alpha'^2 g}\sqrt{\frac{V_4}{Q_1 Q_5}}$, is the effective string tension. Note that it is very small compared to the tension of the original open strings that mediated the interactions between the branes. A very important point that is implicit in the above interaction is the fact that ultra high energy processes in the brane effective theory correspond to low energy processes in the transverse space, which is the space time of the black hole.

Using the above interaction one can calculate the S-matrix element in first order perturbation theory between initial and final states of the black hole constructed out of the states of the SCFT. After this straightforward calculation of the S-matrix, between black hole micro-states we make the statistical assumption that that the black hole is described by a density matrix

$$\rho = \frac{1}{\Omega}\sum_{\{i\}} |i\rangle\langle i| \tag{17}$$

where the sum $\{i\}$ is over all possible distributions keeping N_L and N_R fixed. It is this formula that leads to the entropy in Eq.15, $S = -\text{Tr}\rho \ln \rho = -\sum_{\{i\}} \frac{1}{\Omega} \ln \frac{1}{\Omega} = \ln \Omega$.

Density matrices like the one described above are not unfamiliar in particle physics. They arise, e.g. in calculating the decay rate of an unpolarized particle into unpolarized products. As there, in the present case also, the total "unpolarized" transition probability is given by

$$P_{\text{decay}}(i \to f) = \frac{1}{\Omega} \sum_{\{i\},\{f\}} |\langle f|S|i\rangle|^2 \tag{18}$$

The division by Ω represents averaging over initial states, while the final states are simply summed over. The passage to the decay rate $d\Gamma$ is usual and one can be easily derived. The relevant formulae are

$$\Gamma_H = \sigma_{abs}\rho(\frac{w}{T_H})^{-1}\frac{d^4k}{(2\pi)^4} \tag{19}$$

where σ_{abs} is the absorption cross section given by

$$\sigma_{abs} = \frac{g^2 Q_1 Q_5 \pi^3}{V} w \rho(\frac{w}{T_H}) \rho(\frac{w}{T_L})^{-1} \rho(\frac{w}{T_R})^{-1} \quad (20)$$

and $\rho(x) = (\exp x - 1)^{-1}$ is the Bose factor. $T_L = \frac{r_0 V \exp \frac{\sigma}{2}}{\pi g^2 Q_1 Q_5}$ and $T_R = \frac{r_0 V \exp -\frac{\sigma}{2}}{\pi g^2 Q_1 Q_5}$ are the temperatures of the left and right moving excitations on the effective string and T_H is the Hawking temperature given by $2T_H^{-1} = T_R^{-1} + T_L^{-1}$. This precisely agrees with the same quantities derived using general relativity!

7. Concluding Remarks

It seems clear from the above sketch that the case of the 5-dimensional near extremal black hole of IIb string theory is amenable to a rigorous treatment in terms of the constituent brane model. The low energy processes calculated using his model coincide with the calculations of general relativity. The constituent model produced the answers by calculating the square of the S-matrix between highly degenerate states of the SCFT. After this we assumed the representation of the black hole in terms of a density matrix and calculated the absorption cross section and the corresponding decay rate. This is exactly as in the thermodynamic treatment of a macroscopic body and hence we conclude that within the parameters of the above calculations not only do we have a microscopic model of the black hole degrees of freedom that explain the Bekenstein-Hawking formula, but a resolution of the information paradox itself.

One of the most important and currently active questions under investigation is the bearing or the imprint of the microstates on geometry: holography. There are several inspiring works in this direction [20, 21, 22] which set up a correspondence between the SCFT and the solutions of the wave equations of fields in supergravity in the limit of large 'N', when the metric degenerates into that of anti-De Sitter space. Supergravity and supersymmetric gauge theories have a symbiotic existence and this relationship promises deep insights into the structure of both these theories which are based very different local invariances.

References

1. C.V. Vishveshwara, Phys. Rev. **D1** (1970) 2870.
2. S. Hawking and R. Penrose, *The Nature of Space and Time*, Princeton Univ. Press, (1996); Sci. Am. **275** (1996) 44.
3. L. Susskind, *Some Speculations about Black Hole Entropy in String Theory*, hep-th/9309145.
4. A. Sen, *Extremal Black Holes and Elementary String States*, hep-th/9504147.

5. A. Strominger and C. Vafa, Phys. Lett. **B379** (1996) 99, hep-th/9601029.
6. C. V. Johnson, R. R. Khuri and R. C. Myers, Phys. Lett. **B378** (1996) 78, hep-th/9603061.
7. R. Breckenridge, D. Lowe, R. Meyers, A. Peet, A. Strominger and C. Vafa, Phys. Lett. **B381** (1996) 423, hep-th/9603078;
R. Breckenridge, R. Meyers, A. Peet and C. Vafa, Phys. Lett. **B391** (1997) 93, hep-th/9602065.
8. C. Callan and J. Maldacena, Nucl. Phys. **B475** (1996) 645, hep-th/9602043.
9. J. Maldacena and L. Susskind, Nucl. Phys. **B475** (1996) 679, hep-th/9604042;
10. A. Dhar, G. Mandal and S. R. Wadia, Phys. Lett. **B388** (1996) 51, hep-th/9605234.
11. S. R. Das and S. D. Mathur, Nucl. Phys. **B482** (1996)153, hep-th/9607149.
12. J. Maldacena and A. Strominger, *Black Hole Grey Body Factors and D-brane Spectroscopy*, hep-th/9609026; *Universal Low-Energy Dynamics for Rotating Black Holes*, hep-th/9702015.
13. C. Callan, S. Gubser, I. Klebanov and A. Tseytlin, *Absorption of Fixed Scalars and the D-brane Approach to Black Holes*, hep-th/9610172.
14. I. Klebanov and M. Krasnitz, *Fixed Scalar Grey Body Factors in Five and Four Dimensions* hep-th/9612051; *Testing Effective String Models of Black Holes with Fixed Scalars* htp-th/9703216.
15. S. F. Hassan and S. R. Wadia, Phys. Lett. **B402** (1997) 43, hep-th/9703163.
16. S. F. Hassan and S. R. Wadia, *Gauge Theory Description of D-brane Black Holes: Emergence of the Effective SCFT and Hawking Radiation*, hep-th/9712213.
17. G. t'Hooft, *The Scattering matrix Approach for the Quantum Black Hole*, gr-qc/9607022.
18. J. Polchinski, *TASI Lectures on D-branes*, hep-th/9611050.
19. J. Maldacena, *Black Holes in String Theory*, Ph.D. Thesis, hep-th/9607235.
20. J. Maldacena, *The Large N limit of Superconformal Field Theories and Supergravity*, hep-th/971120.
21. S. Gubser, I. Klebanov and A. Polyakov, *Gauge Theory Correlators from Non-critical String Theory*, hep-th/9802109.
22. E. Witten, *Anti De Sitter Space and Holography*, hep-th/9802150.

Actually, relativity and the geometry associated with it had considerable impact on art and literature. For instance, Jean Metzinger wrote in his essay *Du Cubisme*:

' If we wishd to tie the painter's space to a particular geometry, we should have to study, at some length, certain ones of Riemann's theorems'.

The melted watches in Salvador Dali's painting *Persistence of Memory* represented warped time. He discussed the framework of non-Euclidean geometry and Einstein's theories in his book *The Conquest of the Irrational* and described his watches as the 'extravagant and solitary Camembert of time and space'. In his rather unusual book, *50 Secrets of Magic Craftsmanship*, Dali also spoke of geodesics on the body of a model as equivalent to the volume occupied by the model, a familiar spacetime representation in general relativity:

'The apprentice's Secret Number 22 is that of the drawing of the geodesic lines of his model. Nothing will reveal itself more useful for the understanding of the mysteries of the nude figure than the knowledge to be derived from the assiduous practice of this method. Preferably you must choose a plump model, the curves of whose flesh are as turgescent as possible. The best poses for this are recumbent ones. You need a provision of strings of black cotton which have been previously soaked in linseed oil to which venetian turpentine has been added, in a proportion of five to three. These strings should be hung up the day before using them, so that they may drip off the excess oil, but without drying altogether. Once the model is lying down in the pose which you desire you begin cautiously to lay the strings on the model's body in the places where you wish a clearer indication of the forms. The curve which these strings adopt will naturally be the geodesic lines of the surface which you want made clear. You may then draw your nude,but especially these geodesics which, if they are in sufficient quantity, will suffice - even should you efface the nude - to imprint its absent volume'.

— C.V. VISHVESHWARA

10. QUANTUM GEOMETRY AND BLACK HOLES

ABHAY ASHTEKAR AND KIRILL KRASNOV
*Center for Gravitational Physics and Geometry,
Physics Department, Penn State, University Park,
PA 16802, USA.*

1. Introduction

In his Ph.D. thesis, Bekenstein suggested that, for a black hole in equilibrium, a multiple of its surface gravity should be identified with its temperature and a multiple of the area of its event horizon should be identified with its thermodynamic entropy [1]. In this reasoning, he had to use not only general relativity but also quantum mechanics. Indeed, without recourse to the Planck's constant, $\hbar$, the identification is impossible because even the physical dimensions do not match. Around the same time, Bardeen, Carter and Hawking derived laws governing the mechanics of black holes within classical general relativity [2]. These laws have a remarkable similarity with the fundamental laws of thermodynamics. However, the derivation makes no reference to quantum mechanics at all and, within classical general relativity, a relation between the two seems quite implausible: since nothing can come out of black holes and since their interiors are completely inaccessible to outside observers, it would seem that, physically, they can only have zero temperature and infinite entropy. Therefore the similarity was at first thought to be purely mathematical. This viewpoint changed dramatically with Hawking's discovery of black hole evaporation in the following year [3]. Using an external potential approximation, in which the gravitational field is treated classically but matter fields are treated quantum mechanically, Hawking argued that black holes are not black after all! They radiate as if they are black bodies with a temperature equal to $1/2\pi$ times the surface gravity. One can therefore regard the similarity between the laws of black hole mechanics and those of thermodynamics as reflecting physical reality and argue that the entropy of a black hole is given by 1/4-th its area. Thus, Bekenstein's insights turned out to be essentially correct (although

B. R. Iyer and B. Bhawal (eds.), Black Holes, Gravitational Radiation and the Universe, 149–170.

the precise proportionality factors he had suggested were modified).

This flurry of activity in the early to mid seventies provided glimpses of a deep underlying structure. For, in this reasoning, not only do the three pillars of fundamental physics -quantum mechanics, statistical physics and general relativity- come together but coherence of the overall theory seems to *require* that they come together. It was at once recognized that black hole thermodynamics -as the unified picture came to be called- is a powerful hint for the quantum theory of gravity, whose necessity was recognized already in the thirties.

Let us elaborate on this point. The logic of the above argument is as follows: Hawking's calculations, based on *semi-classical* gravity, lead to a precise formula for the temperature which can then be combined with the laws of black hole mechanics, obtained entirely in the *classical* framework, to obtain an expression of the black hole entropy. Is there a more satisfactory treatment? Can one arrive at the expression of entropy from a more fundamental, statistical mechanical consideration, say by counting the number of 'micro-states' that underlie a large black hole? For other physical systems -such as a gas, a magnet, or the radiation field in a black body- to count the micro-states, one has to first identify the elementary building blocks that make up the system. For a gas, these are atoms; for a magnet, the electron spins; and, for radiation in a black body, the photons. What then are the analogous building blocks of a black hole? They *cannot* be gravitons because the gravitational fields under consideration are static. Therefore, these elementary constituents must be essentially non-perturbative in nature. Thus, the challenges for candidate theories of quantum gravity are: i) isolate these constituents; ii) show that the number of quantum states of these constituents which correspond to large black holes in equilibrium (for which the semi-classical results can be trusted) goes as the exponential of the area of the event horizon; iii) account for the Hawking radiation in terms of quantum processes involving these constituents and matter quanta; and, iv) derive the laws of black hole thermodynamics from quantum statistical mechanics.

These are difficult tasks because the very first step-isolating the relevant constituents- requires new conceptual as well as mathematical inputs. It is only recently, more than twenty years after the initial flurry of activity, that detailed proposals have emerged. One comes from string theory [4]where the relevant elementary constituents are certain objects (D-branes) of the non-perturbative sector of the theory[1]. The purpose of this article is to summarize the situation in another approach, which emphasizes the quantum nature of geometry using non-perturbative techniques from the very

[1]This approach is discussed in the article by Wadia in this volume.

beginning. Our elementary constituents are the quantum excitations of geometry itself and the Hawking process now corresponds to the conversion of the quanta of geometry to quanta of matter.

Although the two approaches are strikingly different from one another, they are also complementary. For example, as we will see, we will provide only an effective quantum description of black holes (in the sense that we first isolate the sector of the theory corresponding to isolated black holes and then quantize it). The stringy description, by contrast is intended to be fundamental. Also, in our approach, there is a one-parameter quantization ambiguity which has the effect that the entropy and the temperature are determined only up to a multiplicative constant which cannot be fixed without an additional, semi-classical input. String theory, by contrast, provides the numerical coefficients unambiguously. On the other hand, so far, most of the detailed work in string theory has been focussed on extremal or near extremal black holes which are mathematically very interesting but astrophysically irrelevant. In particular, a direct, detailed treatment of the Schwarzschild black hole is not available. Our approach, by contrast, is not tied in any way to near-extremal situations and can in particular handle the Schwarzschild case easily. Finally, in the stringy approach, since most actual calculations –such as derivation of the Hawking radiation– are carried out in flat space, the relation to the curved black hole geometry is rather unclear. In our approach, one deals directly with the curved black hole geometry.

The article is organized as follows. Section 2 is devoted to preliminaries. We will first introduce the reader to basic properties of quantum geometry in non-perturbative quantum gravity and then briefly summarize a mathematical model called Chern-Simons theory which is needed in the description of the quantum geometry of the horizon. Main results are contained in section 3. We begin by describing the classical sector of the theory corresponding to black holes and then present the quantum description. By counting the relevant micro-states we are led to the expression of entropy. Finally, we indicate how Hawking evaporation can be regarded as a physical process in which the quanta of the horizon area are converted to quanta of matter. As suggested by the Editors, we shall try to communicate the main ideas at a level suitable for a beginning researcher who is not already familiar with the field. Therefore, the technicalities will be kept to a minimum; details can be found in [5, 6, 7, 8] and references cited therein.

2. Preliminaries

2.1. QUANTUM THEORY OF GEOMETRY

In Newtonian physics and special relativity, space-time geometry is regarded as an inert and unchanging backdrop on which particles and fields evolve. This view underwent a dramatic revision in general relativity. Now, geometry encodes gravity thereby becoming a dynamical entity with physical degrees of freedom and Einstein's equations tell us that it is on the same footing as matter. Now, the physics of this century has shown us that matter has constituents and the 3-dimensional objects we perceive as solids are in fact made of atoms. The continuum description of matter is an approximation which succeeds brilliantly in the macroscopic regime but fails hopelessly at the atomic scale. It is therefore natural to ask: Is the same true of geometry? Is the continuum picture of space-time only a 'coarse-grained' approximation which would break down at the Planck scale? Does geometry have constituents at this scale? If so, what are its atoms? Its elementary excitations? In other words, is geometry quantized?

To probe such issues, it is natural to look for hints in the procedures that have been successful in describing matter. Let us begin by asking what we mean by quantization of physical quantities. Take a simple example -the hydrogen atom. In this case, the answer is clear: while the basic observables -energy and angular momentum- take on a continuous range of values classically, in quantum mechanics their eigenvalues are discrete; they are quantized. So, we can ask if the same is true of geometry. Classical geometrical quantities such as lengths, areas and volumes can take on continuous values on the phase space of general relativity. Can one construct the corresponding quantum operators? If so, are their eigenvalues discrete? In this case, we would say that geometry is quantized and the precise eigenvalues and eigenvectors of geometric operators would reveal its detailed microscopic properties.

Thus, it is rather easy to pose the basic questions in a precise fashion. Indeed, they could have been formulated seventy years ago soon after the advent of quantum mechanics. To answer them, however, is not so easy. For, in all of quantum physics, we are accustomed to assuming that there is an underlying classical space-time. One now has to literally 'step outside' space-time and begin the analysis afresh. To investigate if geometrical observables have discrete eigenvalues, it is simply inappropriate to begin with a classical space-time, where the values are necessarily continuous, and *then* add quantum corrections to it[2]. One has to adopt an essentially

[2]This would be analogous to analyzing the issue of whether the energy levels of a harmonic oscillator are discrete by performing a perturbative analysis starting with a free particle.

non-perturbative approach to quantum gravity. Put differently, to probe the nature of quantum geometry, one should not begin by *assuming* the validity of the continuum picture; the quantum theory itself has to tell us if the picture is adequate and, if it is not, lead us to the correct microscopic model of geometry.

Over the past decade, a non-perturbative approach has been developed to probe the nature of quantum geometry and to address issues related to the quantum dynamics of gravity. The strategy here is the opposite of that followed in perturbative treatments: Rather than starting with quantum matter on classical space-times, one first quantizes geometry and then incorporates matter. This procedure is motivated by two considerations. The first comes from general relativity in which some of the simplest and most interesting physical systems –black holes and gravitational waves– consist of 'pure geometry'. The second comes from quantum field theory where the occurrence of ultraviolet divergences suggests that it may be physically incorrect to quantize matter assuming that space-time can be regarded as a smooth continuum at arbitrarily small scales.

The main ideas underlying this approach can be summarized as follows. One begins by reformulating general relativity as a dynamical theory of connections, rather than metrics[3] [9]. This shift of view does not change the theory classically (although it suggests extensions of general relativity to situations in which the metric may become degenerate). However, it makes the kinematics of general relativity the same as that of $SU(2)$ Yang-Mills theory, thereby suggesting new non-perturbative routes to quantization. Specifically, as in gauge theories, the configuration variable of general relativity is now an $SU(2)$ connection A_a^i on a spatial 3-manifold and the canonically conjugate momentum E_i^a is analogous to the Yang-Mills 'electric' field. However, physically, we can now identify this electric field as a triad; it carries all the information about spatial geometry. In quantum theory, it is natural to use the gauge invariant Wilson loop functionals, $\mathcal{P} \exp \oint_\alpha A$, i.e., the path ordered exponentials of the connections around closed loops α as the basic objects [10, 11]. The resulting framework is often called 'loop quantum gravity'.

In quantum theory, states are suitable functions of the configuration variables. In our case then, they should be suitable functions $\Psi(A)$ of connections. In field theories, the problem of defining 'suitable', i.e., of singling out normalizable states, is generally very difficult and involves intricate

[3]As is often the case, it was later realized that this general idea is not as heretic as it seems at first. Indeed, already in the forties both Einstein and Schrödinger had given such reformulations. However, while they used the Levi-Civita connections, the present approach is based on chiral spin connections. This shift is essential to simplify the field equations and to bring out the kinematical similarity with Yang-Mills theory which in turn is essential for the present treatment of quantum geometry.

functional analysis. In our case, our space-time has no background metric (or any other field). This makes the problem more difficult because the standard methods from Minkowskian field theories cannot be taken over. However, although it is surprising at first, it also makes the problem easier. For, the structures now have to be invariant under the entire diffeomorphism group -rather than just the Poincaré group- and, since the invariance group is so large, the available choices are greatly reduced. This advantage has been exploited very effectively by various authors to develop a new functional calculus on the space of connections. The integral calculus is then used to specify the (kinematical) Hilbert space of states and, differential calculus, to define physically interesting operators, including those corresponding to geometrical observables mentioned above.

For our purposes, the final results can be summarized as follows. Denote by γ a graph in the 3-manifold under consideration with E edges and V vertices. (For readers familiar with lattice gauge theory, γ can be regarded as a 'floating' lattice; 'floating' because it is not necessarily rectangular. Indeed, we don't even know what 'rectangular' means because there is no background metric.) It turns out that the total Hilbert space $\mathcal{H}$ can be decomposed into orthogonal, *finite dimensional* subspaces $\mathcal{H}_{\gamma,\vec{j}}$, where $\vec{j} \equiv \{j_1, \ldots, j_E\}$ stands for an assignment of a non-zero half integer ('spin') -or, more precisely, a non-trivial irreducible representation of $SU(2)$- to each of the edges [12]. To specify a state in $\mathcal{H}_{\gamma,\vec{j}}$, one only has to fix an intertwiner at each vertex which maps the incoming representations at that vertex to the outgoing ones.[4] (For a tri-valent vertex, i.e., one at which precisely three edges meet, this amounts to specifying a Clebsh-Gordon coefficient associated with the j's associated with the three edges.) Each resulting state is referred to as a *spin network* state [13]. The availability of this decomposition of $\mathcal{H}$ in to finite dimensional sub-spaces is a powerful technical simplification since it effectively reduces many calculations in quantum gravity to those involving simple spin systems.

Since the edges of a graph are one dimensional, the resulting quantum geometry is polymer-like and, at the Planck scale, the continuum picture fails completely. It emerges only semi-classically. Recall that a polymer, although one dimensional at a fundamental level, exhibits properties of a three dimensional system when it is in a sufficiently close-knit state.

[4]Given such a specification, we can define a function Ψ on the space of connections as follows. A connection A assigns to each edge e_I an $SU(2)$ element g_I via parallel transport. Consider the $(2j_I+1) \times (2j_I+1)$ matrix M_I associated with g_I in the j_I-th representation. Thus, with each edge we can now associate a matrix. The intertwiners 'tie' the row and column labels of these matrices M_I appropriately to produce a number. This is the value of Ψ on the connection A. Note that $\mathcal{H}_{\gamma,\vec{j}}$ is finite dimensional simply because the space of intertwiners compatible with any spin-assignment $\vec{j}$ on the edges of a graph γ is finite dimensional.

Similarly, a state in which the fundamental one-dimensional excitations of quantum geometry are densely packed in a sufficiently complex configuration can approximate a three dimensional continuum geometry. Individual excitations are as far removed from a classical geometry as an individual photon is from a classical Maxwell field. Nonetheless, these elementary excitations do have a direct physical interpretation. Each edge, carrying a label $j = \frac{1}{2}$ can be regarded as a flux line carrying an 'elementary area': In this state, each surface which intersects only the given edge of γ has an area proportional to ℓ_P^2, where ℓ_P is the Planck length. Thus, given a state in $\mathcal{H}_{\gamma,\vec{j}}$, quantum areas of surfaces are concentrated at points in which they intersect the edges of γ. Similarly, volumes of regions are concentrated at vertices of γ. The microscopic geometry is thus distributional in a precise sense.

In the classical theory, it is the triads that encode all the information about Riemannian geometry. The same is true in the quantum theory. The duals, $\Sigma^i_{ab} = \epsilon_{abc}E^c_i$, of triads are 2-forms and one can show that, when smeared over two dimensional surfaces, the corresponding operators are well-defined and self-adjoint on the Hilbert space $\mathcal{H}$ [14]. Thus, in the technical jargon, triads are operator-valued distributions in two dimensions. Various geometric operators can be constructed rigorously by regularizing the appropriate functions of these triad operators [15, 14, 16, 17, 18]. Each is a self-adjoint operator and one can show that they all have *purely discrete* spectra. Thus, geometry is quantized *in the same sense* that energy and angular momentum are quantized in the hydrogen atom. Properties of these geometric operators have been studied extensively. However, for our purposes, it will suffice to focus only on certain properties of the area operators. We will conclude this section with a brief discussion of these properties.

The *complete* spectrum of area operators is known. The minimum eigenvalue is of course zero. However, the value of the next eigenvalue –the 'area-gap'– is non-trivial; it depends on the topology of the surface. Thus, it is interesting that quantum geometry 'knows' about topology. Although the spectrum is purely discrete, the eigenvalues a_S crowd rather rapidly: for large eigenvalues a_S, the gap Δa_S between a_S and the next eigenvalue goes as $\Delta a_S \leq \exp(-\sqrt{a_S}/\ell_P)$. It is because of this crowding that one can hope to reach the correct continuum limit. More precisely, the detailed behavior of the spectrum is important to recover the correct semi-classical physics. This may seem strange at first: Because the Planck length ℓ_P is so small, one might have thought that even if the spacing between area eigenvalues were uniform, say $\Delta a_S = \ell_P^2$, one would recover the correct semi-classical physics. Detailed analysis shows that is not the case. With an uniform level spacing, for example, one would not recover the Hawking spectrum even for a large black-hole [19]. To obtain even a qualitative agreement with the semi-classical results it is necessary that the eigenvalues crowd and the specific exponential crowding one finds in this approach is also sufficient [14, 20]. Finally, let us describe the spectrum. For our purposes, it is sufficient to display only the eigenvalues that result when edges of a spin network state intersect the surface S under consideration transversely. These are given by [14, 15, 21]:

$$A_S = 8\pi\gamma\ell_P^2 \sum_p \sqrt{j_p(j_p+1)}, \tag{1}$$

where the sum is taken over all points p where edges of a spin network state intersect S, j_p are spin that label the intersecting edges, and γ is an undetermined real number, known as the Immirzi parameter. Thus, the eigenvalues have an ambiguity of an overall multiplicative factor γ which arises from an inherent quantization ambiguity in loop quantum gravity. As mentioned in the Introduction, to fix this ambiguity, one needs additional input, e.g., from semi-classical physics.

2.2. CHERN-SIMONS THEORY

It turns out that our analysis of black hole thermodynamics will require, as a technical ingredient, certain results from a three-dimensional topological field theory known as the Chern-Simons theory. In this sub-section we shall recall these results very briefly.

In this theory, the only dynamical variable is a connection one-form A, which for our purposes can be assumed to take values in the Lie algebra of SU(2). The theory is 'topological' in the sense that it does not need or involve a background (or dynamical) metric. Fix an oriented three manifold Δ (which in our application will be a suitable portion of the black hole horizon), and consider connections A on an SU(2) bundle over it. The action of the theory is given by:

$$S_{CS} = \frac{k}{4\pi}\int_\Delta \mathrm{Tr}\left(A\wedge dA + \frac{2}{3}A\wedge A\wedge A\right). \tag{2}$$

where κ is a coupling constant, also known as the 'level' of the theory, and the trace is taken in the fundamental representation of SU(2). The field equations obtained from the variation of this action simply say that the connection A is flat on Δ. Such connections can still be non-trivial if the manifold Δ has a non-trivial topology. An especially interesting case arises when Δ has a topology $S \times R$ where S is a two-manifold with punctures, i.e., points removed. In this case, holonomies can be non-trivial around punctures; i.e., we can have a distributional non-trivial curvature which 'resides at the deleted points', even though on points included in S, the curvature vanishes everywhere. We will encounter this situation in the next section.

There are several ways to quantize Chern-Simons theory. For our purposes we will need some facts about its canonical quantization. Here, one has to first cast the theory into the Hamiltonian framework. This is achieved by carrying out the canonical (2+1) decomposition of the action, assuming that Δ has the topology of $S \times \mathbb{R}$, where S is a two-dimensional manifold. The phase space consists only of pull-back of A on S; unlike the Yang-Mills theory and general relativity, there are no 'electric fields'. Thus, the components of the connection do not Poisson commute:

$$\left\{A_a^i(x), A_b^j(y)\right\} = \frac{4\pi}{k}\, \varepsilon_{ab} K^{ij} \delta^2(x-y), \tag{3}$$

where K^{ij} is the Cartan-Killing metric on $SU(2)$, and ϵ_{ab} is the (metric-independent) Levi-Civita density on S. As in all generally covariant theories, the Hamiltonian turns out to be a constraint

$$F_{ab} = 0, \tag{4}$$

where F_{ab} is the pull-back of the curvature of A on S.

One is often interested in the quantization of this phase space in the situation in which the two-manifold S has punctures. Then, as mentioned above, even though F_{ab} vanishes on S, the holonomies around these punctures can be non-zero; intuitively, one can now regard the curvature as being 'distributional, residing at the missing points'. To quantize the system, therefore, one has to provide certain additional information –quantum numbers– at these punctures. The resulting Hilbert space of states turns out to be finite dimensional, and the dimension depends on the quantum numbers labelling punctures. (See, e.g., [22].)

3. Application to Black Holes

We are now ready to examine black hole thermodynamics from a fundamental perspective of non-perturbative quantum gravity. We will summarize the overall situation following a systematic approach developed in [5, 7, 8]. (For earlier work, see, e.g., [23].)

In general relativity, supergravity or indeed in any modern classical theory of gravity, using causal structure one can define what one means by a black hole in full generality, without having to restrict oneself to any symmetries. With this notion at hand, one can then restrict attention to specific contexts, such as stationary or axi-symmetric situations. In quantum theory, the situation is not as satisfactory: In none of the approaches available today, does one have an unambiguous notion of a general, quantum black hole. Therefore, within each approach, a strategy is devised to circumvent this problem. The strategy of the present approach is the following: We will first pick out the sector of the classical theory that contains isolated black holes –the analogs of 'equilibrium states' in thermodynamics– and quantize that sector. This will provide an effective description. While this line of attack is sufficient for the analysis of black hole entropy and provides an avenue to understanding the Hawking radiation and laws of black hole mechanics from the perspective of quantum gravity, it is far from being fully satisfactory. For example, since we *begin* with a black hole sector of the classical theory, in a certain sense, the phase of the Hawking process in which the black hole has fully evaporated cannot be encompassed in this approach. Consequently, as it stands, the approach is unsuitable for a comprehensive analysis of issues, such as 'information loss', related to the final stages of the evaporation process. Questions it is best suited to address are of the type: "Given that a space-time contains a black hole of a certain size, what is the associated statistical mechanical entropy and what is the spectrum of the Hawking radiation?" In a more fundamental approach, one would first construct the full theory of quantum gravity, single out, among solutions to the quantum field equations, those states which are to represent a quantum black hole, and analyze their physical properties.

So far, the detailed analysis exists only for non-rotating black holes possibly with electric and dilatonic charges. However, it is expected that the main features of the analysis will carry over also to the rotating case. For simplicity, in the main discussion, we will focus on uncharged black holes and comment on the charged case at the end.

3.1. PHASE SPACE

To single out the appropriate sector of the theory representing isolated black holes, we will consider asymptotically flat space-times with an interior boundary (the horizon) on which the gravitational fields satisfy suitable boundary conditions. These conditions should be strong enough to capture the physical situation we have in mind but also weak enough to allow a large number of space-times. A detailed discussion of the appropriate boundary conditions can be found in [7]. Here, we will only present the underlying ideas.

To motivate our choice of the boundary conditions, let us consider the space-time formed by a collapsing star (see Fig. 1). At some moment, the event horizon $\mathcal{H}$ forms. A part of the gravitational radiation emitted in the process falls in to the horizon along with the matter and the rest escapes to infinity. It is generally expected that the black hole would finally settle

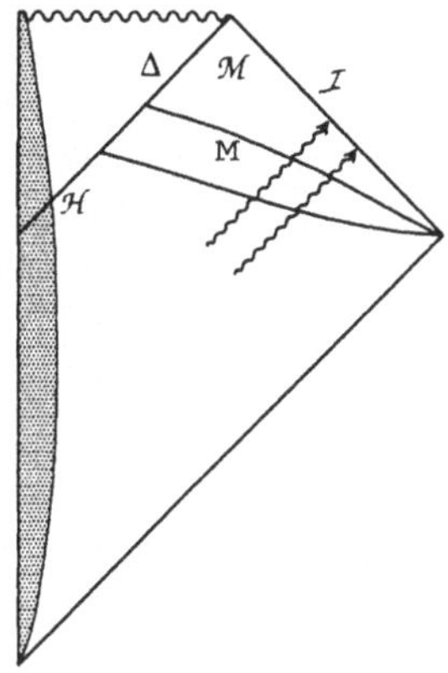

Figure 1. Space-times of interest.

down. Just as in the calculation of entropy in thermodynamics one generally deals with systems that have reached equilibrium, here we will be primarily interested in the final phase in which the black hole has settled down. We will make an idealization and assume that after a certain retarded time no further matter or radiation falls in to the black hole –the black hole is isolated– and focus on the portion Δ of the event horizon to the future of this retarded time. Thus, in this region, the area of any two-sphere cross-section of the horizon will be constant, which we will denote by A_H.

Let us focus on the portion $\mathcal{M}$ of space-time, bounded by Δ and future null infinity $\mathcal{I}$, which can be regarded as the space-time associated with the *isolated* black hole under consideration. For simplicity, we will assume that the vacuum Einstein's equations hold on $\mathcal{M}$, i.e. that all the matter has fallen in to the event horizon in the past of Δ. Note that, as the figure shows, there could still be gravitational waves in $\mathcal{M}$ which escape to $\mathcal{I}$. It is these space-time regions $\mathcal{M}$ that constitute our sector of general relativity corresponding to isolated black holes.

This sector can be isolated technically by specifying boundary conditions which capture the intuitive picture spelled out above [5, 7]. It turns out, however, that, with our choice of boundary conditions at Δ, the standard general relativity action in the connection variables is no longer functionally differentiable: The variation of the action with respect to the connection contains a non-vanishing boundary term at Δ. However, it turns out to be possible to add to the action a boundary term, whose variation exactly cancels the boundary term arising in the variation of 'bulk' action, thus making the total action differentiable. Furthermore, the boundary term turns out to be precisely the Chern-Simons action (2), where the coupling constant κ is now given by: $\kappa = A_H/8\pi\gamma G$. It is quite surprising that there is such a nice interplay between general relativity, the boundary conditions for isolated black holes and the Chern-Simons theory.

To canonically quantize the resulting theory, one has to cast it first in the Hamiltonian form. This can be done in a standard fashion. As in the case without black holes [9], the basic canonical variables are the pull-backs A_a^i of the gravitational connection to a spatial slice, such as M in Fig. 1, and the triads E_i^a on M. However, these fields are now required not only to be asymptotically flat but also to satisfy appropriate boundary conditions on the two-sphere boundary S of M, i.e., the intersection of M and $\mathcal{H}$, which are induced by our boundary conditions on space-time fields. First, the area of S defined by the triads E_i^a is a fixed constant, A_H. Second, the triads and the connections are intertwined by the relation

$$\underline{F}_{ab}^i = -\frac{2\pi\gamma}{A_H}\,\underline{\Sigma}_{ab}^i\,, \tag{5}$$

where, as before, Σ_{ab}^i are the two-forms dual to the triads, F_{ab}^i is the curvature of the connection A_a^i on M and where the under-bars denote pull-backs to the two-sphere boundary S of M. Finally, the boundary conditions also imply that the pull-back of the connection A_a^i to the boundary is reducible, i.e., the (based) holonomies of A_a^i along closed loops within S necessarily belong to an $U(1)$ sub-group of $SU(2)$. Although it is not necessary, for technical simplicity we can fix an internal vector field, r^i, on S along which the curvature of the pulled-back connection is restricted to lie, and partially gauge fix the system. Then the gauge group on the boundary reduces from SU(2) to U(1). Furthermore, now only the r component of (5) is nontrivial.

As one might expect, due to the addition of the surface term to the action, the fundamental Poisson brackets are modified; the symplectic structure also acquires a surface term. This new contribution is just the symplectic structure of the Chern-Simons theory on Δ, where the coupling constant κ is now given by

$$k = \frac{A_H}{8\pi\gamma G}. \tag{6}$$

Note that, up to a numerical coefficient, k is simply the area of the horizon of black hole measured in the units of Planck area ℓ_P^2.

The phase space consisting of the pairs (A, E) satisfying our boundary conditions is infinite dimensional. It is clear, however, that not all the degrees of freedom described by fields A, E are relevant to the problem of black hole entropy. In particular, there are 'volume' degrees of freedom in the theory corresponding to gravitational radiation propagating out to $\mathcal{I}$, which should not be taken into account as genuine black hole degrees of freedom. The 'surface' degrees of freedom describing the geometry of the horizon S have a different status. It has often been argued that the degrees of freedom living on the horizon of black hole are those that account for its entropy. We take this viewpoint in our approach.

Note, however, that in the classical theory that we have described, the volume and surface degrees of freedom cannot be separated: all fields on S are determined by fields in the interior of M by continuity. Thus, strictly speaking, classically there are no independent surface degrees of freedom. However, as we described in subsection 2.1, in the quantum theory the fields describing geometry become distributional, and the fields on S are no longer determined by fields in M; now there are independent degrees of freedom 'living' on the boundary. This striking difference arises precisely because distributional configurations dominate in non-perturbative treatments of field theories and would be lost in heuristic treatments that deal only with smooth fields. Furthermore, as we will see in the next sub-section, it is precisely these quantum mechanical surface degrees of freedom that account for the black hole entropy!

3.2. QUANTIZATION AND ENTROPY

To quantize the theory we proceed as follows. Since in the quantum theory the volume and surface degrees of freedom become independent, we can first quantize volume degrees of freedom by using the well-established techniques of loop quantum gravity summarized in section 2.1. We are interested only in those bulk states which endow the boundary S with an area close to A_H. Boundary conditions (5) then imply that, for each such bulk state, only certain surface states are permissible. To calculate entropy, one then has, in essence, just to count the states satisfying these constraints.

More precisely, the situation is as follows. As discussed in subsection 2.1, in the bulk, we have a 'polymer geometry' with one dimensional excitations. That is, a typical state is associated with a graph γ and the corresponding function $\Psi(A)$ of connections is sensitive only to what the connections do along the edges of γ. Let us take one such state and denote by $p_1, \ldots, p_n$ the intersection points of the edges of γ with the boundary S (see Fig. 2). For such a state to be admissible as a micro-state of our large isolated black-hole –i.e., to belong to 'the quantum micro-canonical ensemble' underlying the classical black hole– two constraints have to be met at the boundary S: i) the quantum analog of the boundary condition (5) has to be satisfied; and, ii) the area assigned by the state to S has to lie in the range given by $(A_H - \ell_P^2,\, A_H + \ell_P^2)$.

As indicated in section 3.1, only the r component of (5) is non-trivial. Hence, in the quantum theory, our states have to satisfy

$$\left(\underline{\hat{F}}^{i}_{ab} r_i + \frac{2\pi\gamma}{A_H}\,\underline{\hat{\Sigma}}^{i}_{ab} r_i\right) \cdot \Psi = 0. \tag{7}$$

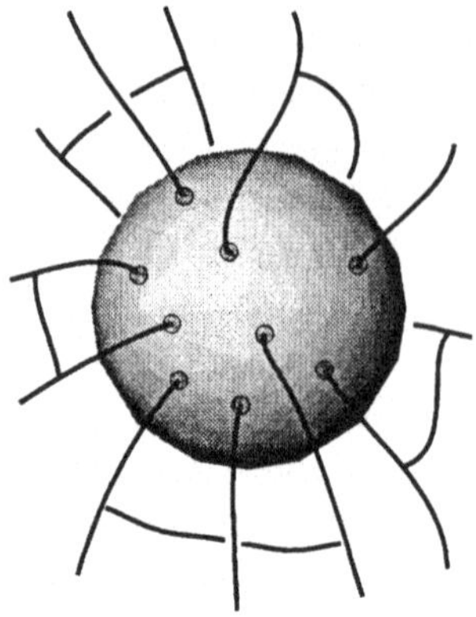

Figure 2. The flux lines of gravitational field pierce the black hole horizon and excite curvature degrees of freedom on the surface. These excitations are described by Chern-Simons theory and account for the black hole entropy in our approach.

Let us first consider $\hat{\underline{\Sigma}}$, the dual of the triad operator. As indicated in section 2.1, this operator is distributional and, on the boundary S, the result of its action turns out to be concentrated precisely at the points $p_1, \ldots, p_n$ where the edges of γ intersect S. Hence, the boundary condition (7) implies that in the quantum theory the state Ψ has a non-trivial dependence only on those connections $\underline{A}$ on S which are flat everywhere except the points $p_1, \ldots, p_n$, where its curvature is distributional. Recall that the Poisson brackets between the connections $\underline{A}$ are precisely those of the Chern-Simons theory. Thus the 'surface part' of the state is precisely a Chern-Simons quantum state associated with the two sphere S with punctures $p_1, \ldots, p_n$. To summarize, the 'volume part' of the state encodes a polymer-type geometry in the bulk, while the surface part is a Chern-Simons state, the two being 'locked together' by (7).

Let us now turn to the second constraint on the states, namely the requirement on area assigned to S by these states. To count the number of states satisfying this constraint, let us examine the eigenstates of the area operator which intersect S, as above, in n points $p_1, \ldots, p_n$. These are precisely the spin network states introduced in section 2.1. Therefore, each intersection point p now inherits a spin label j_p from the edge that intersects S in p. For notational simplicity, from now on, we will refer to the pair (p, j_p) as a puncture and often denote it simply by j_p. The area constraint tells us that we should restrict our attention to the sets $\mathcal{P}$ of punctures,

$$\mathcal{P} = \{j_{p_1}, \ldots, j_{p_n}\}$$

for which the area eigenvalue (associated to S by the corresponding spin-network state)

$$a_{\mathcal{P}} = 8\pi\gamma\ell_P^2 \sum_p \sqrt{j_p(j_p+1)}$$

lies in the interval $A_H - \ell_P^2 \leq a_{\mathcal{P}} \leq A_H + \ell_P^2$. We will refer to these $\mathcal{P}$ as the permissible sets of punctures.

Now, each set $\mathcal{P}$ of punctures gives rise to a Hilbert space of Chern-Simons quantum states of the connection on S. Denote it by $\mathcal{H}_{\mathcal{P}}$. These are the surface states which are 'compatible' with the given permissible set $\mathcal{P}$ of punctures. Now, it follows from standard results in Chern-Simons theory that, for a large number of punctures, the dimension of $\mathcal{H}_{\mathcal{P}}$, goes as

$$\dim\mathcal{H}_{\mathcal{P}} \sim \prod_{j_p \in \mathcal{P}} (2j_p + 1). \tag{8}$$

Intuitively it is obvious that to calculate the entropy it suffices to add up these dimensions, i.e. that the entropy S_{bh} is simply

$$S_{\text{bh}} = \ln \sum_{\mathcal{P}} \dim\left(H_{\mathcal{P}}\right) \tag{9}$$

, where the sum extends over permissible sets of punctures.[5] It is rather straightforward to calculate the sum for large A_H:

$$S_{\text{bh}} = \frac{\gamma_0}{4\ell_P^2\gamma} A_H, \quad \text{where} \quad \gamma_0 = \frac{\ln 2}{\pi\sqrt{3}}.$$

(The appearance of γ can be traced back directly to the formula for the eigenvalues of the area operator, (1)). Thus, in the limit of large area, the entropy is proportional to the area of the horizon. If we fix the quantization ambiguity by setting the value of the Immirzi parameter γ to the numerical constant γ_0 (which is of the order of 1), then the statistical mechanical entropy is given precisely by the Bekenstein-Hawking formula.

Are there independent checks on this preferred value of γ? The answer is in the affirmative. One can carry out this calculation for Reissner-Nordstrom as well as dilatonic black holes. A priori it could have happened that, to obtain the Bekenstein-Hawking value, one would have to re-adjust

[5] More formally, one can proceed as follows. The total Hilbert space carries information about *both*, the bulk and the surface degrees of freedom. We are not interested in this full space since it includes, e.g., states of gravitational waves far away from Δ. Rather, we wish to consider only states of the horizon of a black hole with area A_H. Thus we have to trace over the 'volume' states to construct an density matrix ρ_{bh} describing a maximal-entropy mixture of surface states for which the area of the horizon lies in the range $A_H \pm \ell_P^2$. The statistical mechanical black hole entropy is then given by $S_{\text{bh}} = -\text{Tr}\rho_{\text{bh}} \ln \rho_{\text{bh}}$. As usual, this can be computed simply by counting states, i.e., the right side of (9).

the Immirzi parameter for each value of the electric or dilatonic charge. This does *not* happen. The entropy is still given by (10) and hence by the Bekenstein-Hawking value when $\gamma = \gamma_0$. The way in which the details work out shows that this is quite a non-trivial check.

We conclude with two remarks.

1. An intuitive way to think about the quantum states that underlie an isolated black hole is as follows. As indicated in section 2.1, the edges of spin networks can be thought of as flux lines carrying area. These flux lines endow the horizon S with its area and also pin it (see Fig. 2), exciting the curvature degrees of freedom at the punctures via Eq.(7). With each given configuration of flux lines, there is a finite dimensional Hilbert space $\mathcal{H}_\mathcal{P}$ describing the quantum states associated with these curvature excitations. Heuristically, one can picture the horizon S as a pinned balloon and regard these surface degrees of freedom as describing the 'oscillatory modes' compatible with the pinning.

2. A detailed calculation of the black hole entropy shows that the states that dominate the counting correspond to punctures all of which have labels $j = 1/2$. For, in this configuration, the number of punctures needed to approximate a given A_H is the largest. One can thus visualize each micro-state as providing a yes-no decision or 'an elementary bit' of information. It is very striking that this picture coincides with the one advocated by John Wheeler in his "It from Bits" scenario [24]. Thus, in a certain sense, our analysis can be regarded as concrete, detailed realization of those qualitative ideas.

3.3. BLACK HOLE RADIANCE

Since the number of micro-states of a large black hole grows as the exponential of its area in Planck units, a quantum black hole is system with an astonishingly large number of micro-states.[6] Hence, the statistical mechanical approximations one normally makes, e.g., in treating the emission spectra in atomic physics, are satisfied here very easily. Over twenty years ago, therefore, Bekenstein [25] used analogies from atomic physics to develop a strategy for analyzing black hole radiance from microscopic considerations. Most of the current work in this area uses this strategy in one form or another. Our approach follows this trend.

The mechanism responsible for the radiance in our approach is based on the following qualitative picture. Consider the micro-states of a large

[6]For example, a solar mass black hole has approximately $\exp 10^{92}$ micro-states! Not only is this number extraordinarily large compared to $\exp 10^{23}$, the number we come across in standard statistical systems, but it is very large even on astronomical scales. For example, black body radiation with one solar mass energy and at the same temperature as the sun has about $\exp 10^{60}$ micro-states.

black hole introduced in section 3.2, and let the black hole be initially in an eigenstate $|\Gamma\rangle$ of the horizon area operator. Radiation is emitted when the black hole jumps to a nearby state $|\Gamma'\rangle$ with a slightly smaller area. This change in area corresponds to a change $\Delta M_{\Gamma\Gamma'}$ in the mass of the black-hole which gets radiated away. Suppose, for simplicity, that the emitted particle is of rest mass zero. Then, the frequency $\omega_{\Gamma\Gamma'}$ of the particle when it reaches infinity is given by $\Delta M_{\Gamma\Gamma'} = \hbar\omega_{\Gamma\Gamma'}$. Since the area spectrum (1) is purely discrete, the black-hole spectrum is also discrete at a fundamental level: the emission lines occur at specific discrete frequencies. At first one might worry [19] that such a spectrum would be very different from the continuous, black-body spectrum derived by Hawking using semi-classical considerations. However, as indicated in section 2.1, because the level spacing between the eigenvalues of the area operator decreases exponentially for large areas, the separation between the spectral lines can be so small that the spectrum can be well-approximated by a continuous profile [14, 20].

To determine the intensities of spectral lines and hence the form of the emission spectrum, as in atomic physics, we can use Fermi's golden rule. Thus, the probability of a transition $\Gamma \to \Gamma'$ with the emission of a quantum of radiation is given by

$$W_{\Gamma\to\Gamma'} = \frac{2\pi}{\hbar}\,|V_{\Gamma\Gamma'}|^2\,\delta(\omega - \omega_{\Gamma\Gamma'})\frac{\omega^2 d\omega d\Omega}{(2\pi\hbar)^3}, \tag{10}$$

where $V_{\Gamma\Gamma'}$ is the matrix element of the part of the Hamiltonian of the system that is responsible for the transition, $d\Omega$ is the element of the solid angle in the direction in which the quantum is emitted, and $\omega_{\Gamma\Gamma'}$ is the frequency of the quantum. (For brevity, we have suppressed the dependence of this matrix element on initial and final states of the quantum field describing the radiation.) The total energy dI emitted by the system per unit time in transitions of this particular type can be obtained simply by multiplying this probability by $\hbar\omega$ times the probability $\Pr(\Gamma)$ to find the system in the initial state $|\Gamma\rangle$:

$$dI(\Gamma\to\Gamma') = 2\pi\omega\ \Pr(\Gamma)\ |V_{\Gamma\Gamma'}|^2\,\delta(\omega - \omega_{\Gamma\Gamma'})\frac{\omega^2 d\omega d\Omega}{(2\pi\hbar)^3} \tag{11}$$

The question for us is: For quantum black holes, how do these intensity distributions compare with those of a black-body? To answer the question, one has to calculate the probability distribution $\Pr(\Gamma)$ of the black hole, and the matrix elements $|V_{\Gamma\Gamma'}|$ of the Hamiltonian responsible for transitions. Recall, however, that, in atomic physics, the general form of the spectrum is usually determined by the probability distribution of initial states and by such qualitative aspects of the underlying dynamics as selection rules for quantum transitions. The detailed knowledge of the matrix elements is needed only to determine the finer details of the spectrum. The situation for quantum black holes is analogous.

The overall picture can be summarized as follows. For a large black hole, since the typical energies of the emitted particles are negligible compared to the black hole mass, it is reasonable to appeal to basic principles of equilibrium statistical mechanics and conclude that all accessible micro-states occur with equal probability. Now, the entropy S goes as the logarithm of the number of states. Hence, the probability for occurrence of any one permissible micro-state $|\Gamma\rangle$ is $\exp(-S)$. General considerations involving Einstein's A and B coefficients now imply that the mean number $n_{\omega lm}$ of quanta of frequency ω and angular momentum quantum numbers l, m emitted by the black hole per unit time is given by

$$n_{\omega lm} = \frac{\sigma_{\omega lm}}{e^{\hbar\omega/T} - 1}, \tag{12}$$

where, T is the thermodynamic temperature,

$$\frac{1}{T} = \frac{dS}{dM} \tag{13}$$

and $\sigma_{\omega lm}$ is the absorption cross-section of the black hole in the mode ωlm. (See, e.g., [25, 6]).

Although we have phrased the argument in terms of black holes –system now under consideration– most of the reasoning used so far is quite general. In the case of the black-hole, we know further that the entropy S is given by $A_H/4$, which fixes the thermodynamic temperature in terms of the parameters of the black hole. This turns out to be precisely the temperature that Hawking was led to associate, in the external potential approximation, with the spontaneously emitted radiation. Recall that the initial semi-classical reasoning had an unsatisfactory feature: the Hawking temperature is the property of the emission spectrum at infinity and it is not a priori clear that it is related to the thermodynamic properties of the black hole. The reasoning given above, based on the micro-states of the quantum black hole itself, fills this gap; the temperature in the emission spectrum is indeed the statistical mechanical temperature of the black hole.

Finally, we can obtain the intensity spectrum. Since the s-waves dominate in the emission from non-rotating black holes, we can focus only on the $l = 0$ modes. Then, to obtain the intensity spectrum, one has to multiply $n_{\omega,l=0}$ by the energy $\hbar\omega$ and the density of states $\omega^2 d\omega d\Omega$ in the mode, and integrate out the angular dependence. The result is:

$$I(\omega)d\omega \sim \frac{\hbar\omega^3\sigma_\omega\, d\omega}{e^{\hbar\omega/T-1}}. \tag{14}$$

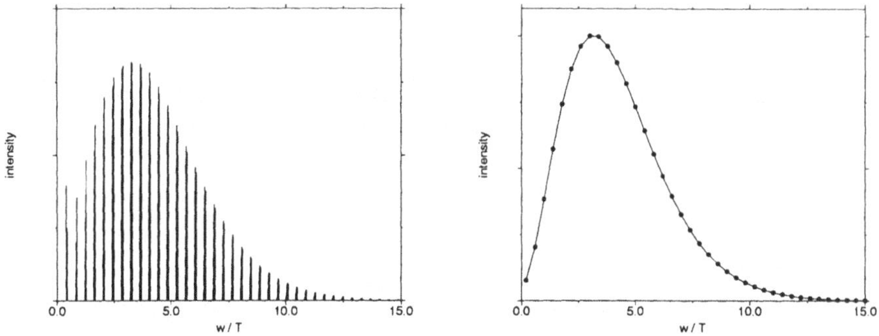

Figure 3. Emission lines with their intensities (left) and the total intensity of the radiation emitted per frequency interval as a function of frequency (right).

Thus, very general arguments lead us to the result that the emission spectrum of the black hole is thermal, in the sense that it has the form (14). The problem now reduces to that of calculating the absorption cross-section σ_ω from first principles. Since the Hawking analysis is semi-classical, there it was consistent to use the classical value of this cross-section. In full theory, on the other hand, we need to compute it quantum mechanically. Thus, the remaining question is: Does the cross-section calculated in the full quantum theory agree with the classical result? If so, one would have a derivation of Hawking radiation from fundamental considerations.

Now, in the classical theory, the cross-section σ_ω in Eq. (14) is an approximately constant function of ω (equal to the horizon area A_H). In our quantum treatment, on the other hand, there is an obvious potential problem. To see this, recall first from section 3.2 that the most likely micro-states are the ones in which each puncture is labelled by $j = \frac{1}{2}$. Indeed they dominate the distribution in the sense that they already provide the leading order contribution to the entropy. Hence, one would expect that, transitions $j = 0 \to j = \frac{1}{2}$ would completely dominate the absorption cross-section. In this case, σ_ω would be peaked extremely sharply at a single frequency and the final spectrum (14) would therefore look *very* different from Hawking's. However, this transition is simply forbidden by a selection rule! More precisely, if the interaction Hamiltonian responsible for the transitions is gauge invariant -which it must be, on general physical grounds- its matrix element between the initial and final states of the above type is identically zero. The situation here is again analogous to that in atomic physics where the selection rules are obtained simply by examining the transformation properties of the interaction Hamiltonian under the rotation group.

With this obstruction out of the way, one can proceed with the calculation of the quantum absorption cross-section. Now one can argue that, for a large class of Hamiltonians, the quantum absorption cross-section is close to the classical one [6]. For instance, if for allowed transitions the dependence on Γ and Γ' of $V_{\Gamma,\Gamma'}$ is negligible, the intensity spectrum is given by Fig. 3. In the first plot, the intensities of lines are computed using Eq.(10). (Here, the overall multiplicative factors are neglected; hence there are no units on the y axis.) The second plot is obtained by first dividing the frequency range into small intervals and then adding the intensities of all the lines belonging to the same interval. It brings out the fact that, although the emission occurs only at certain discrete frequencies, the enveloping curve of the spectrum is thermal.

4. Discussion

The key ideas underlying our approach can be summarized as follows. In the spirit of equilibrium statistical mechanics, we consider large, isolated black holes (in four space-time dimensions). When appropriate boundary conditions at the horizon are imposed, we find that the bulk action of Einstein's theory has to be supplemented by a surface term at the horizon which is precisely the Chern-Simons action. We then quantize the resulting sector of the theory. As one might expect from the general structure of quantum field theories, the fields describing geometry become distributional in the quantum theory. Furthermore, our background independent functional calculus tells us that the fundamental quantum excitations are of a specific type: they are one-dimensional, like polymers. The horizon acquires its area from the points where these one-dimensional excitations pierce it transversely. Associated with each bulk state which endows the horizon with a given area, there are 'compatible' surface states on the horizon itself which come from the quantization of Chern-Simons theory thereon. The total number of these 'compatible' Chern-Simons states grows exponentially with the area, whence the entropy is proportional to the area of the horizon. Finally, using Fermi's golden rule, one can compute the probabilities for transitions among these horizon states. Transitions that decrease the area are accompanied by an emission of particles. From rather general considerations, one can conclude that the emission spectrum is thermal, in agreement with Hawking's semi-classical calculations.

At its core, this is a rather simple and attractive picture of quantum black holes. However, it is far from being complete and further work is being carried out in a number of directions. First, as in other approaches, the calculations leading to the Hawking spectrum are based on a number of simplifying assumptions. Although by and large these are physically motivated, it is important to make sure that the final results are largely insensitive to them. Second, all the detailed work to date has been carried out only for non-rotating black holes. Although the underlying ideas are robust, considerable technical work will be required to extend the results to the non-rotating case.[7] Third, the role played by field equations in this program is yet to be fully understood. Since the construction of the phase space and the associated symplectic structure is delicate, it is clear that, as it stands, the framework is closely tied to general relativity; extension to higher derivative theories, for example, will probably involve significant modifications. (Extension to supergravity, by contrast may not.) Furthermore, in the quantum analysis, the 'kinematic part' of quantum Einstein's equations –the so called Gauss and diffeomorphism constraints– play a significant role. By contrast, however, the role of the Hamiltonian constraint is rather limited and needs to be better understood. Finally, attempts to derive the laws of black hole mechanics in full the quantum theory have just begun.

Perhaps the most puzzling and unsatisfactory aspect of the framework is that there is an inherent quantization ambiguity which leads to a one-parameter family of unitarily inequivalent theories. It shows up in various technical expressions through the Immirzi parameter γ; while the classical theory is completely insensitive to the value of γ, the quantum theory is not. Roughly, this parameter is analogous to the θ angle in Yang-Mills theory and as such its value can be fixed only experimentally. Black hole evaporation can be taken to be an appropriate (thought) experiment for this purpose. Now, it is reasonable to assume that Hawking's semi-classical analysis gives the physically correct answer for large black holes. This answer is recovered in our approach only when the Immirzi parameter is fixed to be $\gamma_0 = \ln 2/\pi\sqrt{3}$. That is, it appears that, only for this value of γ would the non-perturbative quantum theory, with all its polymer geometries and quantized areas, agree, in semi-classical regimes, with the standard quantum field theory calculation in curved space-times. However, the issue is far from being settled. Perhaps a new viewpoint will emerge and it may well cause the present picture to change in certain respects. However, the picture does have a striking coherence. Indeed, it is remarkable that results from three quite different areas –classical general relativity, quantum geometry and the Chern-Simons theory– fit together without a mismatch to provide a consistent and detailed description of the micro-states of black holes. At several points in the analysis, the matching is delicate and consistency

[7]Curiously, as a general rule, the technical problems in extending results from static situations to stationary ones turn out to be 'unreasonably difficult' in general relativity. Examples that readily come to mind are: the definitions of and theorems on multipole moments, the discovery of black hole solutions, the black hole uniqueness theorems, and, the analysis of stability under linearized perturbations.

could easily have failed. Since that does not happen, it seems reasonable to expect that the lines of thought summarized here will continue to serve as main ingredients also in the final picture.

5. Acknowledgements

Much of this review is based on joint work with John Baez and Alejandro Corichi. We are grateful to them and to the participants of the 1997 quantum gravity workshop at the Erwin Schrödinger Institute for many stimulating discussions. The authors were supported in part by the NSF grants PHY95-14240, INT97-22514 and by the Eberly research funds of Penn State. In addition, KK was supported by the Braddock fellowship of Penn State.

References

1. J.D. Bekenstein, *Phys. Rev.*, **D7**, 2333-2346 (1973); **D9**, 3292-3300 (1974).
2. J.W. Bardeen, B. Carter and S.W. Hawking, *Commun. Math. Phys.*, **31**, 161-170 (1973).
3. S.W. Hawking, *Commun. Math. Phys.*, **43**, 199-220 (1975).
4. A. Strominger and C. Vafa, *Phys. Lett.*, **B379**, 99-104 (1996).
   J. Maldacena and A. Strominger, *Phys. Rev. Lett.*, **77**, 428-429 (1996).
5. A. Ashtekar, J. Baez, A. Corichi, K. Krasnov, *Phys. Rev. Lett.*, **80**, No. 5, 904-907 (1998).
6. K. Krasnov, *Quantum geometry and thermal radiation form black Holes*, gr-qc/9710006.
7. A. Ashtekar, A. Corichi, K. Krasnov, *Black hole sector of the gravitational phase space*, (CGPG Pre-print, 1998).
8. A. Ashtekar, J. Baez, K. Krasnov, *Quantum geometry of black hole horizons*, (CGPG Pre-print, 1998).
9. A. Ashtekar, *Phys. Rev. Lett.*, **57**, 2244-2247 (1986); *Phys. Rev.*, **D36**, 1587-1602 (1987).
10. T. Jacobson and L. Smolin, *Nucl. Phys.*, **B299**, 295 (1988).
11. C. Rovelli and L. Smolin, *Nucl. Phys.*, **B331**, 80 (1990).
12. J. Baez, *Spin networks in non-perturbative quantum gravity*, gr-qc/9504036.
13. C. Rovelli and L. Smolin, *Phys. Rev.*, **D52**, 5743-5759 (1995).
14. A. Ashtekar and J. Lewandowski, *Class. Quant. Grav.*, **14**, A55-A81 (1997).
15. C. Rovelli and L. Smolin, *Nucl. Phys.*, **B442**, 593 (1995); Erratum: *Nucl. Phys.*, **B456**, 734 (1995).
16. A. Ashtekar and J. Lewandowski, *Adv. Theo. Math. Phys.*, **1**, 388-429 (1997).
17. T. Thiemann, *A length operator for canonical quantum gravity*, gr-qc/9606092.
18. R. Loll, *Nucl. Phys.*, **B460** 143-154 (1996).
19. J. Bekenstein, V. Mukhanov, *Phys. Lett.*, **B360**, 7 (1995).
20. M. Barreira, M. Carfora, C. Rovelli, *Gen. Rel. and Grav.*, **28**, 1293-1299 (1996).
21. K. Krasnov, *Class. Quant. Grav.*, **15**, L1-L4 (1998).
22. R.K. Kaul and P. Majumdar, *Quantum black hole entropy*, gr-qc/9801080.
23. C. Rovelli, *Phys. Rev. Lett.*, **14** 3288-3291 (1996); *Helv. Phys. Acta.*, **69** 582-611 (1996); K. Krasnov, *Phys. Rev.*, **D55** 3505-3513 (1997); *Gen. Rel. and Grav.*, **30**, No. 1, 53-68 (1998).
24. J. Wheeler, *It from bits and quantum gravity*, Princeton University Report (1992).
25. J.D. Bekenstein and A. Meisels, *Phys. Rev.*, **D15**, 2775-2781 (1977).

11. BLACK HOLES, GLOBAL MONOPOLE CHARGE AND QUASI-LOCAL ENERGY

NARESH DADHICH
Inter-University Centre for Astronomy and Astrophysics, Post Bag 4, Ganeshkhind, Pune - 411 007, India

1. Introduction

Black holes are described by solutions of the Einstein vacuum (or electrovac) equation and these solutions are unique. We wish to consider the question: can we get the same solutions from a slightly different and more general equation, which is not the derivative of the original equation? The answer is yes. Take the case of the Schwarzshild solution. It is the unique spherically symmetric solution of the Einstein vacuum equation, $R_{ik} = 0$. It can also be obtained as the solution of a slightly more general equation than that of vacuum. This happens because the vacuum equation ultimately reduces to two equations: one the good old Laplace equation and the other its first integral [1]. It is therefore sufficient to have only the first integral equation leaving the other free because it would always be implied by the former. This raises an interesting question: what happens when we have the Laplace equation but not its first integral? We shall show that the two sets of equations are dual of each other in a certain sense [2]. In the former case the unique solution is the Schwarzschild black hole while the dual set provides a black hole with a global monopole charge [3]. The latter solution is also unique.

We resolve the gravitational field into electric and magnetic parts relative to a timelike unit vector. The resolution of the Riemann curvature and its double dual relative to a timelike unit vector are termed as the active and passive electric parts respectively while that of the left dual is the magnetic part. The electric and magnetic parts are second rank 3-tensors, the electric parts are symmetric and hence account for 12 components while the remaining 8 components are accounted for by the magnetic part which is trace free, and is composed of the symmetric Weyl magnetic part and

B. R. Iyer and B. Bhawal (eds.), Black Holes, Gravitational Radiation and the Universe, 171–187.

an antisymmetric part representing energy flux. Now the Einstein equation can be written entirely in terms of electromagnetic parts of the curvature, i.e., directly in terms of the Riemann components. Though Ricci is a combination of the Riemann components, the combination implied by the electromagnetic version is more direct and perhaps easier to solve.

The duality that we mentioned earlier refers to interchange of active and passive electric parts. The Einstein vacuum equation is in general symmetric in them and hence it remains invariant under the duality transformation. It however turns out that spherically symmetric vacuum spacetime which is characterized by the Schwarzschild solution can also be obtained as the unique solution of a slightly more general equation. That is, the entire set of vacuum equations is not necessary, rather a subset is sufficient for obtaining the Schwarzschild solution. This subset is not symmetric in the active and passive parts and hence the duality leads to a different set of equations. The dual set of equations, like the original vacuum equations, also admits the general solution that can be interpreted as the static black hole with a global monopole charge [2,3]. This procedure of putting on global monopole charge also works for a charged black hole as well as for vacuum with the cosmological constant, i.e., for a charged black hole embedded in the De Sitter universe.

The vacuum for spherical symmetry can be characterized by a more general equation than vanishing of Ricci, and so also the flat spacetime. The dual set of equations in this case yields a global monopole with zero Schwarzshild mass in the static case and the global texture spacetime for the non-static and homogeneous case. The former massless global monopole spacetime can as well be considered as spacetime of a uniform relativistic potential [1]. In general relativity (GR), the gravitational potential not only determines motion of free particles but it also "curves" the space part of the metric [1]. By relativistic potential we mean the scalar function that satisfies the Laplace equation and also performs this dual role. It is remarkable to note that a uniform relativistic potential produces non-zero curvature and thereby, in contrast to the Newtonian theory, has non-trivial dynamical and physical manifestation. This is one of the most striking and distinguishing features of GR: Its property of producing curvature in space which does not get washed out even when it is constant. This property is in fact related to the non-linear character of the field which is purely a relativistic aspect.

A dual spacetime shares the basic physical character with the original spacetime, for instance radial free fall does not distinguish between the Schwarzschild black hole and the one with the global monopole charge. Since uniform potential and global texture spacetimes are dual to flat spacetime, it would perhaps be justified to envision them as "minimally curved" as they are "dual-flat". The necessary condition for the duality transformation to work is that a subset of the Einstein equation which is not symmetric in active and passive electric parts admits the same solution as the original equation. The sufficient condition would be that the dual set of equations admits a solution different from the original one. Then the two solutions would be duals of each other.

Another issue that we wish to discuss in this article is the characterization of a black hole involving some kind of energy consideration. The most direct and intuitive picture of a black hole is that of a compact object with a very strong gravitational pull, so strong that even photons cannot come out of the bound surface defining its extension in space. It is therefore natural to look for some kind of energy balance argument. The usual relativistic characterization as limit to existence of timelike worldlines and oneway character of compact surface, does not follow directly from any energy consideration. They rather refer to the general spacetime properties. In contrast the classical argument of escape velocity (balance between kinetic and potential energy for a particle travelling with the velocity of light!) is an excellent example of an energy consideration. Though it is intuitively most appealing and does give the right result, the formula employed is certainly not valid for photons. It is an example of a wrong formula giving the right result.

In GR it is not only the matter energy that produces a gravitational field. The former constrains the motion of free particles by pulling them in while the latter constrains through curvature of space. The question is what happens when these two effects become equal? Is then a black hole defined? The field energy influence would be higher order non-linear which is incorporated through space curvature [1]. It is rather very difficult and inherently ambiguous to define its contribution because it is non-localizable. It can therefore be defined only in a quasi-local sense. In view of the inherent ambiguity, there exist several expressions for it in the literature ([4]-[12], in particular see [4] and [6] for other references). The basic idea is to prescribe an integration procedure to compute total energy of spacetime. It is well-known that there is no covariant and unique way of doing it. Another way to look at is to follow the Gauss theorem of classical theory to compute flux of the field (proper red-shifted acceleration experienced by a free particle) through a closed 2-surface to define gravitational charge [2],[10]-[12]. It turns out that these two measures are in general different.

Recently Brown and York [4] have given a definition of quasi-local energy which emerges from the extrinsic curvature of a 2-boundary isometrically embedded in the curved 3-hypersurface. It is the value of the Hamiltonian that generates unit time translations orthogonal to a 3-surface at its boundary. It has some interesting and illuminating features. For instance, for the Schwarzschild black hole it gives in the Newtonian approximation for large r, $E(r) = M + M^2/2r$, and it equals $2M$ at the horizon. It could be understood thus: if the system had mass M in extreme dispersed state at infinity, as it collapses it picks up $-M^2/2r$, field energy which lies outside the radius r. Thus E is the total energy contained inside r and is the sum of energy required to create particles and energy required to assemble them in a sphere of radius r. It is then remarkable to note that the black hole horizon is defined when gravitational field energy $(E(r) - M)$ equals the gravitational charge, M.

That means black hole horizon is in general characterized by equipartition between gravitational charge, which constrains motion by pulling in, and gravitational field energy, which constrains through curvature of space. When these two effects are equal, the body turns into a black hole. This is an energy argument, of the kind we were looking for. In the classical theory, gravitational potential did only one job of pulling in, which could be taken care of by the balance between kinetic and potential energy to define escape velocity. In GR, we have in addition space also curving, and stronger the field gets, stronger should become the curvature contribution. It is the latter that attributes the oneway character, the absolute event horizon, to a black hole. Note that the former alone can never give a oneway surface, it can only prescribe the static limit. Thus in GR we have, rather than equipartition of kinetic and potential energy, equipartition of gravitational charge and gravitational field energy at the horizon. That is black hole's energy at the horizon is equally shared between charge (gravitational attraction) and gravitational field energy (curvature of space). It is an insightful characterization based on the energy consideration.

In Sec. 2, we shall develop the electromagnetic resolution of the Riemann curvature which will be followed by a consideration of the duality transformation and dual spacetimes in Sec.3. In Sec. 4, we shall briefly review the definitions of gravitational charge and the Brown-York quasi-local energy followed by the energetics characterization of black hole in Sec.5. Finally we conclude with a discussion.

2. Riemann electromagnetics

The electromagnetic decomposition of the gravitational field has a long history [13]- [17]. Usually it is the Weyl curvature that is resolved in analogy with the Maxwell field relative to a timelike unit vector into electric and magnetic parts. The Weyl curvature is supposed to describe a free field with its 10 independent components. The electric and magnetic parts are given by second rank symmetric and trace-free 3-tensors. Recently applications of resolution of the Riemann curvature have also been considered [2], [18]-[19]. In this there are two kinds of electric parts coming from resolutions of the Riemann and its double dual, while the magnetic part is given by the resolution of the (single) left dual of Riemann. Again all three are second rank 3-tensors, the electric parts are symmetric but not trace-free, while the magnetic part is trace-free and is composed of the symmetric Weyl magnetic part and an antisymmetric part. Thus electric parts account for 12 and magnetic for 8 of the 20 Riemann components.

Relative to a unit timelike vector u^a, let us resolve the Riemann curvature into electric and magnetic parts as follows:

$$E_{ac} = R_{abcd}u^b u^d, \quad \tilde{E}_{ac} = *R*_{abcd}\, u^b u^d, \tag{1}$$

$$H_{ac} = *R_{abcd}u^b u^d = H_{(ac)} - H_{[ac]} \text{ where,} \tag{2}$$

$$H_{(ac)} = *C_{abcd}u^b u^d, \quad H_{[ac]} = \frac{1}{2}\eta_{abce}R^e_d u^b u^d, \tag{3}$$

$$*R_{abcd} = \frac{1}{2}\eta_{abmn}R^{mn}{}_{cd}.$$

Here C_{abcd} is the Weyl conformal curvature and η_{abcd} is the 4-dimensional volume element. The electric parts are symmetric, $E_{ab} = E_{ba}, \tilde{E}_{ba} = \tilde{E}_{ba}$, and they are all orthogonal to the resolving vector, $(E_{ab}, \tilde{E}_{ab}, H_{ab})u^a = 0$, and $E = E^a_a, \tilde{E} = \tilde{E}^a_a, H = H^a_a = 0$, and $u_a u^a = 1$.

We shall term E_{ab} as active electric part and $\tilde{E}_{ab}$ as passive electric part. The former refer to space-time components of the curvature that are supposed to represent the usual property of gravity - the gravitational attraction, while the latter to space-space components that incorporate non-linear field energy contribution to the field. The magnetic part refers to, as expected, motion of matter and energy flux.

In terms of the electromagnetic parts the Ricci curvature reads as follows:

$$R^b_a = E^b_a + \tilde{E}^b_a + (E + \tilde{E})u_a u^b - \tilde{E}g^b_a + \frac{1}{4}(\eta_{amn}H^{mn}g^b_0 + \eta^{bmn}_{mn}g^0_a). \tag{4}$$

Then the vacuum Einstein equation $R^b_a = 0$ is equivalent to the following set of equations,

$$E \text{ or } \tilde{E} = 0, \quad E^b_a + \tilde{E}^b_a = 0, \quad H_{[ab]} = 0. \tag{5}$$

Note that these equations imply absence of energy flux, and that the sum of the two electric parts equal to zero; i.e. the active and passive parts exactly counterbalance each other. In addition trace of one of the electric parts vanishes. The traces of electric parts give gravitational charge density, $E = -(T_{ik} - \frac{1}{2}Tg_{ik})u^i u^k$ and the energy density, $\tilde{E} = -T_{ik}u^i u^k$ relative to a timelike observer. That means vacuum spacetime is characterized by absence of one of the densities and energy flux, and the two electric parts exactly balancing each-other. The set (5) is symmetric in E_{ab} and $\tilde{E}_{ab}$. Similarly, electrovac equations take the form,

$$E_a^b + \tilde{E}_a^b = -\frac{Q^2}{r^4} g_a^1 g_1^b, \quad E = \tilde{E}, \quad H_{[ab]} = 0, \tag{6}$$

which are also symmetric in active and passive parts and hence will remain invariant when active and passive parts are interchanged.

On the other hand the Einstein space, $R_a^b = \Lambda g_a^b$, is described by

$$E_a^b + \tilde{E}_a^b = 0, \quad \tilde{E} = \Lambda, \quad H_{[ab]} = 0 \tag{7}$$

which are not symmetric in E_{ab} and $\tilde{E}_{ab}$. Here the interchange $E_{ab} \leftrightarrow \tilde{E}_{ab}$ will take De Sitter to anti-De Sitter and the vice-versa.

3. The duality: global monopole and texture

We define the duality transformation by

$$E_{ab} \leftrightarrow \tilde{E}_{ab} \tag{8}$$

and we note that the Einstein equation is invariant under this transformation. The question is: Can there be a situation where a set of equations that are not symmetric, still give the same solution as the symmetric set? The answer is yes, for instance the Schwarzschild solution of Einstein vacuum equation. Let us first demonstrate this.

We claim that the set

$$\tilde{E} = 0, \quad E_a^b + \tilde{E}_a^b = \mu g_a^1 g_1^b, \quad H_{[ab]} = 0, \tag{9}$$

which is not invariant under the duality transformation (8), also admits the Schwarzschild solution as the general solution for spherically symmetric spacetime given by

$$ds^2 = Adt^2 - Bdr^2 - r^2(d\theta^2 + sin^2\theta d\varphi^2). \tag{10}$$

It can be verified that $H_{[ab]} = 0$, leads to time independence and $E_2^2 + \tilde{E}_2^2 = 0$ to $AB = 1$ (note that no boundary condition of asymptotic flatness is used). Then $\tilde{E} = 0$ determines the solution completely, i.e., $B = (1 - 2M/r)^{-1}$, the Schwarzschild solution. Note that the equation $E_1^1 + \tilde{E}_1^1 = \mu$ was not used in determining the solution. That is, of the four independent equations implied by the above set (9), only three were sufficient to determine the solution. Since the Schwarzschild solution is the unique spherically symmetric vacuum solution, the above set (9) also characterizes vacuum for spherical symmetry.

We now employ the duality transformation (8) to write the dual set of equations, which are given by

$$E = 0, \quad E^b_a + \tilde{E}^b_a = \mu g^1_a g^b_1, \quad H_{[ab]} = 0. \tag{11}$$

This also admits the general solution,

$$A = B^{-1} = 1 - k^2 - \frac{2M}{r} \tag{12}$$

which is supposed to describe a Schwarzschild black hole with a global monopole charge k [3]. As before we shall have $AB = 1$, from $E^2_2 + \tilde{E}^2_2 = 0$, and $E = 0$ will give $A' = \text{const}/r^2$ which integrates to give the above solution and $\mu = k^2/r^2$. Like the Schwarzschild solution, this is also unique. It may however be noted that Eq.(11) implies, $A = B^{-1} = 1 + 2\phi$ with $\nabla^2\phi = 0$ [1]. This has the general solution $\phi = -\frac{1}{2}k^2 - M/r$ (The vacuum equation determines $k = 0$). This solution should correspond to uniform potential when $M = 0$.

A global monopole is supposed to be created when global $O(3)$ symmetry is spontaneously broken into $U(1)$ in phase transition in the early Universe. It is modelled by a triplet scalar $\psi^a(r) = kf(r)x^a/r$, $x^a x^a = r^2$ with the Lagrangian

$$L = \frac{1}{2}(\partial\psi^a)^2 - \frac{\lambda}{4}(\psi^a\psi^a - k^2)^2. \tag{13}$$

It would give rise to stresses that at large distance from the core (where $f = 1$) approximate to

$$T^0_0 = T^1_1 = \frac{k^2}{r^2} \tag{14}$$

which are exactly the stresses generated by the solution (12) [3]. Thus the solution (12) describes a black hole with a global monopole charge. We have thus shown that the Schwarzschild black hole with a global monopole charge is dual to the one without it.

Similar to vacuum, flat spacetime in spherical symmetry could be characterized by the set of equations,

$$\tilde{E}^b_a = 0, \quad E^b_a = \mu g^1_a g^b_1, \quad H_{[ab]} = 0. \tag{15}$$

which lead to $A = B = 1$ in the metric (10) giving flat spacetime. Its dual set has $A' = B' = 0$ in place of $A = B = 1$, which leads to $A = 1$ and $B = \text{const.} = (1 - k^2)^{-1}$, say. This is the global monopole solution with $M = 0$, zero mass global monopole. As argued above, it could as well be looked upon as a uniform potential spacetime, which is non-flat with the stresses given by Eq.(14). The relativistic potential performs the dual role of determining free particle motion through A and of accounting for non-linear contribution to field energy through B, curvature of space. The former effect gets washed out as in the Newtonian theory when ϕ is constant. The distinguishing feature of GR is that the latter effect still persists even when $\phi =$ const. $\neq 1$. It is interesting to note that uniform relativistic potential spacetime, which is non-flat with the stresses given by Eq.(14) and zero mass global monopole are described by the same metric. Thus uniform relativistic potential (zero mass global monopole) spacetime is dual to flat spacetime.

For homogeneous, isotropic and non-static case, flat spacetime will be characterized by writing $E^a_b + \tilde{E}^a_b = \mu g^a_b$ in Eq.(15) and then the solution of its dual set is given by the well-known FRW metric

$$ds^2 = dt^2 - S^2(t)[\frac{dr^2}{1-cr^2} + r^2(d\theta^2 + \sin^2\theta d\varphi^2)] \tag{16}$$

with the equation of state $\rho + 3p = 0$. This characterizes spacetime of another topological defect called global texture [20]. The equation of state gives the scale factor,

$$S(t) = \alpha t + \beta \text{ and} \tag{17}$$

$$\rho = 3\frac{\alpha^2 + c}{(\alpha t + \beta)^2}, \quad c = \pm 1, 0. \tag{18}$$

Thus both uniform potential (massless global monopole) are thus dual-flat spacetimes.

The duality transformation could be applied to a charged black hole as well, and then we would have the dual set of equations,

$$E^b_a + \tilde{E}^b_a = (-\frac{Q^2}{r^4} + \mu)g^1_a g^b_1, \quad E = -\frac{Q^2}{2r^4}, H_{[ab]} = 0 \tag{19}$$

which admit the general solution,

$$A = B^{-1} = 1 - k^2 - \frac{2M}{r} + \frac{Q^2}{r^2} \tag{20}$$

where $\mu = k^2/r^2$. This is a charged black hole with global monopole charge.

Note that the equations for the Einstein space (7) imply that De Sitter and anti De Sitter spacetimes are dual of each other. That is, the duality transformation in this case implies change in the sign of Λ. Similarly, a global monopole charge could be added to the De Sitter spacetime. We can, in general, write for a electrovac spacetime with the cosmological constant the dual set of equations as follows:

$$E^b_a + \tilde{E}^b_a = (-\frac{Q^2}{r^4} + \mu)g^1_a g^b_1, \quad E = \wedge - \frac{Q^2}{2r^4}, H_{[ab]} = 0 \tag{21}$$

which give the general solution

$$A = B^{-1} = 1 - k^2 - \frac{2M}{r} + \frac{Q^2}{r^2} - \frac{\wedge}{3}r^2. \tag{22}$$

The duality transformation has also been considered for a fluid distribution [19] and it turns out that under duality transformation pressure anisotropy and heat flux remain unaffected whereas density and pressure change as follows: $(\rho + 3p)/2 \leftrightarrow \rho$, $(\rho - p)/2 \leftrightarrow p$, and the equation of state, $p = (\gamma - 1)\rho \leftrightarrow p = (\bar{\gamma} - 1)\rho, \bar{\gamma} = 2\gamma/(3\gamma - 2)$. This implies that the radiation, $p = \rho/3$, and the De Sitter, $p = -\rho$, universes are self dual. The former is so because when $R = 0$, Ricci is the same as Einstein, while for the latter as shown above De Sitter goes to anti De Sitter but the equation of state remains invariant. As noted earlier, the spacetime with $p = -\rho/3$ is dual to flat spacetime while the stiff fluid $p = \rho$ is dual to dust spacetime.

4. Gravitational Charge and Brown-York quasi-local energy

4.1. GRAVITATIONAL CHARGE

Owing to the non-localizability of gravitational field energy, defining a measure of energy in GR is inherently ambiguous and there are as many or more number of expressions for it as the number of persons who have bothered to address this question [4], [6]. Komar [5] defined the conserved mass of a static black hole by the formula

$$M = \frac{1}{8\pi}\int_s *d\tilde{\xi} = \frac{1}{8\pi}\oint \eta_{abcd} \nabla^a \xi^b dx^c \wedge dx^d, \tag{23}$$

where $\tilde{\xi}$ is the timelike Killing 1-form. This integral will yield a conserved quantity only if the timelike vector is Killing and spacetime is empty. When either of these conditions is violated it will yield a radius dependent expression, a quasi-local quantity as is the case for a charged and/or rotating black hole [10] - [12]. In the case of a rotating black hole we take the corotating timelike vector, $\xi = \frac{\partial}{\partial t} + \omega\frac{\partial}{\partial\varphi}$, which would represent time translation at infinity and would coincide with the horizon generator at the horizon. It thus has proper limits. In between it would give an approximate measure because, unlike the case of spherically symmetry, there does not occur a natural 2-closed surface in this case.

It turns out that the Komar integral is equivalent to the Gauss integral suitably taken over to GR. That means it actually defines gravitational charge, which is responsible for producing acceleration on free particles. Let N be the lapse function for a stationary spacetime. Then the red shifted proper acceleration relative to infinity is given by

$$\mathbf{g} = -N \nabla (\ln N), \tag{24}$$

where $N = |\xi| = (g_{00} - \omega g_{03})^{1/2}$ in the usual notation. This measures effective gravitational field strength and so we apply the Gauss theorem to define effective gravitational charge

$$M_c(r) = -\frac{1}{4\pi} \int_s \mathbf{g}.\mathbf{ds}. \tag{25}$$

It can be easily verified that this is equivalent to the Komar integral (23) [5]. The value of g at the horizon gives the surface gravity of the black hole and hence we have the relation,

$$M_c(r_+) = (\kappa/4\pi)A, \tag{26}$$

where κ is the surface gravity and A is the area of the hole. If we let a test particle tied to a string hang from infinity where the other end is held by an asymptotic observer, the acceleration $\mathbf{g}$ will then measure tension in the string. This is a proper measure of gravitational field strength. We have employed the Gauss theorem of classical theory to measure gravitational charge by evaluating the flux of acceleration across the closed 2-surface.

Gravitational charge for a charged black hole at some radius would be

$$M_c(r) = M - \frac{Q^2}{r} \tag{27}$$

and it would assume the value

$$M_c(r_+) = (M^2 - Q^2)^{1/2} \tag{28}$$

at the horizon. For the rotating black hole for large r it would go as

$$M_c(r) = M - \frac{4M^2a^2}{r^3} + \frac{M^3a^2}{6r^4}, \tag{29}$$

while close to the horizon it assumes the form

$$M_c(r \to r_+) = M - a^2/r \tag{30}$$

which at the horizon yields

$$M_c(r_+) = (M^2 - a^2)^{1/2}. \tag{31}$$

Gravitational charge is obviously a quasi-local measure, which essentially measures the hole's capacity to attract free particles around it. That means it refers directly to the Newtonian aspect (lapse function represents the relativistic potential). This would in general be different from quasi-local energy measures (for a charged black hole, M_c is given by Eq.(27) while quasi-local energy for various definitions is given by $M - Q^2/2r$, see discussion in Ref.[12]).

4.2. BROWN-YORK QUASI-LOCAL ENERGY

Consider a system in the extreme dispersed state at infinity and in this state let its energy be M. Now it contracts on its own without intervention of any external agency under gravity and hence would pick up $-M^2/2r$ field energy. Thus one would expect that the total energy contained inside a radius r should approximate to $M - (-M^2/2r)$, for large r. This approximation is however not expected to be valid in region of very strong field, say near the black hole horizon. The Brown-York definition of quasi-local energy has precisely this intuitively appealing feature [4].

The Brown-York definition [4] is based on the action principle of gravity and matter, and the definition follows from application of the Hamilton-Jacobi analysis. It is defined as *minus* variation in the action in a unit change in proper time separation between B (a closed 2- surface isometrically embedded in a curved 3-hypersurface) and its neighbouring 2-surface. Energy is therefore the value of the Hamiltonian that generates unit time translations orthogonal to 3-hypersurface C at the boundary B. This is clearly a natural measure of energy. It is ultimately defined as follows:

$$E(r) = \frac{1}{8\pi} \int (K - K_0)\sqrt{\sigma} d^2x, \tag{32}$$

where σ is the determinant of the 2-metric and K is the mean extrinsic curvature of the closed 2-boundary. The subscript 0 refers to asymptotic reference spacetime, not necessarily flat. (The symbol $E(r)$ for energy in this and the following section should not be confused with $E = E_a^a$ of Secs. 2 and 3.) The quasi-local energy obeys the additivity property, i.e. energies can be added.

For the spherically symmetric spacetime given by the metric (10) with A and B being functions of r only, the formula (32) then gives

$$E(r) = [r(\frac{1}{\sqrt{B_0}} - \frac{1}{\sqrt{B}})] \tag{33}$$

which, for an asymptotically flat reference where $B_0 = 1$, means

$$E(r) = r[1 - \frac{1}{\sqrt{B}}]. \tag{34}$$

It is obvious that the energy at the horizon is given by the horizon radius, for horizon is characterized by $B^{-1} = 0$. It should read for a charged black hole as

$$E(r) = r[1 - (1 - \frac{2M}{r} + \frac{Q^2}{r^2})^{1/2}] \tag{35}$$

$$\approx M + \frac{M^2}{2r} - \frac{Q^2}{2r} \text{ forlarge} r. \tag{36}$$

This relation can be understood as follows: M stands for the total energy including rest mass and all kinds of interaction energies localizable as well as non-localizable, and energy lying exterior to the radius r will include $Q^2/2r - M^2/2r$, electric and gravitational field energies. The latter though cannot be handled without ambiguity – it will have $-M^2/2r$, the first term in the approximation. Now the energy contained inside a sphere of radius r, will be exactly what is given above. A system is in extreme dispersed state at infinity when all interaction energies vanish and has energy M, and as it collapses under gravity it picks up interaction energies $-M^2/2r$ and $Q^2/2r$. It is natural to expect that energy at any given radius should include all these contributions and this is what the Brown-York definition seems to do.

5. Energetics characterization of a black hole

It is clear from the discussion of the preceding section that gravitational charge hinges on the lapse function $N = \sqrt{A}$ for the metric (10) (though space curvature through g^{11} is involved in computing norm of the acceleration vector **g**) while the Brown-York quasi-local energy entirely arises from the space curvature. It is essentially determined from the mean extrinsic curvature of the 2-surface, which is isometrically embedded in the 3-hypersurface. The asymptotic flat space serves as the reference for measure. The gravitational charge on the other hand is, in the spirit of the Gauss theorem, the measure of flux of the proper acceleration (field strength) across a 2-closed surface.

If we make a black hole from collapse of a dispersed distribution of energy M, then a black hole is characterized by one of the following properties:

(a) When the energy due to all interactions (field energy), localizable or non-localizable, $E(r) - M$, is equal to the gravitational charge enclosed by the radius r.

(b) When matter energy arising from energy-momentum tensor, T_i^k equals the non-T_i^k contribution to the Brown-York energy, $E(r)$. That is the horizon equipartitions the Brown-York energy into matter and non-matter parts.

Gravitational charge of a charged source enclosed by radius r is from Eq.(27) given by

$$M_c(r) = M - Q^2/r, \tag{37}$$

while the Brown-York energy is from Eq.(35) given by

$$E(r) = r[1 - (1 - \frac{2M}{r} + \frac{Q^2}{r^2})^{1/2}]. \tag{38}$$

Setting $E(r) - M$ equal to (37), we get

$$(2Mr - Q^2)(r^2 - 2Mr + Q^2) = 0 \tag{39}$$

which gives the horizon, $r_+ = M + \sqrt{M^2 - Q^2}$, or $r = Q^2/2M$, the hard core radius of naked singularity for $M^2 < Q^2$.

Clearly energy M has the T_i^k -support and so has the contribution $Q^2/2r$, of electric field energy density $Q^2/2r^4$, which lies exterior to r. That means $M - Q^2/2r$ has the T_i^k-support and it is therefore matter energy. Equipartitioning of E into matter and non-matter contributions would then lead to

$$r - (r^2 - 2Mr + Q^2)^{1/2} = 2(M - Q^2/2r) \tag{40}$$

which again implies the same equation (39).

We have thus established the characterizing properties (a) and (b), which are equivalent statements.

Let us further note the following points:

(i) We could in general write

$$E(r_+) - E(\infty) = M_c(r_+) \tag{41}$$

which will be true even for the Kerr black hole as well. Though a 2-surface is not in general closed in the Kerr spacetime, the horizon does provide the closed surface to evaluate the charge (25) and energy (32) integrals. That means the statement (a) will also be true for the Kerr black hole while about (b) we cannot say anything as we do not know how to incorporate rotational energy. If this statement were to be true, rotational energy near the horizon should have the same behaviour as the contribution from electric field energy near the horizon.

(ii) In the spirit of Christodoulou's irreducible mass [21], we define the irreducible energy of a black hole by

$$E_{ir}^2 = r_+^2 + a^2 = E^2(r_+) + a^2 \tag{42}$$

which leads to the equivalent relation

$$M^2 = \frac{1}{4}(E_{ir} + \frac{Q^2}{E_{ir}})^2 + \frac{J^2}{E_{ir}^2} \tag{43}$$

where $E_{ir} = 2M_{ir}$. This can be interpreted in analogy with the special relativistic conservation law, M represents the total energy, J/E_{ir} the momentum and $\frac{1}{2}(E_{ir} + Q^2/E_{ir})$ the rest energy of the hole. The electric field contributes to enhancing the rest energy while rotation as expected contributes explicitly to kinetic energy. E_{ir} will be bounded as: $2M \geq E_{ir} \geq \sqrt{M^2 + Q^2}$ or $\sqrt{2M^2 - Q^2}$. The fraction of energy available for extraction would be given by

$$1 - E_{ir}(\text{extremal})/E_{ir}(\text{sch}). \tag{44}$$

It is 29% for rotating and 5% for charged black hole.
(iii) Note that gravitational charge is a conserved quantity, the same for all r, for the Schwarzschild black hole while the Brown-York energy is conserved, $E = M$ everywhere, for an extremal hole, $M^2 = a^2 + Q^2$. It implies absence of driving force for collapse. Thus an extremal black hole can never be formed through a collapse of a matter distribution. It has to be born as such. It has also been recently shown [22] that as black hole approaches extremality, the allowed window for infalling particle parameters pinches off. It is in agreement with the recent work on black hole involving quantum and topological considerations.
(iv) With reference to the metric (10), note that it is the lapse $\sqrt{A}$ that is responsible for gravitational acceleration, while B measures the curvature caused by gravitational field energy density. A particle at rest only experiences gravitational pull, on the other hand a photon that can never be at rest experiences space curvature. It is the former that determines the limit: how far the gravitational pull is resistable? On the other hand it is the space curvature that determines the limit: when a closed 2-surface turns oneway (limit of irresistability of space curvature)? It is the latter that marks the event horizon. The remarkable thing is that these two considerations give the same answer. This is because photon is the limiting case for ordinary particles.

A black hole results when the measure of gravitational pull is equal to the measure of space curvature. Both constrain free particle motion. Intuitively the former constrains motion in the Newtonian sense while the latter does so through geometry – space curvature. The gravitational charge refers to the former whereas the Brown-York energy to the latter.

6. Discussion

In the above we have discussed two apparently unrelated features of black hole spacetimes. One concerned with the structure of vacuum field equation in its electromagnetic version, which suggested that it could be written in a slightly generalized form, so as to give a different set of equations under the duality transformation. That leads to the dual spacetime of the

Schwarzschild black hole with global monopole. The other concerned measure of gravitational charge and quasi-local energy and their interplay in defining black hole horizon.

There is however a thin thread of concept and principle connecting them at a deeper level. In both of them, the distinctive relativistic feature arising from gravitational field energy plays an important role. In the former case, global monopole charge essentially affects only the space curvature. The acceleration, **g** in Eq.(24), is completely immune to its presence. Global monopole charge is therefore caused by gravitational field energy and hence it is a purely relativistic quantity. Note in particular that spacetime corresponding to uniform relativistic potential is non-flat which represents massless global monopole. That is the potential's contribution to space curvature does not get washed out even when it is uniform. The Brown-York quasi-local energy is entirely determined by the curvature of space (32), its measure is the extrinsic curvature of closed 2-surface. The common thread is thus the non-linear aspect of gravity, arising from field producing field.

By using the measures of gravitational charge and quasi-local energy, we have attempted to formulate an energetics characterization of black hole. Intuitively one can envision a black hole as an object where the field energy's contribution through space curvature becomes equal to that of active gravitational charge. The former constrains motion through space curvature while the latter through gravitational pull. Their equality marks the black hole horizon.

Looking at the vacuum equation (5), we note that the vanishing of the trace of active or passive part signifies absence of active charge density or energy density, while vanishing of $H_{[ab]} = 0$ indicates absence of energy flux, and $E_{ab} + \tilde{E}_{ab} = 0$ means both A and B in (10) are given by the same function. This is why the quantities (M_c in (25) from A and E in (32) from B) arising from them can have a relation between them. It is this relation that characterizes the black hole. Note that the integrals (25, 32) cannot be evaluated at any arbitrary r for the Kerr metric, but they could be at the horizon which is a valid compact 2-surface. Thus the characterizing relation (41) holds good for the rotating black hole as well. It is true in all coordinates. We have verified it for the isotropic coordinates for a charged black hole.

It is interesting to note that gravitational charge (25) is conserved for the Schwarzschild hole while the Brown-York energy (32) is conserved for the extremal hole. That means $E = M$ everywhere, and hence it cannot suffer any gravitational collapse of its own.Thus an extremal hole can never be formed of collapse from a dispersed matter-energy distribution.This agrees with the recent work on black hole dynamics involving classical [22] - [24] and quantum and topological considerations [25]- [26]. In particular it has been demonstrated that as extremality is approached, the parameter window for infall pinches off [22]. Hence extremality is impossible to achieve.

Global monopoles and textures are topological defects which are supposed to be created by spontaneous breaking of global symmetry in phase transitions in the early Universe. Their implications for large scale structure formation and evolution of the Universe is an area of active research. The remarkable fact is that they are dual related to vacuum and flat spacetimes. The physical properties of the Schwarzschild black hole with global monopole has been investigated by several authors [3], [22], [28]-[30]. In particular it has been shown that global monopole's contribution to black hole temperature and particle orbits is to scale the Schwarzschild values by factors $(1-k^2)^2$ and $(1-k^2)^{-3/2}$ respectively [31].

The duality transformation will lead to a distinct dual solution only when the Einstein vacuum (including electrovac with the cosmological constant) solution follows from the subset of the entire set. Then one can introduce some distribution in the free equation, which is not used earlier to break the symmetry between E_{ab} and $\tilde{E}_{ab}$. Now solution of the dual set of equations will incorporate some global parameter in terms of monopole or texture. These spacetimes are dual of each-other, and they do share some basic physical properties. The uniform potential or massless global monopole and the global texture spacetimes are dual to flat spacetime. That is, they are dual-flat and hence are akin to flat spacetime, for instance they are indistinguishable from flat for free-fall. Could they be thought as "minimally" curved? That is to say that dual-flatness is the characterization of "minimality" of curvature. This is a covariant condition which appears reasonable and acceptable on physical grounds.

From Eq.(4) it is clear that the expression for R^b_a is not duality (8) invariant. One may then ask what is the general expression that could be constructed from the Ricci and the metric, which is duality (8) invariant ? It turns out to be

$$R^b_a - \frac{1}{4}Rg^b_a + \wedge g^b_a. \tag{45}$$

Without $\wedge$, it characterizes gravitational instanton which follows from the R^2-action. The instanton equation, $R^b_a - \frac{1}{4}Rg^b_a = 0$ is also conformally invariant. The conformal invariance in fact singles out the R^2- instanton action. Thus conformal invariance automatically includes the duality invariance. On the other hand, the duality invariance of the Palatini action and the requirement that the resulting field equation be valid for all values of R would lead to the conformal invariance and the instanton equation [32], [33].

It is a pleasure to dedicate this work to Vishu on his youthful 60! It was he who had initiated and inspired me to build an intuitively and physically appealing picture of black hole. This is a small attempt on that track.

References

1. N. Dadhich (1997) gr-qc/9704068.
2. N. Dadhich (1997) gr-qc/9712021.
3. M. Barriola and A. Vilenkin (1989) *Phys. Rev. Lett.*, **63**, 341.
4. J.D. Brown and J.W. York (1993) *Phys. Rev.*, **D47**, 1407.
5. A. Komar (1959) *Phys. Rev.*, **113**, 934.
6. G. Bergqvist (1992) *Class. Quantum Grav.*, **9** ,1753.
7. J. Katz, D. Lynden-Bell and W. Israel (1988) *Class. Quantum Grav.*, **5**, 971.
8. A.J. Dougan and L.J. Mason (1991) *Phys. Rev. Lett.*, **67**, 2119.
9. A.N. Petrov and J.V. Narlikar (1996) *Found. Phys.*, **26**, 1201.
10. R. Kulkarni, V. Chellathurai and N. Dadhich (1988) *Class. Quantum Grav.*, **5**, 1443.
11. N. Dadhich (1989) GR-12 Abstracts, p.60.
12. V. Chellathurai and N. Dadhich (1990) *Class. Quantum Grav.*, **7**, 361.
13. C. Lanczos (1938) *Ann. Math.*, **39**, 842.
14. C. Lanczos (1962) *Rev. Mod. Phys.*, **34**, 379.
15. L. Bel (1959) *C.R. Acad. Sci.*, **248**, 1297.
16. J. Ehlers (1993) *Gen. Relativ. Grav.*, **25**, 1255.
17. G.F.R. Ellis (1997) in *Gravitation and cosmology*, eds. S. Dhurandhar and T. Padmanabhan (Kluwer Academic Publishers).
18. M.A.G. Bonilla and J.M.M. Senovilla (1997) *Gen. Relativ. Grav.*, **29**, 91.
19. N. Dadhich, L.K. Patel and R. Tikekar (1998) gr-qc/9801076
20. A. Vilenkin and E.P.S. Shellard (1994) *Cosmic strings and other topological defects* (Cambridge University Press)
21. D. Christodoulou (1970) *Phys. Rev. Lett.*, **25**, 1596.
22. N. Dadhich and K. Narayan (1997) *Phys. Lett.*, **A231**, 335.
23. W. Israel (1986) *Phys. Rev. Lett.*, **57**, 397.
24. R.M. Wald (1974) *Ann. Phys.*, **83**, 548.
25. S.W. Hawking, G.T. Horowitz and S.F. Ross (1995) *Phys. Rev.*, **D51**, 4302.
26. C. Teitelboim (1995) *Phys. Rev.*, **D51**, 4315.
27. D. Harari and C. Lousto (1990) *Phys. Rev.*, **D42**, 2626.
28. G.W. Gibbons, M.E. Ortiz, F. Ruiz Ruiz and T.M. Samols (1992) *Nucl. Phys.*, **B385**, 127.
29. B. Jensen (1995) *Nucl. Phys.*, **B453**, 413.
30. J. Jing, H. Yu and Y. Wang (1993) *Phys. Lett.*, **A178**, 59.
31. N. Dadhich, K. Narayan and U. Yajnik (1996) gr-qc/9703034.
32. K.B. Marathe and G. Martucci (1998) *Geometry of gravitational instantons*, in preparation.
33. N.Dadhich and K.B. Marathe (1998) *Electromagnetic resolution of curvature and gravitational instantons*, in preparation.

It was the best of times, it was the worst of times. For me that is. This was in the year nineteen sixty-nine. I was working at the Institute for Space Studies in New York on a National Research Council fellowship. I was studying the scattering of gravitational waves by black holes through computer simulation in order to find out whether the black hole left its imprint in some way on the scattered wave. This was done by bombarding the black hole with Gaussian wave packets. When the width of the Gaussian was comparable to or less than the radius of the black hole, there emerged a decaying wave pattern later to be called the quasi-normal mode of the black hole. As is well known, a lot of work has been done on the black hole quasi-normal modes since then. Today they are considered to be a means of detecting both black holes and gravitational waves. My original work on the quasi-normal modes was terribly exciting – to me. And that is the best-of-times part of the story. Now for the worst-of-times aspect. Unfortunately, no one at the Institute seemed to be interested in this kind of research. After all, neither black holes nor gravitational radiation had been detected, but only theoretically predicted. Why should any one spend time investigating the interaction between two unobserved entities perhaps of doubtful existence? The consequence of this attitude, which was not at all uncommon in those days, was rather disconcerting, to say the least. My contract was renewed for only three months instead of the normal one year period. As it turned out, however, it was a blessing in disguise, but most effectively disguised at the time as Winston Churchill might have put it.

— C. V. VISHVESHWARA

12. KINEMATICAL CONSEQUENCES OF INERTIAL FORCES IN GENERAL RELATIVITY

A. R. PRASANNA AND SAI IYER
Physical Research Laboratory, Ahmedabad 380 009, India.

1. Introduction

One of the most important concepts of physics, which in fact marked the foundation of physics is the concept of inertia. Inertia as enunciated by Newton is an indicator of the characteristic of a body that determines its motion. Whereas according to Newton's law of motion, a body's state of rest or of uniform motion does in no way characterise its inertia, any change in this state depends upon its inertia. As the change of state can be recorded only through acceleration of the body, the significance of inertia can indeed be understood only by the action of external agencies influencing a body.

However, Newton, while enunciating his famous laws of mechanics, characterised inertia through the mass of a body, which he identified with the quantity of matter contained in it and referred all motion to the so called 'inertial frames' which according to him are the frames at rest or in uniform motion with respect to an 'absolute space'. As a consequence of the notion of absolute space, he had to also assume an absolute time which is same for all observers.

It is a matter of history that the Newtonian concept of absolute space was questioned by Leibnitz and Berkeley on philosophical grounds, whereas as quoted by Weinberg [1], the first constructive attack on this concept was made in 1880 by the Austrian philosopher Ernst Mach who suggested, in a way, that the 'motions of bodies need to be specified in reference to a fixed frame which he identified with that of the fixed stars'.

It is indeed interesting to note that while Mach himself never did put down his notion in the form of a principle, it is only Einstein, who for the first time in 1912 formulated the principle as: '.... the entire inertia of a point mass is the effect of the presence of all other masses, deriving from

B. R. Iyer and B. Bhawal (eds.), Black Holes, Gravitational Radiation and the Universe, 189–206.

a kind of interaction with the latter....' adding a footnote 'this is exactly the point of view which E. Mach urged in his acute investigations on the subject.'

Over the years, several different interpretations of this principle have been put forward, but with the main contention that inertia arises due to interactions of bodies with the rest of the Universe, while the so-called 'fixed stars' arising from averaged motion of the celestial bodies provide the inertial frame in which Newton's laws are applicable.

Einstein was quite enamoured of Mach's idea and he wanted it to be an integral part of his General Relativity. However, there are differing opinions on this as there seem to be as many as twenty different interpretations of the principle itself.

It is well known that while Newton believed that acceleration is absolute, Einstein's theory established clearly that acceleration too is relative and this aspect is well revealed when one considers rotation. One of the simplest ways of appreciating this aspect of Mach's principle is by measuring the spin of the Earth in two different ways. As Narlikar [2] puts it, if one calculates the Earth's rotation by observing the rising and setting of stars or by looking at the rotation of the plane of a Foucault pendulum at a given latitude, one gets the same answer, which shows the relation of rotation with respect to a frame fixed on Earth to the one fixed in space; while in the first method one measures Earth's spin against the background of the fixed distant stars, the second employs the standard Newtonian Mechanics modified in a rotating reference frame.

This idea of relative rotation is inherent in Einstein's general relativity because rotation changes the space-time geometry from that of a static mass, as a consequence of which there always exists a relative drag of inertial frames, which when analysed properly can be ascertained as the effect of inertial forces coming from rotational acceleration. In fact, Einstein himself in 1912 had shown in a preliminary version of a scalar gravitation theory that an accelerated mass shell exerts inertial forces on a test particle near the centre of the shell. However, later, within the framework of General Relativity, Thirring [3] demonstrated that an infinitely thin rotating mass shell (mass M, radius R and angular velocity ω) exerts a Coriolis-type force $\approx (-8Mm/3R)(\omega \times v)$ on a test particle (mass m, velocity v) near the centre of the shell in the approximation $M/R \ll 1$, $\omega R \ll 1$ and an additional force $\approx (-4Mm/15R)\,[\omega \times (\omega \times r) + 2(\omega \cdot r)\omega]$ in the second order of ω, which was interpreted as the centrifugal force.

Unfortunately, as later showed by Lanczos [4], this paper of Thirring [3] violated the energy conservation law at order ω^2 and thus is suspect.

In a classic paper, Brill and Cohen [5] gave an analysis of the question of rotating masses and their effect on inertial frames and obtained the solution for a rotating mass shell with flat space inside and a gravitational field outside, exact upto the first order in ω, which in the weak field limit matched that of Thirring:

$$ds^2 = \psi^4 \left[dr^2 + r^2 d\theta^2 + r^2 \sin^2\theta (d\phi - \Omega(r)dt)^2\right] - V^2 dt^2,$$

where $\Omega(r)$ describes local rotation of inertial frames as seen from infinity.

$$\begin{aligned} U^\mu : U^0 &= \left(1-\sigma^2\right)^{-1/2}, \; U^1 = U^2 = 0, \; U^3 = \sigma/\left(1-\sigma^2\right)^{1/2}, \\ \sigma &= r\psi^2 \sin\theta \left(\omega_s - \Omega\right)/V, \end{aligned}$$

ω_s = angular velocity of the shell, $\alpha = m/2r$.

$$\begin{aligned} \Omega &= \left(r_0\psi_0^2/r\psi^2\right)^3 \omega_s/\left[1 + 3\left(r_0 - \alpha\right)/4m\left(\left(1+\beta_0\right)\right]\right., \quad r > r_0 \\ &= \omega_s/\left(\left(1 + \left[3\left(r_0 - \alpha\right)/4m\left(1+\beta_0\right)\right]\right)\right., \quad r < r_0 \end{aligned}$$

which in the limit $\alpha \ll 1$ reduces to Thirring's result of

$$\Omega = \frac{4m\,\omega_s}{3r_0}.$$

On the other hand, if $r_0 = \alpha$, then inside the shell $\Omega = \omega_s$ which means there is no drag as the rotation rate of the inertial frame gets to be the same as that of the shell. In this spirit the 'distant fixed stars background' is imitated by a thin shell of matter, and as this influences the inertial frame, the effect is Machian.

Twenty years later, Pfister and Braun [6] considered this problem to include the effects of the order ω^2 in deriving the complete solution for a rotating spherical shell with flat space inside, with the metric form

$$ds^2 = -e^{2U} dt^2 + e^{-2U}\left[e^{2K}\left(dr^2 + r^2 d\theta^2\right) + W^2\left(d\phi - \omega A dt\right)^2\right]$$

and obtained a solution which gave the correct centrifugal term in the interior flat space. What is of significance is that the proper boundary fitting at the shell surface $r = r_0$ constrained the shape of the shell to be different from a sphere with the ellipticity of the shell showing an extremum.

Thus a rotating mass shell, with Coriolis and centrifugal forces at the centre has been shown to be an exact solution of Einstein's equation of General Relativity. However, there has been a continued controversy regarding the role of the Machian principle in General Relativity, and this, in our opinion, is due to the language of geometry which takes away the spirit of 'forces' which indeed are physical. Recently, in the last decade, a new way of introducing the concept of Newtonian forces in General Relativity has been introduced by Abramowicz and coworkers, using a 3+1 conformal splitting [7], which has indeed helped in understanding some new features inherent in General Relativity but which were not explicit. The first such feature is the reversal of centrifugal force along the photon circular orbit in static space-times [8] and the kinematical effects associated with this reversal. Further, using the covariant formalism of defining inertial forces in General Relativity [9, 10], Prasanna [11] has defined what is called a "cumulative drag index" for stationary space-time, which helps in seeing that the phenomena of rotational acceleration is inherent to the space-time and thus supports Mach's principle.

In the following, we shall briefly review the work depicting the kinematical effects of rotation in General Relativity which became apparent in the language of "inertial forces".

2. Formalism

Though in principle Abramowicz et al. [9] have defined inertial forces covariantly for space-time without any symmetry, from the point of view of applications one can restrict the treatment to stationary, axisymmetric space-times which have obviously two Killing vectors. It is again useful to use Bardeen's approach of Zero Angular Momentum Observers (ZAMO), which helps in slicing the space-time naturally by their integral hypersurfaces. With this background we consider the space-time [11]

$$ds^2 = g_{tt}dt^2 + 2g_{t\phi}dtd\phi + g_{\phi\phi}d\phi^2 + g_{rr}dr^2 + g_{\theta\theta}d\theta^2, \tag{1}$$

wherein the four-acceleration acting on a particle in circular orbit with constant angular velocity Ω may be decomposed covariantly [10] as given by

$$a_k = -\nabla_k\Phi + \gamma^2 v\left(n^i\nabla_i\tau_k + \tau^i\nabla_i n_k)\right) + (\gamma v)^2\tilde{\tau}^i\tilde{\nabla}_i\tilde{\tau}_k, \tag{2}$$

which has distinctly three parts and they may be easily identified to be [11]:

Gravitational:

$$(Gr)_i = -\nabla_i\Phi = \frac{1}{2}\partial_i\left\{\ln\left[\left(g_{t\phi}^2 - g_{tt}g_{\phi\phi}\right)/g_{\phi\phi}\right]\right\}, \tag{3}$$

Coriolis:

$$\begin{aligned}(Co)_i &= \gamma^2 v n^j\left(\nabla_j\tau_i - \nabla_i\tau_j\right)\\ &= -A^2\left(\Omega-\omega\right)\sqrt{g_{\phi\phi}}\left\{\partial_i\left(g_{t\phi}/\sqrt{g_{\phi\phi}}\right) + \omega\partial_i\sqrt{g_{\phi\phi}}\right\},\end{aligned} \tag{4}$$

and Centrifugal:

$$(Cf)_i = (\gamma v)^2\,\tilde{\tau}^j\tilde{\nabla}_j\tilde{\tau}_i = -\frac{A^2(\Omega-\omega)^2}{2}g_{\phi\phi}\partial_i\left\{\ln\left[g_{\phi\phi}^2/\left(g_{t\phi}^2 - g_{tt}g_{\phi\phi}\right)\right]\right\}. \tag{5}$$

Thus, given any axisymmetric, stationary space-time one can easily evaluate the gravitational, centrifugal and Coriolis components of the acceleration acting on a particle in circular orbit with constant angular velocity Ω.

On the other hand, instead of single particle motion, if we are interested in the kinematics of a rotating fluid configuration with four velocity U^i, one can use the Abramowicz-Carter-Lasota (ACL) formalism [7] of 3+1 slicing as applied to stationary space-time by considering the metric in the locally non-rotating frame. In general, for a perfect fluid distribution, one has the equations of motion as obtained from the conservation laws:

$$(\rho + p)\, U^i_{;j} U^j = h^{ij} p_{,j}, \tag{6}$$

where ρ is the matter density, p the pressure and h^{ij} the projection tensor. Considering a unit element of fluid one can then express the four force acting on it as given by

$$f_i = (\rho + p) \left[U^j \partial_j U_i - \frac{1}{2} U^\ell U^m \partial_i g_{\ell m} \right] + h^j_{\ i} p_{,j}, \tag{7}$$

implying that the vanishing of f_i gives the condition for equilibrium of the fluid element under the influence of forces acting on it.

Following ACL, one can consider the 3+1 slicing of the space-time

$$ds^2 = d\ell^2 - g_{00}\left(dt + 2\omega_\alpha dx^\alpha\right)^2, \tag{8}$$

with $d\ell^2 = \tilde{g}_{\mu\nu} dx^\mu dx^\nu$ representing the positive definite 3-metric, and rewrite the force components to be [12],

$$f_0 = \Phi^{-1}(\rho + p)\tilde{U}^\mu \partial_\mu U_0 + h^\mu_{\ 0} p_{,\mu} \tag{9}$$

and

$$\begin{aligned} f_\alpha \;=\;& \Phi^{-1}(\rho + p)\left[\tilde{U}^\mu \tilde{\nabla}_\mu \tilde{U}_\alpha + 2U^0 \tilde{U}^\mu \omega_{\mu\alpha} + (M_0^2/2\Phi)\partial_\alpha \Phi\right] \\ & + 2\omega_\alpha f_0 + \left(h^\mu_{\ \alpha} - 2\omega_\alpha h^\mu_{\ 0}\right) p_{,\mu}, \end{aligned} \tag{10}$$

with

$$\begin{aligned} M_0^2 &= U_0^2 - \tilde{g}_{\mu\nu}\tilde{U}^\mu \tilde{U}^\nu, \\ \tilde{U}^\mu &= \Phi U^\mu, \\ \tilde{U}_\alpha &= \tilde{g}_{\alpha\beta}\tilde{U}^\beta, \\ \omega_{\mu\nu} &= \partial_\mu \omega_\nu - \partial_\nu \omega_\mu, \\ g_{00} &= -\Phi, \\ g_{\mu 0} &= -2\Phi\omega_\mu, \\ g_{\mu\nu} &= \Phi\left(\tilde{g}_{\mu\nu} - 4\omega_\mu \omega_\nu\right). \end{aligned} \tag{11}$$

In order to consider the discussion of slowly rotating fluid distributions, we shall start with the most general stationary axisymmetric space-time metric, in the locally non-rotating frame as given by Chandrasekhar

$$ds^2 = -e^{2\nu}dt^2 + e^{2\psi}d\hat{\phi}^2 + e^{2\mu_1}dr^2 + e^{2\mu_2}d\theta^2, \tag{12}$$

with ν, ψ, μ_1 and μ_2 being functions of r and θ and $d\hat{\phi} = d\phi - \omega dt$. With $U^{\hat{\phi}} = \Omega - \omega$, being the only non-zero component of spatial velocity, the normalization condition leads to

$$U^0 = \left[e^{2\nu} - e^{2\psi}(\Omega - \omega)^2\right]^{-1/2}, \tag{13}$$

where $\Omega = d\phi/dt$ is the angular velocity of the fluid element and ω is the angular velocity of the locally non-rotating observer (LNRF) as defined by Bardeen, both considered with respect to the observer at infinity. This frame $\left(t,\ \hat{\phi},\ r,\ \theta\right)$ being static, the force equation as obtained above would have to be considered with $\omega_\alpha = 0$ and thus one gets the 3-component of the force acting on the fluid element to be

$$f_\alpha = \Phi^{-1}(\rho + p)\left[\tilde{U}^\mu \tilde{\nabla}_\mu \tilde{U}_\alpha + \frac{M_0^2}{2\Phi} \cdot \partial_\alpha \Phi\right] + h^\mu{}_\alpha p_{,\mu}. \tag{14}$$

3. Kinematical Effects in Static and Stationary Space Times

The introduction of inertial forces in General Relativity reveals some very interesting features associated with space-time, which were not at all obvious while looking at the structure or trajectories in the full four dimensional formalism. In static space-times, as there is no rotation, one has only the centrifugal and the gravitational forces acting on the test particle. It is best to take specific examples of space-times and then following the procedure of conformal splitting, obtain explicitly expressions for forces. As the general expressions have already been derived, given the metric g_{ij} one can read off the expressions for forces from eqs. 3-5.

3.1. SCHWARZSCHILD SPACE TIME

The metric is given by

$$ds^2 = -\left(1 - \frac{2m}{r}\right)dt^2 + \left(1 - \frac{2m}{r}\right)^{-1} dr^2 + r^2 d\Omega^2. \tag{15}$$

As this represents the field outside a static spherical mass m, the forces acting on a test particle of unit mass in circular orbit, in the plane $\theta = \pi/2$

are given by [8],

$$(Gr)_r = -\frac{m}{r^2}\left(1 - \frac{2m}{r}\right)^{-1} \tag{16}$$

and

$$(Cf)_r = \frac{+\Omega^2 r\,(1 - 3m/r)}{(1 - 2m/r - r^2\Omega^2)\,(1 - 2m/r)}. \tag{17}$$

The first consequence that is obvious is that the centrifugal force reverses its sign at $r = 3m$, the photon circular orbit in the Schwarzschild space-time. One of the important consequences of this centrifugal reversal at the photon orbit is that the usual Rayleigh criterion for local stability, with respect to infinitesimal perturbations conserving the specific angular momentum of a fluid in static space-time, also gets reversed.

Consider a collection of particles orbiting a Schwarzschild mass in circular orbits with the angular momentum distribution $\ell = \ell(r)$. The equilibrium condition is given by the balance of forces

$$\zeta\left(r, \ell^2\right) + \chi(r) = 0, \tag{18}$$

where ζ represents the centrifugal force and χ the total force that depends only upon the position and not on any conserved quantity. Assuming now that the force acting on an element of matter displaced from its original position, say r_0, depends only on its new position $r_1 = r_0 + \Delta r$, the centrifugal force in the new position depends upon both the position and the original angular momentum $\ell_0 = \ell\,(r_0)$ resulting in an unbalanced radial force

$$\begin{aligned} \Delta F &= \left[\zeta\,(r_1, \ell_0^2) + \chi\,(r_1)\right] - \left[\zeta\,(r_1, \ell_1^2) + \chi\,(r_1)\right] \\ &= -(\partial\zeta/\partial\ell^2)(d\ell^2/dr)\Delta r. \end{aligned} \tag{19}$$

For a stable situation this unbalanced force would bring the displaced element back towards the equilibrium if

$$\Delta F \cdot \Delta R < 0. \tag{20}$$

This leads to the Rayleigh criterion

$$\frac{d\zeta}{d\ell^2}\frac{d\ell^2}{dr} > 0. \tag{21}$$

In the usual Newtonian theory, as $\zeta \propto \ell^2/r^3$, the criterion demands $(d\ell^2/dr)$ to be always positive, showing that for a stable configuration the specific angular momentum always increases with r. On the other hand, in the present case with ζ as given by eq. 25, the stability criterion outside $r = 2m$ the horizon demands

$$\left(1 - \frac{3m}{r}\right)\frac{d\ell^2}{dr} > 0. \tag{22}$$

Hence for $r > 3m$ one gets the familiar Newtonian result of angular momentum increasing outwards, whereas for $r < 3m$, the angular momentum should decrease with increasing radius. This new effect is purely general relativistic and could be shown explicitly in the language of inertial forces. In fact this effect has been noticed in the full four-dimensional formalism by Anderson and Lemos [13] while studying the dynamics of accretion disks, in the form that very close to black holes the viscous torque transports angular momentum inwards rather than outwards.

3.2. ERNST SPACE TIME

The next example of static space-time that was considered to study the behaviour of inertial forces was the Ernst space-time [14] which represents the field outside a Schwarzschild mass embedded in a uniform magnetic field, B_Γ, as given by

$$\begin{aligned} ds^2 &= -\Lambda^2\left[\left(1-\frac{2m}{r}\right)dt^2-\left(1-\frac{2m}{r}\right)^{-1}dr^2-r^2d\theta^2\right] \\ &\quad +(r^2\sin^2\theta/\Lambda^2)d\phi^2, \\ \Lambda &= 1+B^2r^2\sin^2\theta, \\ B &= B_\Gamma G^{1/2}/c^2. \end{aligned} \tag{23}$$

The two forces acting on a particle in circular orbit on the equatorial plane are:

$$(Gr)_r = -2\Lambda\left[\frac{m}{r^2}+2B^2r\left(1-\frac{3m}{2r}\right)\right] \tag{24}$$

and

$$(Cf)_r = \frac{\ell^2}{r^3}\left[\frac{r^2\Lambda^3(r-2m)}{r^3-\ell^2\Lambda^4(r-2m)}\right]\left[1-\frac{3m}{r}-B^2r^2\left(3-\frac{5m}{r}\right)\right]. \tag{25}$$

Unlike in the pure Schwarzschild geometry, the centrifugal force goes to zero at two different locations which correspond to positive real roots of the cubic

$$-3B^2m^2R^3+5B^2m^2R^2+R-3=0 \tag{26}$$

for values of $Bm < 0.095$. Of the two, one is close to $r = 3m$ and the other very far away. It is easy to see that these two locations exactly coincide with the two circular photon orbits as given by the photon effective potential

$$V_{\text{ph}} = \Lambda^4\ell^2\left(1-\frac{2m}{r}\right)/r^2. \tag{27}$$

This in fact implies that the normal trajectories for test particles as depicted by the equilibrium of forces exist only in between the two photon orbits in this space-time. However, the outer circular photon orbit is very far away and thus becomes significant only for cosmological distances.

3.3. STATIONARY SPACETIMES

As inertial forces are intrinsic to rotation, it is important to understand their behaviour for stationary, axisymmetric space-times like that of Kerr geometry which represents the field outside a rotating black hole. Prasanna and Chakrabarti [15, 16] considered this, mainly with a view to find out the location of the centrifugal reversal radius with respect to the circular photon radius as there are two of them - one prograde and the other retrograde. Using the optical reference geometry approach they obtained the requisite expressions for the Newtonian (inertial) forces in the Boyer-Lindquist coordinates and showed that both the centrifugal and the Coriolis forces reverse sign at several different locations, although none of these linked to the circular photon radii as in the static space-time. They also pointed out the possible relevance of such reversals particularly for the study of the stability properties of compact rotating stars and accretion disks in hydrostatic equilibria. However, there was a lacuna in this study to the extent that as the ergo-surface of the Kerr black hole represents the last static surface in the Boyer-Lindquist frame, the optical reference geometry approach is valid only beyond this and so also the interpretation of the analysis. Sai Iyer and Prasanna [17] overcame this lacuna by studying the forces in Kerr geometry using the locally non-rotating frame for the 3+1 splitting and obtained the expression for the centrifugal force acting on a test particle in circular orbit to be

$$(Cf) = \frac{L^2 r \left(r^5 - 3mr^4 + a^2 r^3 - 3ma^2 r^2 + 6m^2 a^2 r - 2ma^4\right)}{\left(r^3 + a^2 r + 2ma^2\right)^3}. \tag{28}$$

It is clear that this reverses sign at the zeros of the numerator which is a quintic equation. Using Sturm's theorem one can ascertain that for a given value of the Kerr parameter a, there are only three real positive roots of the equation for $0 \leq r < \infty$. Out of these three, two lie within the event horizon while the third is always between $r = 2m$ and $3m$. For $a = 0$ (Schwarzschild) the location is at $3m$, while for the extreme Kerr ($a = 1$) the location is at $2.783m$. Another feature of interest noticed by Iyer and Prasanna is that the centrifugal force shows a clear maximum, whose value decreases with increasing a.

Having seen the behaviour of the centrifugal force on a test particle, we next consider its behaviour in the context of a fluid distribution, which can relate the effects to give a new insight to the change of shape that could occur in rotating fluid configurations.

4. Centrifugal Reversal and Ellipticity of A Slowly Rotating Fluid Configuration

Rotating fluid configurations have been of great interest in the astrophysical context as they are basic for understanding the nature of compact objects and features of gravitational collapse. We have already seen that for a fluid distribution the Rayleigh criterion for stability and the nature of viscous torque both get modified very close to black holes due to the effects of centrifugal reversal. What could be the effect of this feature for a collapsing, rotating fluid sphere as it reaches the status of an ultra compact object which is not a black hole? It is well known that Chandrasekhar and Miller [18] studied slowly rotating homogeneous masses in which the energy density is constant, in the framework of General Relativity and found a result of particular interest, showing that the ellipticity of the configuration, for varying radii but constant mass and angular momentum exhibits a very pronounced maximum. Their interpretation of this result was the hint that it could be mainly due to the effect of dragging of inertial frames. However, after the advent of the 'effect of centrifugal reversal', Abramowicz and Miller [19] considered the behaviour of ellipticity for relativistic Maclaurin spheroids, using the corrected expression for centrifugal force arising from the optical reference geometry approach and obtained the result

$$\bar{\epsilon}\left(\bar{R}\right) = \frac{125}{32}\left[1 - \frac{3}{2\bar{R}}\right]\left(\frac{1}{\bar{R}}\right), \tag{29}$$

where $\bar{R} = R/R_s$ with R_s being the Schwarzschild radius. This is different from the monotonically increasing Newtonian expression of $1/R$ and clearly shows that $\bar{\epsilon}$ has an extremum and particularly a maximum at $R = 3R_s$.

It is important to realise that though the Abramowicz-Miller result explains clearly the existence of an extremum of ellipticity, their use of the Schwarzschild background geometry for a rotating configuration is questionable. In view of this, Gupta, Iyer and Prasanna [12] reconsidered this analysis by taking the metric for the Maclaurin spheroid to be the solution of Hartle [20] and Hartle and Thorne [21], representing the space-time metric of a slowly rotating fluid configuration obtained as a perturbation of the solution for a homogeneous sphere corrected by terms upto second order in the angular velocity $(\mathcal{O}\left(\Omega^2\right))$. Since the idea here was to study the ellipticity, they first obtain an expression for eccentricity through the Newtonian force balance equation

$$g_{\text{equator}} - a\Omega^2 = g_{\text{pole}}\left(1 - e^2\right)^{1/2} \tag{30}$$

and generalise it to

$$F_{ge} - F_{cf} = F_{gp}\left(1 - e^2\right)^{1/2}, \tag{31}$$

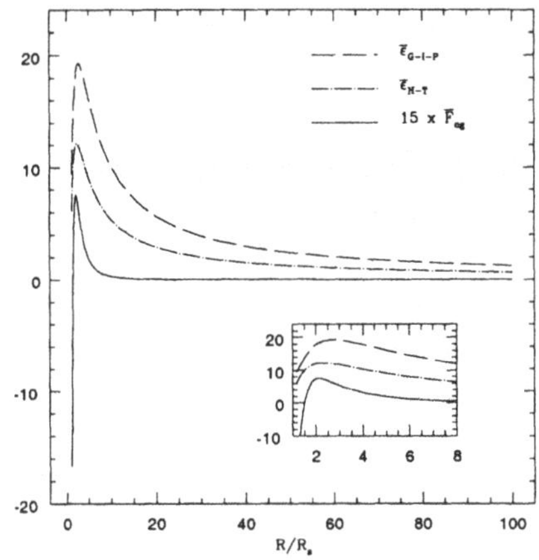

Figure 1. Plots for centrifugal force (in units $(G^2J^2/C^4R_s^5)$) and ellipticity as a function of R/R_s for a homogeneous distribution.

wherein F_g and F_{cf} are the gravitational and centrifugal forces acting on a fluid element as obtained in eqs. 9-11 using the ACL formalism.

For the Hartle metric one can work out these force expressions which turn out to be :

$$\begin{aligned} F_{cf} &= r^2\bar{\omega}^2\left(1/r - \nu_0'/2\right), \\ F_{ge} &= \tfrac{1}{2}e^{\nu_0}\left[\nu_0'\left(1 + 2h_0 - h_2\right) + 2h_0' - h_2'\right], \\ F_{gp} &= \tfrac{1}{2}e^{\nu_0}\left[\nu_0'\left(1 + 2h_0 + 2h_2\right) + 2h_0' + 2h_2'\right], \end{aligned} \tag{32}$$

wherein $\bar{\omega}$ is the angular velocity of the fluid relative to the local inertial frame, h_0 and ν_0 being the potential functions for the non-rotating configuration, while h_2 is the perturbed potential arising from rotation. The subscripts ge and gp indicate the gravitational force at the equator and pole, respectively. The actual equations to determine these quantities are the same as in Hartle and Thorne [21]. Using the relevant expressions in eq. 39 and noting that in the limit of slow rotation ($e \ll 1$) $\epsilon = e^2/2$, one gets from eq. 38 for the ellipticity of the configuration

$$\epsilon = 3\left(h_2 + \frac{h_2'}{\nu_0'}\right) + r^2\bar{\omega}^2 e^{-\nu_0}\left[(2/r\nu_0') - 1\right]. \tag{33}$$

Fig. (1) shows the plots of centrifugal force and of ellipticity for a sequence of homogeneous configurations of varying radii, keeping the mass and angular momentum constant.

The centrifugal force reaches a maximum at $R \approx 2.1R_s$ and turns *negative* for $R \sim 1.45R_s$ while the ellipticity reaches a maximum at $R \sim 2.75R_s$, a value closer to the exact result of Chandrasekhar and Miller [18]

($R \sim 2.3R_s$) as compared to that of Abramowicz and Miller [19] ($R \sim 3R_s$) who did not consider the changes in the metric potential arising from rotation.

Gupta, Iyer and Prasanna [22] extended this study to consider the effect of centrifugal reversal for inhomogeneous distributions by considering the fluid distribution with different equations of state. The results show that while centrifugal force attains a maximum somewhere between $R/R_s = 2.1$ and 2.3, its reversal of sign can be seen only in one case (EOS B), whereas in the other cases the equilibrium configuration becomes unstable before reaching the value $R/R_s \approx 1.5$.

On the other hand a completely new situation arises as far as the ellipticity is concerned in the sense that the ellipticity keeps decreasing and reaches *negative maximum*, indicating that the configurations turn to prolate shape.

It is in fact not a completely new situation, for Pfister and Braun [6], while considering the solution for a rotating mass shell with correct Coriolis and centrifugal forces at the centre, have found that a proper matching of the solutions across the shell surface would be possible only when the ellipticity is negative indicating that the configuration has to be prolate. This feature, however, needs a more careful looking into as it is very important to understand the change in shape of ultra compact objects, which would have a bearing on possible gravitational radiation emission as well as structural stability of the configuration.

5. Frame Dragging and Cumulative Drag Index

As General Relativity provides an intricate relation between the space-time geometry and physics, it is important to identify what is intrinsic to space-time structure that can influence physical characteristics arising from matter distribution in the space-time. As all motion is relative, what is rotation relative to? Is there a force that induces rotation or is rotation intrinsic to space-time structure? The best way to understand this is to see the influence of a given space-time on a test particle. In fact, it is well known that whereas a static space-time offers only the gravitational and centrifugal forces for balancing a test particle in its trajectory, a stationary space-time has Coriolis force too along with the other two. It is also well known that the stationary space-time induces inertial frame dragging, normally referred to as the Lense-Thirring effect. The drag when viewed in the full four dimensional formalism is a net effect due to the total acceleration induced by the curvature. As we now have a covariant decomposition of the total acceleration into various inertial and gravitational parts, it is worth examining the 'drag' as an effective consequence of the relative acceleration components. With this in view, a new parameter has been defined called the 'cumulative drag index' [11, 23] as given by

$$\mathcal{C} = \frac{(Cf + Co - Gr)}{(Cf + Co + Gr)}, \tag{34}$$

wherein Cf is the centrifugal, Co is the Coriolis and Gr the gravitational accelerations acting on the particle in circular orbit in the given space-time. Obviously the index is finite for all non-geodesic orbits and thus characterises the nature of the drag.

Prasanna [11] considered the nature of the drag coefficient for orbits along which the centrifugal force is zero, in Kerr and Gödel space-time and showed the remarkable similarity both for co-rotating and counter-rotating particles in circular orbits.

In fact these two space-times are ideal for studying the rotational aspects, as one describes the space-time outside a rotating black hole, while the other represents a homogeneous rotating Universe. Also, as the orbit considered is the one where centrifugal force is zero, the only forces acting on the particle are the Coriolis and the gravitational forces. The index is positive only for a very small range of Ω values, $|\, \Omega \,| \leq \Omega_{\pm}$, where $\Omega_{\pm}$ correspond to geodesic angular velocity for a given r and a. This behaviour is same both in the black hole space-time as well as the cosmological space-time for both co- and counter-rotating particles.

On the other hand, if we consider the drag index $\mathcal{C}$ at any other value of r, then its behaviour shows a characteristic difference for orbits inside the photon orbit and the ones outside the photon orbit. In fact, as depicted in Fig. 2, the drag index is positive beyond the prograde photon orbit for all values of Ω, whether the particle is co-rotating or counter-rotating. This feature clearly shows that except very close to black holes, the rotational drag does not distinguish co- and counter-rotation of test particles, indicating that the feature is basically intrinsic to the space-time or Machian. Further, as shown by Prasanna and Iyer [23], this feature of rotational drag is almost independent of approximations as the behaviour of the drag index is almost the same in the full Kerr geometry or linearised Kerr geometry or in the Newtonian limit. Prasanna and Iyer have also considered the behaviour of $\mathcal{C}$ for the Neugebauer-Meinel metric which represents the space-time of a rotating disk of dust as discussed by Meinel and Kleinwächter [24]. The significant result obtained from the study of the drag index is the exact similarity of the constraint on the retrograde orbits of particles in the disk. The remarkable similarity of the index behaviour for the prograde geodesic for $\mu = 0.5$, with that of the behaviour for the retrograde photon orbit in the Kerr space-time clearly indicates that $\mu = 0.5$ corresponds to the last possible counter-rotating orbit in the disk. As μ is related to z, the redshift

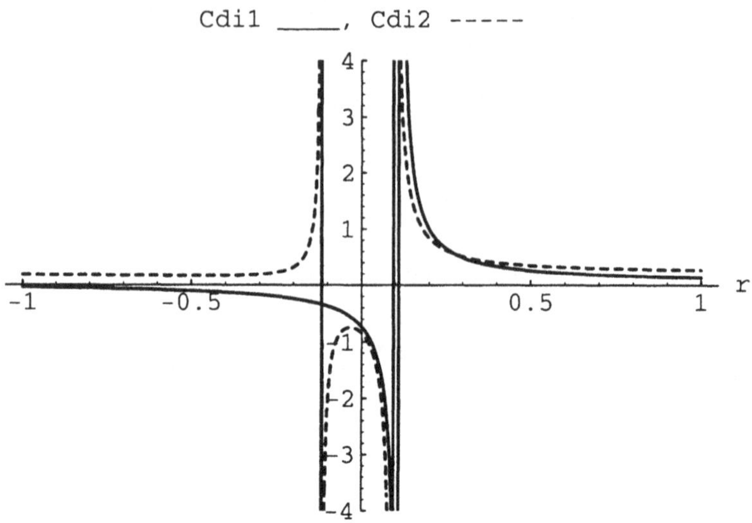

Figure 2. Plot of $\mathcal{C}(\Omega)$ for $a = 1$, $r = 4$ and $a = 0.9$, $r = 4.5$ for Kerr space-time when $Cf \neq 0$.

parameter, one can clearly see that counter-streaming particles would have to have $z < 0.285$. This may indeed act as a possible test for understanding the significance of the drag index, when tested through the redshift measurement of counter-rotating streams in the Virgo cluster. We shall return to the discussion of the drag index again, while considering the gyroscopic precession in the black hole space-time.

6. Gyroscopic Precession and Inertial Forces

It is well known that the structure of space-time is best understood through the dynamics of test particles, which are geodesics when no force is acting on them. For particles with spin which may or may not be following geodesics, there will be a spin precession, arising from the spin angular velocity precession relative to the fixed frame at infinity, an effect often referred to as arising from inertial frame dragging. The best way to test this effect is through a gyroscope, whose spin vector follows a Fermi-Walker transport so that there is no precession in the local frame. Though there are different ways to consider the gyroscopic precession, the most elegant way is due to Iyer and Vishveshwara [25], who use the Frenet-Serret formalism that characterises the trajectory of the gyroscope in an invariant geometric manner. We do not discuss the formalism here as it has been discussed elsewhere in this volume [26].

What is of interest is to see the relation of the gyroscopic precession with the inertial forces, which could give a better understanding of the geometrical features of space-time in terms of more tangible physical effects. Nayak and Vishveshwara [27] have investigated this aspect both in static and stationary space-times, using the Frenet-Serret formalism. As examples for static space-times they consider the Ernst space-time and the Reissner-Nordström space-times and in both cases find that exactly at the photon circular orbit the gyroscopic precession *reverses* sign, just as the centrifugal force does.

On the other hand, in the Kerr-Newman space-time, there is no such correlation between gyroscopic precession reversal and the centrifugal force reversal. While the location of the orbit where the gyroscopic precession changes sign depends on the constant angular speed Ω of the gyroscope (particle), the location of the orbit where the centrifugal force reverses sign is independent of Ω. Further, neither of these orbits coincide with the photon circular orbit as in the case of static space-times. This should not be too surprising as in the stationary space-time the presence of Coriolis force would influence the gyroscopic precession and thus the behaviour of this effect depends on Ω, the angular velocity, explicitly. Specifically, for the Kerr metric the gyroscopic precession $\tau_1(r)$, for a gyroscope confined to the equatorial plane of the black hole is given by Prasanna [11] and Nayak and Vishveshwara [27],

$$\tau_1 = \frac{\Omega r^3 - 3m\Omega(1-\Omega a)r^2 + ma(1-\Omega a)^2}{r^2\left[-\Omega^2 r^3 + (1-\Omega^2 a^2)\,r - 2m\,(1-\Omega a)^2\right]}, \tag{35}$$

which when $a = 0$, is zero at $r = 3m$, as in the static Schwarzschild geometry. Fig. 3 gives the behaviour of τ_1 as a function of Ω at different r for $a = 1$, the extreme Kerr case. In general, one can see that the gyroscopic precession rate is positive only for a small range of values of Ω ($|\,\Omega\,| < 0.2$) and is negative for all other values, for locations $r \leq r_{php}$ the prograde photon orbit. However, for $r \geq 6$ the precession rate is positive for counter-rotating orbits and negative for co-rotating orbits for $|\,\Omega\,| > \Omega_\pm$, the values corresponding to the location of infinite redshift orbit and is vice versa for $|\,\Omega\,| < \Omega_\pm$.

Considering both the cumulative drag index and the gyroscopic precession for the same set of parameters one finds that sufficiently far away from the black hole for the same values of a and Ω, the drag index and the gyroscopic precession have exactly opposite behaviour. However as a function of Ω for a given a the behaviour of $\mathcal{C}$ and τ_1 differ depending upon the location, whether the particle is inside the prograde photon orbit or outside, as depicted in Figs. 4 and 5. When the drag index is infinite, as the sum of the forces acting on the particle is zero, the gyroscope will be

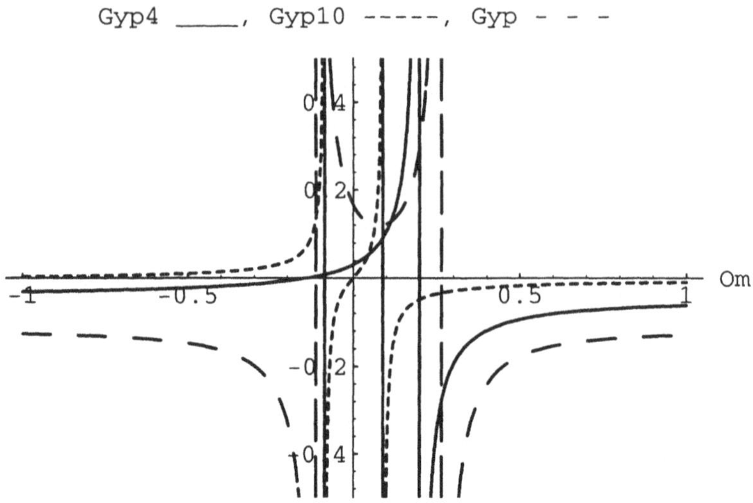

Figure 3. Plots for τ_1 as a function of Ω for $a = 1$ at $r = r_{php}$ the prograde photon orbit, at 10 and where the centrifugal force is zero.

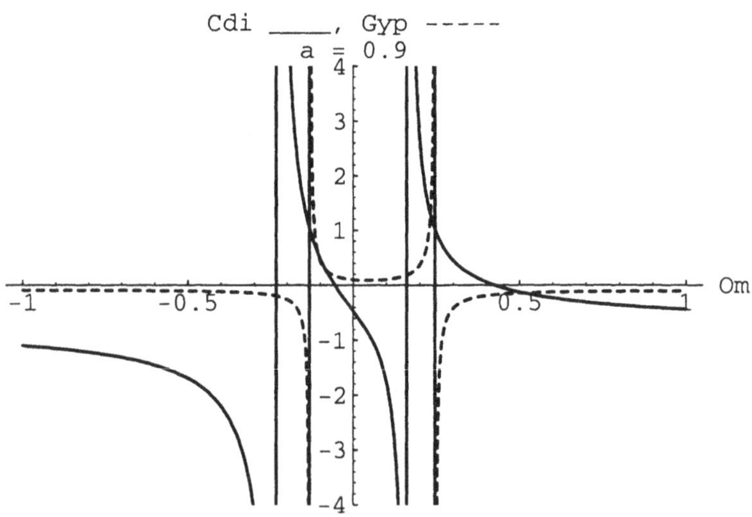

Figure 4. Plots for C and τ_1 as a function of Ω for $a = 0.9$ and $r = 3$.

on a geodesic and the precession rate is positive if it is co-rotating and negative if it is counter-rotating. As this measurement is made by the locally non-rotating observer, it clearly shows that there exists a cosmic rotation which influences the gyroscope rotating around a black hole, to increase the precession if it is co-rotating and decrease it if it is counter-rotating.

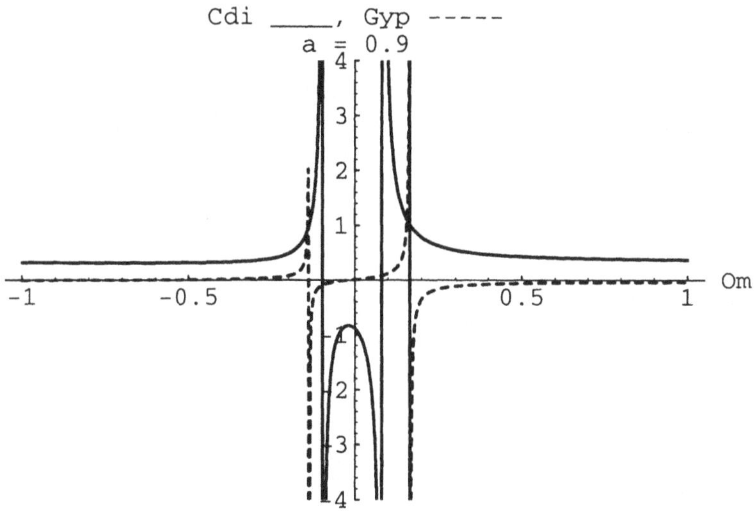

Figure 5. Same as Fig. 4 for $r = 5$ (outside r_{php}).

In fact, another aspect that signifies the global feature of rotation is that the behaviour of neither the drag index nor that of the precession rate of the gyroscope changes across the location where the centrifugal force goes to zero.

7. Concluding Remarks

Defining the inertial (Newtonian) forces in general relativity, it has been seen that there are several interesting features in the general relativistic formalism of particle and fluid behaviour in strong gravitational fields, which could be of significance both for astrophysical and fundamental physics phenomena. The reversal of centrifugal force and the behaviour of ellipticity (which shows a possible change in the shape of the ultra compact fluid configuration) can be of significance while considering the dynamics of ultra compact objects resulting from continued gravitational collapse. Existence of prolate configurations in the case of inhomogeneous fluid distributions can be of particular interest while considering hybrid stars and further in calculating the final burp of gravitational radiation that the collapsing star may emit.

On the other hand, with the help of the covariant definition of inertial forces we have been able to define a new index whose behaviour for different space-times, exhibits features to indicate that the rotation is an intrinsic character of space-time and has a global influence on particles in circular motion independent of whether their sense of rotation is co- or counter-with respect to that of either a black hole or that of a rotating Universe. The clear presence of this effect within the framework of general relativity, which becomes apparent when analysed through the introduction of inertial forces explicitly, is in our view a manifestation of the Machian view of the influence of cosmic motion of matter distribution on local phenomena.

We are very happy to dedicate this article to the occasion of the sixtieth birthday of our dear friend VISHU, wishing him a HAPPY BIRTHDAY and many many more years of fruitful research to leave behind a trail for generations to follow.

References

1. Weinberg, S. (1972) *Gravitation and Cosmology*, p. 16.
2. Narlikar, J. V. (1977) *The Structure of the Universe*, p. 168.
3. Thirring, W. (1918) *Physicalische Zeitschrift* **19**, 33.
4. Lanczos, K. (1923) *Zeit. f. Physik* **14**, 204.
5. Brill, D. R. and Cohen, J. (1966) *Phys. Rev.* **143**, 1011.
6. Pfister, H. and Braun, K.H. (1985) *Class. & Quan. Grav.* **2**, 909.
7. Abramowicz, M. A., Carter, B. and Lasota, J.P. (1988) *G.R.G. Journal* **20**, 1173.
8. Abramowicz, M. A. and Prasanna, A. R. (1990) *M.N.R.A.S.* **245**, 720.
9. Abramowicz, M. A., Nurowski, P. and Wex, N. (1993) *Class. & Quan. Grav.* **10**, L183.
10. Abramowicz, M. A., Nurowski, P. and Wex, N. (1995) *Class. & Quan. Grav.* **12**, 1467.
11. Prasanna, A. R. (1997) *Class. & Quan. Grav.* **14**, 227.
12. Gupta, A., Iyer, S. and Prasanna, A. R. (1996) *Class. & Quan. Grav.* **13**, 2675.
13. Anderson, M. R. and Lemos, J. P. S. (1988) *M.N.R.A.S* **233**, 489.
14. Prasanna, A. R. (1991) *Phys. Rev.* **D43**, 1418.
15. Prasanna, A. R. and Chakrabarti, S. K. (1990) *Gen. Rel. Grav.* **22** 9987.
16. Chakrabarti, S. K. and Prasanna, A. R. (1990) *J. Astrophys. Astron.* **11**, 29.
17. Iyer, S. and Prasanna, A. R. (1993) *Class & Qua. Grav.* **10**, L13.
18. Chandrasekhar, S. and Miller, J. C. (1974) *M.N.R.A.S.* **167**, 63.
19. Abramowicz, M. A. and Miller, J. C. (1990) *M.N.R.A.S.* **245**, 729.
20. Hartle, J. (1967) *Astrophys. J.* **150**, 1005.
21. Hartle, J. and Thorne, K. S. (1968) *Astrophys. J.* **153**, 807.
22. Gupta, A., Iyer, S. and Prasanna, A. R. (1997) *Class. & Quan. Grav.* **14**, L143.
23. Prasanna, A. R. and Iyer, S. (1997) *Phys. Letts.* **A233**, 17.
24. Meinel, R. and Kleinwächter, A., in *Mach's Principle from Newton's Bucket to Quantum Gravity* (1993), Eds. Barbour and Pfister, p. 339.
25. Iyer, B. R. and Vishveshwara, C. V. (1993) *Phys. Rev.* **D48**, 5706.
26. See chapter 13 by Nayak in this volume.
27. Nayak, K.R. and Vishveshwara, C.V. (1996) *Class. & Quan. Grav.* **13**, 1783.

13. GYROSCOPIC PRECESSION AND INERTIAL FORCES

The Frenet-Serret Approach

K. RAJESH NAYAK
Indian Institute of Astrophysics,
Bangalore 560 034, India.

I wish to express my deep gratitude to C. V. Vishveshwara for his encouragement. It is a great honour to be his student. It is also a pleasure to write this article in honor of him.

The phenomenon of gyroscopic precession is of interest in various situations in physics. In the context of the general theory of relativity, gyroscopic precession involves kinematic effects, contributions from spacetime curvature and the effect of inertial frame dragging when the spacetime possesses inherent rotation. It has been proposed as a test of the theory itself. In their recent work, Iyer and Vishveshwara [1] have presented a detailed study of this phenomenon in stationary spacetimes. Their approach involved the use of the elegant and covariant Frenet-Serret formalism for the general description of precession and for computations pertaining to specific examples. Such examples would be of interest not only from conceptual point of view but also for astrophysical applications. This forms the basis for a covariant description of gyroscopic precession. At the same time, there has been considerable interest in the study of the general relativistic equivalents of inertial forces, especially the centrifugal force, pursued notably by Abramowicz and coworkers. Abramowicz, Nurowski and Wex [2] have presented the most general formalism for inertial forces. The motivation for this work stemmed from the earlier studies of centrifugal force in static spacetimes and its reversal at the photon orbits. This occurs in the Schwarzschild spacetime at $r = 3M$ where a circular null geodesic exists. This phenomenon is therefore relevant to compact objects with M/R(compactness) smaller than $1/3$. Objects of this nature termed ultracompact objects were first considered by Iyer, Vishveshwara and Dhurandhar [9] and were shown to exist as stable configurations. This is in fact a generic property of static spacetimes in which particles following circular orbits experience only gravitational and centrifugal forces [5, 6, 7]. In the case of stationary spacetimes, the situation alters significantly. Additional

B. R. Iyer and B. Bhawal (eds.), Black Holes, Gravitational Radiation and the Universe, 207–218.

inertial forces such as the coriolis force can now exist. We have now, on the one hand, gyroscopic precession which is influenced not only by spacetime curvature but also by the rotation inherent to the spacetime and on the other hand, the general relativistic equivalents of inertial forces in whose description too the rotational effects enter. Furthermore, one would like to study the possible reversal of both gyroscopic precession and the centrifugal force in a stationary spacetime. Taking the Kerr-Newman spacetime as a specific example of stationary spacetimes Nayak and Vishveshwara [5] have shown that, the situation in the case of stationary spacetimes is much more complicated than in the case of static spacetimes. Neither centrifugal force nor gyroscopic precession reverses at the photon orbit.

The above studies raise some interesting questions. Is gyroscopic precession directly related to centrifugal force in all static spacetimes? If so, do they both necessarily reverse at the photon orbit? In the case of stationary spacetimes is it possible to make a covariant connection between the gyroscopic precession on the one hand and the inertial forces on the other, not necessarily just the centrifugal force? Does such a connecting formula reveal the individual non-reversal of gyroscopic precession and centrifugal force at the photon orbit? In this article we consider these and related questions.

1. Frenet-Serret Formalism

The Frenet-Serret(FS) formalism is one of the elegant methods for investigating the geometric properties of the curves. In this formalism the curves are studied by assigning an orthonormal frame called the 'Frenet-Serret frame' at each point. Rates of changes of these frames are expressed by Frenet-Serret formulae in term of scalar parameters called Frenet-Serret parameters . These parameters along with the Frenet-Serret frames describe the fundamental geometric properties of the curve.

The Frenet-Serret equations in general are given by

$$\begin{aligned} \dot{e}^a_{(0)} &= \kappa e^a_{(1)}, \\ \dot{e}^a_{(1)} &= \kappa e^a_{(0)} + \tau_1 e^a_{(2)}, \\ \dot{e}^a_{(2)} &= -\tau_1 e^a_{(1)} + \tau_2 e^a_{(3)}, \\ \dot{e}^a_{(3)} &= -\tau_2 e^a_{(2)}. \end{aligned} \tag{1}$$

The parameters κ,τ_1 and τ_2 are respectively the curvature, and the first and second torsions while $e^a_{(i)}$ form an orthonormal tetrad. These six quantities not only describe the geometry of the worldline completely but also elucidate the interlink between the physical quantities and geometric properties.

When the formalism is applied to the Killing trajectories of space-times many interesting features emerge. These considerations apply to a single trajectory in any specific example. However, additional geometric insight may be gained by identifying the trajectory as a member of one or more congruences generated by combining different Killing vectors. For this purpose the FS formalism is applied to quasi-Killing trajectories.

Let us consider a spacetime that admits a timelike Killing vector ξ^a and a set of spacelike Killing vectors $\eta_{(A)}$ (A=1,2,... m). Then a quasi-Killing vector may be defined as

$$\chi^a \equiv \xi^a + \omega_{(A)}\eta^a_{(A)}, \tag{2}$$

if the Lie derivative of the functions $\omega_{(A)}$ with respect to χ^a is assumed to vanish, $\mathcal{L}_\chi \omega_{(A)} = 0$. In the above index (A) is summed over. A congruence of quasi-Killing trajectories is generated by the integral curves of χ^a. As a special case we obtain a Killing congruence when $\omega_{(A)}$ are constants.

Assuming χ^a to be timelike, we may define the four velocity of a particle following χ^a by

$$e^a_{(0)} \;\equiv\; u^a \;\equiv\; e^\psi \chi^a. \tag{3}$$

e^ψ is the redshift factor and,

$$\dot{e}^a_{(0)} \;\equiv\; e^a_{(0);b}e^b_{(0)} \;=\; F^a_{\ b}e^b_{(0)}, \tag{4}$$

where

$$F_{ab} \;\equiv\; e^\psi \left(\xi_{a;b} + \omega_{(A)}\eta_{(A)a;b}\right). \tag{5}$$

The derivative of $\omega_{(A)}$ drops out of the equation. The Killing equation and the equation $\xi_{a;b;c} \equiv R_{abcd}\xi^d$ satisfied by any Killing vector lead to

$$F_{ab} \;=\; -F_{ba} \text{ and } \dot{F}_{ab} \;=\; 0. \tag{6}$$

In the case of the quasi-Killing trajectories one can show that κ, τ_1 and τ_2 are constants and that each of $e^a_{(i)}$ satisfies a Lorentz-like equation:

$$\dot{\kappa} = \dot{\tau}_1 = \dot{\tau}_2 = 0, \tag{7}$$

$$\dot{e}^a_{(i)} = F^a_{\ b}e^b_{(i)}. \tag{8}$$

Further, κ, τ_1, τ_2 and $e^a_{(\alpha)}$ can be expressed in terms of $e^a_{(0)}$ and $F^n{}_{ab} \equiv F_a{}^{a_1}F_{a_1}{}^{a_2}\cdots F_{a_{n-1}b}$. These relations were first derived by Honig, Schücking and Vishveshwara [10] to describe charged particle motion in a homogeneous electromagnetic field. Interestingly, they are identical to those that arise in the case of quasi-Killing trajectories [1].

One can also show that

$$\alpha \equiv \frac{1}{2}F^a_{\ b}F^b_{\ a} = \kappa^2 - \tau_1^2 - \tau_2^2. \tag{9}$$

The Eq.(9) can be written as,

$$\frac{1}{4}n^a n_a = \tilde{\omega}^a \tilde{\omega}_a + \frac{1}{2}(\chi_{b;c}\chi^{b;c})\chi_a\chi^a, \tag{10}$$

where n^a, the normal to the family of surfaces $\chi_a\chi^a$ = constant is given by $n_a = (\chi_b\chi^b)_{;a} = -2e^{-2\psi}\kappa e_{(1)a}$ and $\tilde{\omega}^a\tilde{\omega}_a = \tau_1^2 + \tau_2^2$. Equation (10) was first derived by Vishveshwara [11] to define the Killing horizon in black hole solutions like Schwarzschild and Kerr solutions [12].

The Frenet-Serret formalism can be generalized in a straightforward manner to higher dimensional spacetimes [13]. This would be an useful technique in investigating the black holes in higher dimensions.

2. Frenet-Serret Description of Gyroscopic Precession

The Frenet-Serret (FS) formalism offers a covariant method for treating gyroscopic precession. It turns out to be a convenient and elegant description of the phenomenon when the worldlines along which the gyroscopes are transported follow spacetime symmetry directions or Killing vector fields. In fact, in most cases of interest, orbits corresponding to such worldlines are considered for simplicity. As we shall see, the torsions τ_1 and τ_2 are directly related to gyroscopic precession.

Let us consider an inertial tetrad $\left(e^a_{(0)}, f^a_{(\alpha)}\right)$ which undergoes Fermi-Walker (FW) transport along the worldline. The triad $f_{(\alpha)}$ may be physically realized by a set of three mutually orthogonal gyroscopes. Then, the angular velocity of the FS triad $e^a_{(\alpha)}$ with respect to the FW triad $f^a_{(\alpha)}$ is given by [1]

$$\omega^a_{FS} = \tau_2 e^a_{(1)} + \tau_1 e^a_{(3)}. \tag{11}$$

Thus the gyroscopes precess with respect to the FS frame at a rate given by $\Omega_{(g)} = -\omega_{FS}$. In case of the Killing congruence ω_{FS} is identical to the vorticity of the congruence.

An axially symmetric stationary metric admits a timelike Killing vector ξ^a and a spacelike Killing vector η^a with closed circular orbits around the axis of symmetry. We assume orthogonal transitivity. The quasi-Killing vector field

$$\chi^a = \xi^a + \omega\eta^a \tag{12}$$

generates closed circular orbits around the symmetry axis with constant angular speed ω along each orbit. The FS parameters and the tetrad can be determined either by the direct substitution of χ^a or by transforming to a rotating coordinate frame as discussed in [1]. They can be written in terms of the Killing vectors and their derivatives as follows.

$$\kappa^2 = -g^{ab}a_a a_b \tag{13}$$

$$\tau_1^2 = \left[g^{ab} a_a d_b\right]^2 \tag{14}$$

$$\tau_2^2 = \left[\frac{\varepsilon^{abcd}}{\sqrt{-g}} n_a \tau_b a_c d_d\right]^2 . \tag{15}$$

In the above,

$$\begin{aligned} d_a &= \left(\frac{\mathcal{B}}{2\sqrt{-\Delta_3 \kappa}}\right)\left[\frac{\mathcal{B}_a}{\mathcal{B}} - \frac{\mathcal{A}_a}{\mathcal{A}}\right] \\ \mathcal{A} &= (\xi^a \xi_a) + 2\omega\,(\eta^a \xi_a) + \omega^2\,(\eta^a \eta_a) \\ \mathcal{B} &= (\eta^a \xi_a) + \omega\,(\eta^a \eta_a) \\ \mathcal{A}_a &= (\xi^b \xi_b)_{,a} + 2\omega(\eta^b \xi_b)_{,a} + \omega^2(\eta^b \eta_b)_{,a}; \quad a = 1, 2. \\ \mathcal{B}_b &= (\eta^a \xi_a)_{,b} + \omega(\eta^a \eta_a)_{,b}; \quad b = 1, 2. \\ \Delta_3 &= (\xi^a \xi_a)(\eta^b \eta_b) - (\eta^a \xi_a)^2, \end{aligned} \tag{16}$$

where n^a is the unit vector along $\zeta_a = \xi_a - \left(\xi^b \eta_b\right) / (\eta^c \eta_c)\, \eta_a$ and τ^i is the unit vector along the rotational Killing vector η^a. We may note that all the above equations can be specialized to a static spacetime by setting $\xi^a \eta_a = 0$ and $\zeta^a \equiv \xi^a$.

3. Inertial Forces

As has been mentioned earlier, in a recent paper Abramowicz *et. al.* [2] have formulated the general relativistic analogues of inertial forces in an arbitrary spacetime. The particle four velocity u^a is decomposed as

$$u^a = \gamma\,(n^a + v\tau^a)\,. \tag{17}$$

In the above, n^a is a globally hypersurface orthogonal timelike unit vector, τ^a is the unit vector orthogonal to it along which the spatial three velocity v of the particle is aligned and γ the normalization factor that makes $u^a u_a = 1$.

Then the forces acting on the particle are written down as

$$\begin{aligned} \text{Gravitational force } G_k &= \phi_{,k} \\ \text{Centrifugal force } Z_k &= -(\gamma v)^2\, \tilde{\tau}^i \tilde{\nabla}_i \tilde{\tau}_k \\ \text{Euler force } E_k &= -\dot{V} \tilde{\tau}_k \\ \text{Coriolis – Lense – Thirring force } C_k &= \gamma^2 v X_k \end{aligned} \tag{18}$$

where

$$\begin{aligned} \dot{V} &= (v e^{\phi} \gamma)_{,i} u^i \\ X_k &= n^i\,(\tau_{k;i} - \tau_{i;k}) \\ \phi_{,k} &= -n^i n_{k,i}. \end{aligned} \tag{19}$$

Here $\tilde{\tau}^i$ is the unit vector along τ^i in the conformal space orthogonal to n^i with the metric

$$\tilde{h}_{ik} = e^{-2\phi}\left(g_{ik} - n_i n_k\right) \tag{20}$$

As has been shown by Greene, Schücking and Vishveshwara [12], axially symmetric stationary spacetimes with orthogonal transitivity admit a globally hypersurface orthogonal timelike vector field

$$\zeta^a = \xi^a + \omega_0 \eta^a, \tag{21}$$

where the fundamental angular speed of the irrotational congruence is

$$\omega_0 = -(\xi^a \eta_a)/(\eta^b \eta_b). \tag{22}$$

The unit vector along ζ^a is identified with n^a. Further, if u^a follows a quasi-Killing circular trajectory, then τ^i is along the rotational Killing vector η^a. In this case it is easy to show that $\dot{V} = 0$ and hence the Euler force does not exist. From the above relations, we can write down the inertial forces from their definitions as follows.
Gravitational force

$$G_k = \phi_{,k}, \tag{23}$$

Centrifugal force

$$Z_k = \frac{1}{2} e^{2(\psi+\phi)} \tilde{\omega}^2 \left(\frac{\eta^a \eta_a}{\zeta^b \zeta_b}\right)_{,k}, \tag{24}$$

Coriolis-Lense-Thirring force

$$C_k = e^{2(\psi+\alpha)} \tilde{\omega} \left(\frac{\xi^a \eta_a}{\eta^b \eta_b}\right)_{,k}, \tag{25}$$

where $\phi = \frac{1}{2}\ln(\zeta^a \zeta_a)$, $\alpha = \frac{1}{2}\ln(-\eta^a \eta_a)$, $\psi = \frac{1}{2}\ln(\chi^a \chi_a)$ and $\tilde{\omega} = (\omega - \omega_0)$.

In a static spacetime the global timelike Killing vector ξ^a itself is hypersurface orthogonal. The unit vector n^a is now aligned along ξ^a,

$$n^a = e^{-\phi} \xi^a. \tag{26}$$

Then we have the inertial forces as follows:
Gravitational force

$$G_k = \phi_{,k}, \tag{27}$$

Centrifugal force

$$Z_k = -\frac{\omega^2}{2} e^{2(\psi+\alpha)} \left[\ln\left(\frac{\eta^i \eta_i}{\xi^j \xi_j}\right)\right]_{,k}, \tag{28}$$

Coriolis-Lense-Thirring force is identically zero.

4. Covariant Connections

In the preceding section we have derived expressions for τ_1 and τ_2 which give gyroscopic precession rate in terms of the Killing vectors. Similarly, inertial forces in an arbitrary axisymmetric stationary spacetime have also been written down in terms of the Killing vectors. All these quantities have been defined in a completely covariant manner. We shall now proceed to establish covariant connections between gyroscopic precession, *i.e.* the FS torsions τ_1 and τ_2, on the one hand and the inertial forces on the other. First, we shall consider the simpler case of static spacetimes.

4.1. STATIC SPACETIMES

We have derived in Eqs. (14) and (15), the FS torsions τ_1 and τ_2 for a stationary spacetime. As has been mentioned earlier, for a static spacetime $\xi^a\eta_a = 0$ and $\zeta^a = \xi^a$ in the above equations as well as in the expressions for inertial forces. With this specialization, centrifugal force can be written from Eq.(28) as

$$Z_b = e^{-(\phi-\alpha)}\omega\kappa d_b \tag{29}$$

Substituting Eq.(29) in Eqs.(14) and (15) we arrive at the relations

$$\tau_1^2 = \frac{\beta^2}{\omega^2}\left[a^b Z_b\right]^2, \tag{30}$$

and

$$\tau_2^2 = \frac{\beta^2}{\omega^2}\left[\frac{\varepsilon^{abcd}}{\sqrt{-g}}n_a\tau_b a_c Z_d\right]^2, \tag{31}$$

where

$$\beta = \frac{e^{(\phi-\alpha)}}{\kappa}. \tag{32}$$

The equations above relate gyroscopic precession directly to the centrifugal force. The two torsions τ_1 and τ_2, equivalent to the two components of precession, are respectively proportional to the scalar and cross products of acceleration and the centrifugal force. We shall discuss the consequences of these relations later on.

4.2. STATIONARY SPACETIMES

We decompose the angular speed ω with reference to the fundamental angular speed of the irrotational congruence $\omega_0 = -(\xi^a\eta_a)/(\eta^a\eta_a)$,

$$\omega = \tilde{\omega} + \omega_0. \tag{33}$$

From Eq.(16) and from Eq.(14), for τ_1^2 we can show

$$\tau_1^2 = \frac{\beta^2}{\tilde{\omega}^2}\left[g^{ab}a_a\left(Z_b + \beta_1 C_a\right)\right]^2 \tag{34}$$

where

$$\beta = \frac{e^{(\phi-\alpha)}}{\kappa}$$
$$\beta_1 = -\frac{1}{2}\left[1+\tilde{\omega}^2 e^{2(\alpha-\phi)}\right] \tag{35}$$

Again, from Eq.(15), we obtain the expression

$$\tau_2^2 = \frac{\beta^2}{\tilde{\omega}^2}\left[\frac{\varepsilon^{abcd}}{\sqrt{-g}}n_a\tau_b a_c\left(Z_d + \beta_1 C_d\right)\right]^2 . \tag{36}$$

These relations are more complicated than those we have derived in the static case. Nevertheless, they closely resemble the latter with the centrifugal force replaced by the combination of centrifugal and Coriolis forces $(Z_a + \beta_1 C_a)$. The static case formulae are obtained from those of stationary case by setting the Coriolis force to zero.

A formula for gyroscopic precession in the case of circular orbits in axially symmetric stationary spacetimes was derived by Abramowicz, Nurowski and Wex [15] within a different framework. We note that gyroscopic precession does not involve the gravitational force. In case of geodetic orbits, total force is zero but not the centrifugal and Coriolis force individually. Therefore gyroscopic precession is also nonzero even for geodetic orbits.

We may also note that the definition of inertial force is not unique. For instance, De Felice [16], Semerak [17] Barrabes, Boisseau and Israel [18] have suggested approaches different from the one employed in the present paper. The gyroscopic precession was solved in an alternative formalism by Semerak [19, 20] (see also De Felice [21]). A recent paper by Bini, Carini and Jantzen [22] has established the geometrical basis and interrelation of various modes of force splitting and applied their general results to circular orbits in axially symmetric stationary spactimes [23].

5. Reversal of Gyroscopic Precession and Inertial Forces:

The condition for the reversal of gyroscopic precession is given by

$$\omega^a_{FS} = \tau_1 e^a_{(3)} + \tau_2 e^a_{(1)} = 0. \tag{37}$$

Since $e^a_{(1)}$ and $e^a_{(3)}$ are linearly independent vector fields at each point, this condition is the same as requiring

$$\tau_1 = \tau_2 = 0. \tag{38}$$

By considering the actual structure of τ_2, it is easy to show that τ_2 becomes zero on a plane about which the metric components are reflection invariant. The equatorial plane in the black hole spacetime is an example of this.

We shall now examine the vanishing of the FS torsions expressed as above and in relation to the inertial forces.

In what follows, we shall prove a theorem that relates the simultaneous reversal of centrifugal force and gyroscopic precession along a time like circular orbit to the existence of a circular null geodesic with common orbit in three space orthogonal to n^a. We start from the condition for gyroscopic precession reversal, *i.e.* $\tau_1 = \tau_2 = 0$, and show that at the point where this occurs a null circular geodesic must exist along the same spatial three orbit.

We shall now assume that the gyroscope is transported along a circular orbit which is not a geodesic, *i.e.* $\kappa \neq 0$. This we do in anticipation of the result that a null geodesic, not a timelike one, exists with its spatial trajectory identical to that of this timelike orbit. Now $\kappa \neq 0$ implies $\mathcal{A}_1 \neq 0$ and $\mathcal{A}_2 \neq 0$ from Eq.(16). We arrive at

$$\mathcal{A}\mathcal{B}_a - \mathcal{B}\mathcal{A}_a = 0.$$

Then, Eq.(16) reduces this condition to

$$(\xi^b\xi_b)(\eta^c\eta_c)_{,a} - (\xi^b\xi_b)_{,a}(\eta^c\eta_c) = 0. \tag{39}$$

With the help of this equation we show below that, if a circular geodesic exists where precession reverses, then it has to be null.

The condition for circular geodesic is

$$(\xi^b\xi_b)_{,a} + \omega^2(\eta^b\eta_b)_{,a} = 0. \tag{40}$$

This can be proved from the geodesic equation, assuming that the four velocity u^a is proportional to $\xi^a + \omega\eta^a$. Using condition (39), this reduces to

$$\frac{(\xi^b\xi_b)_{,a}}{(\xi^c\xi_c)}\left[(\xi^d\xi_d) + \omega^2(\eta^d\eta_d)\right] = 0. \tag{41}$$

Since $(\xi^b\xi_b)_{,a}/2(\xi^c\xi_c)$ is the gravitational force, which is assumed to be nonzero, this is equivalent to

$$(\xi^d\xi_d) + \omega^2(\eta^d\eta_d) = 0. \tag{42}$$

This means that the geodesic, if one exists, is null. Now we shall show that in fact a geodesic must exist at the point of precession reversal.

If a geodesic does not exist at the point of reversal, then

$$(\xi^b\xi_b)_{,a} + \omega^2(\eta^b\eta_b)_{,a} \neq 0 \tag{43}$$

for all values of ω. However, Eq.(39) may be recast as

$$(\xi^b\xi_b)_{,a} - \left(\frac{\xi^c\xi_c}{\eta^c\eta_c}\right)(\eta^b\eta_b)_{,a} = 0. \tag{44}$$

This shows that the geodesic condition is satisfied for $\omega^2 = -(\xi^c\xi_c/\eta^c\eta_c)$. Therefore there does exist a geodesic and we have already shown that it has to be null. We shall now prove the converse, *i.e.* if a circular null geodesic exists then τ_1 and τ_2 are zero at the null geodesic.

The condition for a circular null geodesic is given by Eq.(39). Dividing this equation by $(\xi^a\xi_a)(\eta^b\eta_b)$, we see that it reduces to $\left[\ln\left(\eta^b\eta_b/\xi^c\xi_c\right)\right]_{,k}$ which is proportional to Z_a from Eq.(28) and is equal to zero. Further, from the dependence of τ_1 and τ_2 on Z_a from Eqs.(30) and (31) we see that $\tau_1 = \tau_2 = 0$. We may note the fact that both gyroscopic precession and centrifugal force reverse simultaneously as is evident from Eqs.(30) and (31). We have therefore proved the following theorem.

Theorem: In the case of circular orbits in static spacetimes reversal of gyroscopic precession and centrifugal force takes place at some point, if and only if a null geodesic exists at that point.

6. Gravi-electric and Gravi-magnetic fields

Gravielectric and gravi-magnetic fields are closely related to the idea of inertial forces. These fields with respect to observers following the integral curves of n^a can be defined as follows.
Gravi-electric field:

$$E^a = F^{ab}n_b \tag{45}$$

Gravi-magnetic field:

$$H^a = \tilde{F}^{ab}n_b \tag{46}$$

where $\tilde{F}^{ab}$ is the dual of F^{ab}. In the above, as before, $F^{ab} = e^{\psi}(\xi_{a;b}+\omega\eta_{a;b})$. Projecting onto the space orthogonal to n^a with $h_{ab} = g_{ab} - n_a n_b$ and decomposing u_a as given in (17), the force acting on the particle can be written as follows

$$\dot{u}_{\perp a} = \gamma\left[E + v\times H\right] = f_{GEa} + f_{GHa}. \tag{47}$$

We can therefore define
Gravi-electric force:

$$f_{GEa} = \gamma F_{ac}n^c \tag{48}$$

Gravi-magnetic force:

$$f_{GHa} \;=\; \gamma v\sqrt{-g}\varepsilon_{abcd}n^b\tau^c H^d. \tag{49}$$

In the stationary case, n^a is unit vector along irrotational vector field. As before we decompose $\omega = \tilde{\omega} + \omega_0$, where ω_0 is given by (22). Then a straightforward computation gives the expression for the gravi-electric force.

$$f_{GEa} \equiv -e^{2(\psi+\phi)}G_a \;+\; C_a. \tag{50}$$

This shows the relation of gravi-electric field or force to both gravitational and centrifugal forces.

$$f_{GHa} \;\equiv\; \left[\frac{C_a}{2} \;+\; e^{2(\psi+\alpha)}\tilde{\omega}^2 G_a \;-\; Z_a\right]. \tag{51}$$

Hence gravi-magnetic force is related to all the three inertial forces – gravitational, centrifugal and Coriolis.

Several authors have discussed gravi-electromagnetism in earlier papers with application to gyroscopic precession. We may cite as examples the papers by Embacher [24], Thorne and Price [25], Jantzen, Carini and Bini [26], and Ciufolini and Wheeler [27].

7. Conclusion

The main purpose of the present article was to establish covariant connection between gyroscopic precession on the one hand and the analogues of inertial forces on the other. This has been accomplished in the case of axially symmetric stationary spacetimes for circular orbits. In the special case of static spacetimes gyroscopic precession can be directly related to the centrifugal force. From this we have been able to prove that both precession and centrifugal force reverse at a photon orbit, provided the latter exists. In the case of stationary spacetimes, the corresponding relations are more complicated. The place of centrifugal force is now taken by a combination of centrifugal and Coriolis-Lense-Thirring forces. As a result, gyroscopic precession and centrifugal force do not reverse in general at the photon orbit. We have also studied some of the above aspects in the spacetime conformal to the original static spacetime [28]. In this approach part of the gravitational effect is factored out thereby achieving certain degree of simplicity and transparency in displaying interrelations and the reversal phenomenon. Closely related to these considerations is the idea of gravi-electric and gravi-magnetic fields. We have covariantly defined these with respect to the globally hypersurface orthogonal vector field that constitutes the general relativistic equivalent of Newtonian rest frame. In this instance, these fields can be related to the inertial forces. When these fields are formulated with respect to the orbit under consideration, they lead to a striking similarity to the corresponding physical quantities that arise for a charge moving in an actual, constant electromagnetic field. We have thus established connections and correspondences among several interesting general relativistic phenomena.

Acknowledgement :

I wish to thank B.R. Iyer for his critical reading and for his useful remarks on this article.

References

1. B.R. Iyer and C.V. Vishveshwara, *Phys. Rev.*, **D48**, 5706 (1993).
2. M.A. Abramowicz, P.Nurowski and N. Wex, *Class Quantum. Grav,*, **10**, L183 (1993).
3. M. A. Abramowicz and A.R. Prasanna, *Mon. Not. R. Astr. Soc*, **245**, 720 (1990).
4. A.R. Prasanna, *Phys. Rev,*, **D43**, 1418 (1991).
5. K. Rajesh Nayak and C. V. Vishveshwara, *GRG*, **29**, 291(1997).
6. M. A. Abramowicz, *Mon. Not. R. Astron. Soc*, **245**, 733 (1990).
7. W. Rindler and V. Perlick, *GRG*, **22**, 1067(1990).
8. K. Rajesh Nayak and C. V. Vishveshwara, *Class Quantum. Grav,*, **13**, 1173(1996).
9. B.R. Iyer, C.V. Vishveshwara and S.V. Dhurandhar, *Class. Quantum. Grav,*, **2**, 219(1985).
10. E. Honig, E.L. Schücking and C.V. Vishveshwara, *J. Math. Phys.*, **15**, 774 (1974).
11. C. V. Vishveshwara,*J. Math. Phys.*, **9**, 1319 (1968).
12. R. D. Greene, E. L. Schücking and C.V. Vishveshwara, *J. Math. Phys.*, **16**, 153 (1975).
13. B.R. Iyer and C.V. Vishveshwara *Class. Quantum. Grav,*, **5**, 961(1988).
14. M. A. Abramowicz, B. Carter and J. P. Lasota, *Gen. Rel. Grav.*, **20** 1173 (1988).
15. M.A. Abramowicz, P.Nurowski and N. Wex, *Class Quantum. Grav,*, **12**, 1467 (1995).
16. F. de Felice, *MNRAS*, **252**, 197 (1991).
17. O. Semerak, *N. Cimento*, **B 110**, 973 (1995).
18. C. Barrabes, B. Boisseau and W. Israel, *MNRAS*, **276**, 432 (1995).
19. O. Semerak, *CQG*, **13**, 2987 (1996).
20. O. Semerak, *GRG*, **29**, 153 (1997).
21. F. de Felice, *CQG*, **11**, 1283 (1994).
22. D. Bini, P. Carini, R.T Jantzen, *Int. J. Mod. Phys.*, **D 6** 1 (1997).
23. D. Bini, P. Carini, R.T Jantzen, *Int. J. Mod. Phys.*, **D 6** 143 (1997).
24. F. Embacher, *Found. Phys.*, **14**, 721 (1984).
25. K.S. Thorne, R.H. Price, D.A. Macdonald (Eds.): *Black Holes: The Membrane Paradigm*, Yale Univ. Press, New Haven (1986).
26. R.T. Jantzen, P.Carini and D. Bini, *Ann. Phys. (N.Y.)*, **215**, 1 (1992).
27. I. Ciufolini and J. A. Wheeler, *Gravitation and Inertia*, Princeton Univ. Press, Princeton (1995).
28. K. Rajesh Nayak and C. V. Vishveshwara, *Gyroscopic Precession and Inertial Forces in Axially Symmetric Stationary Spacetimes, GRG*, (in press) 1998.

14. ANALYSIS OF THE EQUILIBRIUM OF A CHARGED TEST PARTICLE IN THE KERR-NEWMAN BLACK HOLE

J.M. AGUIRREGABIRIA, A. CHAMORRO AND J. SUINAGA
Física Teórica, Universidad del País Vasco, Apdo. 644, 48080 Bilbao, Spain.

1. Introduction

It is well known that, contrary to what happens in prerelativistic physics, there is no general exact solution in general relativity to the problem of two extended charged bodies in equilibrium under the combined action of the gravito-electromagnetic forces due to their masses and charges. Approximate particular solutions for a static axisymmetric configuration with an arbitrary number of sources were obtained in 1947 by Papapetrou [1] and Majumdar [2] who found that, in appropriate units, the equilibrium condition was

$$m_i^2 = q_i^2, \qquad i = 1, \ldots, n, \tag{1}$$

m_i and q_i being the mass and electric charge of particle i. Kramer [3] obtained in 1988 an exact particular static axisymmetric solution for two charged masses which had to satisfy

$$m_1 m_2 = q_1 q_2, \tag{2}$$

a condition compatible with but not equivalent to (1). Within the post-Newtonian approximation, and in the case of two non-rotating arbitrary bodies, Barker and O'Connell [4] had already arrived at the equilibrium condition (2), that, according to them, had to be supplemented by

$$m_1 q_2 (q_2 - q_1) = m_2 q_1 (q_2 - q_1) \,. \tag{3}$$

If $q_1 = q_2$ one gets from Eq.(2) $q_i = \pm (m_1 m_2)^{1/2}$, $i = 1, 2$, while for $q_1 \neq q_2$ Papapetrou and Majumdar's condition (1) is recovered. Kimura

B. R. Iyer and B. Bhawal (eds.), Black Holes, Gravitational Radiation and the Universe, 219–234.

and Ohta [5] in 1977 and Ohta and Kimura [6] in 1982 approached the same problem, again for static axisymmetric dispositions of two charged bodies, in the post-post-Newtonian approximation concluding that Papapetrou and Majumdar's result (1) was the equilibrium condition.

All the above mentioned conditions were obtained as sufficient for equilibrium and obviously separation-independent. It was however shown by Bonnor [7] in 1993 that, when the Reissner-Nordstrom black hole is substituted for one of the particles, a second charged test particle may sit at certain separation dependent equilibrium points.

In this contribution we shall present a fairly complete analysis of the equilibrium of a spinless charged test particle in the Kerr-Newman black hole. Section 2 is a summary of Bonnor's results for the Reissner-Nordstrom black hole case. In Section 3 we generalize Bonnor's study to the Kerr-Newman black hole. By using the formulation of inertial forces in general relativity, as given by Abramowicz et al [8], we analyze in Section 4 the equilibrium of the test particle in the Kerr-Newman black hole in terms of the different forces. Finally, in Section 5, we make some comments about how the present work could be further generalized to a spinning test particle and about some open problems relating to the dynamical behaviour of spinning particles in gravito-electromagnetic fields.

It is a pleasure to acknowledge here that the contents of Sections 3 and 4 are the outcome of our ongoing collaboration with Professor C. V. Vishveshwara. They have been published in more detail elsewhere [9, 10].

2. A charged test particle in the Reissner-Nordstrom black hole

Bonnor conjectured in 1981 [11] that the condition for equilibrium of two spherically symmetric charged particles in general relativity was $m_1 m_2 = q_1 q_2$, that is the same as in classical physics. As was mentioned in Section 1 Kramer arrived at the same conclusion for the case of a particular static axisymmetric solution of the Einstein-Maxwell field equations. But in Kramer's solution the particles departed from spherical symmetry as they exhibited non-null multipolar moments. To throw light on the problem Bonnor [7] analyzed the case in which one of the particles was replaced by the Reissner-Nordstrom black hole, whose metric is

$$\begin{aligned} ds^2 &= -\left(1-\frac{2M}{r}+\frac{Q^2}{r^2}\right)dt^2+\left(1-\frac{2M}{r}+\frac{Q^2}{r^2}\right)^{-1}dr^2+ \\ & \quad r^2\left(d\theta^2+\sin^2\theta\, d\varphi^2\right), \end{aligned} \tag{4}$$

and the other by a charged test particle of charge q and mass m. The motion equation–ignoring all kinds of radiation back reaction–for the test particle is

$$ma^\alpha = qF^{\alpha\beta}u_\beta, \tag{5}$$

where u^α and a^α respectively are the 4-velocity and 4-acceleration of the particle and $F^{\alpha\beta}$, the electromagnetic field, is given by

$$F_{\alpha\beta} = \phi_{\beta;\alpha} - \phi_{\alpha;\beta} \tag{6}$$

with

$$\phi_\alpha = -\delta^0_\alpha \frac{Q}{r}. \tag{7}$$

From the previous expressions Bonnor obtained as the unique condition for the equilibrium of the test particle the following equation:

$$m\left(M - Q^2r^{-1}\right) = qQ\left(1 - 2Mr^{-1} + Q^2r^{-2}\right)^{1/2}. \tag{8}$$

Two cases can be distinguished:

1. *Separation-independent equilibrium*: It follows from (8) that the necessary and sufficient conditions are

$$m^2 = q^2, \qquad M^2 = Q^2, \tag{9}$$

provided that r be such that $g_{00} < 0$.

2. *Separation-dependent equilibrium*: This case is somewhat more complicated since it includes two subcases. In fact (8) leads to

$$r = \frac{Q^2}{M - q\,\mathrm{sign}\,Q\sqrt{(Q^2 - M^2)\,/\,(m^2 - q^2)}} \tag{10}$$

that makes apparent the two possibilities:

2.1. $M^2 < Q^2$, $q^2 < m^2$. There is no event horizon and the only requirement is that r be positive, that is

$$q\,\mathrm{sign}\,Q\sqrt{(Q^2 - M^2)\,/\,(m^2 - q^2)} < M. \tag{11}$$

The equilibrium is possible whenever q and Q have opposite signs, or if they have the same sign when $qQ < mM$, that easily follows from (11).

2.2. $M^2 > Q^2$, $q^2 > m^2$. Now there is event horizon. Equilibrium outside the horizon is possible only if $mM < qQ$. This can be seen from Eq.(10) by observing that q and Q must share the same sign since otherwise one gets $r < M$. We see therefore that, as was to be expected, the analysis of equilibrium in general relativity is more complex and richer than in non-relativistic classical theory. Clearly the equilibrium conditions are stronger than in the Newtonian theory. What could be surprising is that the separation-independent equilibrium still is possible in the relativistic framework.

3. A charged test particle in the Kerr-Newman black hole

3.1. THE EQUILIBRIUM POSITIONS

When the charged test particle is placed in the Kerr-Newman black hole field it is to be expected that the angular momentum of the source should make a difference with respect to the equilibrium conditions in absence of rotational effects.

The metric is

$$\begin{aligned} ds^2 &= -\frac{\Delta}{\rho^2}\left[dt - a\sin^2\theta\, d\varphi\right]^2 + \frac{\sin^2\theta}{\rho^2}\left[\left(r^2+a^2\right)d\varphi - a\,dt\right]^2 + \\ &\quad \frac{\rho^2}{\Delta}dr^2 + \rho^2\, d\theta^2, \end{aligned} \tag{12}$$

where

$$\Delta \equiv r^2 - 2Mr + Q^2 + a^2, \tag{13}$$

$$\rho^2 \equiv r^2 + a^2\cos^2\theta, \tag{14}$$

a being the angular momentum per unit mass of the source.

The electromagnetic field F of the source is given by the 2-form $F = 2\,dA$, where A, the 4-potential 1-form, is

$$A = -\frac{Qr}{\rho^2}\left(dt + \frac{r^2+a^2}{\Delta}dr - a\sin^2\theta\, d\varphi\right). \tag{15}$$

As in the previous section the motion equation for the test particle is taken to be

$$mu^{\alpha}_{;\beta}u^{\beta} = qF^{\alpha\beta}u_{\beta}. \tag{16}$$

Since we are interested in the rest equilibrium positions of the particle in the frame defined by the space coordinates of the metric (12), the 4-velocity u^α, in that situation, will be along the global timelike Killing vector, ξ^α, of the metric:

$$u^\alpha = e^\chi \xi^\alpha = e^\chi \delta^\alpha_0, \tag{17}$$

with

$$u^\alpha u_\alpha = -1, \qquad e^{-2\chi} = -\xi^\beta\xi_\beta = -g_{00}. \tag{18}$$

With the help of (17), (18) and the Killing equation, (16) becomes, for the desired equilibrium positions,

$$\left(me^{-\chi} - qA_0\right)_{,\alpha} = 0, \tag{19}$$

where the comma stands for the ordinary derivative. Eq. (19) is the necessary condition for equilibrium. Obviously it is also sufficient since, if it is fulfilled for all t, (16) will also be verified for all t with u^α given by (17). But if (19) holds at some point for some t it will hold for any t at the same point because neither e^χ nor A_0 depend on t. The bracket in (19) only depends on r and θ which leads to only two equations:

$$m\frac{\Delta\left(Q^2r - Mr^2 + Ma^2\cos^2\theta\right)}{\rho^4\left(r^2 - 2Mr + Q^2 + a^2\cos^2\theta\right)} + qQ\frac{\Delta\left(r^2 - a^2\cos^2\theta\right)}{\rho^5\left(r^2 - 2Mr + Q^2 + a^2\cos^2\theta\right)^{1/2}} = 0, \tag{20}$$

$$ma^2\frac{\left(2Mr - Q^2\right)\sin\theta\cos\theta}{\rho^4\left(r^2 - 2Mr + Q^2 + a^2\cos^2\theta\right)} - 2qQa^2\frac{r\sin\theta\cos\theta}{\rho^5\left(r^2 - 2Mr + Q^2 + a^2\cos^2\theta\right)^{1/2}} = 0. \tag{21}$$

These equations are invariant under symmetry about the equatorial plane $\theta = \pi/2$. Thus only the closed interval $[0, \pi/2]$ needs to be considered.

Case 1: Arbitrary values for θ, excepting $\theta = 0, \pi/2$. In both instances (20) and (21) must be satisfied. It is easy to see that this is equivalent to the requirement $\rho^2 = 0$, which is impossible. Therefore there is no equilibrium in the region $0 < \theta < \pi/2$.

Case 2: Equatorial plane $\theta = \pi/2$. Eq. (21) is automatically satisfied and the rotational parameter a dissapears in Eq. (20). Thus the analysis and results of Bonnor, considered in Section 2, apply to this case.

Case 3: The axis, $\theta = 0$. Again Eq. (21) holds trivially. The remaining equation, (20), takes the form

$$m\left(Mr^2 - Ma^2 - Q^2r\right)\sqrt{r^2 + a^2} = qQ\left(r^2 - a^2\right)\sqrt{r^2 - 2Mr + Q^2 + a^2}. \tag{22}$$

By using the notations $A \equiv q/m$, $B \equiv Q/M$, $C \equiv a/M$, $u \equiv r/M$ and squaring Eq. (22), that really does not introduce new spurious equilibrium positions, one gets

$$\begin{aligned} &-\left(A^2B^4C^4\right) + \left(1 - A^2B^2\right)C^6 + 2\left(1 + A^2\right)B^2C^4u \\ &+ \left[B^4C^2 + 2A^2B^4C^2 - \left(1 - A^2B^2\right)C^4\right]u^2 - 4A^2B^2C^2u^3 \\ &+ \left[\left(1 - A^2\right)B^4 - \left(1 - A^2B^2\right)C^2\right]u^4 - 2\left(1 - A^2\right)B^2u^5 \\ &+ \left(1 - A^2B^2\right)u^6 = 0. \end{aligned} \tag{23}$$

Separation-independent equilibrium on the axis demands the vanishing of all the coefficients of the above polynomial in u. It straight-forwardly follows that this is not possible if $a \neq 0$. Eq. (22) also yields that the solution $r = 0$, $\theta = 0$ exists only if $mMa = qQ\sqrt{Q^2 + a^2}$. To find the remaining separation-dependent equilibrium positions–and only those which are outside the horizon when the latter exists–one has to obtain the roots of a sixth-degree polynomial in the interval $(0, \infty)$ or, if $Q^2 + a^2 < M^2$ in $(1 + \sqrt{1 - B^2 - C^2}, \infty)$. In order to ascertain whether or not there are equilibrium points it suffices to use Sturm's theorem [12] to count the number of roots in those intervals. Numerical computation shows that equilibrium points exist indeed for some ranges of the parameters.

Finally, equilibrium is possible on the axis for $r = a$ when $Q = 0$ and no horizon exists ($M^2 < a^2$) as Eq. (22) makes it clear. It seems as if the rotation of the source induced a repulsive force on the test particle. The consideration of the radial motion, along the axis, supports this observation: in that case the force, $du^\alpha/d\tau$, consists of the addition of two terms; one corresponds to the Schwarzchild field, the other is always positive and proportional to a^2.

3.2. STABILITY OF THE EQUILIBRIUM POSITIONS

Separation-independent equilibrium, that only takes place on the equatorial plane, is obviously neutral, like in the Newtonian case. Thus we only need to pay attention to the separation-dependent equilibrium positions on the equatorial plane and the axis. The appropriate analysis starts deriving the linearized perturbative equations, for small displacement of the equilibrium positions, from the motion equation (16). It is well known that the time behaviour, $e^{\lambda t}$, of the perturbations is relevant for the stability question [13]. In brief: if all the real parts of the characteristic exponents λ are negative, the equilibrium is asymptotically stable. If there exists at least one characteristic exponent with a positive real part, the equilibrium is unstable. Finally, if the largest real part is null, the equilibrium is stable but not asymptotically stable for the linear system, while for the complete non-linear system the analysis is inconclusive. In the last case the system is non-hyperbolic and the non-linear terms may stabilize or unstabilize the equilibrium point. So, more information is needed in that case.

Let us consider first the stability on the axis, $\theta = 0$. There the possible values for the parameter λ are

$$\lambda = 0, 0, \qquad \pm a^3 \sqrt{\frac{Q^2 \Delta (m^2 + q^2)}{m^2 (a^2 - r^2)\ (a^2 + r^2)^5} P\left(\frac{r^2}{a^2}\right)}, \tag{24}$$

$$\pm i a \sqrt{\frac{Q^2}{(a^2 - r^2)(a^2 + r^2)^2}}, \tag{25}$$

where the following polynomial has been used:

$$P(x) \equiv \mu x^3 + 3x^2 + 3\mu x + 1, \quad \mu \equiv \frac{m^2 - q^2}{m^2 + q^2}. \tag{26}$$

If the equilibrium position is such that $a < r$, one of the last two values is positive and therefore the equilibrium is unstable. The same happens for the other two values of λ if $r < a$ since $P(x)$ is positive for $0 < x < 1$ and $|\mu| < 1$. If $r = a$ the equilibrium is possible only for $Q = 0$, as was previously stated. In this case one readily sees that the equilibrium is unstable as the values for λ are

$$\lambda = 0, 0, \pm\sqrt{\frac{M(M-a)}{2a^4}}, \pm\sqrt{\frac{M}{4a^3}}. \tag{27}$$

Consequently, the equilibrium points on the axis are always unstable. For the equatorial plane, $\theta = \pi/2$, the characteristic exponents are

$$\lambda = 0, 0, \pm i\frac{a|Q|}{r^3}, \pm\sqrt{(M^2 - Q^2)\, r^2 - a^2Q^2}. \tag{28}$$

If $M^2 \geq Q^2 + a^2$ we have at least one positive λ and the equilibrium is unstable outside the horizon, $M + \sqrt{M^2 - Q^2 - a^2} < r$, because of the inequalities

$$\left(M^2 - Q^2\right) r^2 - a^2Q^2 \geq a^2\left(r^2 - Q^2\right) > a^2\left(M^2 - Q^2\right) \geq a^4 > 0. \tag{29}$$

Therefore if there exists equilibrium on the equatorial plane it is unstable, save in the following cases:

Case 1: Separation-independent equilibrium: $m^2 = q^2$, $M^2 = Q^2$ and $qQ > 0$.
Case 2: Separation-dependent equilibrium:
Case 2.1: $q^2 < m^2$, $M^2 < Q^2$ and $qQ < mM$.
Case 2.2: $m^2 < q^2$, $Q^2 < M^2 < Q^2 + a^2$, $mM < qQ$ and $(M^2 - Q^2)r^2/Q^2 \leq a^2$, with r obtained from (20) for $\theta = \pi/2$:

$$r = \frac{Q^2}{M - |q|\sqrt{(Q^2 - M^2)/(m^2 - q^2)}}. \tag{30}$$

In the last two cases the equilibrium is stable for the linearized system. To be sure that the same is true for the complete system, one has to study the behaviour on the non-linear terms. Perhaps in that respect, two properties of the four-dimensional system (16) may be useful, namely that it is Hamiltonian and separable. There are no stable equilibrium points outside the horizon if the latter exists, like in the absence of rotation situation discussed by Bonnor.

4. Force analysis

Although we have studied the equilibrium of a test particle in the Kerr-Newman metric in Section 3, nothing has been said there about the forces acting on the particle. In order to analyse these forces we shall now use the concept of inertial forces in general relativity after the formulation given by Abramowicz et al [8]. Nayak and Vishveshwara [14, 15, 16] have studied these forces in detail, their possible direction reversal and their relation to gyroscopic precession in general relativity. A brief summary of the formulation of inertial forces and the application to our metric will be presented first. Then a similar procedure will be followed for the electromagnetic forces. It will be always assumed that the test particle moves along a circular orbit normal to the symmetric axis with uniform angular speed ω. The particle at rest is treated as a special case by setting ω to zero. Finally we shall interpret the equilibrium in terms of the inertial and electromagnetic forces.

4.1. INERTIAL FORCES

Abramowicz et al [8] gave a formulation for the general relativistic analogues of the inertial forces. The particle 4-velocity is written as

$$u^{\alpha} = \gamma\left(n^{\alpha} + v\tau^{\alpha}\right), \tag{31}$$

where n^{α} is a globally hypersurface orthogonal timelike unit vector, τ^{α} is a unit vector orthogonal to the former and along which the spatial 3-velocity of the particle is aligned and γ is a normalization factor such that $u^{\alpha}u_{\alpha} = -1$. Projecting the acceleration of the particle onto the space orthogonal to n^{α} by means of the projection tensor $h_{\alpha\beta} = g_{\alpha\beta} + n_{\alpha}n_{\beta}$ one can write

$$a_{\perp\alpha} = \phi_{,\alpha} + (\gamma v)^2 \tilde{\tau}^{\beta}\tilde{\nabla}_{\beta}\tilde{\tau}_{\alpha} + \gamma^2 v X_{\alpha} + \dot{V}\tilde{\tau}_{\alpha}. \tag{32}$$

Here $\tilde{\tau}^{\beta}$ is the unit vector along τ^{β} in the conformal space orthogonal to n^{β} with the metric $\tilde{h}_{\alpha\beta} = e^{-2\phi}(g_{\alpha\beta} + n_{\alpha}n_{\beta})$. It can also be shown that the covariant derivatives in the two spaces are related by

$$\tilde{\tau}^{\beta}\tilde{\nabla}_{\beta}\tilde{\tau}_{\alpha} = \tau^{\beta}\nabla_{\beta}\tau_{\alpha} - \tau^{\beta}\tau_{\alpha}\nabla_{\beta}\phi + \nabla_{\alpha}\phi. \tag{33}$$

Also we have:

$$\begin{aligned} \dot{V} &= \left(e^{\phi}\gamma v\right)_{,\beta} u^{\beta}, \\ X_{\alpha} &= n^{\beta}\left(\tau_{\alpha;\beta} - \tau_{\beta;\alpha}\right), \\ \phi_{,\alpha} &= n^{\beta} n_{\alpha;\beta}. \end{aligned} \tag{34}$$

Then from Eq.(32) the forces acting on the particle are written down as

$$\begin{aligned} \text{Gravitational force:} \quad & G_\alpha = -m\phi_{,\alpha}\,, \\ \text{Coriolis–Lense–Thirring force:} \quad & C_\alpha = -m\gamma^2 v X_\alpha, \\ \text{Centrifugal force:} \quad & Z_\alpha = -m\,(\gamma v)^2\,\tilde{\tau}^\beta \tilde{\nabla}_\beta \tilde{\tau}_\alpha, \\ \text{Euler force:} \quad & E_\alpha = -m\dot{V}\tilde{\tau}_\alpha. \end{aligned} \tag{35}$$

The above formalism will now be applied to axially symmetric stationary spacetimes. As was shown by Greene et al [17], axially symmetric stationary spacetimes with orthogonal transitivity admit a globally hypersurface orthogonal vector field, ζ^α, such that

$$\zeta^\alpha = \xi^\alpha + \omega_0 \eta^\alpha, \tag{36}$$

where the fundamental angular speed of the irrotational congruence is

$$\omega_0 = -\frac{\xi^\alpha \eta_\alpha}{\eta^\beta \eta_\beta}. \tag{37}$$

In the above, ξ^α is a global timelike Killing vector and η^α is the rotational Killing vector. The unit vector along ζ^α is identified with n^α. Further, if n^α follows a circular trajectory, then τ^α is along η^α. In this case it is easy to see that the Euler force vanishes. More specifically

$$u^\alpha = e^\psi\left(\xi^\alpha + \omega\eta^\alpha\right) = \gamma\left(n^\alpha + v\tau^\alpha\right) = e^\psi \chi^\alpha. \tag{38}$$

The former relations yield the following expressions for the gravitational, Coriolis–Lense–Thirring and centrifugal forces respectively:

$$G_\alpha = -m\phi_{,\alpha}, \tag{39}$$

$$C_\alpha = me^{2\psi}\,\frac{\omega-\omega_0}{\eta^\sigma\eta_\sigma}\left[\left(\eta^\beta\eta_\beta\right)\,\left(\xi^\rho\eta_\rho\right)_{,\alpha} - \left(\xi^\beta\eta_\beta\right)\left(\eta^\rho\eta_\rho\right)_{,\alpha}\right], \tag{40}$$

$$Z_\alpha = \frac{1}{2}me^{2\psi}\,\frac{(\omega-\omega_0)^2}{\zeta^\sigma\zeta_\sigma}\,\left[\left(\zeta^\beta\zeta_\beta\right)\left(\eta^\rho\eta_\rho\right)_{,\alpha} - \left(\eta^\beta\eta_\beta\right)\left(\zeta^\rho\zeta_\rho\right)_{,\alpha}\right]. \tag{41}$$

4.2. ELECTROMAGNETIC FORCES

If $F_{\alpha\beta}$ is the electromagnetic tensor field the electric and magnetic fields, E^α and H^α, can respectively be defined with respect to an irrotational observer with unit velocity vector n^α as

$$E^\alpha = F^{\alpha\beta} n_\beta, \qquad H^\alpha = \tilde{F}^{\alpha\beta} n_\beta, \tag{42}$$

where $\tilde{F}_{\alpha\beta} = \frac{1}{2}\sqrt{-g}\epsilon_{\alpha\beta\rho\sigma}F^{\rho\sigma}$ is the dual of $F^{\alpha\beta}$. The Lorentz equation of motion (16) can be projected onto the space orthogonal to n^α and decomposing the velocity according to Eq. (31) we get

$$m\dot{u}_{\perp\alpha} = q\gamma\left[E_\alpha + vn^\sigma\sqrt{-g}\,\varepsilon_{\sigma\rho\beta\alpha}\,\tau^\rho H^\beta\right]. \tag{43}$$

Comparing this equation with that of classical electrodynamics:

$$\mathbf{F} \equiv q\left(\mathbf{E} + \mathbf{v}\times\mathbf{H}\right) \equiv \mathbf{f}_{\mathrm{E}} + \mathbf{f}_{\mathrm{H}} \tag{44}$$

we may define $\mathbf{f}_{\mathrm{E}}$ and $\mathbf{f}_{\mathrm{H}}$ as the electric and magnetic forces by

$$\begin{array}{rl}\text{Electric force:} & f_{\mathrm{E}\alpha} = q\gamma E_\alpha, \\ \text{Magnetic force:} & f_{\mathrm{H}\alpha} = q\gamma v n^\sigma\sqrt{-g}\,\varepsilon_{\sigma\rho\beta\alpha}\,\tau^\rho H^\beta.\end{array} \tag{45}$$

4.3. EQUILIBRIUM IN TERMS OF THE FORCES

The dynamical equation for the circular orbits we are considering in terms of the forces is

$$G_\alpha + Z_\alpha + C_\alpha + f_{\mathrm{H}\alpha} + f_{\mathrm{E}\alpha} = 0. \tag{46}$$

According to Eqs.(32) and (43), Eq.(46) above may be explicitly written as

$$m\left[\phi_{,\alpha} + (\gamma v)^2\tilde{\tau}^\beta\tilde{\nabla}_\beta\tilde{\tau}_\alpha + \gamma^2 v X_\alpha\right] = q\gamma\left[E_\alpha + vn^\sigma\sqrt{-g}\,\varepsilon_{\sigma\rho\beta\alpha}\,\tau^\rho H^\beta\right]. \tag{47}$$

4.3.1. *Case* $v = 0$.

When $v = 0$ the particle will travel along the irrotational trajectory determined by ζ^α. Then the centrifugal, Coriolis and magnetic forces all vanish and the equation of motion is reduced to

$$m\phi_{,\alpha} = q\gamma F_{\alpha\beta}n^\beta. \tag{48}$$

4.3.2. *Case* $\omega = 0$.

For a stationary test particle, its 4-velocity will be along the Killing vector ξ^α. The equilibrium conditions are now expressed by putting $\omega = 0$ in Eqs. (38) and (47) after taking into account the fact that Eq.(38) implies that $v = \left[\eta^\alpha\eta_\alpha/\left(-\zeta^\beta\zeta_\beta\right)\right]^{1/2}(\omega - \omega_0)$.

We note a curious fact here. The centrifugal force can be decomposed into three terms, $Z^\alpha = Z_1^\alpha + Z_2^\alpha + Z_3^\alpha$, where

$$Z_1^\alpha = -\; m(\gamma v)^2\nabla^\alpha\phi, \tag{49}$$

$$Z_2^\alpha = -\; m(\gamma v)^2\tau^\beta\nabla_\beta\tau^\alpha, \tag{50}$$

$$Z_3^\alpha = m(\gamma v)^2\tau^\alpha\tau^\beta\nabla_\beta\phi. \tag{51}$$

As a consequence of the symmetry with respect to τ^α, which is along the Killing vector η^α, we have $Z_3^\alpha = 0$. Then, if we define

$$F_1^\alpha = G^\alpha + \frac{1}{2}C^\alpha + Z_1^\alpha + f_{\rm E}^\alpha, \tag{52}$$

$$F_2^\alpha = \frac{1}{2}C^\alpha + Z_2^\alpha + f_{\rm H}^\alpha, \tag{53}$$

the sum of the inertial and electromagnetic forces, $F^\alpha = G^\alpha + C^\alpha + Z^\alpha + f_{\rm E}^\alpha + f_{\rm H}^\alpha$, is split into two terms $F^\alpha = F_1^\alpha + F_2^\alpha$. The intriguing fact is that, though the equilibrium conditions are

$$F^\alpha = F_1^\alpha + F_2^\alpha = 0, \tag{54}$$

we may impose separately

$$F_1^\alpha = 0, \tag{55}$$

and

$$F_2^\alpha = 0, \tag{56}$$

since Eq.(55) is equivalent to Eq.(54) and Eq.(56) is satisfied identically on the axis ($\theta = 0$) and is again equivalent to Eq.(54) on the equatorial plane ($\theta = \pi/2$). This follows directly from the fact that the components of F^α are proportional to those of the individual parts of F_1^α and F_2^α for a static charge ($\omega = 0$).

In the above, inertial and electromagnetic forces have been defined with respect to the globally hypersurface orthogonal, irrotational congruence n^α. This is a satisfactory way to treat forces and interactions, since the congruence n^α defines the closest analogue of the Newtonian rest frame as has been discussed in Ref.[17]. The vector field n^α is orthogonal to $t = constant$ surfaces. Thus, observers with world lines along n^α share three-dimensional spaces determined by these surfaces and for them t is the common synchronous time. Nevertheless, as viewed by stationary observers with world lines along ξ^α, they follow circular orbits with angular speed ω_0.

On the axis, $\omega_0 = 0$ and the two sets of observers become identical. Consequently, on the axis only the gravitational force survives with both the centrifugal and the Coriolis forces going to zero. Similarly, only the electric force is non-zero on the axis and the two forces balance each other. In Figure 1 we have plotted these two forces for a typical set of mass, charge and angular momentum parameters. One can see their general behaviour, points of reversal and location of the equilibrium positions where they balance each other. It can be easily shown that both forces reverse if there is no ergosphere, $M^2 < Q^2 + a^2$, and point always in the same direction outside the ergosphere if this exists.

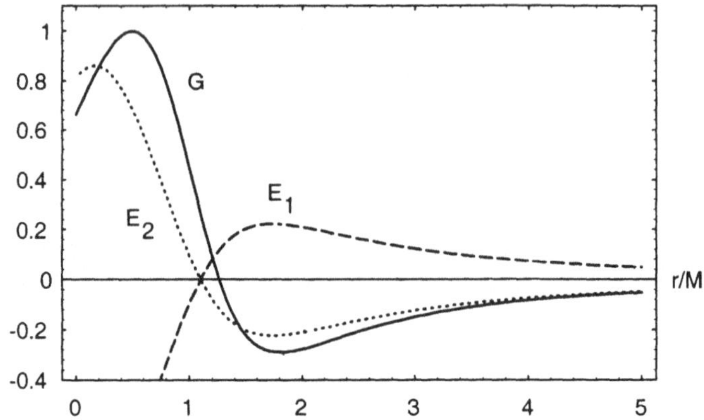

Figure 1. Gravitational force (G) on the axis for $|Q| = 0.55M$ and $a = 1.1M$. The electric forces corresponding to $|q| = 2m$ are also displayed for $qQ > 0$ (E_1) and $qQ < 0$ (E_2). All forces are multiplied by the scale factor M/m. The intersection of curves G and E_1 corresponds to the equilibrium point for $qQ < 0$ and the equilibrium for $qQ > 0$ occurs at the points where G and E_2 intersect.

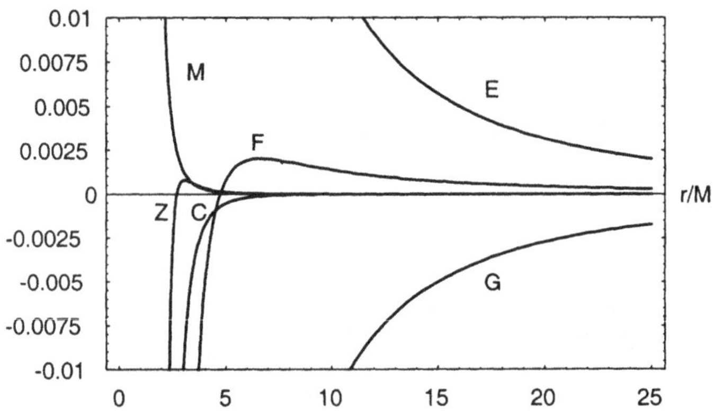

Figure 2. Gravitational (G), Coriolis-Lense-Thirring (C), centrifugal (Z), electric (E) and magnetic (M) forces on the equatorial plane for $|q| = 2m$, $|Q| = 0.6M$, $a = 0.3M$ and $qQ > 0$. The sum of these five forces (F) is also displayed and its intersection with the horizontal axis corresponds to the equilibrium position.

The situation on the equatorial plane is more complicated. The stationary particle world-line along ξ^α rotates with respect to the irrotational observer following n^α. Therefore all the forces–save the Euler force–are nonzero. Figures 3, and 4 show plots of all these individual forces as well as

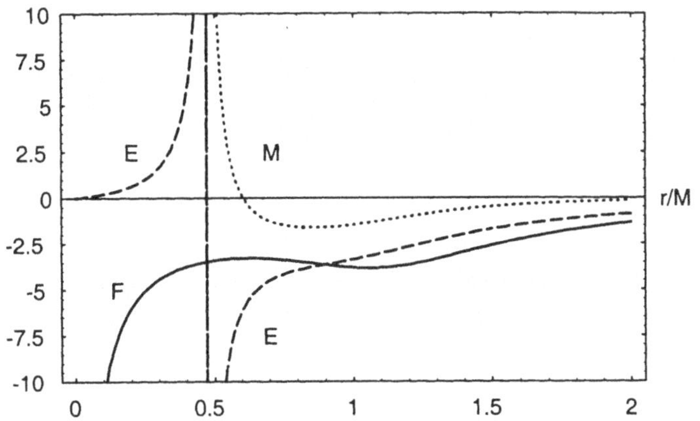

Figure 3. Electric (E) and magnetic (M) forces for $|q| = 2m$, $|Q| = 1.1M$, $a = 1.1M$ and $qQ < 0$. The sum of inertial and electromagnetic forces (F) does not cross the horizontal axis and, in consequence, there is no equilibrium position.

their total sum. Only points located outside the ergosphere are considered when the latter exists. Again one can see and the equilibrium positions where the individual forces reverse. Unlike on the axis, now the electric force has a definite sign. In Table 1 we have displayed the conditions under which the remaining forces can reverse. In some cases the conclusions in this table have been easily drawn from analytical considerations, but the analysis of the gravitational and centrifugal forces requires dealing with polynomials of fifth and sixth order, respectively. In the cases indicated in Table 1, numerical analysis has been used.

TABLE 1. On the equatorial plane ($\theta = \pi/2$) the electric force always points in the same direction but the inertial and magnetic forces reverse their direction under the following conditions when (1) both the ergosphere and the horizon exist, (2) the ergosphere exists but there is no horizon and (3) there is neither ergosphere nor horizon.

Case	Parameters	G_r	C_r	Z_r	f_{H_r}
(1)	$Q^2 + a^2 < M^2$	never†	never	always†	never
(2)	$Q^2 < M^2 < Q^2 + a^2$	sometimes†	never	sometimes†	never
(3)	$Q^2 > M^2$	always	always	sometimes†	always

† This is a numerical result.

5. Conclusions and further remarks

This work has aimed at revealing the new effects that could appear in the context of the equilibrium of a charged test particle, when the background Reissner-Nordstrom metric is replaced by the Kerr-Newman spacetime. We have seen that one of the effects is that equilibrium becomes impossible except on the axis and the equatorial plane. It is interesting that the introduced rotation does not alter the equilibrium conditions on the equatorial plane since the angular momentum parameter a does not appear in Eq. (20), nor obviously in Eq. (21) that is identically satisfied. Thus Bonnor's results remain valid on that plane, save for a point: The parameter a is involved in the stability analysis in such a way that it precludes stable equilibrium outside the horizon if the latter exists ($Q^2 + a^2 < M^2$). We note, however, that this absence of stability was also present in the Reissner-Nordstrom case when $Q^2 < M^2$ and $m^2 < q^2$. The equilibrium condition on the axis leads to a sixth-degree polynomial in the radial coordinate r. Numerical computations show that there are equilibrium positions for certain ranges of the mass, charge and angular momentum whether $M < |Q|$ or $|Q| < M$; but all these equilibrium points are unstable. Therefore the introduction of rotation in spacetime completely modifies the characteristics of static equilibrium except on the equatorial plane.

A comprehensive analysis of the forces involved in the equilibrium of a charged test particle in the Kerr-Newman spacetime has been presented. It has been made manifest that for each force –not only for the centrifugal one–points can be found at which its direction is reversed.

A natural generalization of the work presented here would be to assume that the charged test particle also possesses spin. This considerably increases the complexity of the problem, both technically and conceptually. To describe the motion of that sort of particle in an external field one needs two sets of equations: one set determining the spin dynamics proper and a second one corresponding to the trajectory of the particle, which should contain some terms accounting for the spin influence on such trajectory. There is a wide consensus that the dynamical spin equation for a charged dipole test particle in a gravito-electromagnetic field can be taken as

$$\frac{DS_\alpha}{d\tau} = u_\alpha \frac{Du^\beta}{d\tau} S_\beta + \frac{gq}{2m}(F_{\alpha\lambda} + u_\alpha u^\beta F_{\beta\lambda})S^\lambda, \tag{57}$$

where S_α is the spin 4-vector and g is the gyromagnetic ratio of the particle [18]. The situation with the trajectory dynamical equation is different. Interestingly enough, even neglecting any possible radiation reaction on the particle, different authors have been maintaining conflicting views on the matter. The origin of the discrepancy is related, in some cases, to what should be the form of a further constraint between the 4-velocity and the spin that has to be imposed to make the system determinate [18, 19, 20, 21]. If one starts from the special relativistic equation for the trajectory given by Teitelboim et al [18] for a dipole charged test particle to find its simplest extension to general relativity the result is,

$$\begin{aligned} m\frac{Du^\alpha}{d\tau} = & \quad qF^{\alpha\beta}u_\beta + \eta^{\alpha\beta\gamma\delta}S_\beta u_\gamma \frac{D^2u_\delta}{d\tau^2} + \frac{1}{2}R^\alpha{}_{\lambda\mu\nu}u^\lambda\eta^{\mu\nu\rho\sigma}S_\rho u_\sigma \\ & + \frac{gq}{2m}\tilde{F}_{\beta\gamma}{}^{;\alpha}u^\beta S^\gamma \\ & + \frac{gq}{2m}\frac{D}{d\tau}(S_\beta \tilde{F}^{\beta\alpha}) - \frac{gq}{2m}\eta^{\alpha\beta\gamma\delta}S^\sigma F_{\sigma\beta}u_\gamma \frac{Du_\delta}{d\tau}, \end{aligned} \tag{58}$$

where $R^\alpha{}_{\lambda\mu\nu}$ is the Riemann curvature tensor of the metric with the sign convention used in Ref.[22], $\eta_{\alpha\beta\gamma\delta} = \sqrt{-g}\epsilon_{\alpha\beta\gamma\delta}$ is the Levi-Civita tensor (note that in the square root $-g$ is the absolute value of the determinant of the metric, not the negative of the gyromagnetic ratio). Again for the Kerr-Newman spacetime we have published elsewhere [23] an analysis of the stationary equilibrium of the spinning particle based on Eq. (57), the simplification that results from neglecting the coupling of the magnetic moment to the electromagnetic field in Eq.(58) –that is setting $g = 0$–and the constraint,

$$S^\alpha u_\alpha = 0. \tag{59}$$

Naturally new effects appear, such as the possibility of equilibrium off the equatorial plane and the axis, or points on the equatorial plane and the axis at which the particle remains at rest while its spin precesses.

In a future work we intend to repeat the analysis by using the full equation (58). But much more important than that is the standing question: Would it be possible to advance reasons to single out a self-consistent complete set of dynamical and constraint equations describing to the "best" approximation a classical charged dipole point-like test particle? It might even be that this question has to be rephrased. Hopefully we shall also address this problem in the near future.

6. Acknowledgements

Most of this work was done in collaboration with Prof. C.V. Vishveshwara as already mentioned. K. Rajesh Nayak contributed very effectively to the results presented in Section 4. The whole contribution has been partially supported by The University of the Basque Country under contract UPV/EHU 172.310-EB036/95.

References

1. Papapetrou, A. (1947) *Proc. R. Irish Acad.* **51**, pp. 248–268.

2. Majumdar, S.M. (1947) *Phys. Rev.* **72**, pp. 390–398.
3. Kramer, D. (1988) *Class. Quantum Grav.* **5**, pp. 1435–1442.
4. Barker, B.M. and O'Connell, R.F. (1977) *Phys. Lett.* **61 A**, pp. 297–298.
5. Kimura, T. and Ohta, T. (1977) *Phys. Lett.* **63 A**, pp. 193–195.
6. Ohta, T. and Kimura, T. (1982) *Prog. Theor. Phys.* **68**, pp. 1175–1195.
7. Bonnor, B.W. (1993) *Class. Quantum Grav.* **10**, pp. 2077–2082.
8. Abramowicz, M.A., Nurowski, P. and Wex, N. (1993) *Class. Quantum Grav.* **10**, L183.
9. Aguirregabiria, J.M., Chamorro, A., Suinaga, J. and Vishveshwara, C.V. (1995) *Class. Quantum Grav.* **12**, pp. 699–705.
10. Aguirregabiria, J.M., Chamorro, A., Nayak, R.K., Suinaga, J. and Vishveshwara, C.V. (1996) *Class. Quantum Grav.* **13**, pp. 2179–2189.
11. Bonnor, B.W. (1981) *Phys. Lett.* **83 A**, pp. 414–416.
12. Stoer, J. and Burlisch, R. (1980) *Introduction to Numerical Analysis* (Berlin: Springer), p. 281.
13. Hirsch, M.W. and Smale, S. (1974) *Differential Equations, Dynamical Systems and Linear Alagebra* (New York: Academic), p. 180.
14. Nayak, R.K. and Vishveshwara, C.V. (1997) *Gen. Rel. Grav.* **29** pp. 291–306.
15. Nayak, R.K. and Vishveshwara, C.V. (1996) *Class. Quantum Grav.* **13**, pp. 1783-1795.
16. Nayak, R.K. and Vishveshwara, C.V. (1997) *Gyroscopic precession and inertial forces in axially symmetric stationary spacetimes*, to be published.
17. Greene, R.D., Schücking, E.L. and Vishveshwara, C V. (1975) *J. Math. Phys.* **16**, p. 153–157.
18. See, for instance, Teitelboim, C., Villarroel, D. and van Weert, Ch.G. (1980) *Riv. Nuovo Cimento* **3**, pp. 1–64.
19. Dixon, W.G. (1970) *Proc. R. Soc. London* **A 314**, pp. 499–527; (1970) *Proc. R. Soc. London* **A 319**, pp. 509–547; (1974) *Phil. Trans. R. Soc. London* **A 277**, pp. 59–119.
20. Yee, K. and Bander, M. (1993) *Phys. Rev. D* **48**, pp. 2797–2799.
21. Khriplovich, I.B. and Pomeransky, A.A. (1996) *Phys. Lett. A* **216**, pp. 7–14; gr-cr/9710098, 21 Oct. 1997, and references therein.
22. Misner, C.W., Thorne, K.S. and Wheeler, J.A. (1973) *Gravitation* (San Francisco: W.H. Freeman and Company).
23. Aguirregabiria, J.M., Chamorro, A., Suinaga, J. and Vishveshwara, C.V. (1996) *Class. Quantum Grav.* **13**, pp. 417–424

15. NEUTRON STARS AND RELATIVISTIC GRAVITY

M. VIVEKANAND
National Centre for Radio Astrophysics
Tata Institute of Fundamental Research
Pune University Campus, P. O. Box 3
Ganeshkhind, Pune 411007, INDIA.

1. Introduction

Relativistic gravity shapes neutron stars; they in turn display some of the fascinating predictions of relativistic gravity. In simple terms relativistic gravity can be understood as the warping of space-time near any object that has mass. Consider the following examples. In Newtonian gravity, a planet orbits around a star in an ellipse. In relativistic gravity the orbit is not a closed curve; it is approximately an ellipse, whose major axis precesses in the plane of the orbit (the planet mercury displays this motion). Consider another example. In Newtonian gravity a particle (with zero angular momentum) that is attracted towards a rotating object will fall radially inwards. In relativistic gravity the particle will be deflected azimuthally as it falls inwards. The explanation is that the rotating object drags along with it the local space-time. Consider yet another example. In Newtonian gravity a gyroscope in free fall does not naturally precess, while it does in relativistic gravity. This is supposed to be due to an additional gravitational torque felt by the gyroscope. It is difficult to observe such effects in our solar system, where the gravitational field is relatively weak. However one can observe them in different astrophysical situations such as binary systems, globular clusters, etc., by monitoring the varying clock rate of pulsars.

This article describes how relativistic gravity determines the structure and stability of neutron stars, and how they in turn are useful to study relativistic gravity. It begins with a brief introduction to neutron stars.

At the end of its evolution, a star of mass ≥ 8 solar masses ($M_\odot$) is supposed to consist of a core of incombustible iron, surrounded by shells

B. R. Iyer and B. Bhawal (eds.), Black Holes, Gravitational Radiation and the Universe, 235–248.

of burning silicon, magnesium, oxygen, neon, carbon, helium and hydrogen (Burbidge et al, 1957; Schwarzschild, 1965; Clayton, 1983; Woosely & Weaver, 1986). The core with a radius of several thousand kilometers (km) is supported against gravitational collapse by the pressure of non-relativistic degenerate electrons (Fowler, 1926); it is surrounded by the so called red supergiant of radius 100 million km. Further crushing by gravity makes the electrons relativistic. The pressure decreases for the same density, since the dependence upon density of pressure in a relativistic degenerate gas of electrons is weaker than that in a non-relativistic degenerate gas of electrons. The electrons also get captured by protons in the nucleus (inverse beta decay) to reduce the electron pressure still further. If the mass of the core remains below the Chandrasekhar limit of 1.4 $M_{\odot}$, it settles down to be a white dwarf. If the mass grows beyond that limit, it collapses to a star consisting mainly of neutrons, supported against gravity by their degeneracy pressure. This is a neutron star; see Meszaros (1992) and Glendenning (1997) and references therein.

A Neutron star is supposed to be born with internal temperature in excess of 10^{11} Kelvin (K). Within a few days it cools to 10^9 K, its surface temperature at this point being about 5×10^6 K. So far the cooling is by emission of neutrinos. When the internal temperature falls to around 10^8 K the cooling is mainly by emission of photons (Nomoto & Tsuruta, 1987; Ogelman, 1995). Its radius is about 10 km and its density is about 10^{15} g.cm^{-3} (Harrison, Thorne, Wakano & Wheeler, 1965). Its magnetic field is supposed to be greater than 10^8 Gauss (Manchester & Taylor, 1977). It is supposed to be rapidly rotating on account of the angular momentum of the stellar core. Some neutron stars are supposed to emit coherent radio waves in the frequency range 10 MHz to 10 GHz, from a localised region on the surface, beamed into a narrow, hollow cone centered about the axis of the dipole magnetic field (Ruderman & Sutherland, 1975). If and when this beam sweeps past an observer(s), he sees a pulse of radiation, at the period of rotation of the star. The period increases continuously due to the loss of rotational energy, supposedly in the form of relativistic electrons and positrons, and very low frequency (inverse of the period) un-pulsed incoherent electromagnetic radiation. Such neutron stars are known as pulsars. There are two classes of pulsars – normal pulsars with periods P between 0.1 and 5.0 sec, period derivative dP/dt ($\dot{P}$) around 10^{-15}, and magnetic field B around 10^{12} Gauss; and milli second (ms) pulsars with $P \approx 0.001$ to 0.1 sec, $\dot{P} \approx 10^{-18}$, and $B \approx 10^8$ Gauss (Manchester, 1993). The pulsed coherent radiation is only about one part in 10^5 of the total energy lost by the pulsar (Manchester & Taylor, 1977).

The extreme stability of the periods of (particularly) ms pulsars makes them some of the best clocks in the universe (Ryba & Taylor, 1991). This is in spite of two well known perturbations to pulsar periods. One is called timing noise, which is the residual error in arrival time of the average pulse; it is $\leq 1\%$ of the pulsar period. The second is called period glitches, which is the abrupt change in P and $\dot{P}$, of magnitudes $\Delta P/P \approx 10^{-9}$ to 10^{-6}, and $\Delta \dot{P}/\dot{P} \approx 10^{-3}$ to 10^{-1}, that occur once in several years (Lyne, 1993). However, both the timing noise and period glitches occur in young pulsars (Lyne et al, 1995), so the much older ms pulsars can be considered to be excellent clocks.

2. Inside Neutron Stars

2.1. NON-ROTATING NEUTRON STARS

The important difference between Newtonian and relativistic gravities is that in the latter, pressure contributes to gravitational force. A non-rotating neutron star is assumed to a spherically symmetric configuration, the space-time around it being described by the Schwarzschild metric (Weinberg, 1972; Misner, Thorne & Wheeler, 1973). The equations of hydrostatic equilibrium for a spherically symmetric star are known as the Oppenheimer-Volkoff equations (Oppenheimer & Volkoff, 1939). By solving these one obtains the pressure and density profiles within the star. An order of magnitude estimate of some parameters of neutron stars can be obtained by considering a configuration in which the nucleons are packed at maximum density, i.e., next to each other. Then the number of baryons A_* in the star is approximately 10^{57}, its radius R_* is 7 km, and its mass M_* is 2.3 $M_\odot$. The average density is about 10^{15} g.cm^{-3}, which is higher than nuclear densities. The gravitational energy is about 18% of the rest mass energy of the star (Glendenning, 1997).

The strength of interaction between nucleons plays an important role in determining the structure and stability of neutron stars. Ideally the interaction should be computed in curved space-time, but one is allowed to compute it in flat space-time if a local inertial frame (in which the laws of special relativity are valid) can be defined over the mean distance between a pair of interacting nucleons. It can be shown that the components of the metric $g_{\mu\nu}$ change by negligible amounts over this distance in a neutron star. So the flat space-time approximation is sufficient. Similarly it can be shown that relativistic gravity will not affect the internal working or the structure of an atom or a nucleus.

However, relativistic gravity changes the conditions of chemical and thermal equilibrium in neutron stars. It can be shown that the baryon chemical potential μ, defined as the change in baryon energy density for a unit change in baryon number density, varies with radial distance r from the center of the star as

$$\mu(r)\left[1-2\frac{GM_*(r)}{c^2r}\right]^{1/2}=\text{constant},\tag{1}$$

where $M_*(r)$ is the mass within the radius r and c is the velocity of light; temperature varies similarly (Zeldovich & Novikov, 1971). So the condition for thermal equilibrium in a neutron star is not a constant temperature throughout the star. This arises because the conserved quantity in relativistic gravity is not the local energy E but the quantity $E \times (-g_{00})^{1/2}$; the condition for thermal equilibrium is the requirement that net thermal flux is zero.

In the original solution of Oppenheimer and Volkoff the limiting radius and mass of neutron stars was approximately 9.6 km and 0.72 $M_\odot$. This is a lower bound since nucleon-nucleon repulsions were ignored. The nuclear force stiffens the equation of state, i.e., it increases the pressure at a given density. In the stiffest equation of state the speed of sound is equal to the speed of light; this gives a limiting mass of 3.2 $M_\odot$ and radius of 13.4 km (Glendenning, 1997). The mass referred to here is the sum of the rest mass of the baryons (M_A) and the gravitational mass defect due to the gravitational potential energy (Zeldovich & Novikov, 1971). The baryon mass M_A is higher than the gravitational mass M_* for all but the least massive neutron stars ($\leq$ 0.2 $M_\odot$). In such stars the binding energy per baryon defined as $(M_A - M_*)/A_*$, turns out to be negative from the model calculations. Clearly such neutron stars are difficult to manufacture (Glendenning, 1997).

While the Oppenheimer-Volkoff equations describe the equilibrium configuration of neutron stars, they do not assure stability, for which the basic criterion is the condition $dM_*/d\epsilon_c > 0$; ϵ_c is the central density. There are two regions of stability, one corresponding to white dwarfs for $\epsilon_c < 10^{10}$ g.cm^{-3}, and another for neutron stars for which ϵ_c lies in a very narrow range around 10^{15} g.cm^{-3} (Harrison, Thorne, Wakano & Wheeler, 1965). Gravitational collapse of stars occurs in relativistic gravity at finite central densities, while it occurs at infinite central density in Newtonian gravity (Harrison, Thorne, Wakano & Wheeler, 1965).

2.2. EFFECT OF ROTATION

In rotating neutron stars the dragging of the local space-time, or the so called Lense-Thirring effect (Weinberg, 1972), has to be incorporated into the model calculations. In brief, an observer influenced by a rotating star will also rotate, by an amount that depends upon the distance from the center of the star. The angular velocity $\omega(r)$ of the observer (or, equivalently, that of the local inertial frame) varies as r^{-3} with the distance r from the center of the star, as measured by an observer at infinity (Weinberg, 1972).

For a typical neutron star of limiting mass the ratio $\omega(r)/\Omega$ (where Ω is the angular velocity of the star assuming rigid rotation) ranges from about 0.6 at the centre of the star, to about 0.2 at its edge; see (Glendenning, 1997), (Weber, 1992) and (Hartle, 1967).

Rotation will change the Kepler frequency Ω_c, at which the centripetal force equals the gravitational force, making the star unstable. In Newtonian gravity $\Omega_c = [GM_*/R_*^3]^{1/2}$; in relativistic gravity Ω_c is less than this value. This is contrary to expectation since the centripetal force on an element of stellar fluid depends upon the differential rotation $\Omega - \omega(r)$. While the exact calculations of Ω_c for a rotating star in relativistic gravity are difficult, an empirical formula is $\zeta^2 GM_*/R_*^2 = \Omega_c^2 R_*$, where ζ is about 0.625 (Weber & Glendenning, 1992). The minimum rotation period for a neutron star varies from 0.26 ms to 0.58 ms (Glendenning, 1997). An absolute upper limit to Ω_c is 0.167 $\times M_*/M_\odot$ ms (Glendenning, 1997), obtained from the absolute upper limit allowed for M_*/R_* for any star obeying Einstein's equations (Weinberg, 1972).

The limiting rotation period of neutron stars may differ significantly if gravitational radiation is emitted by the star. A vibrating, rotating neutron star has a changing quadrupole moment, which produces gravitational radiation. If the loss of this radiation enhances the vibrational instability, the star is unstable; see Glendenning (1997) and references therein. Detailed calculations are available for the minimum period of a neutron star as a function of its mass, for several temperatures and several modes of instability. However, these have not resolved the issue. According to Glendenning (1997), for hot (and therefore young) neutron stars with low viscosity, this mechanism sets a limit higher than the above mentioned numbers. For a neutron star having the limiting mass with $T = 10^{10}$ K, the limiting rotation period is 1.0 ms, while the Kepler period is only 0.63 ms. For $T = 10^6$ K, the two periods are much closer. However this subject has several uncertainties which are not yet resolved and requires much more detailed analysis (Glendenning, 1997). For example, Lindblom and Mendell (1995) claim that gravitational radiation does not limit the rotation rate of neutron stars, due to the viscosity of the interior superfluid.

Rotation also makes the moment of inertia of neutron stars an increasing function of the angular frequency Ω; the average value increases by about 15% to 20%. Rotation increases the minimum allowed mass of neutron stars by about 10% and their radius by about 10% to 20%; it decreases their central densities by about 20% (Friedman et al, 1986; Weber & Glendenning, 1992).

2.3. CRUST-CORE INTERACTION

Neutron stars are supposed to consist of an outer crust and a superfluid core (Baym et al, 1969; Irvine, 1978). The spin axis of the superfluid core might get misaligned with respect to that of the crust, due to repeated glitches (Blandford, 1995). The crust will then undergo Lens-Thirring precession about the direction of total angular momentum of the system, with a period of 3 to 8 pulsar periods. Since the observed radio radiation of pulsars is tied to the crust, one can expect to see the above effect in the timing data of pulsars (Blandford, 1995).

3. Outside Neutron Stars

3.1. GRAVITATIONAL REDSHIFT

The ratio ν/ν_0 of the frequency of electromagnetic radiation that is emitted at point r_0 and received at point r, is given by $[g_{00}(r_0)/g_{00}(r)]^{1/2}$, in general theory of relativity (Weinberg, 1972). In weak gravitational fields: $g_{00} = (1 + 2\phi/c^2)$, where ϕ is the gravitational potential. The fractional change in frequency is $z = (\nu_0 - \nu)/\nu = \Delta\nu/\nu \approx [\phi(r) - \phi(r_0)]/c^2$. Thus a spectral line emitted at distance r_0 from the center of the star suffers the gravitational redshift $z \approx GM_*/(c^2 r_0)$, by the time it is observed at infinity. This effect is negligible for our sun, and is 10 to 100 times larger for white dwarfs. It has been measured for the white dwarf 40 Eridani (Popper, 1954) to be $z = 7\ (\pm 1) \times 10^{-5}$. Theoretically, any star obeying the Oppenheimer-Volkoff equations satisfies the stability condition $2GM_*/(c^2 R_*) < 8/9$, which implies that $z < 2$ for such stars (Weinberg, 1972). The redshift is 0.13 for a typical non-rotating neutron star (Weinberg, 1972). For rotating neutron stars it is calculated to lie between 0.184 and 0.854 for various models (Lindblom, 1984).

Gravitational redshift is believed to be seen in X-ray pulsars, which consist of a neutron star accreting matter from a normal stellar companion in a binary system. They were discovered by the UHURU satellite in the early seventies (Giacconi et al, 1971); since then they have been studied by other satellites such as EINSTEIN, GINGA, ROSAT, etc (Meszaros, 1992). Several X-ray pulsars show emission lines in the 10 kilo electron volt (KeV) to 40 Kev band, which are thought to be gravitationally redshifted cyclotron lines from strong magnetic fields. For $B = 10^{12}$ gauss the fundamental cyclotron frequency is at 11.6 $/(1 + z)$ Kev. In Her X-1 a spectral feature at 40 Kev is best interpreted as a higher harmonic of such a cyclotron line. Spectral lines at similar energies in objects such as 4U 0115+63, 4U 1538-52, X 0331+53, etc., all point to magnetic fields of the order of 10^{12} gauss (Mihara, 1995; Meszaros, 1992). Damen et al (1990) have explored a method to estimate the gravitational redshift from neutron stars using X-ray bursts.

Gravitational redshift is also believed to be seen in γ-ray burst sources, which are supposed to be neutron stars accreting matter impulsively. They were first detected by the VELA satellite which was set up to monitor nuclear explosions on earth (Klebesadel et al, 1973). Two kinds of spectral lines were claimed to have been seen: (1) absorption lines in the 20 Kev to 60 Kev band, and (2) emission lines around 400 Kev; both are supposed to be gravitationally redshifted spectral features; see Meszaros (1992) and references therein. However, the BATSE detector aboard the Compton Gamma-Ray Observatory (CGRO) failed to confirm these findings (Hartmann, 1995; Fishman, 1996). So these results are controversial.

Gravitational redshift modifies the observed spectrum from the surface of neutron stars (Pavlov et al, 1995); it reduces the observed surface temperatures by the factor $(1 + z)$ (Becker & Aschenbach, 1995).

3.2. TIME DELAY

In general relativity the extra time of travel of electromagnetic signals from a distance r to the closest point of approach r_0 to a star is given by

$$t(r, r_0) = 2\frac{M_* G}{c^3} \ln\left[\frac{r + (r^2 - r_0^2)^{1/2}}{r_0}\right] + \frac{M_* G}{c^3}\left[\frac{r - r_0}{r + r_0}\right]^{1/2} \tag{2}$$

(Weinberg, 1972). Thus the round trip time delay for a radar signal to travel from earth to the planet mercury past the sun, and back is

$$\Delta t = 2\left[t(r_\oplus, R_\odot) + t(r_m, R_\odot)\right] \approx 4\frac{M_\odot G}{c^3}\left[1 + \ln\left(\frac{4 r_m r_\oplus}{R_\odot^2}\right)\right] \tag{3}$$

where $r_\oplus$ and r_m are the distances of earth and mercury from the sun, and $R_\odot$ is the radius of the sun, the closest distance of passage of the radar signal from the center of the sun. In 1964 Shapiro proposed (Shapiro, 1964) and later observed the delay of 240 micro seconds. In 1968 Shapiro (Shapiro, 1968) proposed that such time delay be observed from pulsar signals passing close to the sun.

The Shapiro delay has been best measured in the binary pulsar PSR B1855+09, whose pulsar period is 5.36 ms, orbital period is 12.327 days, and whose plane of orbit is seen almost edge on (Taylor, 1993). The time delay Δt is a function of the minimum distance between the line of sight and the companion, which is a low mass star ($0.26\ M_\odot$), as well as upon the phase ϕ of the pulsar in the binary orbit:

$$\Delta t = -2\,p \ln\left[1 - q \cos\left\{2\pi\left(\phi - \phi_0\right)\right\}\right], \tag{4}$$

where p and q are the size and the shape factors of Shapiro delay (q is the sine of the angle of inclination of the binary orbit). The observed time delay in this pulsar closely fits the expected curve; the maximum delay in the orbit is about 20 μsec. In fact, the determination of p and q in this pulsar provided the additional equations required to determine the mass of the pulsar and its binary companion (Taylor, 1995).

Stochastic Shapiro delay of pulsar signals on account of background stars or dark matter passing close by the line of sight has been investigated (Fargion & Conversano, 1997; Walker, 1996; Ohnishi et al, 1995). The effect was found to be very small. Therefore pulsars can in principle be used as the clocks for the civil time standard.

3.3. BENDING OF LIGHT

To first order the deflection $\Delta\phi$ in the angle of a ray of light passing at the closest distance r_0 to a star is given by $4GM_*/(c^2 r_0)$. This is 1.75 arc seconds ($''$) for starlight passing by the sun, which was verified during the total solar eclipse of 1919 (Dyson et al, 1920). Radio interferometry was also used to measure the change in position of several quasars as their radio waves passed past the sun (Fomalont & Sramek, 1977); the results supported Einstein's theory.

Bending of light near neutron stars has been discussed in several diverse contexts. In the X-ray pulsar Her X-1, the variation of the cyclotron line energy with the pulse phase favours a pencil beam; on the other hand the observed total X-ray luminosity favours a fan beam. This discrepancy can possibly be resolved by invoking gravitational bending of the X-rays at the surface of the neutron star (Meszaros, 1992; Pechenick et al, 1983). Bending of light has also been invoked in the analysis of timing data from binary pulsars whose light has been bent by their companion (Doroshenko & Kopeikin, 1995; Churchill, 1992; Schneider, 1990; Gorham, 1986). Gravitational lensing by neutron stars has been invoked to amplify the light from a binary companion (Schneider, 1989), and for detection of isolated pulsars in our galaxy (Horvath, 1996).

3.4. PRECESSION OF PERIHELIA

In general relativity, the precession of the perihelion of the orbit of a planet around a star is $\Delta\phi = 6\pi M_* G/(c^2 L)$ radians in each revolution, where L is the semi-latus rectum of the orbit (Weinberg, 1972). For mercury (the planet closest to the sun) this is predicted to be $43.03''$ per century, which has been verified by the analysis of data available since 1765 (Clemence, 1947). Although the accuracy of measurement diminishes for the rest of the planets, the general relativistic precession of periastron has also been verified for the planets Venus and earth and the asteroid Icarus (Clemence, 1947).

In binaries containing neutron stars the magnitude of the precession is expected to be much higher. Indeed, the binary pulsar PSR B1913+16, which contains two neutron stars, has a measured precession of 4.23 degrees per year (Taylor, 1993). The rate of periastron advance has also been measured in the binary pulsars PSR B1534+12, PSR B2127+11C and PSR B2303+46 (Taylor, 1993). In these binaries general relativity is used as a tool for astronomy, to obtain the masses of neutron stars.

3.5. PRECESSION OF ORBITING GYROSCOPES

In relativistic gravity, a gyroscope orbiting the earth will have two components of angular velocity of precession. One is known as "geodetic" precession, and depends purely upon the mass of earth. Another depends upon the spin angular momentum of earth, due to dragging of the local inertial frame; this is known as the Lens-Thirring effect or the "hyperfine" term (Weinberg, 1972).

Schiff (1960) suggested that a gyroscope placed in orbit around the earth could be used to measure the precession predicted by general relativity. Assuming a circular orbit, the ratio of the hyperfine to the geodetic precession rate is 6.5 $\times 10^{-3}$ (Weinberg, 1972). So the dominant effect is geodetic precession of the gyroscope. The average value of its angular frequency $< \Omega_g >$ is

$$< \Omega_g > \approx 1.5 \frac{[GM_{\oplus}]^{3/2}}{r^{5/2}} \approx 8.4 '' \left[\frac{R_{\oplus}}{r}\right]^{5/2} \text{ per year,} \tag{5}$$

where r is the radius of orbit of the gyroscope and $R_{\oplus}$ is the radius of earth (Weinberg, 1972). While this is measurable, the hyperfine term alone is

$$< \Omega_h > = 0.5 \frac{GJ_{\oplus}}{r^3} \approx 0.055 '' \left[\frac{R_{\oplus}}{r}\right]^{3} \text{ per year,} \tag{6}$$

(Weinberg, 1972); $J_{\oplus}$ is the orbital angular momentum of earth. To measure this comparatively tiny precession, the spin axis of the gyroscope should be perpendicular to both the direction of $J_{\oplus}$ as well as to the plane of its own orbit. The optimal experiment for this is to place a gyroscope in polar orbit around earth, with its spin axis perpendicular to the plane of its orbit (Weinberg, 1972). There are several problems in conducting such experiments (see references in Will, 1993). However, the earth-moon system can be considered to be a gyroscope with its spin axis perpendicular to the orbital plane; $< \Omega_g >$ is predicted by general relativity to be about 2″ per century, which has been measured by lunar laser ranging data (Will, 1993).

Efforts are under way to detect the predicted precession of the rotation axis of the binary pulsar PSR B1913+16, by means of long term changes in its pulse shape (Damour, 1993). The relative powers in the two components of the pulse profile of this pulsar are changing systematically, by about one percent every year. This is apparently consistent with the expected geodetic precession of 1.21 degrees per year of the pulsar spin axis (Weisberg et al, 1989).

4. Gravitational Radiation

It is well known that Maxwell's equations in electromagnetism have radiative or wavelike solutions. In general relativity Einstein's equations also have radiative solutions. If one considers a small perturbation $h_{\mu\nu}$ on a general metric $g_{\mu\nu}$ (not necessarily the Minkowski or the flat space-time metric $\eta_{\mu\nu}$), the result is a wave equation for $h_{\mu\nu}$, after imposing a gauge condition (gravitational waves are more complicated than electromagnetic waves because they also contribute to the gravitational field, unlike electromagnetic waves). Just like electromagnetic waves, $h_{\mu\nu}$ also has two independent states of polarisation. However, electromagnetic waves have helicity 1 while gravitational waves have helicity 2. In the language of field theory, photons have spin 1 and gravitons have spin 2 (Weinberg, 1972; Misner, Thorne & Wheeler, 1973). In the low velocity limit, gravitational radiation occurs due to accelerating mass quadrupole tensor $\ddot{I}_{ij} = d^2 I_{ij}/dt^2$ where

$$I_{ij} = \int d^3x \rho (3x_i x_j - \delta_{ij} r^2), \tag{7}$$

ρ being the mass density (Landau & Lifshitz, 1975).

An ellipsoid rotating with angular frequency ω around one of its principal axis of inertia, will emit gravitational waves at frequency 2ω. The total power emitted is

$$\frac{dE}{dt} = \frac{32G}{5c^5} \omega^6 I^2 e^2 \quad \text{ergs/sec}, \tag{8}$$

where $I = I_{11} + I_{22}$, and $e = (I_{11} - I_{22})/I$ is the equatorial ellipticity (Weinberg, 1972). A newly born pulsar is supposed to have $I = 10^{45}$ g.cm^2 and $\omega/2\pi \approx 10^4$ sec^{-1}; it is expected to emit gravitational radiation at the rate of $10^{55} e^2$ ergs/sec (Weinberg, 1972). Since the total rotational energy is only about 10^{53} ergs, most of the initial rotational energy is expected to be carried away by gravitational radiation provided $e \geq 10^{-4}$ (Thorne, 1969; Ostriker & Gunn, 1969).

For two stars of masses M_1 and M_2 in an elliptical orbit, the average power emitted per orbit is

$$\frac{dE}{dt} = \frac{32G^4}{5c^5a^5}M_1^2M_2^2(M_1+M_2) \times f(e), \tag{9}$$

where a is the semi-major axis and $f(e)$ is a function of the eccentricity e of the orbit (Landau & Lifshitz, 1975; Peters & Mathews, 1963). Since the binary is losing energy, the orbital period P_b will decrease at the rate

$$\dot{P}_b = \frac{dP_b}{dt} = -\frac{192\pi}{5c^5}\left[\frac{2\pi G}{P_b}\right]^{5/3}\frac{M_1M_2}{(M_1+M_2)^{1/3}} \times f(e), \tag{10}$$

(Taylor, 1995; Wagoner, 1975). The binary pulsar PSR B1913+16 consists of two neutron stars in an orbit with $e = 0.62$ and $P_b = 7.75$ hours. The predicted $\dot{P}_b$ for this binary is -2.403×10^{-12} while the observed value is -2.425×10^{-12}, verifying the existence of gravitational radiation (Taylor, 1993). For more details see (Iyer, 1998).

Direct detection of gravitational waves using resonant quadrupole antennas (Weber, 1968) is currently one of the foremost challenges to physicists. Modern gravitational wave detectors consist of four masses arranged along the two arms of an L; laser interferometers monitor the relative motion of these masses due to the passage of a gravitational wave (Abramovici et al, 1992). For more details see (Bhawal, 1998). Gravitational waves from some models of the early universe occur over a very broad band of frequencies. However, gravitational waves due to the motion of compact stars in binaries, including their coalescence, fall within the band 10^{-4} to 10^4 Hz, while gravitational waves from rotating compact stars or from supernova explosions fall in the range 1 to 10^4 Hz (Thorne, 1996). Compact, dynamical concentrations of large amounts of mass are supposed to be strong emitters of gravitational radiation. Currently the best astrophysical scenario for direct detection of gravitational waves is supposed to be mergers of neutron stars or black holes of a binary systems, or collisions between these compact stars (Thorne, 1996). Such systems will obviously be rare in the universe, and gravitational wave emission from them is clearly an episodic phenomenon. For more details see (Sathyaprakash, 1998). The high frequency gravitational waves can be studied using earth based detectors, but the low frequency waves might require space based antennas (Thorne, 1996).

The gravitational radiation from a new born neutron star is supposed to have a characteristic downward sweep in frequency, from a few hundred Hz to zero (Lai & Shapiro, 1995). Possibilities have been discussed and unsuccessful attempts have been made to detect gravitational radiation from single pulsars (Dhurandhar et al, 1996; New et al, 1995; Hough et al, 1983; Hirakawa, 1980; Zimmermann & Szedenits, 1979). Gravitational radiation from asymmetric supernova explosions is also a possibility (Nazin & Postnov, 1997; Burrows & Hayes, 1996).

The binary pulsar PSR B1913+16 is supposed to be emitting gravitational waves at roughly 10^{-4} Hz, which is twice its orbital frequency. With the passage of time the orbit will shrink. As the two neutron stars spiral in the frequency will increase; In the last fifteen minutes before collision it sweeps from 10 Hz to 10^3 Hz (Thorne, 1996). The details of the coalescence of such systems, the ensuing gravitational radiation, and the optimal signal processing techniques are currently being worked out (Ruffert et al, 1997; Thorne, 1996; Ruffert et al, 1996). The above calculations are normally done using Newtonian gravitational for the orbital motion and the quadrupole formula for gravitational wave emission. Incorporation of relativistic gravity more completely predicts several interesting phenomena, such as (1) the backscatter of gravitational waves from the warped space-time of the binary pulsar, (2) the Lens-Thirring effect modifying the observed waveform, etc., (Thorne, 1996). For more details see (Sathyaprakash, 1998; Dhurandhar, 1998).

The expected rate of occurrence of such events was until recently a few mergers per year within 200 mega parsec of the universe (Phinney, 1991; Narayan et al, 1991). This number is currently being revised downwards due to input from pulsar searches (Bailes, 1996).

Acknowledgements: I thank T. Padmanabhan for a critical reading of this article and useful discussions. This article has made use of NASA's Astrophysics Data System Abstract Service.

References

Abramovici, A., Althouse, W. E., Drever, R. W. P., Gursel, Y., Kawamura, S., Raab, F. J., Shoemaker, D., Sievers, L., Spero, R. E., Thorne, K. S. (1992) *Science*, **256**, 325.

Bailes, M. (1996) *Compact Stars in Binaries*, I. A. U. Symposium no. 165, Kluwer Academic Publishers, Netherlands, eds. J. van Paradijs, E. P. J. van den Heuvel, E. Kuulkers, 213.

Baym, G., Pethik, C., Pines, D., Ruderman, M. (1969) *Nature*, **224**, 872.

Bhawal, B. (1998) See chapter 27 in this volume.

Becker, W., Aschenbach, B. (1995) *The Lives of the Neutron Stars*, Kluwer Academic Publishers, NATO ASI Series, eds. M. A. Alpar, U. Kiziloglu, J. van Paradijs, 47.

Blandford, R. D. (1995) *J. Astrophys. Astr.*, **16**, 191.

Burbidge, E. M., Burbidge, G. R., Fowler, W. A., Hoyle, F. (1957) *Rev. Mod. Phys.*, **29**, 547.

Burrows, A., Hayes, J. (1996) *Phys. Rev. Lett*, **76**, 352.

Churchill, C. W. (1992) *Bull. Am. Astron. Soc.*, **181**, 9909.

Clayton, D. D. (1983) *Principles of Stellar Evolution and Nucleosynthesis*, University of Chicago Press, Chicago.

Clemence, G. M. (1947) *Rev. Mod. Phys.*, **19**, 361.
Damen, E., Magnier, E. Lewin, W. H. G., Tan, J., Pennix, W., van Paradijs, J. (1990) *Astron. Astrophys.*, **237**, 103.
Damour, T. (1993) *Pulsars as Physics Laboratories*, Oxford University Press, Oxford, eds R. D. Blandford, A. Hewish, A. G. Lyne and L. Mestel, 135.
Dhurandhar, S. V., Blair, D. G., Costa, M. E. (1996) *Astron. Astrophys.*, **311**, 1043.
Dhurandhar, S.V. (1998) See chapter 28 in this volume.
Doroshenko, O. V., Kopeikin, S. M. (1995) *Mon. Not. Roy. Astron. Soc.*, **274**, 1029.
Dyson, F. W., Eddington, A. S., Davidson, C. (1920) *Mem. Roy. Astron. Soc.*, **62**, 291.
Fargion, D., Conversano, R. (1997) *Mon. Not. Roy. Astron. Soc.*, **285**, 225.
Fishman, G. J. (1996) *Compact Stars in Binaries*, I. A. U. Symposium no. 165, Kluwer Academic Publishers, Netherlands, eds. J. van Paradijs, E. P. J. van den Heuvel, E. Kuulkers, 467.
Fomalont, E. B., Sramek., R. A. (1977) *Comments on Astrophys.*, **7**, 19.
Fowler, R. H. (1926) *Mon. Not. Roy. astron. Soc.*, **87**, 114.
Friedman, J. L., Ipser, J. R., Parker, L. (1986) *Astrophys. J.*, **304**, 115.
Giacconi, R., Gursky, H., Kellogg, E., Schreier, E., Tananbaum, H. (1971) *Astrophys. J.*, **167**, L67.
Glendenning, N. K. (1997) *Compact Stars*, Springer Verlag, New York.
Gorham, P. W. (1986) *Astrophys. J.*, **303**, 601.
Harrison, B. K., Thorne, K. S., Wakano, M. and Wheeler, J. A., (1965) *Gravitation Theory and Gravitation Collapse*, University of Chicago Press, Chicago.
Hartle, J. B. (1967) *Astrophys. J.*, **150**, 1005.
Hartmann, D. (1995) *The lives of the Neutron Stars*, Kluwer Academic Publishers, NATO ASI Series, eds. M. A. Alpar, U. Kiziloglu, J. van Paradijs, 495.
Hirakawa, H. (1980) *Gravitational Radiation, Collapsed Objects and Exact Solutions*, Proceedings of Einstein Centenary Summer School, Perth, Springer Verlag, Berlin, 331.
Horvath, J. E. (1996) *Mon. Not. Roy. Astron. Soc.*, **278**, L46.
Hough, J., Drever, R. W. P., Ward, H., Munley, A. J., Newton, G. P., Meers, B. J., Hoggan, S., Kerr, G. A., (1983) *Nature*, **303**, 216.
Irvine, J. M. (1978) *Neutron Stars*, Oxford University Press, Oxford.
Iyer, B. R. (1998) See chapter 26 in this volume.
Klebesadel, R. W., Strong, I. B., Olson, R. A. (1973) *Astrophys. J. Lett*, **182**, L85.
Lai, D., Shapiro, S. L. (1995) *Astrophys. J.*, **442**, 259.
Landau, L. D., Lifshitz, E. M (1975) *The Classical Theory of Fields*, Pergamon Press, New York.
Lindblom, L. (1984) *Astrophys. J.*, **278**, 364.
Lindblom, L., Mendell, G. (1995) *Astrophys. J.*, **444**, 804.
Lyne, A. G. (1993) *Pulsars as Physics Laboratories*, Oxford University Press, Oxford, eds R. D. Blandford, A. Hewish, A. G. Lyne and L. Mestel, 29.
Lyne, A. G., Pritchard, R. S., Shemar, S. L. (1995) *J. Astron. Astrophys.*, **16**, 179.
Manchester, R. N., Taylor, J. H. (1977) *Pulsars*, W. H. Freeman and Co., San Fransisco.
Manchester, R. N. (1993) *Pulsars as Physics Laboratories*, Oxford University Press, Oxford, eds R. D. Blandford, A. Hewish, A. G. Lyne and L. Mestel, 3.
Meszaros, P. (1992) *High Energy Radiation from Magnetized Neutron Stars*, University of Chicago Press, Chicago.
Mihara, T., Makishima, K., Nagase, F. (1995) *Bull. Am. Astron. Soc.*, **187**, 10403.
Misner, C. W., Thorne, K. S., Wheeler, J. A. (1973) *Gravitation*, W. H. Freeman and Co., San Fransisco.
Narayan, R., Piran, T., Shemi, A. (1991) *Astrophys. J.*, **379**, L17.
Narlikar, J. V. (1978) *General Relativity and Cosmology*, The Macmillan Company of India Ltd., New Delhi.
Nazin, S. N., Postnov, K. A. (1997) *Astron. Astrophys.*, **317**, L79.

New, K. C. B., Chanmugam, G., Johnson, W. W., Tohline, J. E. (1995) *Astrophys. J.*, **450**, 757.
Nomoto, K., Tsuruta, S. (1987) *Astrophys. J.*, **312**, 711.
Ogelman, H. (1995) *The Lives of the Neutron Stars*, Kluwer Academic Publishers, NATO ASI Series, eds. M. A. Alpar, U. Kiziloglu, J. van Paradijs, 101.
Ohnishi, K., Hosokawa, M., Fukushima, T., Takeuti, M. (1995) *Astrophys. J.*, **448**, 271.
Oppenheimer, J. R., Volkoff, G. M. (1939) *Phys. Rev.*, **55**, 374.
Ostriker, J. P., Gunn, J. E. (1969) *Astrophys. J.*, **157**, 1395.
Pavlov, G. G., Shibanov, Yu. A., Zavlin, V. E., Meyer, R. D. (1995) *The lives of the Neutron Stars*, Kluwer Academic Publishers, NATO ASI Series, eds. M. A. Alpar, U. Kiziloglu, J. van Paradijs, 71.
Pechenick, K. R., Ftaclas, C., Cohen, J. M. (1983) *Astrophys. J.*, **274**, 846.
Peters, P. C., Mathews, J. (1963) *Phys. Rev.*, **131**, 435.
Phinney, S. E. (1991) *Astrophys. J.*, **380**, L17.
Popper, D. M. (1954) *Astrophys. J.*, **120**, 316.
Ruderman, M. A., Sutherland, P. G., *Astrophys. J.*, **196**, 51 (1975).
Ruffert, M. Janka, H. T., Schafer, G. (1996) *Astronom. Astrophys.*, **311**, 532.
Ruffert, M. Rampp, M., Janka, H. T. (1997) *Astronom. Astrophys.*, **321**, 991.
Ryba, M. F., Taylor, J. H. (1991) *Astrophys. J.*, **371**, 739.
Sathyaprakash, B. S. (1998) See chapter 25 in this volume.
Schiff, L. I. (1960) *Phys. Rev. Lett*, **4**, 215.
Schneider, J. (1989) *Astron. Astrophys.*, **214**, 1.
Schneider, J. (1990) *Astron. Astrophys.*, **232**, 62.
Schutz, B. (1985) *A First Course in General Relativity*, Cambridge University Press, Cambridge.
Schwarzschild, M. (1965) *Structure and Evolution of the Stars*, Dover, New York.
Shapiro, I. I. (1964) *Phys. Rev. Lett.*, **13**, 789.
Shapiro, I. I. (1968) *Science*, **162**, 352.
Shapiro, S. L. and Teukolsky, S. A. (1983) *Black Holes, White Dwarfs, and Neutron Stars*, John Wiley & Sons, Inc., New York.
Taylor, J. H. (1993) *Pulsars as Physics Laboratories*, Oxford University Press, Oxford, eds R. D. Blandford, A. Hewish, A. G. Lyne and L. Mestel, 117.
Taylor, J. H. (1995) *J. Astron. Astrophys.*, **16**, 307.
Thorne, K. S. (1969) *Astrophys. J.*, **158**, 1.
Thorne, K. S. (1996) *Compact Stars in Binaries*, I. A. U. Symposium no. 165, Kluwer Academic Publishers, Netherlands, eds. J. van Paradijs, E. P. J. van den Heuvel, E. Kuulkers, 153.
Wagoner, R. V., (1975) *Astrophys. J. Lett*, **196**, L63.
Walker, M. A., (1996) *Pub. Astron. Soc. Austral.*, **13**, 236.
Weber, F. (1992) *Hadron Physics and Neutron Star Properties*, Habilitation Thesis, University of Munich.
Weber, F., Glendenning, N. K. (1992) *Astrophys. J.*, **390**, 541.
Weber, J,. (1968) *Physics Today*, **21**, 34.
Weinberg, S. (1972) *Gravitation and Cosmology*, John Wiley & Sons Inc., New York.
Weisberg, J. M., Romani, R. W. Taylor, J. H. (1989) *Astrophys. J.*, **347**, 1030.
Will, C. M. (1993) *Theory and Experiment in Gravitational Physics*, Cambridge University Press, Cambridge.
Woosely, S. E., Weaver, T. A. (1986) *Ann. Rev. Astron. Astrophys.*, **24**, 205.
Zeldovich, Y. B. and Novikov, I. D. (1971) *Relativistic Astrophysics Vol I Stars and Relativity*, University of Chicago Press, Chicago.
Zimmermann, M., Szedenits, E. Jr. (1979) *Phys. Rev. D.*, **20**, 351.

16. ACCRETION DISKS AROUND BLACK HOLES

PAUL J. WIITA
Department of Physics & Astronomy,
Georgia State University,
Atlanta, Georgia 30303, U.S.A.

AND

Physics Department,
Indian Institute of Technology,
Powai, Mumbai 400 076, India.

1. Introduction

It is patently impossible to provide a thorough review of a topic as vast as accretion disks around black holes in only 15 relatively small pages, and so this chapter will mention only a few topics, and concentrate on fewer still. In 1996 alone, 250 papers listing accretion or accretion disks as subject headings appeared in *The Astrophysical Journal* and 80 more were published in the *Monthly Notices of the Royal Astronomical Society*, with other leading journals containing much additional research. In that accretion onto white dwarfs, neutron stars, and young stellar objects are also active research areas, only roughly half of the above flood of calculations and data was directly related to accretion onto black holes (hereafter, BHs). Nonetheless, the literature since the early 1970's, when this phenomenon was first considered seriously (e.g., Shakura & Sunyaev 1973; Novikov & Thorne 1973), amounts to several thousand papers. Roughly one international conference per year concentrates on accretion disk phenomena (for a recent one, see Abramowicz et al. 1997a), and entire texts (e.g., Frank et al. 1992), monographs (e.g., Chakrabarti 1990a), and many far more extensive reviews than this current effort (substantial recent ones include Chakrabarti 1996a; Lin & Papaloizou 1996; Papaloizou & Lin 1995), have appeared on this subject. This topic has now become so mainstream that it would be amazing if a modern text on either general relativity theory or

B. R. Iyer and B. Bhawal (eds.), Black Holes, Gravitational Radiation and the Universe, 249–263.

basic astronomy did not discuss accretion disks around black holes.

In a volume celebrating the contributions of C.V. Vishveshwara, it seems appropriate to begin by noting some collaborative efforts that began our scientific acquaintance. Vishu had already performed computations of electromagnetic fields and orbits of particles in the vicinity of BHs back in the 1970's (Chitre & Vishveshwara 1975; Prasanna & Vishveshwara 1978). During my first scientific visit to India, Naresh Dadhich and I examined the effect of strong magnetic fields on the efficiency with which matter would be accreted onto non-rotating BHs (Dadhich & Wiita 1982). Using the Ernst (1976) metric, we located the marginally stable and minimum binding energy orbits (r_{ms} and r_{mb}, respectively) for accretion disks around a static BH immersed in a uniform magnetic field. When I met him in Bangalore, Vishu proposed that we generalize this effort to the more difficult case of Kerr BHs, using the effective potential for trajectories worked out by Prasanna and himself (Prasanna & Vishveshwara 1978). We first considered uniform magnetic fields (Wiita et al. 1983) around rotating BHs and later analyzed the situation for dipolar magnetic fields (Iyer et al. 1985). Charged test particles could see the efficiency with which their rest mass is converted to energy rise dramatically in the presence of magnetic fields; however, the increase of accretion efficiency with magnetic field strength is lower for black holes with higher specific angular momenta. Nonetheless, when full-fledged hydrodynamical accretion is considered, the effects of the magnetic field on accretion efficiency are invariably quite small.

Vishu encouraged me to give a series of lectures on accretion disks in Bangalore in the summer of 1981 and then arranged to distribute those lecture notes (Wiita 1981). These talks inspired additional works on black hole accretion disks by Indian astrophysicists, the first of which was an analysis of the maximum luminosity that could emerge from the "funnels" along the rotation axes of optically and geometrically thick accretion disks (§2.2) around black holes (Nityananda & Narayan 1982). This paper was followed by an estimate of the strength of the flow of ejected matter from these funnel regions (Narayan et al. 1983). We shall not further consider the launching of jets from accretion disks here, other than to mention the probable importance of magnetic fields and refer the reader to other reviews (e.g., Wiita 1990, 1996a; Chakrabarti 1996a).

In §2 the major classes of accretion disks (thin and thick) around black holes will be outlined. Over the past few years, models in which much of the energy produced in the disk is advected inwards to the black hole before all the radiation can escape from the surface of the disk have become the focus of intense investigation, and a brief discussion of the properties of these disks will be given in §3; it is worth noting that two of the leading contributors to the study and popularity of such "unified" or "advection dominated" accretion flows are Indians: Sandip Chakrabarti and Ramesh Narayan. Observable variations produced by fluctuations on accretion disks around black holes will be briefly discussed in §4.

2. Standard Accretion Disk Models

Accretion disks around black holes were first envisaged to occur within a binary stellar system where a very massive primary star had evolved and collapsed into a black hole; the secondary would feed matter onto the BH either through a wind or through Roche lobe overflow. In either case, the accreted matter would clearly possess substantial angular momentum with respect the BH, and would form a ring around it, where centripetal and gravitational forces balance. Viscosity within the fluid, presumably produced by turbulence or magnetic fields, would allow most of the mass to filter inwards, while the bulk of the angular momentum would be transported outwards (e.g. Shakura & Sunyaev 1973). Clear evidence for such accretion disks around white dwarfs in binaries was quickly provided by analysis of cataclysmic variables (e.g., Robinson 1976).

The tremendous powers required to energize quasars, Seyfert galaxies, BL Lacertae objects, and other active galactic nuclei (hereafter AGN) demanded an engine that could very efficiently convert matter into electromagnetic radiation. The rapid variability in luminosity arising from AGN also implied very compact central engines. Accretion from a disk around a Schwarzschild (non-rotating) BH produces efficiencies of $\eta \simeq 0.057$, where the total power liberated is related to the mass accretion rate by $L = \eta \dot{M} c^2$, roughly an order of magnitude above that attainable through nuclear fusion. Accretion onto a maximally rotating Kerr BH is even more efficient, yielding $\eta \simeq 0.42$, though most models provide limits of $\eta \simeq 0.32$ for a Kerr BH with $a/M \simeq 0.998$ (Thorne 1974). These computations are based upon the binding energy of a test particle at the last stable circular orbit around the BH; $r_{ms} = 6GM/c^2 \equiv 6r_g \equiv 3R_s$ for a Schwarzschild BH. In that the bulk of this luminosity is found to emerge from distances of $< 20R_s$ for accretion disks around black holes, both the efficiency and compactness desiderata are beautifully satisfied, and this picture, with $M_{BH} \sim 10^6$–$10^9 M_\odot$, quickly became the dominant model for the AGN powerhouses (e.g., Rees 1984). The more powerful quasars still required substantial mass influxes ($\sim 10 M_\odot \mathrm{yr}^{-1}$) which are difficult, but not impossible, to find from mass loss processes in dense stellar clusters. The uncertainty in the net angular momentum of the gas falling into a central mass implies that the disk only might form at relatively small radii, where rotation finally becomes important enough to balance gravity.

2.1. THIN ACCRETION DISKS

The original accretion disk models (Shakura & Sunyaev 1973; Novikov & Thorne 1973) assumed that the mass accretion rate was substantial, but that $L \ll L_E$, where $L_E = 4\pi c G M_{BH} m_p/\sigma_T = 1.3\times10^{38} \mathrm{erg\,s^{-1}} (M_{BH}/M_\odot)$ is the Eddington luminosity, or the maximum luminosity that can escape from a spherical system and still allow matter to be accreted inwards (i.e., when radiation pressure on electrons is just balanced by gravitational attraction on protons). The other key assumptions were that the disk had a Keplerian angular momentum distribution so that $v_\phi \propto r^{-1/2}$ (in the Newtonian approximation), and that the radial velocity of infall, $v_r \ll v_\phi$. Under these circumstances, the radial equations of the steady state disk structure decouple from the vertical ones, and complete solutions can be obtained based on the relativistic conservation laws for mass, energy and angular momentum. Define: the surface mass density as $\Sigma(r) = \int_{-h}^{+h} \rho(r,z)dz$, with ρ the mass density and $h(r)$ the disk half thickness; $F(r)$ is the flux radiated from the upper disk surface; and $W(r) \equiv \int_{-h}^{+h} \langle T_{\phi r}\rangle dz$ is the integrated stress between two disk annuli. Then, in terms of non-dimensional radii, $r_* \equiv r/M$, and angular momentum parameter of the BH, $a_* \equiv a/M$, one can find the equations of radial structure to be (Novikov & Thorne 1973):

$$\dot{M} = -2\pi r \Sigma v_r (1 - 2/r_* + a_*^2/r_*^2)^{1/2}, \tag{1}$$

$$W = \frac{\dot{M}}{2\pi}\left(\frac{M}{r^3}\right)^{1/2} \frac{(1 - 3/r_* + 2a_*/r_*^{3/2})^{1/2} Q}{r^2(1 - 2/r_* + a_*^2/r_*^2)(1 + a_*/r_*^{3/2})}, \tag{2}$$

$$F = \frac{3\dot{M}}{8\pi r^2}\left(\frac{M}{r}\right) \frac{Q}{(1 + a_*/r_*^{3/2})(1 - 3/r_* + 2a_*/r_*^{3/2})^{1/2}}, \tag{3}$$

where Q is an extremely messy integral, which, however, $\approx 1 - (r_{ms}/r_*)^{1/2}$ for small a_*.

The Newtonian approximation shows that the total luminosity emerging from both surfaces of the disk, $L_{\rm disk} = GM\dot{M}/2r_{in}$ and this is one-half of the maximum total luminosity contained in the infalling matter. Therefore this matter retains as kinetic energy half the potential energy it lost while in-spiralling, implying that an equivalent amount of additional energy could be radiated in a boundary layer, where the disk "rubs up against" the star (e.g. Frank et al. 1992). This phenomenon is possible for central objects with hard surfaces, such as white dwarfs and neutron stars (though strong magnetic fields may prevent the disk from reaching the stellar surface); however, BHs do not possess hard surfaces, and once inside the marginally stable orbit the infalling matter is trapped and very little additional radiation will escape; instead it will be swallowed by the BH, along with the accreted gas.

Standard thin disk models approximate the shear as proportional to the pressure in the disk, $T_{\phi r} = \alpha P$, with $\alpha < 1$. With this simplification, values of F, Σ, h, and T are found as algebraic functions of $r, M, \dot{M}, a$ and α (Novikov & Thorne 1973). Small corrections to some of these results, particularly for h, were recently demonstrated (Riffert & Herold 1995). Such disks are usually considered to be divided into three regions: at large radii (typically $r_* > 100$) the 'outer region' has gas pressure much greater than radiation pressure and the opacity is mostly free-free; a 'middle region', from which most of the luminosity arises, still has gas pressure dominating, but the opacity is mostly due to electron scattering; the 'inner region' (typically $r_* < 4$), is dominated by radiation pressure and electron scattering. While the flux is essentially independent of the viscosity (shear) prescription, the disk thickness, surface density and central temperatures all do depend on α. Much effort has gone into producing self-consistent models for the viscosity and this may finally be bearing fruit (cf. §3.3), with the possibility that $\alpha \sim 0.01$–0.3 is generally appropriate.

For the disk to remain thin ($h/r \ll 1$), it is easily seen that the local Keplerian velocity is highly supersonic, $c_s \ll v_\phi$, in that vertical pressure balance implies that $h \approx c_s(r/GM)^{1/2}r$. In this case, the circular matter velocity will be very close to the Keplerian value, but this will not be true very close to r_{ms}. However, the radial velocity must be substantially subsonic in the bulk of the disk, since $v_r \approx \alpha c_s(h/r) \ll c_s$.

2.2. THICK RADIATION SUPPORTED ACCRETION DISKS

Of course the assumptions implying accretion disks are always thin can break down, and they almost certainly do in the innermost regions when careful consideration of the effects of general relativity show that the flow must go supersonically through a cusp (e.g., Abramowicz et al. 1978). If the accretion rate is very high and/or the viscosity is very low then the matter will pile up in the vicinity of the central BH. The accreting material forms a thick, radiation pressure supported, accretion disk, which is often called an accretion torus. Lynden-Bell (1978) first showed the basic shape such flows should have, and pointed out that there would be funnels along the BHs rotation axis in which high radiation pressures could drive matter outwards, possibly producing jets. Paczyński & Wiita (1980) produced the first thick disk models that connected smoothly to more standard thin accretion disks at larger radii, and pointed out several important features of these configurations. These computations approximated the effects of the Schwarzschild metric in a very simple way, by using a pseudo-Newtonian potential of the form, $\Phi_{PW} = -GM/(r - 2r_g)$ in lieu of the usual $\Phi_N = -GM/r$. This potential gives the correct values of $r_{ms} = 6r_g$ and $r_{mb} = 4r_g$, and closely approximates the efficiency, with $\eta(r_{ms}) = 0.0625$. A recent study (Artimova et al. 1996) has shown that this potential remains better in nearly every regard than others that have been employed.

These disks are characterized by accretion rates so high that more than the Eddington luminosity should be produced, i.e., $\dot{M} > \dot{M}_E$. Under these circumstances the radiation produced in the accreting matter exerts large pressures on the gas and forces significant changes in the disk structure, so that $h(r) \sim r$, at least for the inner 10–100r_g. When the pressure gradient term is important (basically when $GM/r \approx P/\rho \approx c_s^2$), it modifies the angular momentum distribution. In the inner portion (out to a few r_{in}) the specific angular momentum will be above the Keplerian value (typically close to constant), while outside of that distance (corresponding to the densest portion of the disk on the midplane) the angular momentum distribution will be below the Keplerian value, though it becomes very close to it at large r. The inner edge of the disk is pushed closer to the BH than r_{ms}, and in the limit of extremely high $\dot{M}$, it approaches r_{mb} (Paczyński & Wiita 1980); this also means that the structure in the inner region is nearly independent of the viscosity. This inward motion of the disk's inner edge implies that the efficiency of accretion drops dramatically, so that, while in principle, super-Eddington luminosities can be produced, the growth of the luminosity is far below that of the accretion rate. Further, these early models involved only barytropic equations of state, and did not properly compute how much of the generated luminosity would be advected with the matter and swallowed by the BH.

More sophisticated models of radiation supported thick disks included limits imposed by the disk's self-gravity and nuclear fusion (Wiita 1982). Later work was able to include self-gravity in a reasonable approximation and make full general relativistic computations of the angular momentum distribution (Chakrabarti 1988; Lanza 1992). The spectra emitted from thick disks are similar to those of thin disks in the optical, but thick disks produce much more radiation at high frequencies (EUV and soft X-ray), if viewed essentially face-on; however, the spectra are much weaker and softer if viewed at high inclination angles (so that the hot, radiation drenched funnels are blocked from view by the rest of the disk) (Madau 1988). Computation of fusion reactions indicate that nucleosynthesis is probably unimportant for thick disks around supermassive BHs in AGN, but that it might be of real importance around solar mass BHs, and the ejected mass might have significantly enhanced metallicity (Chakrabarti et al. 1987).

Despite being able to generate jets (albeit, probably not highly relativistic ones, cf. Melia & Königl 1989), super-Eddington luminosities, sometimes better fitting spectra than thin disks, and nucleosynthesis, the study of thick disks fell from favor in the late 1980's. This was mainly because of the demonstration that thin tori were dynamically and globally unstable to non-axisymmetric perturbations (Papaloizou & Pringle 1984; Goldreich et al. 1986; Hawley & Blaes 1987). However, other calculations showed that this instability grows much more slowly for more realistic thick disks when compared to slender tori (Kojima 1986) and that actual accretion reduces the reflection from the inner boundary that allows the modes to grow quickly (e.g., Blaes 1987).

Therefore, this radiation supported thick disk picture may remain viable, if high accretion rates involving both low angular momentum and low radial infall velocities can be achieved. Accomplishing both simultaneously is difficult in the simplest astrophysical situations, but could well occur through the production of shocks in initially thin disks (e.g., Chakrabarti 1990a, 1996a). Inside of that shock zone, probably at 10–20 r_g, the flow is likely to puff up and resemble the thick disk models we have just summarized.

3. Composite Disk Models

Over the past few years, a great deal of attention has been focused on "advection dominated accretion flows" (or ADAFs, as they are often abbreviated). These more complex models now appear to be necessary to match the spectral and variability properties of observed black hole binaries, and they may also be relevant to AGN powerhouses. The possibilities that electrons and protons do not necessarily equilibrate before being accreted and that disks can form with separate temperatures for each were pointed out by Lightman & Eardley (1974) and Shapiro et al. (1976), but they emphasized the thermal instability of such disks. Ichimaru (1977) studied two-temperature disks where the proton cooling time exceeded the accretion time and he showed that these could be thermally stable; this work probably qualifies as the first discussion of ADAFs (Svensson 1997). The idea that geometrically-thick, but optically thin, two-temperature disks (with $T_{\rm ion} \sim T_{\rm virial} \sim 10^{11} K \gg T_e$) could be the natural state for very low accretion rates was stressed by Rees et al. (1982). Such ion-supported accretion disks (or tori) would radiate very little directly, but could tap the rotational energy of the central BH through magnetic fields, and thereby rather easily produce relativistic jets with small mass fluxes. Rees et al. argued that such central engines were good models for radio galaxies as opposed to quasars, where the tremendous optical brightness probably originates from thin or radiation supported thick disks. Optically thick "slim accretion disk" models allowing for advection were produced (e.g., Abramowicz et al. 1988), but may be unstable.

Two basic approaches to a detailed analysis of the more complex accretion disk models have been undertaken. One approach to these unified models stresses the topologies of the solutions to steady-state models for the radial structure, and can naturally include shock transitions; however, to solve these equations, one must integrate over the vertical structure and must make approximations concerning the heating and cooling rates (e.g., Matsumoto et al. 1984; Abramowicz & Chakrabarti 1990; Chakrabarti 1990a,b 1996a,b; Chakrabarti & Titarchuk 1995). The other route to studying these flows includes a more explicit treatment of vertical structure and heating and cooling processes, but makes other simplifications, such as self-similarity, which preclude recognition of shocks in the flows (e.g., Narayan & Yi 1994, 1995; Abramowicz et al. 1995; Chen et al. 1995; Esin et al. 1996; Lasota et al. 1996; Narayan 1996). We now turn to brief descriptions of some key results obtained from these studies to date.

3.1. FLOW TOPOLOGY AND UNIFIED DISK MODELS

An easy way to think about the topology of rotating accretion flows is to first consider the specific energy of an inviscid, thin, isothermal flow, which can be shown to be (e.g., Chakrabarti 1996a)

$$e = \frac{1}{2}v_r^2 + c_s^2 \ \log(\Sigma) + \frac{1}{2}\frac{\lambda^2}{x^2} - \frac{1}{2(x-1)}, \tag{4}$$

where $x = r/2r_g$, λ is the assumed constant specific angular momentum, c_s is the constant sound speed, and the pseudo-Newtonian potential is employed for simplicity. Close to the BH ($x \sim 1$), the potential energy term dominates the rotational kinetic energy term, and $e \approx 0.5[v_r^2 - (x-1)^{-1}]$. The negative sign connecting the two terms indicates that contours of constant energy in the (v_r, x) plane are hyperbolic and the asymptotes cross to form a saddle type ('X') inner sonic point. Similarly, at great distances from the BH, since the rotational energy term falls as $\propto x^{-2}$ while the potential falls as $\propto x^{-1}$, the contours of constant e also have a hyperbolic character and there is an outer saddle type sonic point. But, for intermediate distances, where $x \sim \lambda$, the rotational energy term dominates the potential term, and $e \approx 0.5[v_r^2 + (\lambda/x)^2]$; the positive sign indicates elliptical contours for e, implying the presence of an 'O' type sonic point (i.e., as for a harmonic oscillator). In the presence of viscosity, this 'O' type point changes to a spiral, or 'S', type, as for a damped harmonic oscillator (Chakrabarti 1996a). A general relativistic treatment for the Kerr metric obviously changes many details but not the fundamental topologies and number of sonic points (Chakrabarti 1990a), so this type of analysis is generally useful.

Now the actual presence of all three possible sonic points is determined by the angular momentum of the flow and the heating and cooling processes that yield the internal energy of the fluid, which was neglected in the above simple approximation. For no angular momentum in the accreting matter, the three critical points merge into a single transonic radius, as for simple spherical accretion (Bondi 1952). Up to two stable shocks can occur in these accretion flows for higher λ, but even the outer one typically occurs at $x < 30$. The inner sonic point exists only for black hole accretion (Chakrabarti 1990a), while the outer two can exist around neutron stars. Therefore the flow onto the BH is supersonic in going through the horizon, while the fluid should strike a neutron star's surface subsonically. This difference in flow topologies can have observable consequences. If the accretion rate is high enough, the flow onto a BH through an inner shock will be sufficiently convergent to Comptonize soft photons into a power-law spectrum with fairly steep spectral index (e.g., Chakrabarti & Titarchuk 1995), while this spectral component should be absent from flows onto neutron stars. Some spectral characteristics of AGN can be well fit by flows including shocks (e.g., Chakrabarti & Wiita 1992).

When viscosity is included in the analysis, not only does the central 'O' type sonic point become an 'S' type but the outer stable shock also becomes weaker (Chakrabarti 1990b; Chakrabarti and Molteni 1995). Once the viscosity rises above a critical value, α_{cr}, the outer shock disappears, and the topology changes to allow two smooth solutions. The one with higher dissipation passes through the inner sonic point, while the one with lower dissipation passes through the outer sonic point. The necessary viscosity depends on the disk temperature, sonic point location and angular momentum at that point, but typically corresponds to $\alpha_{cr} \sim 0.02$.

By allowing for different assumptions about heating and cooling mechanisms, these topological results permit the development of a 'unified model' for accretion disks (Chakrabarti 1989, 1996a,b). As long as α viscosity provides the heating, $Q_+ \propto \alpha P x d\Omega/dx$. For an optically thick flow, the cooling is reasonably approximated by the diffusive case, so $Q_- = 4acT^4/3k\Sigma$; but if the flow is optically thin, then Q_- must be examined more carefully, with terms for bremsstrahlung, Comptonization, and synchrotron losses included. The idea of this unification scheme is that while α_{cr} $(0 < \alpha_{cr} < 1)$ depends on the relative importance of Q_- to Q_+ (or how much energy is advected inwards), the location of the inner sonic point, and the specific angular momentum at the inner edge of the disk (Chakrabarti 1996a), the fundamental possible solution topologies do not change. The key point is that if $\alpha < \alpha_{cr}$ the flow can sustain a shock, but if $\alpha > \alpha_{cr}$ a shock will not form and the fluid can accrete smoothly through the inner sonic point.

3.2. ADAFS AND CRITICAL ACCRETION RATES

The other approach to multi-phase disk models has stressed the change in structure that can be expected as the accretion rate varies. Most models for BH powerhouses in AGN were concerned with the need to generate high luminosities and therefore emphasized models with high efficiencies and relatively high accretion rates; the appropriate models were therefore the standard thin and radiation supported thick disks summarized in §2. But the possibility of starved accretion flows, with $\dot{m} \ll 1$ ($\dot{m} \equiv \dot{M}/\dot{M}_E$) must be considered, particularly in binary systems with the secondary underfilling its Roche lobe, or in ex-(or future) AGN where there is little gas available near the central massive BH.

The basic idea can be summarized as follows (e.g., Ichimaru 1977; Svensson 1997): there are two stable branches in the as $\dot{m}$ vs. Σ plane. The high value of Σ branch corresponds to having $t_{\rm cool} < t_{\rm heat}$ at the outer edge of the disk ($r_{\rm out}$), and yields the Shakura–Sunyaev disks with gas pressure support; as $\dot{m}$ rises the radiation pressure grows, leading towards the more unstable thick disks. The low value of Σ branch is selected when $t_{\rm cool} > t_{\rm heat}$ at $r_{\rm out}$ and corresponds to advection dominated flows; there exists a maximum value of $\dot{m}$ (and Σ) for which such flows can exist. An actual accretion flow can also be a mixture of these two physical branches; typically the outer portion would correspond to a standard thin disk, but the inner portion could go over to an ADAF.

In these ADAF models, the key consideration is the nature of the heating and cooling processes, for $Q_+ = Q_- + Q_{\rm adv}$ must be considered separately for protons and electrons. While the electrons can almost always radiate efficiently, in the innermost portions of accretion flows with low $\dot{m}$, the protons will not, as long as Coulomb processes are the only thing that share energy between electrons and protons. This produces a two-temperature disk, with $T_i \sim 10^{12}{\rm K}/x$ and $T_e \sim 10^9$–10^{10}K (Shapiro et al. 1976; Rees et al. 1982). At these temperatures the electrons are mildly relativistic, electron scattering dominates the opacity, and the models appear to be both thermally and viscously stable. Depending upon the angular momentum, this inner portion could resemble either the ion-supported torus (Rees et al. 1982) or a nearly spherical shape (Narayan & Yi 1995). It is important to realize that here 'advection' does not mean that photons are trapped in an optically thick fluid and dragged into the BH; rather, not many photons are produced in the first place, and matter is swallowed without emitting much radiation.

The critical accretion rate below which these two-temperature models become important is given by $\dot{m}_{\rm cr} \sim 0.3\alpha^2$ (e.g., Narayan & Yi 1994, 1995; Abramowicz et al. 1995; Chen et al. 1995). These underluminous models have $\eta \sim 0.1(\dot{m}/\dot{m}_{\rm cr})$ so that the total power is only $L \approx (\dot{m}/\dot{m}_{\rm cr})^2 L_E$. The spectra of these disks are very hard, and inverse Compton scattering of synchrotron emission (based upon reasonable approximations to magnetic field strength) allows quite remarkable fits to spectra of various objects over about 15 decades of frequency. In particular, fits to soft X-ray transients such as A0620−00 and V404 Cygni in their quiescent states (Narayan et al. 1996, 1997), Sagittarius A* at the center of the Milky Way (Narayan et al. 1995) and the Seyfert AGN NGC 4258 (Lasota et al. 1996) have all been quite good, and require setting very few parameters; these are M_{BH} (which can often be independently estimated), $\dot{m}$, α, $P_{\rm mag}/P_{\rm gas}$, and $r_{\rm tr}$, the radius at which the standard disk converts to the two-temperature flow. The possibility that ADAF models could also work for luminous accreting black holes has been explored by Narayan (1996) who found that good spectral fits were possible, but that an unlikely high value of α (≈ 1) was necessary to obtain such fits. Advection dominated one-temperature models have also been considered (Esin et al. 1996). When the effects of non-local radiation on the heating and cooling of such accretion flows is included then they are apparently only possible for extremely low accretion rates, $\dot{m}_{\rm cr} < 0.003\alpha^2$ (Esin 1997), and have not been able to fit the spectral of any particular astrophysical object (Narayan 1997).

One particularly interesting application of these ADAF models is to the spectral changes that should occur as the accretion rate changes. Esin et al. (1997) have argued that the differing spectra seen for of Nova Muscae in the X-ray can be explained in terms of differing accretion rates. The idea is that if $\dot{m} \approx 0.01$ the transition radius is very large, corresponding to the quiescent state and a hard spectrum; as the accretion rate rises $r_{\rm tr}$ decreases. For $\dot{m} \approx 0.05$ a low state with a flatter spectrum is achieved and for $\dot{m} \approx 0.08$ a transition to an intermediate state occurs, with a peak in the hard X-rays. For slightly larger $\dot{m}$, the ADAF portion of the flow disappears and the spectrum is the soft one of a standard thin disk. Narayan (1997) has argued that the excellent fit of ADAF spectra to X-ray transient sources that appear to have massive primaries is strong evidence that one is directly detecting a BH in those cases. This is because these solutions are only available for central objects possessing event horizons, and the X-ray spectra and L/L_E ratios of neutron star candidates differ systematically from those of BH candidates.

3.3. CAN α BE DETERMINED?

The hope that the viscosity parameter, α, could be determined from either *ab initio* calculations or observations has long been held by astrophysicists modeling accretion systems. Values of $\alpha \sim 0.1$ seem to give good fits to dwarf nova outbursts (e.g., Cannizzo et al. 1988), and nearly everyone accepted the arguments of Shakura and Sunyaev (1973) that $\alpha < 1$. Further constraints from observations relevant to BHs or convincing theoretical models have been lacking until very recently. The ADAF models discussed in §3.2 fit the observed spectra best if $\alpha \approx 0.2$–0.3 (Narayan 1997). The critical value of α in the unified scheme is about 0.02 (§3.1).

Three-dimensional numerical simulations involving magnetic fields in sheared flows have strongly indicated that MHD turbulence does indeed transport angular momentum and that the magnetic fields are actually amplified (Brandenberg et al. 1995; Stone et al. 1996; Hawley et al. 1996). The values of α deduced from these simulations are typically quite low, ranging from 0.001–0.01. However, these results depend both on the resolution of the grid used in the simulation and at what values the magnetic fields saturate. Stronger angular momentum transport is found for higher resolution simulations (Brandenberg 1997). Therefore the results from the simulations to date are probably lower limits that could be raised somewhat when bigger computers allow for smaller grids; nonetheless, it will be important to see if direct simulations can actually produce the substantially higher values of α apparently favored by spectral modeling.

4. Variability from Accretion Disks

In view of space constraints, only a couple of topics can be mentioned here; for a more extensive reviews, see Chakrabarti (1996a) and Wiita (1996b). Although most strong variability, particularly from radio loud AGN, is attributed to shocks in relativistic jets, there is no question that instabilities of various types exist on accretion disks. Observations of rapid X-ray variability in Seyfert galaxies (e.g., Lawrence & Papadakis 1993), and intra-night optical variability in radio quiet QSOs (e.g., Sagar et al. 1996) are explicable in terms of bright spots on accretion disks (e.g., Abramowicz et al. 1991; Mangalam & Wiita 1993).

The key idea of these models is that any intrinsic variability produced in or just above an accretion disk will be strongly modulated by the orbital motion of the radiating flare. Both Doppler effects and purely general relativistic light bending effects come into play, and are particularly important for fluctuations produced in the inner portions of accretion disks. Phenomenological bright spot models were first applied to X-ray time series and were shown to produce excellent fits to generic light curves and their power spectral densities (Abramowicz et al. 1991). They were also shown to be viable models for the smaller optical fluctuations in BL Lacertae objects and other AGN (Mangalam & Wiita 1993). The anti-correlation between X-ray luminosity and rapid variability for Seyfert galaxies (Lawrence &

Papadakis 1993) is at least partially explained by these models (Bao & Abramowicz 1996). The high temporal coherence between different X-ray bands recently seen for galactic black hole candidates (Vaughn & Nowak 1997) is difficult to explain in most models, but can be fitted by reasonable bright spot scenarios (Abramowicz et al. 1997b).

Although various physical causes for such fluctuations have been considered, none has been worked out to the extent to provide fully convincing pictures. Vortices forming within a disk (Abramowicz et al. 1992) could be relevant, as could plasma dominated events just above the disk surface (e.g., Krishan & Wiita 1994). Spiral shocks produced in the accretion disk around a BH by passing massive stars, molecular clouds or companion BHs could also yield suitable luminosity fluctuations from the disks (e.g., Chakrabarti & Wiita 1993). The possibility that self-organized criticality plays a role in producing many flares of different strengths has also been seriously considered (e.g., Mineshige et al. 1994; Wiita & Xiong 1997).

An interesting feature of these models involves the variations in polarization that should be produced. The X-rays originating from the hot inner region are expected to be partially linearly polarized as electron scattering dominates the opacity. If strong magnetic fields are present on or above the disk then synchrotron radiation from electrons may also contribute to the polarization. Rapid polarization variability arises from these orbiting regions and the variability of both the net degree of polarization and the net angle of the polarization plane are energy-dependent, i.e., the polarization variability amplitudes are larger at higher energy (Bao et al. 1996). This trend depends only weakly on the local physics such as the specific polarization mechanism or optical depth of the sources (Bao et al. 1997). Energy-dependent polarization variability is a direct result of near field bending of light rays by the central BH and it is unique to black hole systems involving accretion disks. Since accretion disks around BHs are expected in both active galactic nuclei and X-ray binaries, future X-ray polarimetry missions could confirm the prediction of this phenomenon.

5. Conclusions

Tremendous effort continues on models for accretion disks around black holes. These models have progressed to the point that detailed comparisons with the spectra and variability characteristics of many astrophysical systems are being made. Effects arising from the strong gravitational fields in the immediate environs of black holes are possibly being observed, and convincing evidence for such effects should be available in the near future.

This work is supported in part by NASA grant NAG 5-5098, and by Research Program Enhancement funds at Georgia State University.

References

1. Abramowicz, M.A., Bao, G., Lanza, A. and Zhang, X.-H. (1991) *A.& A.*, **245**, 454.
2. Abramowicz, M.A., Bao, G., Larsson, S. and Wiita, P.J. (1997b) *Ap. J.*, (in press).
3. Abramowicz, M.A., Björnsson, G., and Pringle, J.E., eds., (1997a) *Theory of Black Hole Accretion Disks*, Cambridge University Press, Cambridge, in press.
4. Abramowicz, M.A. and Chakrabarti, S.K. (1990) *Ap. J.*, **350**, 281.
5. Abramowicz, M.A., Chen, X., Kato, S., Lasota, J.-P., and Regev, O. (1995) *Ap. J.*, **438**, L37.
6. Abramowicz, M.A., Jarosyński, M. and Sikora, M. (1978) *A.& A.*, **63**, 221.
7. Abramowicz, M. A., Lanza, A., Spiegel, E. A., Szuszkiewicz, E. (1992) *Nature*, **356**, 41.
8. Abramowicz, M.A., Czerny, B., Lasota, J.P. and Szuszkiewicz, E. (1988) *Ap. J.*, **332**, 646.
9. Artimova, I.V., Björnsson, G., and Novikov, I.D. (1996) *Ap. J.*, **461**, 565.
10. Bao, G., and Abramowicz, M. (1996) *Ap. J.*, **465**, 646.
11. Bao, G., Hadrava, P., Wiita, P.J., and Xiong, Y. (1997) *Ap. J.*, 20 Sept., in press.
12. Bao, G., Wiita, P. J., and Hadrava, P. (1996), *Phys. Rev. Lett.*, **77**, 12.
13. Blaes, O.M. (1987) *M. N. R. A. S.*, **227**, 975.
14. Bondi, H. (1952) *M. N. R. A. S.*, **112**, 195.
15. Brandenberg, A. (1997) in *Theory of Black Hole Accretion Disks*, Abramowicz, M.A. et al. eds., Cambridge University Press, Cambridge, in press.
16. Brandenberg, A., Nordlund Å., Stein, R.F., Torkelson, U. (1995) *Ap. J.*, **446**, 741.
17. Cannizzo, J.K., Shafter, A.W., and Wheeler, J.C. (1988) *Ap. J.*, **333**, 227.
18. Chakrabarti, S.K. (1988) *J. Ap. Astron.*, **9**, 49.
19. Chakrabarti, S.K. (1989) *Ap. J.*, **347**, 365.
20. Chakrabarti, S.K. (1990a) *Theory of Transonic Astrophysical Flows*, World Scientific, Singapore.
21. Chakrabarti, S.K. (1990b) *M. N. R. A. S.*, **243**, 610.
22. Chakrabarti, S.K. (1996a) *Phys. Reports*, **266**, 229.
23. Chakrabarti, S.K. (1996b) *Ap. J.*, **464**, 664.
24. Chakrabarti, S.K., Jin, L. and Arnett, W.D. (1987) *Ap. J.*, **313**, 674.
25. Chakrabarti, S.K. and Molteni, D. (1995) *M. N. R. A. S.*, **272**, 80.
26. Chakrabarti, S.K. and Titarchuk, L.G. (1995) *Ap. J.*, **455**, 623.
27. Chakrabarti, S.K. and Wiita, P.J. (1992) *Ap. J.*, **387**, L21.
28. Chakrabarti, S.K. and Wiita, P.J. (1993) *Ap. J.*, **411**, 602.
29. Chen, X., Abramowicz, M.A., Lasota, J.-P., Narayan, R., and Yi, I. (1995) *Ap. J.*, **443**, L61.
30. Chitre, D.M. and Vishveshwara, C.V. (1975) *Phys. Rev.*, **D12**, 1538.
31. Dadhich, N. and Wiita, P.J. (1982) *J. Phys. A.: Math. Gen.*, **15**, 2645.
32. Ernst, F.J. (1976) *J. Math. Phys.*, **17**, 54.
33. Esin, A.A. (1997) *Ap. J.*, **482**, 400.
34. Esin, A.A., McClintock, J.E. and Narayan, R. (1997) *Ap. J.*, in press.
35. Esin, A.A., Narayan, R., Ostriker, E., and Yi, I. (1996) *Ap. J.*, **465**, 312.
36. Frank, J., King, A. and Raine, D. (1992) *Accretion Power in Astrophysics*, Cambridge University Press, Cambridge.
37. Goldreich, P., Goodman, J. and Narayan, R. (1986) *M. N. R. A. S.*, **221**, 339.
38. Hawley, J.F. and Blaes, O.M. (1987) *M. N. R. A. S.*, **225**, 677.
39. Hawley, J.F., Gammie, C.F., and Balbus, S.A. (1996) *Ap. J.*, **464**, 690.
40. Ichimaru, S. (1977) *Ap. J.*, **214**, 840.
41. Iyer, B.R., Vishveshwara, C.V., Wiita, P.J. and Goldstein, J.J. (1985) *Pramāna*, **25**, 135.
42. Kojima, Y. (1986) *Prog. Theor. Phys.*, **75**, 251.
43. Krishan, V. and Wiita, P.J. (1994) *Ap. J.*, **423**, 172.
44. Lanza, A. (1992) *Ap. J.*, **389**, 141.

45. Lasota, J.-P., Abramowicz, M.A., Chen, X., Krolik, J., Narayan, R. and Yi, I. (1996) **462**, 142.
46. Lawrence, A. and Papadakis, I. (1993) *Ap. J.*, **414**, L85.
47. Lightman, A.P. and Eardley, D.E. (1974) *Ap. J.*, **187**, L1.
48. Lin, D.N.C. and Papaloizou, J.C.B. (1996) *Ann. Rev. Astr. Ap.*, **34**, 703.
49. Lynden-Bell, D. (1978) *Physica Scripta*, **17**, 185.
50. Madau, P. (1988) *Ap. J.*, **327**, 116.
51. Mangalam, A.V. and Wiita, P.J. (1993) *Ap. J.*, **406**, 420.
52. Matsumoto, R., Kato, S., Fukue, S., and Okazaki, A.T. (1984) *Pub. Astron. Soc. Japan*, **36**, 71.
53. Melia, F. and Königl, A. (1989) *Ap. J.*, **340**, 162.
54. Mineshige, S., Ouchi, N.B., and Nishimori, H. (1994) *Pub. Astr. Soc. Japan*, **46**, 97.
55. Narayan, R. (1996) *Ap. J.*, **462**, 136.
56. Narayan, R. (1997) in *Theory of Black Hole Accretion Disks*, Abramowicz, M.A. et al. eds., Cambridge University Press, Cambridge, in press.
57. Narayan, R., Barret, D., and McClintock, J.E. (1997) *Ap. J.*, **482**, 448.
58. Narayan, R., McClintock, J.E. and Yi, I. (1996) *Ap. J.*, **457**, 821.
59. Narayan, R., Nityananda, R., Wiita, P.J. (1983) *M. N. R. A. S.*, **205**, 1103.
60. Narayan, R. and Yi, I. (1994) *Ap. J.*, **428**, L13.
61. Narayan, R. and Yi, I. (1995) *Ap. J.*, **444**, 231.
62. Narayan, R., Yi, I., and Mahadevan, R. (1995) *Nature*, **374**, 623.
63. Nityananda, R. and Narayan, R. (1982) *M. N. R. A. S.*, **201**, 697
64. Novikov, I. and Thorne, K.S. (1973) in *Black Holes*, eds. C. DeWitt and B. DeWitt, Gordon and Breach, New York, p. 343.
65. Paczyński, B. and Wiita, P.J. (1980) *A.& A.*, **88**, 23.
66. Papaloizou, J.C.B. and Lin, D.N.C. (1995) *Ann. Rev. Astr. Ap.*, **33**, 505.
67. Papaloizou, J.C.B. and Pringle, J.E. (1984) *M. N. R. A. S.*, **208**, 721.
68. Prasanna, A.R. and Vishveshwara, C.V. (1978) *Pramāna*, **11**, 359.
69. Rees, M.J. (1984) *Ann. Rev. Astr. Ap.*, **22**, 471.
70. Rees, M.J., Begelman, M.C., Blandford, R.D., and Phinney, E.S. (1982) *Nature*, **295**, 17.
71. Riffert, H. and Herold. H. (1995) *Ap. J.*, **450**, 508.
72. Robinson, E.L. (1976) *Ann. Rev. Astr. Ap.*, **14**, 119.
73. Sagar, R., Gopal-Krishna, and Wiita, P.J. (1996) *M. N. R. A. S.*, **281**, 1267.
74. Shakura, N.I. and Sunyaev, R.A. (1973) *A.& A.*, **24**, 337.
75. Shapiro, S.L., Lightman, A.P. and Eardley, D.M. (1976) *Ap. J.*, **204**, 187.
76. Stone, J.M., Hawley, J.F., Gammie, C.F., Balbus, S.A. (1996) *Ap. J.*, **463**, 656.
77. Svensson, R. (1997) in *Theory of Black Hole Accretion Disks*, Abramowicz, M.A. et al. eds., Cambridge University Press, Cambridge, in press.
78. Thorne, K.S. (1974) *Ap. J.*, **191**, 507.
79. Vaughn, B.A. and Nowak, M.A. (1997) *Ap. J.*, **474**, L43.
80. Wiita, P.J. (1981) *Accretion Disks in Astrophysics*, RRI, Bangalore.
81. Wiita, P.J. (1982) *Ap. J.*, **256**, 666.
82. Wiita, P.J. (1990) in *Beams and Jets in Astrophysics*, ed. P.A. Hughes, Cambridge University Press, Cambridge, p. 379.
83. Wiita, P.J. (1996a) in *Energy Transport in Radio Galaxies and Quasars*, P. Hardee, A. Bridle, and A. Zensus, eds. **ASP Conf. Series, Vol. 100**, Astron. Soc. Pacific, San Francisco, p. 395.
84. Wiita, P.J. (1996b) in *Blazar Continuum Variability*, H.R. Miller, J.R. Webb, and J.C. Noble, eds. **ASP Conf. Series, Vol. 110**, Astron. Soc. Pacific, San Francisco, p. 42.
85. Wiita, P.J., Vishveshwara, C.V., Siah, M.J. and Iyer, B.R. (1983) *J. Phys. A.: Math. Gen.*, **16**, 2077.
86. Wiita, P.J. and Xiong, Y. (1997) in *Theory of Black Hole Accretion Disks*, Abramowicz, M.A. et al. eds., Cambridge University Press, Cambridge, in press.

An important contribution of the general theory of relativity to cosmology has been to keep out theologians by a straightforward application of tensor analysis.

— ENGELBERT SCHUCKING

There is a lovely passage in William Faulkner's *The Sound and the Fury* where he invokes the possibility of light returning to the earth after traversing a curved space allowing people to see the past. Quentin, who wishes to escape time, muses:
'It was a while before the last stroke ceased vibrating. Like all bells that ever rang still ringing in the long dying light rays....Like father said, down the long and lonely light rays you might see Jesus walking'.

— C. V. VISHVESHWARA

Astronomers often try to make important discoveries by combining the observational techniques of ESP with the theoretical methods of science fiction.

— ENGELBERT SCHUCKING

17. ASTROPHYSICAL EVIDENCE FOR BLACK HOLE EVENT HORIZONS

K. MENOU, E. QUATAERT AND R. NARAYAN
Harvard-Smithsonian Center for Astrophysics,
60 Garden Street, Cambridge, MA 02138, USA. ‡§

1. Introduction

Astronomers have identified potential black holes in a variety of astrophysical objects. In binary star systems, consisting of two stars revolving around each other under the influence of their mutual gravitational attraction, black holes with masses of order several solar masses ($M \sim 5 - 20M_{\odot}$) are thought to exist as the dead remnant of the initially more massive of the two stars.

Occasionally, the black hole in such a system accretes (receives) matter from the outer layers of its companion and the binary becomes a powerful emitter of high energy radiation (hence the name: X-ray binary, XRB) [1]. The radiation is ultimately due to the conversion of gravitational potential energy: matter heats up as it falls into the deep potential well of the black hole and the hot gas radiates [2].

In addition, supermassive black holes ($M \sim 10^6 - 10^{10}M_{\odot}$) probably exist at the centers of most galaxies. By accreting matter (stars or gas) in their vicinity, these black holes may produce the intense emission that we observe from quasars and other Active Galactic Nuclei (AGN) [3].

Since the discovery of black hole solutions in Einstein's theory of general relativity [4, 5], no definitive proof of their existence has been found. The main issue is that a black hole remains, by its very nature, concealed to the electromagnetic observer and can only be observed through its indirect

‡Narayan has fond memories of his association with Vishu during their years together in Bangalore. Vishu was generous with his support and encouragement as Narayan was beginning his career as an astrophysicist.

§To appear also in the Proceedings of the Eighth Marcel GROSSMANN Meeting on General Relativity.

B. R. Iyer and B. Bhawal (eds.), Black Holes, Gravitational Radiation and the Universe, 265–287.

gravitational influence on the world beyond its event horizon [6]. Mass estimates of compact objects in some systems argue for the existence of black holes. In addition, several observational signatures have been proposed as suggestive of a black hole environment. These lines of evidence do not, however, constitute definitive proof of the reality of black holes.

Recently, a class of accretion solutions called advection-dominated accretion flows (ADAFs) has been extensively applied to a number of accretion systems; for these models, most of the gravitational energy released in the infalling gas is carried (or advected) with the accretion flow as thermal energy, rather than being radiated as in standard solutions [7]. Since a large amount of energy falls onto the central object, there is an important distinction between compact objects with an event horizon and those without; for those with an event horizon, the energy is "lost" to the outside observer after it falls onto the central object, while for those with a hard surface, the energy is re-radiated and ultimately observed [8]. This new and potentially powerful method of discriminating between black holes and other kinds of compact objects is the subject of this review.

In §2, we list the best black hole candidates, and briefly review the classical techniques and theories used to detect potential black holes. In §3, we describe various solutions to the astrophysical problem of gas accreting onto a central object, emphasizing the specific characteristics of ADAFs. In §4, we describe how ADAF models have been applied to X-ray binaries and galactic nuclei and we explain why these studies strengthen the evidence for the existence of black holes.

2. "Classical" Evidence for Black Holes

2.1. EVIDENCE BASED ON MASS IN X-RAY BINARIES

It is well known that neutron stars possess a maximum mass above which they collapse to a black hole. The precise mass limit is not known due to uncertainties in the physics of neutron star interiors; like white dwarfs, neutron stars are supported by degeneracy pressure [9], but unlike white dwarfs, nuclear forces play an important role in their structure [10]. The equation of state of matter at ultra-nuclear densities is not well known. The maximum mass for the stiffest equation of state is $\sim 3\ M_\odot$ [11], and the limit goes up by no more than 25 % if one includes the effects of rotation [12, 13]. Even without detailed knowledge of the equation of state, it is possible from first principles to put a limit on the maximum mass. With an equation of state satisfying causality (sound speed less than the speed of light) and assuming that general relativity is correct, the limit is $\sim 3.4\ M_\odot$ [14]. Without causality, it reaches $5.2\ M_\odot$ [15, 13]. Although exotic compact stars with very large masses could theoretically exist [16, 17], it is generally accepted by the astronomical community that compact stellar objects with masses above $\sim 3\ M_\odot$ are strong black hole candidates.

In various X-ray binary systems, one can constrain the mass of the central object by observations of the spectral lines of the secondary star (usually a main-sequence star). The Doppler shifts of these spectral lines give an estimate of the radial velocity of the secondary relative to the observer. When combined with the orbital period of the binary, this can be translated (through Kepler's 3rd law) into a "mass function" for the compact object,

$$f(M_X) = \frac{M_X \sin^3 i}{(1+q)^2}. \tag{1}$$

The mass function does not directly give the mass M_X of the compact star because of its dependence on the unknown inclination i between the normal to the binary orbital plane and the line of sight, and the ratio of the secondary mass to the compact star mass, $q = M_2/M_X$. However, the mass function is a firm lower limit to M_X. Therefore, mass functions above $3\ M_\odot$ suggest the presence of black holes. Additional observational data (absence or presence of eclipses, for instance, or information on the nature of the secondary star) can help to constrain i or q, so that a likely value of M_X can often be determined. The best stellar mass black hole candidates currently known are summarized in Table 1. A detailed account of dynamical mass estimates can be found in *X-ray Binaries* [1].

2.2. EVIDENCE BASED ON MASS AND VELOCITY PROFILES OF GALACTIC CENTERS

Lynden-Bell's conjecture [33] that AGN are powered by accreting supermassive black holes is supported by the intense and energetic emission seen in these systems (implying a high efficiency process like accretion onto a compact object, cf. *3.1*). Observations of superluminal jets in some sources (implying a relativistic source) confirm the hypothesis. In addition, rapid variability of the luminosity, in particular the X-ray luminosity, is often observed. For a time-scale of variability Δt, causality implies a source size less than $R \sim c\Delta t$, where c is the speed of light. The estimates of R from the observed variability are typically small, thus supporting the black hole hypothesis [3].

In recent years, new lines of evidence have emerged, thanks to improved observational techniques. Detailed spectroscopic studies of the central regions of nearby galaxies provide valuable information on the line-of-sight velocities of matter (gas or stars). Since the presence of a massive black hole influences the dynamics of matter orbiting around it [34], these studies can help discriminate between black hole and non black hole models.

TABLE 1. The best black hole candidates among XRBs [20,133]. Type refers to Low Mass X-ray Binary (relatively low mass companion star) vs. High Mass X-ray Binary. Mass functions $f(M_X)$ and likely masses M_X are given in solar units ($M_\odot$).

Object	Type	$f(M_X)$	Likely M_X	References
GRO J0422+32 (XN Per 92)	LMXB	1.21 ± 0.06	≥ 9	[18, 19, 20]
A 0620-00 (XN Mon 75)	LMXB	2.91 ± 0.08	$4.9 - 10$	[21, 22]
GRS 1124-683 (XN Mus 91)	LMXB	3.01 ± 0.15	$5 - 7.5$	[23]
4U 1543-47	LMXB	0.22 ± 0.02	$1.2 - 7.9$	[24]
GRO J1655-40 (XN Sco 94)	LMXB	3.16 ± 0.15	7.02 ± 0.22	[25, 26]
H 1705-250 (XN Oph 77)	LMXB	4 ± 0.8	4.9 ± 1.3	[27]
GS 2000+250 (XN Vul 88)	LMXB	4.97 ± 0.1	8.5 ± 1.5	[28]
GS 2023+338 (V404 Cyg)	LMXB	6.08 ± 0.06	12.3 ± 0.3	[29, 30]
0538-641 (LMC-X-3)	HMXB	2.3 ± 0.3	$7 - 14$	[31]
1956+350 (Cyg X-1)	HMXB	0.24 ± 0.01	$7 - 20$	[32]

Current results favor the presence of black holes, but alternative scenarios still remain possible [35]. In Table 2 we list the best supermassive black hole candidates revealed by these techniques [36, 37].

2.3. EVIDENCE BASED ON X-RAY SPECTRAL LINES

X-ray emission lines are observed in many Seyfert 1 galaxies (a subclass of AGN). The lines are probably due to fluorescence by iron atoms in a relatively cold gas irradiated by X-ray photons [58, 59]. As matter nears a black hole, the velocity of the gas becomes large. Doppler shifts, as well as gravitational redshifts, are expected in the observed radiation. In the galaxy MCG-6-30-15, a very broad emission line has been observed and is believed to originate from gas rotating close to the central mass at a speed comparable to the speed of light [57, 56]. It has been claimed that the Kerr rotation parameter of the postulated black hole in MCG-6-30-15 can be constrained to a value close to unity [60], though this has since been challenged [61].

TABLE 2. Summary of supermassive black hole candidates at the centers of nearby galaxies. The first three are the best candidates. The likely mass is given in solar units ($M_\odot$).

Galaxy	Likely Mass	References
Milky Way (Sgr A*)	2.5×10^6	[38, 39, 42, 40]
NGC 4486 (M87)	3.2×10^9	[43, 44]
NGC 4258	3.6×10^7	[45]
NGC 224 (M31)	7.5×10^7	[36, 35]
NGC 221 (M32)	3×10^6	[46, 47, 48]
NGC 1068	10^7	[41]
NGC 3115	2×10^9	[49]
NGC 3377	8×10^7	[35]
NGC 3379	$0.5 - 2 \times 10^8$	[50]
NGC 4261	4.9×10^8	[51]
NGC 4342	3.2×10^8	[36]
NGC 4374 (M84)	1.5×10^9	[52]
NGC 4486B	6×10^8	[53]
NGC 4594	10^9	[54]
NGC 4945	10^6	[55]
NGC 6251	7.5×10^8	[36]
MCG-6-30-15	?	[56, 57]

2.4. BETTER EVIDENCE ?

Due to the nature of a black hole, it is difficult to prove its existence directly. Arguably, the best proof would be based on strong field general relativistic effects. However, quite generally, these effects can be well imitated by a neutron star or a more exotic compact object with a radius of a few Schwarzschild radii. The exception is the event horizon, which cannot be mimicked by any other object, since it is constitutive of black holes. Advection-dominated accretion flows may provide unique evidence for this intrinsic feature of black holes [62, 63, 64, 8]

3. Accretion solutions

3.1. WHY ACCRETION ?

Astrophysics deals with a variety of sources that show either persistent or intermittent high energy (typically X-ray) emission. Many of these sources involve enormous amounts of released energy. Accretion onto a neutron star or a black hole is the most efficient known process for converting matter into energy (excluding the obvious matter-antimatter annihilation process), far more efficient than the fusion of hydrogen into helium [2, 3]. Therefore, one is naturally inclined to explain high-energy astrophysical observations through the accretion paradigm.

In these models, the accreting gas radiates (in part or totally) the energy that it acquires in falling into the gravitational potential well of the central object. For a given accretion rate $\dot{M}$ (quantity of matter per unit time), the gravitational energy available during accretion onto a spherical object of radius R_X and mass M_X is

$$L_{\rm acc} = \frac{GM_X \dot{M}}{R_X}, \tag{2}$$

where G is the gravitational constant. The radiative efficiency of an accretion flow is defined as its ability to convert the rest mass of particles at infinity into radiated energy,

$$\eta = \frac{L_{\rm rad}}{\dot{M}c^2}, \tag{3}$$

where $L_{\rm rad}$ is the luminosity radiated by the flow (less than $L_{\rm acc}$) and c is the speed of light. For a compact object, i.e. an object with a large ratio M_X/R_X, we see from Eq. 2 that the efficiency is potentially very high. For a neutron star or a black hole, η can be ~ 0.1 (in contrast to the ~ 0.007 efficiency of hydrogen fusion).

It is convenient to scale the luminosity and the mass accretion rate in terms of the so-called Eddington limit. The Eddington luminosity is that luminosity at which the radiation pressure on a spherical flow balances the gravitational attraction of the central object:

$$L_{\rm Edd} = 1.3 \times 10^{38} \left(\frac{M_X}{M_\odot} \right) \ {\rm erg\ s^{-1}}. \tag{4}$$

This is the theoretical upper limit to the luminosity of a steady spherical accretion flow (but it is usually meaningful for other geometrical configurations as well). Another useful quantity is the Eddington accretion rate, defined here as the accretion rate at which a 0.1 efficiency accretion flow emits the Eddington luminosity:

$$\dot{M}_{\rm Edd} = \frac{L_{\rm Edd}}{0.1c^2} = 1.4 \times 10^{18} \left(\frac{M_X}{M_\odot}\right) \ {\rm g\ s^{-1}}. \tag{5}$$

We use the following notation for the scaled accretion rate:

$$\dot{m} = \frac{\dot{M}}{\dot{M}_{\rm Edd}}. \tag{6}$$

3.2. STANDARD SOLUTIONS

The equations describing the accretion of gas onto a central object are the standard conservation laws of mass, momentum and energy, coupled with detailed equations for the microscopic radiation processes. These equations are highly nonlinear, and allow multiple solutions. One is generally interested only in sufficiently stable solutions, since these are the ones that can be observed in nature.

Historically, the first solution to be considered was Bondi spherical accretion [65], in which matter without angular momentum accretes radially onto the central object. This solution is usually not relevant close to the central object (where the bulk of the observed radiation originates), since accreting matter in astrophysical settings invariably possesses a finite amount of angular momentum and cannot adopt a purely radial accretion configuration.

The second well known accretion solution is the "thin disk" solution in which matter revolves around the central object in nearly Keplerian orbits as it slowly accretes radially [66, 67, 68]. To reach the central object, the accreting matter must get rid of its angular momentum through viscous transport processes. Associated with the transport of angular momentum, viscous dissipation heats the accreting gas, and controls, along with the cooling processes, the energetics of the accretion flow. In the thin disk solution, the viscously dissipated gravitational energy is radiated very efficiently. In terms of the efficiency defined in Eq. 3, a thin disk has $\eta \sim 0.06$ for a Schwarzschild black hole, and $\eta \sim 0.42$ for a maximally rotating Kerr black hole [69]. (In the case of a neutron star star, $\eta \sim 0.2$.) Since the matter cools efficiently, it adopts a vertically thin configuration where $H/R \ll 1$, R being the distance from the central object and H the height of the disk as measured from the orbital plane. The thin disk solution has been the common paradigm in accretion theory for many years, and has been successfully applied to disks around young stars, disks in binaries and AGN disks [2].

The thin disk model assumes that, in steady-state (constant $\dot{m}$ through the disk), the local viscous dissipation rate is everywhere exactly balanced by the locally emitted flux. The disk is optically thick, which means that each annulus of the disk emits a blackbody spectrum (to first approximation). The overall emission from the disk is the superposition of the emission from all annuli, which means a superposition of blackbody spectra at different temperatures. However, since the emission is dominated by the innermost annuli, what emerges is similar to a single temperature blackbody spectrum. For a detailed review of thin accretion disk theory and its applications, see *Accretion Power in Astrophysics* [2].

The thin disk model has serious difficulties explaining the properties of some observed systems. In particular, it cannot account for the fact that radiation from many systems often spans the entire electromagnetic spectrum, with significant emission from radio to gamma rays; this is very different from a simple blackbody type spectrum.

3.3. ADVECTION-DOMINATED ACCRETION FLOWS

The thin disk paradigm describes the structure of an accretion flow when the gas can (locally) radiate all of the viscously generated energy. When the gas does not radiate efficiently, the viscously generated energy is stored as thermal energy and advected by the flow. In this case, one has an ADAF, whose structure and properties are markedly different from a thin disk.

ADAFs can exist in two regimes. The first is when the gas is optically thin and the cooling time exceeds the inflow time (defined as the time-scale for matter to reach the central object) [70, 71, 7, 8, 72]. The second is when the gas is so optically thick that a typical photon diffusion time out of the flow exceeds the inflow time [73, 74, 75] In both situations, the accretion flow is unable to cool efficiently and advects a significant fraction of the internally dissipated energy [76]. We focus on optically thin ADAFs, since they have seen the most applications to observed systems (see Narayan [77] and Narayan, Mahadevan & Quataert [78] for reviews of ADAFs and their applications).

3.3.1. *Optically thin, two-temperature ADAFs: Basic Assumptions*

Important studies of two-temperature accretion flows and advection-dominated accretion flows have been carried out in the past [71, 72, 79], but a detailed and consistent picture has only emerged recently [70, 7, 80, 8], offering the opportunity to apply specific models to astrophysical systems.

A standard assumption in the theory of hot accretion flows is that the gas is two temperature, with the ions significantly hotter than the electrons. In two-temperature ADAFs (2-T ADAFs), the viscous dissipation is assumed to preferentially heat the ions. The fraction of the viscous heating that goes directly to the electrons is parameterized by a factor δ, usually set to 10^{-3} ($\sim m_e/m_p$). In addition, the electrons receive energy from the ions via Coulomb collisions. When the density is low, however, this process is not very efficient. Therefore, only a small fraction of the viscous energy reaches the electrons; this energy is usually radiated (though not always [81, 82]). The rest of the energy remains in the ions, which are unable to cool in an inflow time. The energy in the ions is thus advected with the flow and is finally deposited on the central object. If the object is a black hole, the energy disappears; if it has a surface, the advected energy is re-radiated.

The efficiency of the turbulent viscous transport and dissipation in the flow are parameterized, as in the standard thin disk theory, by the "α-prescription" [68]: the viscosity is written as $\nu = \alpha c_s^2/\Omega_K$, where α is the viscosity parameter, c_s is the sound speed in the plasma and Ω_K is the local Keplerian angular velocity. If we assume that the Balbus-Hawley instability [83] is the origin of the MHD turbulence in the flow, we expect the magnetic fields and the gas in an ADAF to be in equipartition (i.e. magnetic pressure comparable to gas pressure), and the viscosity parameter α to be close to 0.3 [84].

3.3.2. *2-T ADAFs: Properties of the Flow and the Emitted Radiation*

The dynamical properties of ADAFs were first described by analytical self-similar solutions [7]; these were soon replaced by global numerical solutions [85, 86, 87, 88, 89, 90]. The flows are geometrically thick ($H/R \sim 1$) and are well approximated as spherical flows with most of the gas properties roughly constant on spherical shells [80]. Matter falls onto the central object at radial speeds less than, but comparable to, the free fall velocity and with angular velocities significantly less than the local Keplerian value. ADAFs are radially convectively unstable, but the implications of this for the structure and dynamics of the flows are still unclear [8].

A self-consistent determination of the thermal structure of an ADAF requires detailed numerical calculations [62, 8]. The poor energetic coupling between ions and electrons, and the advection of heat by the ions, is reflected in their temperature profiles: the ion temperature goes as $T_i \sim 10^{12}K/r$ (r is the distance from the central object in units of the Schwarzschild radius) while the electron temperature saturates at $T_e \sim 10^{9-10}$ K in the inner $10^2 - 10^3$ Schwarzschild radii. Above a critical accretion rate, $\dot{m}_{\rm crit}$, the density in the plasma becomes sufficiently high that Coulomb collisions efficiently transfer energy from the ions to the electrons. The gas then radiates most of the viscously dissipated energy and is no longer advection-dominated. ADAFs therefore only exist for accretion rates below $\dot{m}_{\rm crit} \sim \alpha^2$ [70, 71, 8, 72]. For $\dot{m} > \dot{m}_{\rm crit}$, accretion occurs as a cool thin disk.

Since ADAFs are optically thin and most of the viscously released energy is advected, they are significantly underluminous compared to a thin accretion disk with the same accretion rate; in fact, the radiative efficiency can be as low as $\eta \sim 10^{-3} - 10^{-4}$ at low $\dot{m}$. The emission by the hot plasma comes almost entirely from the electrons. Bremsstrahlung is an important emission process. In addition, since the electrons are in the presence of significant magnetic fields and are marginally relativistic, synchrotron emission and inverse Compton radiation are also very important. Synchrotron emission is the dominant mechanism at radio, infrared and optical wavelengths, with the peak emission occurring at a wavelength that depends on the black hole mass ($\lambda_{\rm peak} \propto M_X^{1/2}$) [91]. Comptonization of soft photons by hot electrons contributes from infrared to hard X-rays and is a strong function of the accretion rate. Bremsstrahlung emission generally dominates in the X-ray band at low mass accretion rates, but Comptonization takes over at higher accretion rates. A significant amount of γ-rays is also emitted, resulting from neutral pions (created by proton-proton collisions) which decay into very hard photons [92]. Pair creation processes are not very important in ADAFs since the radiation energy densities in the flow are low [93, 94].

3.3.3. *2-T ADAFs : Stability*

The stability of accretion solutions is an important issue in accretion physics since an accretion solution is not viable if it is unstable (by that one generally means linearly unstable). For instance, the important 2-T (non advective) solution of Shapiro et al. [79] is known to be violently unstable [95] and for this reason will probably never be observed in nature. The stability of 2-T ADAF solutions has been investigated in the long and short wavelength limits [96, 97, 8, 98, 99]. These studies conclude that, since the only unstable modes have sufficiently slow growth rates, ADAFs constitute a viable solution.

4. Applications of ADAFs

ADAF models have been applied to a number of low luminosity systems. They give a satisfying description of the spectral characteristics of the source Sgr A* at the center of our Milky Way Galaxy [100, 64], the weak AGN NGC 4258 [101] and M87 [102], and of several quiescent black hole XRBs [103, 62, 63]. All of these systems are known to experience low efficiency accretion and the thin disk solution encounters serious difficulties in explaining the observed spectral properties. In addition, the ADAF model also fits brighter systems with higher efficiencies [104, 105, 106].

It is worth mentioning here that previous studies have shown that the spectra of some of these systems can be explained by constructing models of optically thin accreting plasmas. However, these models are not necessarily dynamically consistent and do not explicitly attempt to satisfy mass, angular momentum and energy conservation. The ADAF solutions solve the observational and theoretical problems, for the first time, in a reliable and dynamically consistent way.

4.1. MODELING TECHNIQUES

Popham & Gammie [90], among others [85, 89], have computed the steady-state dynamical structure of ADAFs in the Kerr geometry. Their numerical solutions have been used in the spectral models presented here. The dynamical model provides the radial profiles of quantities such as radial speed, angular speed, density, pressure and viscous dissipation. In these models, the structure is determined by 3 parameters : α (viscosity parameter), γ (adiabatic index of the gas), and f (advection parameter, i.e. the fraction of the viscously dissipated energy which is advected). Usually, models assume that a constant fraction $(1-\beta) = 0.5$ of the total pressure comes from the magnetic field pressure and that $\alpha \sim 0.6(1-\beta) = 0.3$. The adiabatic index is given by $\gamma = (8-3\beta)/(6-3\beta) = 1.44$ [107]. The parameter f is solved self-consistently by calculating the radiation processes in detail and feeding back the information to the dynamical solution [105]. So far, studies have been restricted to the Schwarzschild geometry, but ultimately, in full generality, the Kerr rotation parameter a of the black hole will constitute an additional parameter.

The steady-state energetic structure of the flow is found by solving the electron and proton energy equations. The emission spectrum is obtained as a byproduct of this procedure, in which synchrotron, bremsstrahlung and Compton processes are all taken into account. Since soft photons can be scattered off electrons more than once before escaping the flow, the Compton problem has to be solved by a global "iterative scattering method" for consistency [62]. In the models presented here, the photon transport has been treated as Newtonian, but including gravitational redshift effects. Generic ADAF spectra with photon transport in the Kerr metric have been computed by Jaroszynski & Kurpiewski [108].

The basic ADAF model has one adjustable parameter: the mass accretion rate $\dot{m}$. The X-ray flux emitted by the ADAF is very sensitive to the density and therefore the accretion rate. For this reason, $\dot{m}$ is usually adjusted so that the resulting spectrum fits the available X-ray data.

The spectral characteristics of some systems are better explained by mixed ADAF + thin disk models. The accretion proceeds via an ADAF

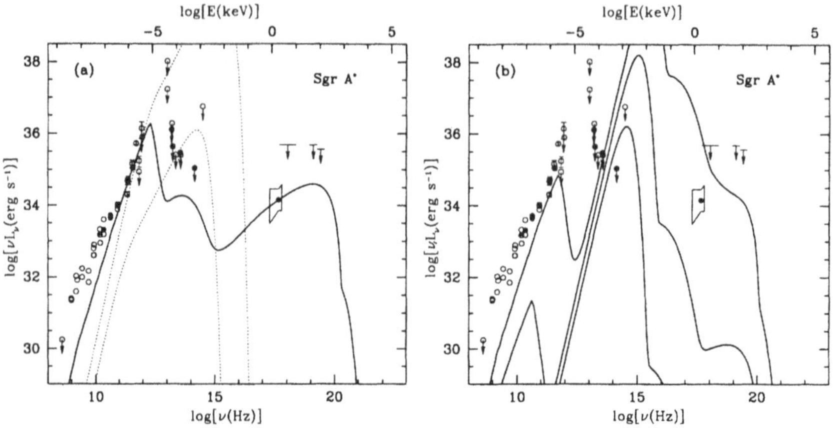

Figure 1. (a) Spectrum of an ADAF model of Sgr A* (solid line, based on [92]). The mass accretion rate inferred from this model is $\dot{m} = 1.3 \times 10^{-4}$ in Eddington units, in agreement with an independent observational determination. Dotted lines show the spectra of thin accretion disks, at the same accretion rate (upper) and at $\dot{m} = 1 \times 10^{-8}$ (lower). Observational data are shown as circles with error bars, and upper limits are indicated by arrows. The box indicates the constraints on the flux and the spectral index in soft X-rays. (b) Spectra of ADAF models of Sgr A* where the central mass is taken to have a surface at 3 $R_{\rm Schw}$ and the advected energy is assumed to be re-radiated as a blackbody. From top to bottom, the three spectra correspond to $\dot{m} = 10^{-4}, 10^{-6}, 10^{-8}$. All three models violate the infrared limit.

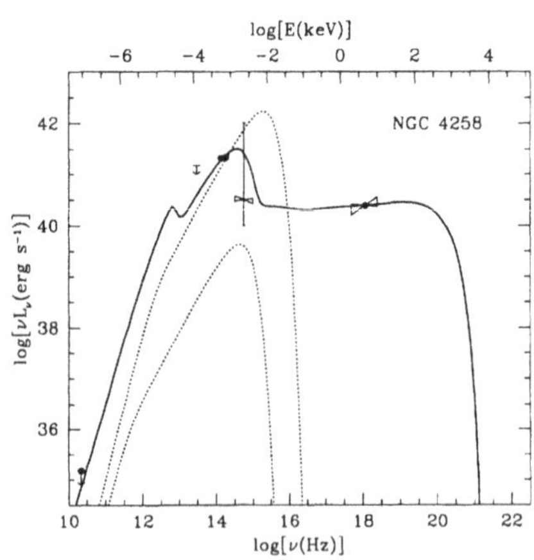

Figure 2. Spectrum of an ADAF model of NGC 4258 (solid line), at an accretion rate $\dot{m} = 9 \times 10^{-3}$, with a transition radius $R_{\rm trans} = 30 R_{\rm Schw}$. Dotted lines show the spectra of thin accretion disks, at an accretion rate $\dot{m} = 4 \times 10^{-3}$ (upper, adjusted to fit the infrared points) and $\dot{m} = 10^{-5}$ (lower).

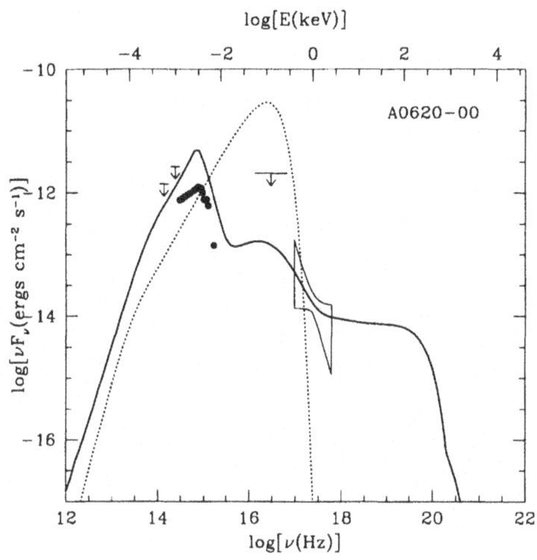

Figure 3. Spectrum of an ADAF model of A0620-00 (solid line, based on [88]) at an accretion rate $\dot{m} = 4 \times 10^{-4}$, compared with the observational data. The dotted line shows the spectrum of a thin accretion disk with an accretion rate $\dot{m} = 10^{-5}$ (adjusted to fit the optical flux).

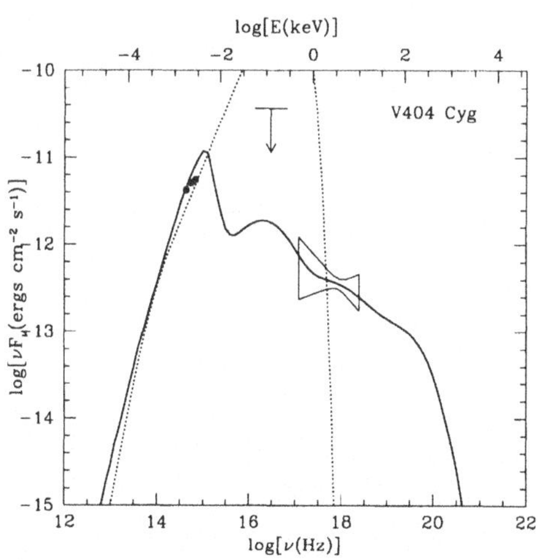

Figure 4. Spectrum of an ADAF model of V404 Cyg (solid line, based on ref. [88]) at an accretion rate $\dot{m} = 2 \times 10^{-3}$, compared with the observational data. The dotted line shows the spectrum of a thin accretion disk with $\dot{m} = 1.8 \times 10^{-3}$ (adjusted to fit the optical flux).

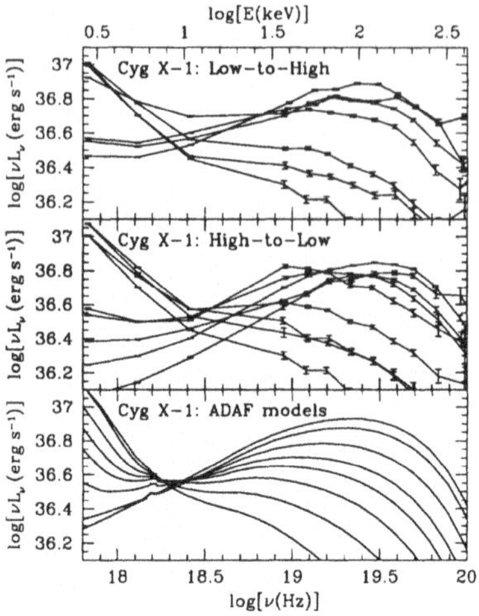

Figure 5. The broadband simultaneous RXTE (1.3-12 keV) and BATSE (20-600 keV) spectra of Cyg X-1 observed during the 1996 low-high (upper panel) and high-low (middle panel) state transitions. The bottom panel shows a sequence of ADAF models which are in good agreement with the observations [29].

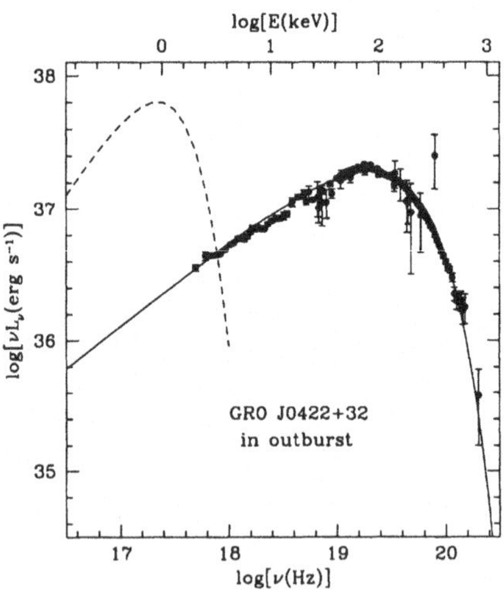

Figure 6. An ADAF model of J0422+32 (solid line) in the so-called low state (which is actually during outburst), compared with the observational data (dots and errorbars) [29]. The dashed line shows a thin disk model at the same accretion rate $\dot{m} = 0.1$.

in the inner regions of the accretion flow, and via a thin disk beyond a so-called transition radius ($R_{\rm trans}$). The contribution of the thin disk to the overall spectrum is a strongly decreasing function of $R_{\rm trans}$, so that $R_{\rm trans}$ is a relevant parameter (and therefore is well determined) only in those systems where the thin disk contribution to the emitted flux is significant (see NGC 4258 and Cyg X-1, for instance).

ADAF models and the corresponding observational data points and constraints are shown in Fig. 1 to Fig. 6 for a number of well-known black hole systems. Spectra of thin accretion disks are also shown. While the ADAF models are in qualitative agreement with the observations, the thin disk models are generally ruled out quite convincingly.

4.2. SPECTRAL MODELS OF LOW LUMINOSITY SYSTEMS

4.2.1. *Sgr A**

Sgr A* is an enigmatic radio source at the center of our Galaxy. Dynamical mass estimates show that a dark mass $M \sim 2.5 \times 10^6 M_\odot$ resides in a region of less than 0.1 pc in extension around this source [38, 42, 40]. Based on this, it is assumed that Sgr A* is a supermassive black hole. Observations of gas flows and stellar winds in its vicinity suggest that Sgr A* accretes matter at a rate in the range $\dot{M} \sim$ a few $\times 10^{-6} - 10^{-4} M_\odot$ yr^{-1} ($\dot{m} \sim 10^{-4} - 10^{-2}$) [109, 110]. A standard thin disk accreting with this $\dot{m}$ at a typical efficiency $\eta \sim 0.1$ is far too luminous to account for the observed integrated flux. Such a model also disagrees with the observed shape of the spectrum and is ruled out (by orders of magnitude) by an upper limit on the Sgr A* flux in the near infrared (Fig. 1a).

The first suggestion that Sgr A* may be advecting a significant amount of energy was made by Rees [111], and the first ADAF spectral model of this source appeared in Narayan, Yi & Mahadevan [112]. Since then, the modeling techniques [82, 113] and observational constraints have improved, allowing the construction of a refined model of Sgr A* [100, 64]. The model is consistent with the *independent* estimates of M and $\dot{M}$ mentioned above, and it fits the observed fluxes in the radio, millimeter and X-ray bands and upper limits in the sub-millimeter and infrared bands (see Fig. 1a).

In evaluating the quality of the fit in Fig. 1a, it should be kept in mind that only one parameter, $\dot{m}$, has been adjusted; this parameter has been optimized to fit the X-ray flux at 2 keV. The position of the synchrotron peak at 10^{12} Hz and its amplitude are not fitted but are predicted by the model; the agreement with the data is good. There is, however, a problem at lower radio frequencies, $\lesssim 10^{10}$ Hz, where the model is well below the observed flux. This is currently unexplained.

An important feature of the ADAF model of Sgr A* is that the observed low luminosity of the source is explained as a natural consequence of the advection of energy in the flow (the radiative efficiency is very low, $\eta \sim 5 \times 10^{-6}$) and the disappearance of this energy through the event horizon of the black hole. The model will not work if the central object has a hard surface, as demonstrated in Fig. 1b.

4.2.2. *NGC 4258*

The mass of the central black hole in the AGN NGC 4258 has been measured to be $3.6 \times 10^7 M_\odot$ [45]. Highlighting the fact that the observed optical/UV and X-ray luminosities are significantly sub-Eddington ($\sim 10^{-4}$ and $\sim 10^{-5}$ respectively), Lasota et al. [101] proposed that accretion in NGC 4258 proceeds through an ADAF. They found that the emission spectrum and the low luminosity of this system can be explained by an ADAF model, provided that most of the viscously dissipated energy in the flow is advected into a central black hole. Since then, new infrared measurements have been made [114] that constrain the transition radius to be $R_{\rm trans} \sim 30$ Schwarzschild radii. The ADAF produces the X-ray emission while the outer thin disk accounts for the newly observed infrared emission (see Fig. 2, unpublished). The refined model is also in agreement with a revised upper limit on the radio flux [115]. Maoz & McKee [116] and Kumar [117] find, via quite independent arguments, an accretion rate for NGC 4258 in agreement with the ADAF model. Neufeld & Maloney [118] estimate a much lower $\dot{m}$ ($\leq 10^{-5}$) in order to explain the maser emission in the source, but it is hard to explain the observed spectrum with such an $\dot{m}$ (Fig. 2).

4.2.3. *Other Low Luminosity Galactic Nuclei*

Quasars are luminous high redshift AGN, which are believed to be powered by supermassive black holes with masses around $10^8 - 10^9\ M_\odot$ [119]. Most nearby bright elliptical galaxies are believed to host dead quasars, i.e. quasars which have become inactive through evolution [120, 121]. As pointed out by Fabian & Canizares [122], however, the nuclei of these elliptical galaxies are much too dim, given the mass accretion rates inferred from independent methods. Fabian & Rees [123] suggested that the problem could be resolved if the accretion in these nuclei is occurring through an ADAF. This proposition has been confirmed by Mahadevan [91] and models have been developed for specific galaxies such as M87 [102] and M60 [124]. Lasota et al. [101] have similarly argued that low-luminosity LINER and Seyfert galaxies have supermassive black holes in their nuclei accreting via ADAFs.

4.2.4. *A0620-00 and V404 Cyg*

Soft X-ray Transients (SXTs) are a class of mass transfer X-ray binaries in which the accreting star is often a black hole candidate. Episodically, these systems enter high luminosity "outburst" phases, but for most of the time they remain in a very low luminosity "quiescent" phase. One of the main issues in modeling black hole SXTs in quiescence is that a thin accretion disk cannot explain both the observed low luminosity and the X-ray flux in these s systems : at an accretion rate low enough to fit the observed luminosity, a thin disk will not emit any significant flux in the X-ray band. Moreover, the X-ray spectrum will have the wrong shape. Narayan, McClintock & Yi [63] argued that the dilemma can be solved by means of the ADAF model and demonstrated this by carrying out spectral fits of A0620-00, V404 Cyg and Nova Mus 91 in quiescence. Narayan, Barret & McClintock [62] have recently refined these models for V404 Cyg and A0620-00 (see Fig. 3 and Fig. 4). The models are in good agreement with the observed shapes of the X-ray spectra (recall that the X-ray flux level is fitted by adjusting $\dot{m}$, but the spectral slope is unconstrained in the fit) and they satisfy other observational constraints such as optical measurements and extreme ultraviolet upper limits. In the optical, the models are somewhat too luminous to fit the data; the shapes of the spectra, however, are predicted well (see especially Fig. 3, where the position of the peak is reproduced well by the model). In contrast, the thin disk model deviates very significantly from the data and is easily ruled out. Hameury et al. [103] showed that observations of another SXT, GRO J1655-40, are consistent with the presence of an ADAF in quiescence. In addition, their model provides a convincing explanation for the time delay that was observed between the optical and X-ray light curves during a recent outburst [125].

4.3. MORE LUMINOUS SYSTEMS

4.3.1. *Spectral States of XRBs*

Narayan [106] suggested that the various spectral states (from quiescence to the luminous outburst phase) of black hole binaries (especially SXTs) can be understood in terms of a sequence of thin disk + ADAF models, where the thin disk accounts for most of the luminosity in outburst and is gradually replaced by an ADAF during the transition to quiescence. Recently, Esin et al. [105] have developed these ideas with detailed calculations, and succeeded in explaining the observed spectral states of the system Nova Mus 91 surprisingly well. The same models also appear to explain other luminous black hole systems such as Cyg X-1 (Fig. 5) and J0422+32 (in outburst, Fig. 6) [104]. These studies, for the first time, unify a fair fraction of the phenomenology of black hole XRBs. Once again, the existence of a

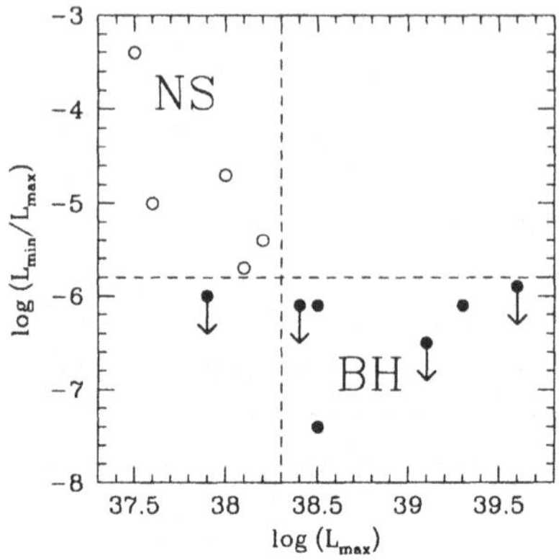

Figure 7. A comparison between black hole (BH, filled circles) and neutron star (NS, open circles) SXT luminosity variations [41]. The ratio of the quiescent luminosity to the peak outburst luminosity is systematically smaller for BH systems than for NS systems. This confirms the presence of event horizons in BH candidates.

central black hole, into which energy is lost, is an essential feature of the models.

4.3.2. *Galaxies and Quasars*

A similar unification is also possible in our understanding of the various degrees of activity in galactic nuclei. Yi [126] has proposed an explanation for the apparent evolution of quasars from a large population of luminous objects at redshift ≥ 2 to a much smaller population in the nearby universe. He postulates that the mass accretion rate in galactic nuclei has decreased over cosmic time. Supermassive black holes in galactic nuclei had $\dot{m} > \dot{m}_{\rm crit}$ at early times and accreted through thin disks with a high radiative efficiency. As $\dot{m}$ decreased with time, the accretion flow switched to a low efficiency ADAF. This explains the absence of very luminous AGN in the local neighborhood of our Galaxy.

Di Matteo & Fabian [127] proposed that a significant fraction of the hard X-ray (≥ 2 keV) background could be due to emission from a population of galaxies undergoing advection-dominated accretion in their nuclei. Yi & Boughn [128] discuss a test of the ADAF paradigm in this context; they argue that measurements of the radio and X-ray fluxes could provide direct estimates of the central black hole masses in many galaxies.

4.3.3. *Neutron Star SXTs vs. Black Hole SXTs*

As mentioned earlier, SXTs are XRBs that exhibit large luminosity variations. Most of the time they remain in a quiescent state, in which their luminosities are low and the transferred mass is partially stored in a non-steady thin disk. Episodically, they experience an outburst event, during which their luminosities become very high [129]. During outbursts, which correspond to a sudden accretion of the mass stored in the disk, these sources reach luminosities of order the Eddington luminosity (cf. Eq. 4). From dynamical mass estimates, and observations of "X-ray bursts" (very probably thermonuclear explosions on the surface of neutron stars), one can distinguish between neutron star and black hole candidate systems [1]. Since black holes are more massive than neutron stars (and the Eddington luminosity scales with the mass of the accreting object), black hole SXTs are naturally observed in outburst at a higher maximum luminosity than neutron star SXTs.

Assuming that the ADAF model of SXTs in quiescence [62, 63] is generic, Narayan, Garcia & McClintock [130] argued that black hole SXTs should experience more substantial luminosity changes from quiescence to outburst than neutron star SXTs. In quiescence, the central object accretes matter through an ADAF : most of the viscously dissipated energy is advected by the flow onto the central object. In the neutron star case, this advected energy has to be re-radiated from the stellar surface, whereas in the black hole case, it is lost through the horizon. Therefore, black hole SXTs in quiescence must be much less luminous (relative to their maximum luminosity) than neutron star SXTs, which is exactly what is observed in nature (see Fig. 7). This systematic effect confirms the presence of an event horizon in black hole SXTs [130, 131].

5. Conclusion

Advection-dominated accretion flow models of black holes in various XRBs and AGN are in qualitative agreement with the best current observations. They provide a satisfying explanation for both the spectral characteristics and the extremely low luminosities observed in these systems. The models require that a very large fraction of the dissipated energy in the flow is advected into the central black hole. If a standard star with a surface, and not an event horizon, were present at the center of these systems, changes in the spectra and luminosities by orders of magnitude are predicted (e.g. compare Fig. 1a and Fig. 1b). The success of ADAF models thus constitutes strong evidence for the existence of stellar mass and supermassive black holes in the Universe.

We conclude with an interesting observation. If one considers the candidate black hole XRBs listed in Table 1, apart from LMC X-3, all the other systems spend the bulk of their time in the ADAF state and only occasionally switch to the more efficient thin accretion disk state. Similarly, in Table 2, the three best candidates, Sgr A*, NGC 4258 and M87, appear to have ADAFs, and it is plausible that many of the other systems have ADAFs as well, except for NGC 1068, MCG-6-30-15, and possibly NGC 4945. If ADAFs are, as the evidence suggests, very common, there is in principle no difficulty finding systems in which to test for the presence of event horizons. The challenge at the moment is primarily observational — ADAFs by their nature are very dim and one needs instruments with superior sensitivity to achieve adequate signal-to-noise.

Acknowledgments

We are grateful to Vicky Kalogera, Chris Kochanek, Avi Loeb, Rohan Mahadevan, Jeff McClintock and Doug Richstone for useful discussions. Fig. 5 and Fig. 6 were kindly provided by Ann Esin. The electronic preprints referenced below can be found at the Los Alamos National Laboratory e-print service : http://xxx.lanl.gov/. This work was supported in part by NSF grant AST 9423209 and NASA grant NAG 5-2837. EQ was supported by a NSF graduate research fellowship.

References

1. *X-ray Binaries*, eds. W. H. G. Lewin, J. Van Paradijs and E. P. J. Van den Heuvel (1995), (Cambridge University Press, Cambridge).
2. Frank, J., King, A., and Raine, D. (1992), *Accretion Power in Astrophysics*, 2nd Ed. (Cambridge University Press, Cambridge).
3. *Physics of Active Galactic Nuclei*, eds. W. J. Duschl and S. J. Wagner (1992), (Springer, Berlin).
4. Kerr, R.P. (1963), *Phys. Rev. Lett.* **11**, 237.
5. Oppenheimer, J.R., and Snyder, H. (1939), *Phys. Rev.* **56**, 455.
6. Misner, C.W., Thorne, K.S., and Wheeler, J.A. (1973), *Gravitation* (Freeman and Co., New York).
7. Narayan, R., and Yi, I. (1994), *Ap. J. Lett.* **428**, L13.
8. Narayan, R., and Yi, I. (1995), *Ap. J.* **452**, 710.
9. Chandrasekhar, S. (1931), *M. N. R. A. S.* **91**, 456.
10. Hartle, J.B. (1978), *Phys. Rep.* **46**, 201.
11. Kalogera, V. and Baym, G. (1996), *Ap. J. Lett.* **470**, L61.
12. Friedman, J.L., and Ipser, J.R. (1987), *Ap. J.* **314**, 594.
13. Shapiro, S.L., and Teukolsky, S.A. (1983), *Blackholes, White Dwarfs and Neutron Stars*, (Wiley Interscience, New York).
14. Rhoades, C.E., and Ruffini, R. (1974), *Phys. Rev. Lett.* **32**, 6.
15. Hartle, J.B. and Sabbadini, A.G. (1977), *Ap. J.* **213**, 831.
16. Bahcall, S., Lynn, B.W., and Selipsky, S.B. (1990), *Ap. J.* **362**, 251.

17. Miller, J.C., Shahbaz, T. and Nolan, L.A. (1997), *M. N. R. A. S.* submitted, *Astro-ph*/9708065.
18. Beekman, G., et al. (1997), *M. N. R. A. S.* **290**, 303.
19. Casares, J., et al. (1995), *M. N. R. A. S.* **276**, L35.
20. Filippenko, A.V., Matheson, T., and Ho, L.C. (1995), *Ap. J.* **455**, 614.
21. McClintock, J.E., and Remillard, R.A. (1986), *Ap. J.* **308**, 110.
22. Shahbaz, T., Naylor, T., and Charles, P.A. (1994), *M. N. R. A. S.* **268**, 756.
23. Orosz, J.A., et al. (1996), *Ap. J.* **468**, 380.
24. Orosz, J.A., et al. (1997), *Ap. J.* submitted.
25. Bailyn, C.D. et al. (1995), *Nature* **378**, 157.
26. Orosz, J.A, and Bailyn, C.D. (1997), *Ap. J.* **477**, 876.
27. Remillard, R.A. (1996), *Ap. J.* **459**, 226.
28. Callanan. P.J. et al. (1996), *Ap. J. Lett.* **470**, L57.
29. Sanwal, D., et al. (1996), *Ap. J.* **460**, 437.
30. Shahbaz, T., et al. (1994), *M. N. R. A. S.* **271**, L10.
31. Cowley, A.P. et al. (1983), *Ap. J.* **272**, 118.
32. Gies, D.R., and Bolton, C.T. (1982), *Ap. J.* **260**, 240.
33. Lynden-Bell, D. (1969), *Nature* **223**, 690.
34. Binney, J., and Tremaine, S. (1987), *Galactic Dynamics* (Princeton University Press, Princeton).
35. Kormendy, J., and Richstone, D. (1995), *Annu. Rev. Astron. Astrop.* **33**, 581.
36. Ford, H.C., Tsvetanov, Z.I., Ferrarese, L., and Jaffe, W. (1997), in proc. IAU Symp. 184, *The Central Region of the Galaxy and Galaxies*, ed. Y. Sofue, *Astro-ph*/9711299.
37. Richstone, D. (1997), in proc. IAU Symp. 184, *The Central Region of the Galaxy and Galaxies*, ed. Y. Sofue.
38. Eckart, A., and Genzel, R. (1997), *M. N. R. A. S.* **284**, 576.
39. Genzel, R., Eckart, A., Ott, T., and Eisenhauer, F. (1997), *M. N. R. A. S.* **291**, 219.
40. Haller, J.-W., et al. (1996), *Ap. J.* **468**, 955.
41. Greenhill, L.J., Gwinn, C.R., Antonucci, R., and Barvainis, R. (1996), *Ap. J. Lett.* **472**, L21.
42. Haller, J.-W., et al. (1996), *Ap. J.* **456**, 194.
43. Harms, R.J., et al. (1994), *Ap. J. Lett.* **435**, L35.
44. Macchetto, F., et al. (1997), *Ap. J.* **489**, 579.
45. Miyoshi, M., et al. (1995), *Nature* **373**, 127.
46. Bender, R., Kormendy, J., and Dehnen, W. (1996), *Ap. J. Lett.* **464**, L123.
47. van der Marel, R.P., de Zeeuw, P.T., Rix, H.-W., and Quinlan, G.D. (1997), *Nature* **385**, 610.
48. van der Marel, R.P., Cretton, N., de Zeeuw, P.-T., and Rix, H.-W. (1997), *Ap. J.* submitted, *Astro-ph*/9705081.
49. Kormendy, J., et al. (1996), *Ap. J. Lett.* **459**, L57.
50. Gebhardt, K., et al. (1996), *Bull. Am. Astron. Soc.* **28**, 1422.
51. Ferrarese, L., Ford, H.C., and Jaffe, W. (1996), *Ap. J.* **470**, 444.
52. Bower, G.A., et al. (1997), *Ap. J. Lett.* in press, *Astro-ph*/9710264.
53. Kormendy, J., et al. (1997), *Ap. J.* in press, *Astro-ph*/9703188.
54. Kormendy, J., et al. (1996), *Ap. J. Lett.* **473**, L91.
55. Greenhill, L.J., Moran, J.M., and Herrnstein, J.R. (1997), *Ap. J. Lett.* **481**, L23.
56. Fabian, A.C., et al. (1995), *M. N. R. A. S.* **277**, L11.
57. Tanaka, Y., et al. (1995), *Nature* **375**, 659.
58. George, I.M. and Fabian, A.C. (1991), *M. N. R. A. S.* **249**, 352.
59. Matt, G., Fabian, A.C. and Reynolds, C.S. (1997), *M. N. R. A. S.* **289**, 175.
60. Dabrowski, Y., et al. (1997), *M. N. R. A. S.* **288**, L11.
61. Reynolds, C.S., and Begelman, M.C. (1997), *Ap. J.* in press, *Astro-ph*/9705136.
62. Narayan, R., Barret, D., and McClintock, J.E. (1997), *Ap. J.* **482**, 448.

63. Narayan, R., McClintock, J.E., and Yi, I. (1996), *Ap. J.* **457**, 821.
64. Narayan, R., et al. (1997), *Ap. J.* in press, *Astro-ph/*9706112.
65. Bondi, H. (1952), *M. N. R. A. S.* **112**, 195.
66. Novikov, I.D., and Thorne, K.S. (1973), in *Blackholes*, ed. C. DeWitt and B. DeWitt (Gordon and Breach, New York).
67. Pringle, J.E. (1981), *Annu. Rev. Astron. Astrop.* **19**, 137.
68. Shakura, N.I., and Sunyaev, R.A. (1973), *A.& A.* **24**, 337.
69. Eardley, D.M. and Press, W.H. (1975), *Annu. Rev. Astron. Astrop.* **13**, 381.
70. Abramowicz, M.A., et al. (1995), *Ap. J. Lett.* **438**, L37.
71. Ichimaru, S. (1977), *Ap. J.* **214**, 840.
72. Rees, M.J., Begelman, M.C., Blandford, R.D., and Phinney, E.S. (1982), *Nature* **295**, 17.
73. Abramowicz, M.A., Czerny, B., Lasota, J.-P., and Szuszkiewicz, E. (1988), *Ap. J.* **332**, 646.
74. Begelman, M.C. (1978), *M. N. R. A. S.* **184**, 53.
75. Katz, J. (1977), *Ap. J.* **215**, 265.
76. Chen, X., et al., (1995), *Ap. J. Lett.* **443**, L61.
77. Narayan, R. (1997), in Proc. IAU Colloq. 163, ASP Conf. Series vol. 121, *Accretion Phenomena and Related Outflows*, eds. D. T. Wickramasinghe. L. Ferrario and G. V. Bicknell, p. 75.
78. Narayan, R., Mahadevan, R., and Quataert, E. (1997), in Proc. Reykjavik Symp. on *Non-linear Phenomena in accretion Disks around Black Holes*, eds. M. A. Abramowicz, G. Bjornsson and J. E. Pringle, in press.
79. Shapiro, S.L., Lightman, A.P., and Eardley, D.M. (1976), *Ap. J.* **204**, 187.
80. Narayan, R., and Yi, I. (1995), *Ap. J.* **444**, 231.
81. Mahadevan, R., and Quataert, E. (1997), *Ap. J.* **490**, 605.
82. Nakamura, K.E., Kusunose, M., Matsumoto, R., and Kato, S. (1997), *Publ. Astron. Soc. Jap.*, **49**, 503.
83. Balbus, S.A., and Hawley, J.H. (1991), *Ap. J.* **376**, 214.
84. Hawley, J.F., Gammie, C.F., and Balbus, S.A. (1996), *Ap. J.* **464**, 690.
85. Abramowicz, M.A., Chen, X.-M., Granath, M., and Lasota, J.-P. (1996), *Ap. J.* **471**, 762.
86. Chen, X., Abramowicz, M.A., and Lasota, J.-P. (1997), *Ap. J.* **476**, 61.
87. Gammie, C.F., and Popham, R. (1997), *Ap. J.* submitted, *Astro-ph/*9705117.
88. Narayan, R., Kato, S., and Honma, F. (1997), *Ap. J.* **476**, 49.
89. Peitz, J. and Appl, S. (1997), *M. N. R. A. S.* **286**, 681.
90. Popham, R. and Gammie, C.F. (1997), *Ap. J.* submitted.
91. Mahadevan, R. (1997), *Ap. J.* **477**, 585.
92. Mahadevan, R., Narayan, R., and Krolik, J. (1997), *Ap. J.* **486**, 268.
93. Bjornsson, G., Abramowicz, M.A., Chen, X.-M., and Lasota, J.-P. (1996), *Ap. J.* **467**, 99.
94. Kusunose, M., and Mineshige, S. (1996), *Ap. J.* **468**, 330.
95. Piran, T. (1978), *Ap. J.* **221**, 652.
96. Kato, S., Yamasaki, T., Abramowicz, M.A., and Chen, X.-M. (1997), *Publ. Astron. Soc. Jap.*, **49**, 221.
97. Manmoto, T., et al. (1996), *Ap. J. Lett.* **464**, L135.
98. Wu, X.-B. (1997), *Ap. J.* **489**, 222.
99. Wu, X.-B. (1997), *M. N. R. A. S.* in press, *Astro-ph/*9707329.
100. Manmoto, T., Mineshige, S., and Kusunose, M. (1997), *Ap. J.* **489**, 791.
101. Lasota, J.-P., et al. (1996), *Ap. J.* **462**, 142.
102. Reynolds, C.S., et al. (1996), *M. N. R. A. S.* **283**, L111.
103. Hameury, J.-M., Lasota, J.-P., McClintock, J.E., and Narayan, R. (1997), *Ap. J.* in press, *Astro-ph/*9703095.
104. Esin, A.A., et al. (1997), *Ap. J.* submitted, *Astro-ph/*9711167.
105. Esin, A.A., McClintock, J.E., and Narayan, R. (1997), *Ap. J.* **489**, 865.

106. Narayan, R. (1996), *Ap. J.* **462**, 136.
107. Esin, A.A. (1997), *Ap. J.* **482**, 400.
108. Jaroszynski, M., and Kurpiewski, A. (1997), *A.& A.* **326**, 419.
109. Genzel, R., Hollenbach, D., and Townes, C.H. (1994), *Rep. Prog. Phys.* **57**, 417.
110. Melia, F.(1992), *Ap. J. Lett.* **387**, L25.
111. Rees, M.J. (1982), in *The Galactic Center*, eds G. Riegler and R. D. Blandford.
112. Narayan, R., Yi, I., and Mahadevan, R. (1995), *Nature* **374**, 623.
113. Nakamura, K.E., Matsumoto, R., Kusunose, M., and Kato, S. (1996), *Publ. Astron. Soc. Jap.*, **48**, 761.
114. Chary, R., and Becklin, E.E. (1997), *Ap. J. Lett.* **485**, L75.
115. Herrnstein, J.R. et al. (1997), private communication.
116. Maoz, E. and McKee, C.F. *Ap. J.* in press, *Astro-ph/*9704050 (1997).
117. Kumar, P. (1997), *Ap. J.* submitted, *Astro-ph/*9706063.
118. Neufeld, D.A., and Maloney, P.R. (1995), *Ap. J. Lett.* **447**, L17.
119. Rees, M.J. (1984), *Annu. Rev. Astron. Astrop.* **22**, 471.
120. Chokshi, A., and Turner, E.L. (1992), *M. N. R. A. S.* **259**, 421.
121. Soltan, A. (1982), *M. N. R. A. S.* **200**, 115.
122. Fabian, A.C., and Canizares, C.R. (1988), *Nature* **333**, 829.
123. Fabian, A.C., and Rees, M.J. (1995), *M. N. R. A. S.* **277**, L55.
124. Di Matteo, T., and Fabian, A.C. (1997), *M. N. R. A. S.* **286**, 50.
125. Orosz, J.A., et al. (1997), *Ap. J.* **478**, 830.
126. Yi, I. (1996), *Ap. J.* **473**, 645.
127. Di Matteo, T., and Fabian, A.C. (1997), *M. N. R. A. S.* **286**, 393.
128. Yi, I., and Boughn, S.P. (1997), *Ap. J.* submitted, *Astro-ph/*9710147.
129. Lasota, J.-P. (1996), in *Compact Stars in Binaries*, ed. J. van Paradijs et al.
130. Narayan, R., Garcia, M.R., and McClintock, J.E. (1997), *Ap. J. Lett.* **478**, L79.
131. Garcia, M.R., McClintock, J.E., Narayan, R., and Callanan, P.J. (1997), in Proc. 13th NAW on CVs, Jackson Hole, eds S. Howell, E. Kuulkers, and C. Woodward, *Astro-ph/*9708149.

How can one detect a black hole? The philosophy of detection is that of Sherlock Holmes: When you have excluded the impossible, whatever remains, however improbable, must be the truth. If the existence of an invisible object has been established by inference and if the possibility of its being a collapsed object like a black dwarf or a neutron star is ruled out, then it must be a black hole. Of course, a black hole in isolation does not betray its existence; it has to be spotted through the company it keeps.

— C. V. VISHVESHWARA

If general relativity were devoid of applications in physics or astronomy, it would be a closed subject, and would soon return to the stagnation and sterility that characterized it before the relativistic renaissance of recent decades. It is precisely those applications in black-hole searches, gravitational-wave astronomy, and cosmology that have imbued the subject with a vitality and vigor that have made it an open-ended field, not a closed book, despite the fact that the structure of the theory has not changed in seventy years

– CLIFFORD M. WILL

18. BLACK HOLES IN ACTIVE GALACTIC NUCLEI

AJIT K. KEMBHAVI
Inter-university Centre for Astronomy and Astrophysics, Pune 411 007, India.

1. Introduction

There are two completely different astrophysical contexts in which black holes are believed to play an important role: (1) As members of stellar-mass binary systems and (2) as supermassive black holes in the nuclei of active galaxies and quasars. A black hole most vividly manifests itself when it accretes matter from a companion star or the surroundings. As the matter moves under the spell of the black hole's gravity, it heats up to high temperatures and emits radiation. Rotational energy can also be extracted from a black hole through the medium of a magnetic field. Much of the energy released can appear in non-thermal form, and can be transported through relativistic beams, which give rise to startling phenomena like apparent faster than light motion, rapid variability and flux amplification.

While the presence of a black hole can lead to very interesting consequences, none of the observations actually made so far can be taken to be an unambiguous proof of the existence of a black hole. The observed circumstances could very well be due to the presence of a compact object which has a radius greater than the Schwarzschild limit. In the case of binary systems, X-ray emission from accreting matter are produced whether the accreting object is a black hole, or a neutron star with a radius of $\sim 10\,\mathrm{km}$. Any observed time variation in the emission, which can be traced to features in the compact object, immediately rules out a black hole. But, so far, the only way available to establish that the object is a black hole is to show that it has a mass in excess of the accepted limiting mass of a neutron star. Unfortunately, mass determinations are made ambiguous due to our ignorance about the angle of inclination of the orbit with the line of sight. Only mass upper limits are therefore available, and while there

B. R. Iyer and B. Bhawal (eds.), Black Holes, Gravitational Radiation and the Universe, 289–307.

are several promising candidates, including the original black hole binary candidate Cygnus X-1, one still cannot say that a stellar mass black has indeed been found.

In the case of active galaxies and quasars, the inference that a $\sim 10^6 - 10^9 M_\odot$ black hole is the prime mover rests mainly on the high efficiency of energy release, rapid variability at X-ray energies and observations which can be traced to relativistic bulk motion. As in the case of binaries, none of the observed phenomena unequivocally demand the existence of a supermassive black hole. Bursts of star formation, involving $\sim 10^8 M_\odot$ of gas, can reproduce many of the observations, with the continuum being contributed by the many massive stars and supernova remnant involved in the burst. But while star bursts probably make important contributions to the overall spectrum and observed morphology of the active galaxy (see e. g. Terlevich 1992, Perry 1994), they are unlikely to be able to be the sole explanation, except in the objects with modest luminosity. It is possible to substitute a black hole with a dense star cluster or supermassive star to create a deep potential well for the energy release. But it can be argued (Rees 1984) that much of the total mass of such structures will inevitably collapse to a black hole.

While the existence of black holes in active galaxies and quasars remains to be established, they have become more or less accepted as the primary energy providers in active galaxies. Energy is extracted from the black hole through accretion or other means and is reprocessed to produce the observed continuum and line spectrum. While the basic processes are always the same, a variety of objects is observed because of variation in the ratio of the power output of the central engine to the host galaxy, environmental effects and orientation relative to the observer. This point of view, which could be termed as the "standard model", is hardly likely to be sufficient to explain the observed variety. But it provides a framework with which observations can be compared, and relative to which successes and failures, in developing an overall understanding, may be judged.

In the following we will review, very briefly and somewhat incompletely, the phenomenology of active galaxies and quasars and the physical processes associated with the "standard model", and compare observations with theoretical models. It will be left to the reader to decide whether we have actually seen a singularity in action, and if we have or are soon going to, what the philosophical implications are. As for the other setting for black holes, the binary X-ray sources, which are in fact more likely than the active galaxies to lead to a positive detection in the near future, we shall say nothing more.

One important topic which is intentionally skipped in this article is a detailed discussion on accretion disks and especially the thick disks. These topics have been reviewed in detail in chapters 16 and 17 by Wiita and Menou, Quataert & Narayan respectively, in this volume.

Further material on the subjects covered here may be found in the excellent reviews by Rees (1984) and Blandford (1990). A detailed review of the phenomenology and models at the pedagogical level has been provided by Kembhavi and Narlikar (1998, hereafter KN98). This work elaborates on most of the points considered in the present article and includes an exhaustive list of references.

2. Taxonomy

2.1. QUASARS

The first quasars to be identified, 3C 273 and 3C 48 were discovered as the optical counterparts of bright radio sources. When accurate position of the radio source 3C 273 became available through Lunar occultation, it was found that the optical object within the small error circle appeared to be star-like with apparent magnitude ~ 13. This could not, however, be a normal star, since it had strong emission lines in its spectrum, with redshift $z = \Delta\lambda/\lambda = 0.16$, where $\Delta\lambda$ is the increase in the wavelength of the emission lines relative to their rest frame values. If this redshift was interpreted as being due to cosmological expansion, then the distance at which the object was situated was so large that its luminosity would be $\sim 10^{47}\ \mathrm{erg\ s^{-1}}$, i. e. , a few orders of magnitude more than the luminosity of our entire galaxy. The star-like object associated with 3C 48 was found to have a redshift of 0.37. In the process of further optical identifications of radio sources, a number of objects similar to the radio quasars in their optical properties, but not having discernible radio emission, were also found. In fact most of the ~ 8000 quasars[1] now known, with redshift as large as ~ 5, have been discovered in optical surveys which have been designed to discover quasars with specific optical properties, or in the identification of X-ray sources. Only a few percent of all quasars have significant radio emission.

A quasar is generally defined to be an object with a starlike appearance in the discovery survey, which exhibits variability in the observed flux, broad emission lines and high redshift. The degree to which any of these criteria is met varies across the very large sample now available. Closer examination with deep exposures have shown that a nebulosity, which is

[1]In the literature the term quasar is sometimes used only for objects discovered as optical counterparts of radio sources, while similar objects not having significant radio emission are called quasi-stellar objects. We use the term quasar for the whole class, and use the term radio quasar if the discussion refers specially to sources with high radio luminosity.

presumably due to a host galaxy, is associated with at least the nearby quasars. While many surveys look for excessively blue objects, quasars with redshift ~ 4 can in fact appear to be red in colour.

In the early days of quasar research, attempts were made to interpret the redshifts as being due to to the Doppler effect or being gravitational in origin. These interpretations are now no longer in favour, the former because of the excessively high velocities required and the complete absence of blue shifted objects, and the latter because of the difficulties in making a deep enough gravitational potential well within reasonable astrophysical constraints. The cosmological interpretation of the redshift is now almost universal, and an integral feature of the "standard model", but the tantalizing possibility that at least a part of the redshift could be due to new physical effects still remains (see KN98 for a discussion).

2.2. ACTIVE GALAXIES

Active galaxies are those which have emission from non-stellar sources, at a level comparable to, or in excess of, the emission from the stars in the galaxy. The first active galaxies to be discovered were *radio galaxies*, which are galaxies identified with bright radio sources. These exhibit powerful radio emission from large structures which often extend to sizes many times larger than the size of the galaxy itself. Also present are emission lines and other signs of quasar-like activity from the nucleus.

Seyfert galaxies form a class of galaxies with bright star-like nuclei and peculiar morphology. These galaxies have strong emission lines produced in their nuclear regions, and a subdivision of the class of Seyfert galaxies can be made on the basis of their emission line properties. Some of the emission lines from a *Seyfert type 1* galaxy are broad, with a full width at half maximum (FWHM) greater than a few thousand kilometres per second, while other lines are relatively narrow, with widths greater than a few hundred kilometres per second. The nuclei of Seyfert galaxies also have strong continuum emission, and are very much like quasars in their properties, though not quite as luminous. A *Seyfert Type 2* galaxy has a weak continuum and only strong narrow emission lines. Seyfert galaxies of intermediate properties are also known.

BL Lac objects have strong nuclear continuum radiation that shows high polarization and rapid variability. Emission lines are, however, absent or weak. BL Lacs are found mainly in the optical identifications of radio or X-ray sources, and all known examples are strong radio sources. BL Lacs are sometimes found to be the nuclei of elliptical galaxies.

The nuclei of active galaxies have properties similar to those of quasars. Their presence causes disturbance to the morphology of the extended re-

TABLE 1. Local space densities

Type of object	Number density Mpc^{-3}
field galaxies	10^{-1}
luminous spirals	10^{-2}
Seyfert galaxies	10^{-4}
radio galaxies	10^{-6}
quasars	10^{-7}
radio quasars	10^{-9}

gions of the galaxy, but these largely leave the galaxy intact. We are concerned here only with the active galactic nuclei (AGN), where the black hole is to be found. In the standard model the difference between a quasar and an AGN is only quantitative: quasars are active galaxies with high nuclear-to-galaxy luminosity ratio $L_{\rm n}/L$. The high brightness of the nucleus swamps the rest of the galaxy making it difficult to detect, especially because of their small angular size and low surface brightness produced due to cosmological effects.

Amongst the quasars are objects which have very rapid and high amplitude flux variability and highly polarized radiation. These always have high radio luminosity, and are known as *blazars*. BL lac objects too belong to this class. The spectrum of blazars is non-thermal in nature (see below), and it is believed that the emission is dominated by a relativistically beamed component.

Active galaxies constitute only a small fraction of all galaxies, and quasars which are possibly the high luminosity end of the active galaxy population, are even rarer. The approximate space densities of these objects in our local neighbourhood are given in Table 1(Osterbrock 1989). The change in these comoving densities with redshift is discussed below.

3. Continuum Radiation

The radiation from an AGN or quasar has a continuum component, which stretches over an extremely wide frequency band from the radio to the high energy $\gamma -$ ray region. Along with the continuum are found emission and absorption lines. The emission lines can be broad or narrow, as mentioned in Section 2.2 and are produced in numerous small clouds of gas surrounding the nucleus. The broad lines are produced in dense clouds spread over a region smaller than $\sim 1\,\mathrm{pc}$, while the narrow lines are produced in less dense clouds spread over a region $\lesssim 1\,\mathrm{kpc}$ in size. The emission lines provide important diagnostics of the physical conditions and the kinematics of the line producing regions. The absorption lines are produced either in the object itself, or by intervening matter. We shall not elaborate on the emission and absorption spectra, and only consider the continuum, which is produced close to the black hole

In the optical band, quasars and AGN have continuum colours which are bluer than those of most stars. It was believed from early observations that the continuum in the optical could be represented by a power-law of the form $L(\nu) \propto \nu^{-\alpha}$, with $\alpha \simeq 0.5$. Power-law forms also seemed to apply to other regions of the spectrum, like the radio and X-ray regions, at least over limited sections of the frequency band. The simple power-law nature indicated that the continuum radiation must be predominantly non-thermal in origin, produced possibly by the synchrotron emission of relativistic electrons, with a power-law distribution of energy, spiraling in a magnetic field. The higher energy photons in the continuum could also be produced by the inverse Compton scattering of low energy photons by the relativistic electrons.

The present picture of the spectrum is far more complex than the early observations indicated. It appears that that there are basically two kinds of continuum shape.

3.1. BLAZAR SPECTRA

Blazars have a smooth, non-thermal continuum spectrum from the radio to the ultraviolet band, possibly produced by synchrotron emission. The spectrum in this region lacks features which are present in the thermally dominated objects discussed below. The non-thermal emission continues in the high energy region, upto the highest energy $\gamma-$ ray band explored. The luminosity per decade of frequency, $\nu L(\nu)$, increases from the radio region towards higher frequencies and generally peaks somewhere in the submillimetre to infrared regions.

At X-ray energies blazars have an approximately power-law spectrum, which is generally flatter than the X-ray spectra of radio-quiet quasars. Recent observations of blazars by $\gamma-$ray detectors on the Comptel satellite have shown that a significant fraction of blazars are $\gamma-$ray emitters at GeV energies, and that the power in the $\gamma-$ ray region can exceed by an order of magnitude or more the peak power at lower frequencies. In one case, photons with energy in the TeV range have been detected.

The radio cores of blazars are bright and rapidly variable, and the flux is polarized. All these facts point to a synchrotron origin for the emission at radio wavelengths. The radio emission from extended regions, whether associated with blazars or other kinds of objects, has a simple power-law form with index $\alpha \lesssim 1$. The radio emission from blazars, on the other hand, is often dominated by a compact component which has a spectrum which is flat, with $\alpha \simeq 0$, or inverted with $\alpha < 0$ or of complex form. This is traced to synchrotron self-absorption of the radio photons, which occurs when the source is very compact and the density of photons is high. The absorption increases towards lower frequencies and there is a turnover in the spectrum at a frequency $\nu_a \propto \theta^{-4/5}$, where θ is the angular size of the source. For frequencies $\nu < \nu_a$ the optical depth to self-absorption exceeds unity and it can be shown that the spectrum is of the form $L(\nu) \propto \nu^{5/2}$. Different parts of the emitting region can get self-absorbed at different frequencies, and the observed spectrum is a superposition, so that it can have a flat or a complex shape.

The brightness temperature $T_b = c^2 I(\nu)/(2k\nu^2)$, where $I(\nu)$ is the intensity of radiation at frequency ν and k is the Boltzmann constant in the infrared and optical regions is $\gtrsim 10^6$ K. This is much higher than the black body temperature that would be associated with emission peaks at these wavelengths. The emission is therefore non-thermal and the continuity of the spectrum between the radio and ultraviolet wavelengths points to a synchrotron origin for the emission in all these regions. The emission at higher energies could also be due to the same process, but this would require the emitting electrons to have very high energies. The high energy photons can, however, be straightforwardly produced from inverse Compton scattering of low energy synchrotron photons by the electrons that produced them in the first place, i. e., synchrotron self-Compton emission. This boosts photons of energy ϵ to $\sim \gamma^2 \epsilon$.

3.1.1. *Relativistic effects*

Superluminal motion has been observed in a number of blazars. The first quasar in which such an effect was observed was 3C 273: observations using very long base line interferometry at a radio frequency in the gigahertz range, which provided an angular resolution of ~ 1 milliarcsecond, showed that radio structures in the quasar were moving apart at a speed exceeding that of light. Superluminal motion at speeds $\lesssim 10c$ has now been observed in several tens of quasars.

The most elegant and universally applicable explanation is that a blob which appears to be moving faster than light is the projection of a structure which is moving at a highly relativistic speed in the rest frame, making a small angle of inclination $\psi \sim 1/\gamma$ to the observer's line of sight, where γ is the Lorentz factor of the blob's rest frame motion.

Another effect associated with the relativistic motion is that the intensity of the blob appears to be enhanced by a factor $\propto (1 - \beta \cos \psi)^{-(3+\alpha)}$, where $\beta = v/c$ and α is the power-law spectral index of the flux emitted by the blob. It has been found that the radio jets in quasars are always one-sided, while there is no reason to believe that the basic mechanism for the production of the jets is asymmetric. If one of the beams in twin relativistic jets points close to the observer's line of sight, and the other away from it, then the flux of the incoming jet is significantly amplified relative to the outgoing jet. The surface brightness of this jet can be several thousand times higher than that of the jet moving away. Because of the limited brightness range over which observations can be made, the difference in brightness makes the jets appear to be one sided. This explanation of course is not an established fact, in the sense that it has not yet been shown that the large scale one-sided jets are relativistic, and that the visible jet is directed towards the observer. Superluminal motion has only been observed in jets close to the AGN which have sizes on the parsec scale. But there are many indications that large scale jets on the kiloparsec scale are also relativistic. Bulk relativistic motion can also account for the rapid variability which is found in many blazars.

3.2. THERMALLY DOMINATED OBJECTS

In radio-quiet quasars, Seyfert galaxies and radio galaxies, i. e., the non-blazar sector of the population, the dominant emission is in the $\sim 0.01 - 100\,\mu$m region. Characteristic features of these spectra are (1) a peak in emitted power in the infra-red region between 10 and 100 microns called the *infrared bump*, and another prominent peak in the ultraviolet region called the *big blue bump* or the *UV excess*. The continuum in the far ultraviolet region is responsible for ionization of the broad line and narrow line emitting regions, and therefore some idea of its shape can be obtained from the study of emission lines. Thermally dominated objects can have detectable radio emission, with power much higher than found in normal galaxies. But except in the powerful radio galaxies, the radio emission is only a fraction of the total power.

Observations of a large number of quasars in the ~ 0.3 nm–6 cm (5 GHz–10^{19} Hz) range (Sanders et al. 1989) have shown that the continuum shapes of the thermally dominated objects are remarkably similar. The average continua for quasars which are significant radio emitters, and for those which are not, are shown in Figure 1. The average spectrum clearly shows the infrared and big blue bumps, which are also apparent in individual spectra.

Beyond the extreme ultraviolet is the soft X-ray region at ~ 0.1 keV and then there is the X-ray region, followed by the γ − ray region. The

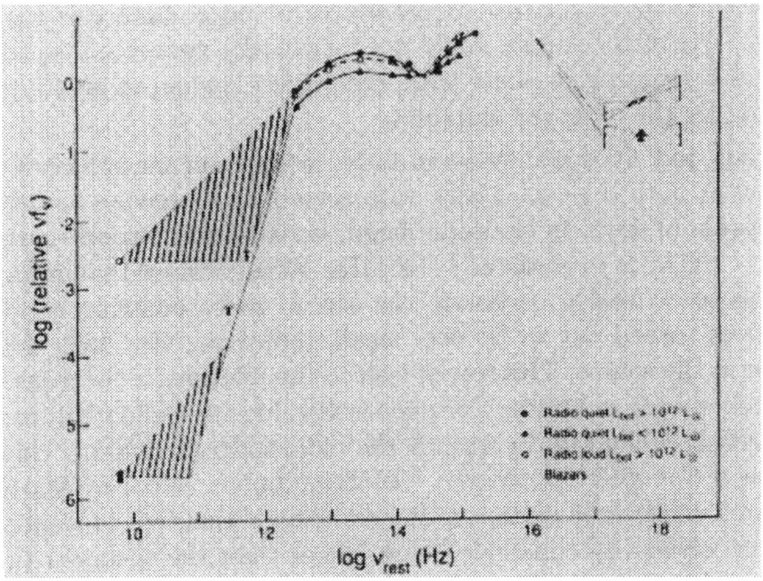

Figure 1. Average continuum distributions of radio-quiet and radio-loud quasars. Each data point represents the mean over the corresponding sample. The hatched region in the radio band indicates the spread of radio spectral indices for each sample. νf_ν, where f_ν is the flux density, is plotted as the ordinate. The infrared bump and big blue bump are clearly seen. Reproduced from Sanders et al. (1989)

spectra up to a few keV are of power-law form, with radio-loud quasars having flatter spectra than the radio-quiet ones. Beyond this range and up to a few tens of kiloelectronvolts the spectra have an approximate power-law shape with a canonical spectral index of ~ 0.7. The simple power law is often modified by X-ray absorption at low energy, a soft X-ray excess and the effects of pair creation due to photon–photon interactions in the most compact sources. In addition there are modifications to the X-ray spectrum, including emission lines, that are interpreted to be the consequences of reflection by cold matter. Observations with COMPTEL have shown that Seyferts and radio-quiet quasars are not strong emitters above some hundreds of keV, while the radio-loud blazars often show high energy γ – ray emission.

4. Variability

Variability studies are very important in identifying the physical processes and the size of the region in which the radiation, in a given wavelength range, is produced. If variability is observed on a timescale of Δt_{var} in the source frame, then the radiation must be produced in a region with size

R constrained by $R \lesssim c\Delta t_{var}$. If the source is bigger than this limit, then different parts of the source would not be causally connected, so that they would not be varying in phase with each other, leading to greatly reduced amplitude for the observed variability.

Quasars and AGN are observed to be variable on timescales of months and years at optical wavelengths, with some blazars having variability on the timescale of days. In the radio band, variability timescales range from years at $\sim 100\,\text{MHz}$ to weeks at $\sim 100\,\text{GHz}$. When observed radio variability timescales were used to estimate the size of radio emitting regions, the source sizes turned out to be very small, indicating very high density of radiation in the source. This would lead to the relativistic electrons quickly losing their energy in inverse Compton scattering the radio photons to high energy, which would rapidly quench the radio source in what is colourfully termed as a *Compton catastrophe.* The situation was saved by Martin Rees by invoking relativistic motion of the emitting region, which results in the rest frame variability timescale being longer than the observed timescale by a factor γ^3. This implies a larger source size and much lower radiation density in the source frame, thus avoiding the catastrophe. The pattern of variability in different wavelength regions can have important implications for the modelling of the radiation processes and understanding the geometry of the emitting region.

The fastest observed variability occurs in the X-ray band, with timescales of minute and hours being quite common in non-blazar type of AGN. Strong X-ray variability is universal to blazars. The X-ray timescale shows that the high energy emission arises deep inside the AGN, and should provide the best probe for the region closest to the black hole.

Ultraviolet variability timescales in blazars show a negative correlation with their ultraviolet luminosity, as well as optical polarization. This is consistent with the relativistic beaming of a significant fraction of the UV and optical emission, since beaming amplifies observed fluxes, shortens observed variability timescales and increases the observed polarization. In the case of Seyfert galaxies, the objects with higher ultraviolet luminosity show less variation. This is consistent with the emission arising from an accretion disk or other isotropic emitter: the more luminous such a source is, the greater is its light crossing time. The variability timescale then increases and the amplitude decreases. Accretion disk models, however, receive a setback from the observation that in some sources, ultraviolet and optical variability occur in phase. This is not expected in accretion disks, since there the more energetic ultraviolet radiation must be emitted from the hotter inner parts of the disk, while the optical emission comes from a region which is about seven time larger in radius. Since in the disk model the two layers can only communicate at sound speed, the optical variation should lag the ultraviolet variation by several years.

In the far infrared band, variability has been observed in blazars with high brightness temperature, which is consistent with a non-thermal origin. The other kinds of AGN do not show such variability, indicating that in this case the far infrared radiation is thermal emission from dust in an extended region.

5. Luminosity Functions

We have seen in Section 2 that the local number density of quasars and AGN is just a small fraction of all galaxies. The proportion however is not constant in time. It was suspected from the early quasar surveys that their number density *increases* at earlier epochs. This evolution of number density with number density is now an established fact, and it is only the form of the evolution, and its physical interpretation, which remain to be unambiguously determined.

The luminosity function of quasars is defined to be the comoving number density $\Phi(L, z)$, per unit luminosity and redshift range, of quasars as a function of luminosity and redshift. One could of course consider luminosity in different bands like the optical, radio, X-ray and so on, as well as other properties of the objects concerned, as independent variables of the function. We will only consider the optical luminosity in the blue band. To determine this function observationally, it is first necessary to determine the surface density of objects as a function of their flux (apparent magnitude) and redshift. This is done by counting *all* the objects in a small area of the sky which satisfy specific criteria, like having excess blue colour. Well defined surveys which include several hundreds of quasars are now available, but even these are not sufficient to obtain the luminosity function by dividing the observed sample into luminosity and redshift bins. A variety of techniques are used to deproject the observed surface density into a number density, and simple analytical forms of $\Phi(L, z)$ are fitted to the distribution. Theoretical models of quasar evolution have not reached a stage where physically motivated models can be tested against observation. The functions used are those which provide the simplest representation of the data in terms of a few parameters.

From the early surveys it appeared that the dependence of $\Phi(L, z)$ on luminosity and redshift could be separated, and the distribution of luminosity had the simple power-law form $\Phi(L) \propto L^{-\beta}$. The increase in density at earlier epochs could be accommodated by assuming that the comoving number density increased with redshift as $(1+z)^{\delta}$, with $\delta =$ constant. Other functions of redshift, like an exponential could also be used. This simple scheme of evolution was known as *pure density evolution*. In an alternate scheme, known as *pure luminosity evolution*, it is assumed that the luminosity of the objects increases with redshift. Then in any luminosity bin there are more objects per unit volume at higher redshift. For a featureless power-law luminosity function, it is not possible to distinguish between the two simple schemes of evolution as they are mathematically equivalent.

The luminosity functions which have been determined from large surveys which became available in the late 1980s showed that a simple power-law is not valid over the entire range of luminosity sampled. Adequate fits have been obtained with the luminosity function distributed in a broken power-law form, with the the function becoming steeper at higher luminosity. The evolution is obtained by a making the luminosity redshift dependent i.e., through luminosity evolution. This in effect shifts the break towards higher luminosity. Pure density evolution is no longer permitted, because of the break in the luminosity distribution.

A broken-power law fit to the quasar luminosity function obtained by Boyle et al. (1991) is shown in Figure 2. The plot shows the comoving number density of quasars as a function of absolute blue magnitude, for several redshifts. It is clear that at a given absolute magnitude the number density increases with redshift. The form of the function remains the same at all redshifts, and the function at a higher redshift can be obtained from a lower redshift bin by simply translating every luminosity to a higher value. The power-laws, in terms of luminosity[2] are $\Phi(L) \propto L^{-1.6}$ for $L < L_*$ and $\Phi(L) \propto L^{-3.8}$ for $L > L*$. The break luminosity is redshift dependent, with $L* \propto (1+z)^{3.5}$. The numerical values are somewhat dependent on the cosmological and spectral parameters assumed, but the result that there is a broken power-law, and that there is luminosity evolution, appear to be fairly robust within the range of parameter space explored. It is obvious from Figure 2 that the evolution slows down beyond a redshift of $z \sim 2$. It is not yet clear whether the comoving density of quasars remains constant beyond this point, or whether it actually decreases. Knowing the true situation is of course critical to the theories of quasar and galaxy formation. We shall examine in Section 7 simple models which attempt to reproduce the observed luminosity function.

The luminosity function at low redshifts have not been well determined. At low redshift the host galaxy of a quasar is easy to detect, and it is possible to confuse quasars with AGN. Quasars with redshift $z < 0.3$ are therefore often excluded from luminosity function determinations. We therefore do not have a local luminosity function for quasars. By the same token, Seyferts at high redshift would appear to be star-like, and may therefore be classified as quasars. We therefore have a luminosity function for Seyferts

[2]Absolute magnitude M and luminosity L are related through $M = -2.5 \log L +$ constant.

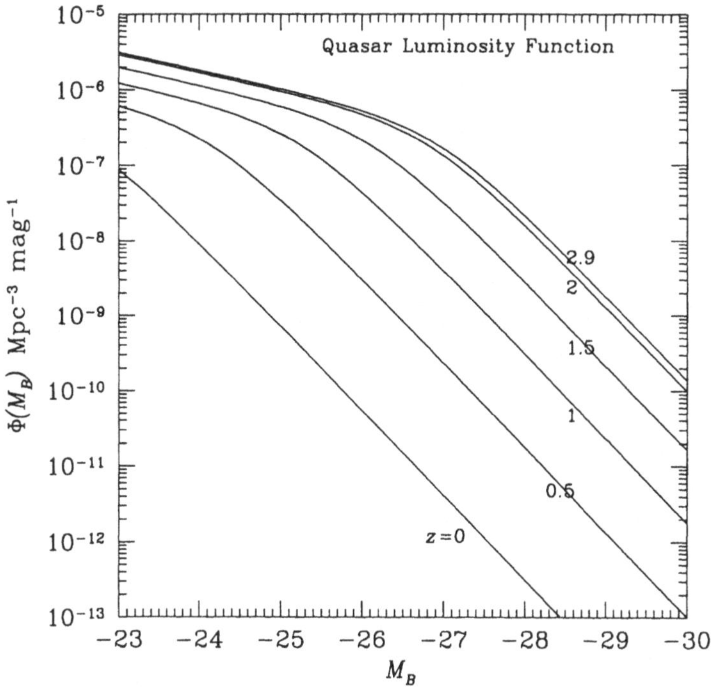

Figure 2. A broken power-law representation of the luminosity function of quasars (Boyle et al. 1991). The plot for $z = 0$ is obtained by extrapolation of the data for $z > 0.3$.

at low redshift, and one for quasars at high redshift, but it is not clear what the relationship between the two is. It is possible that all species of AGN and quasars belong to the same basic population, and the appearance is governed by the ratio of the AGN to host galaxy luminosity. If the luminosity evolution extends to low luminosities, then local AGN could evolve into high luminosity quasars at high redshift. The alternative is that evolution is the prerogative of the more luminous objects, and therefore an unevolved population AGN, exists also at higher redshift.

Recently Köhler et al. (1997) have obtained the luminosity function for a sample of low luminosity quasars/AGN at low redshift, and find that it can be represented by a single power-law $\Phi \propto L^{-2.1}$, with no evolution occuring to $z = 0.3$. If this form is correct, then either the broken power-law found at higher luminosity and redshift must be incorrect, or pure luminosity evolution is not a correct model over the entire range of luminosity.

6. Black Holes

The high efficiency with which AGN generate energy imply a gravitational origin for the energy. The short variability timescales which are observed require a very compact emitting region, while bulk relativistic motion calls for a deep gravitational potential well. These are all pointers to the presence of a black hole hole at the center of an AGN. An advantage in having a black hole is that in contrast to the complexities associated with other possible compact structures, a black hole is a very simple object, thanks to the no-hair theorems: it is fully characterized by its mass and spin, and the geometry around it by the Kerr metric. The maximal specific angular momentum permitted for a spinning black hole of mass M is GM/c: if this is exceeded, the event horizon would disappear, which would go against cosmic censorship. The matter from which the black hole is formed is very likely to have specific angular momentum greater than this, so it is not unrealistic to assume that at formation the spin angular momentum of the black hole is close to the maximal value.

The minimum characteristic timescale associated with a black hole of Schwarzschild radius $r_{\rm S}$ is $t_{\rm S} = r_{\rm S}/c \sim 10^3 M_8\,\text{sec}$, where M_8 is the mass of the black hole in units of $10^8 M_\odot$a nd $M_\odot$ is one Solar mass. A variability timescale $\gtrsim 10^2\,\text{sec}$, as observed in the case of X-rays requires a mass $\gtrsim 10^6 M_\odot$. If a black hole accretes matter spherically symmetrically, it can radiate at most at the Eddington limit of $L_{\rm E} = 4\pi c G M m_{\rm p}/\sigma_{\rm T} \simeq 1.2\times10^{46} M_8\,\text{erg}\,\text{s}^{-1}$, where $m_{\rm p}$ is the proton mass and $\sigma_{\rm T}$ the Thomson scattering cross section. For luminosities exceeding this limit, the pressure of the radiation on the infalling matter would prevent further accretion. In spite of the very specific assumptions which go into deriving its value, the Eddington limit provides a characteristic luminosity. If AGN and quasars are emitting close to the Eddington limit, the corresponding black hole masses would be in the range of $10^6 - 10^9 M_\odot$.

6.1. ENERGY EXTRACTION FROM BLACK HOLES

In addition to an event horizon, the Kerr black hole has a critical radius called the *static limit*. Particles from inside this radius can escape to infinity, but they always corotate with the black hole. The region between the event horizon and the static limit is called the *ergosphere*.

The ergosphere has the property that particles moving in it with orbits crossing the event horizon can have negative total energy, which in the Newtonian approximation includes the gravitational, kinetic and rest-mass energy. These orbits are contained within the static limit. When a particle on such an orbit crosses the event horizon, the mass of the hole decreases. This can be used in principle to extract energy from the black hole via the

Penrose process (Penrose 1969), in which a particle with positive energy enters the ergosphere and splits into two particles, one of which has negative energy and is in an orbit that takes it into the event horizon. The other particle acquires more positive energy than the original particle had, and leaves the ergosphere. The net result is that energy is extracted from the black hole, which then has less angular momentum than it did before the event.

The Kerr black hole has two kinds of energy: that due to the spin of the hole, and that due to an irreducible mass. The latter is equal to $m/\sqrt{2}$, so that the fraction which can be extracted is $(\sqrt{2}-1)/\sqrt{2}$, i. e., $\sim$ 29 percent of the rest-mass of the black hole.

Though the Penrose process is very efficient and could provide immense sources of energy, it has not been possible to imagine realistic scenarios in which the energy can actually be extracted. Some simplified processes involve the scattering of photons by charged particles inside the ergosphere. In the absence of an electromagnetic field the Penrose process requires that for a fragment to attain negative energy it should have a speed exceeding $c/2$ relative to the other fragment; this translates to unacceptable requirements and low efficiency for the energy extraction. Dadhich and his co-workers (see Wagh and Dadhich 1989) have considered the Penrose process in the presence of a magnetic field associated with the accretion disk. The twisting magnetic field lines give rise to a quadrupole electric field. The threshold energy required to get a particle into a negative energy orbit can now come from electromagnetic interaction, so that an exorbitant amount of kinetic energy is not required. It has been demonstrated that the process can have very high efficiency when the accretion of discrete particles is considered, but details of the complex magnetohydrodynamic processes have not been worked out.

Another process for extracting black hole rotational energy using magnetic fields is the Blandford–Znajek mechanism (see Blandford 1990), which depends on the fact that the event horizon of a black hole behaves like a spinning conducting surface, though not a perfect one, with surface resistivity 377 ohms. A spinning hole embedded in a magnetic field acquires a quadrupole distribution of electric charge and a corresponding poloidal electric field. In analogy with a unipolar inductor, power can be extracted by having a current flow between the spinning hole's equator and poles. When the angular velocity of the field lines at infinity is zero, the maximum power extracted is $\sim B^2a^2c$, where B is the uniform magnetic field in which the hole is embedded. The efficiency of energy extraction in slowing down a hole which starts off spinning maximally is $\sim$ 9 percent.

7. Evolutionary scenarios

We have seen in Section 5 that the space density of quasars as a function of redshift, as inferred from observations, is consistent with luminosity evolution. In this picture, the luminosity of the quasar population as a whole increases with redshift up to $z \simeq 2$, after which the evolution slows down, and there could even be a decline in luminosity towards higher redshift. This behaviour of the population can be reproduced by different patterns of activity in individual quasars. These have been discussed most clearly by Cavaliere and Padovani (1989). It is assumed in these models that quasar luminosity is the result of accretion of matter onto a black hole with an efficiency η for converting the accreted mass into energy.

In one scenario, it is assumed that luminosity evolution refers to the dimming of individual quasars. All quasars dim continuously on the same timescale $t_{\rm L}$ following a short formation phase at $z \simeq 2$. It is assumed for simplicity that this dimming is exponential. The epoch at $z = 2$ from which quasars have been dimming corresponds to a look-back time of 5.4 Gyr for a $q_0 = 0$ universe. For the luminosity evolution of Boyle et al. (1988), the dimming from redshift z is by a factor $(1 + z)^{3.5}$. For this model the luminosity at the present epoch is given by

$$\frac{L}{L_{\rm Edd}} = 2\times10^{-4}\left(\frac{\eta}{0.1}\right). \tag{1}$$

In this scenario the quasar phenomenon is limited to a very small fraction of all galaxies. The dimmed counterparts could be Seyfert galaxies, but then the local space density of Seyferts would be required to be consistent with the space density of high redshift quasars. The observational situation here is not clear because of the uncertainties in the Seyfert luminosity function and the evolution of the quasar luminosity function at the faint end.

Another scenario is where all galaxies have a single, short event of quasar-like activity, corresponding mainly to the black hole formation phase. If during the episode the accretion rate is $\dot{M} \sim \eta^{-1}\dot{M}_{\rm Edd}$, then $L \sim L_{\rm Edd}$. If, further, the luminosity (i. e., the mass of the black hole formed in the episode) depends inversely on the cosmological epoch, then the episodes will on average present the appearance of luminosity evolution, with $L \sim L_{\rm Edd}$ at all redshifts. This scenario requires that galaxies with $L \gtrsim L_*$, where L_* is the characteristic luminosity of the Schechter luminosity function of galaxies, would all have passed through a phase of quasar activity. Since the average luminosity increases with redshift, at any epoch the inactive galaxies harbour nuclei that are more massive on the average than those of the active galaxies.

Intermediate to the two extreme alternatives above is the scenario where quasar activity occurs intermittently in each of a set of galaxies. The duration of each outburst is short compared to the timescale over which the luminosity of the quasar population is found to evolve. In this model the ratio $L/L_{\rm Edd}$ increases from a local value $\sim 5\times10^{-2}$ to $\gtrsim 1$ at $z \simeq 2$. The number of host galaxies required in this scenario scales with $L/L_{\rm Edd}$ and for the value 5×10^{-2} approaches the same number as in the scenario with $L \sim L_{\rm Edd}$.

Examining the observational data available to them, Cavaliere and Padovani (1989) have estimated that $L/L_{\rm Edd} \sim 5\times10^{-2}$ (it is assumed here that $H_0 = 50\,\rm km\,sec^{-1}\,Mpc^{-1}$). In spite of the considerable uncertainty in the determination of this ratio, the nominal observed value rules out the model in which quasars are to be found in a very small fraction of galaxies and are subject to continuous dimming. The model with $L \sim L_{\rm Edd}$ is also ruled out, which leaves the scenario involving recurrent episodes of quasar activity.

If quasar activity is long lived, it is limited to the small fraction of all galaxies which are specially suitable for sustaining it. In the standard picture, these galaxies would have a high-mass black hole at their centres. The situation is quite different for the other patterns we have discussed. In these cases a considerable fraction of all galaxies would have passed through single or recurrent episodes of activity. A super-massive black hole is therefore expected in many galaxies, with the presently inactive bright galaxies having more-massive black holes than the presently active ones. These alternatives need to be examined further, both theoretically as well as observationally, because of their important implications for quasar and galaxy formation and evolution.

Small and Blandford (1992) have considered a model that is somewhat different from the scenarios described above. This model involves the quick formation of a black hole in newly formed galaxies, with the hole radiating at the Eddington limit. At later epochs, when the supply of fuel is no longer plentiful, the holes accrete only intermittently at an average rate that is the same for all quasars and is given by $\dot{M} \propto M^{-1.5}t^{-6.7}$, t being the cosmological time. The number of relic black holes per decade of mass is $\propto M^{-0.4}$, for black hole masses in the range $3\times10^7 \lesssim M/M_\odot \lesssim 3\times10^9$. In this model all sufficiently massive galaxies pass through a quasar phase, the mass of the black hole formed being a monotonic function of the host galaxy mass. It is therefore possible to test the model by comparing the local density of galaxies having known central masses with the density of high redshift quasars. Haenhelt and Rees (1993) have developed a model for the evolution of the luminosity function, using a superposition of ~ 100 generations of quasars with lifetime $\sim 10^8$ yr and an evolutionary timescale for the population of $\sim 10^9$ yr. In this model quasars are assumed to be the first phase in the formation of a galaxy, in the potential well of a dark-matter halo. The number of newly forming dark-matter haloes at successive cosmological epochs is estimated using the Press–Schechter formalism in the cold dark matter (CDM) scenario. The luminosity functions calculated from this model are in good agreement with observation.

8. Discussion

We have considered in this brief review a few observational aspects of AGN and quasars, and the standard model which involves a black hole. It is apparent even from the sketchy account given here that there is a considerable gap between the available models, and their application to the interpretation of the rather extensive data. There are many compelling arguments in favour of the black hole model. There are also many interesting processes available, to either extract energy from the rotating black hole and to release energy from accreting matter. But from the observations it is not clear whether any or all of these processes is in operation, or whether some completely novel mechanisms remain to be discovered. Energy extracted from the black hole, with the help perhaps of a torus around it, has to be channeled into twin relativistic beams. These have to carry energy from the region close to the black hole to regions which are hundreds of kiloparsec away. The beams have been observed in radio sources, and there is ample evidence that they are relativistic on the parsec scale, while the nature of the beams on the large scale still remains to be established. The beams seem to occur only in a fraction of quasars and AGN, while the rest have all the attributes of being active, except that they do not manifest the beams.

The radiation spectrum extends over a very broad range of frequencies, and though the basic processes for the production of photons are very well understood, it is still a challenge to put them together to reproduce the observed spectrum. The form of the spectrum depends on the distribution of matter around the AGN, which provides emitting, absorbing and reflecting surfaces. The distribution of matter is not known and has to be guessed from the spectrum. There are many types of AGN, and it would be very convenient to be able to reduce them to very few basic types, and ascribe the observed variety to the observed effect of orientation relative to the observer. Such orientation becomes important because of the relativistic beams, as well as the presence of an obscuring torus which has dimensions larger than the parsec scale. Only when the observer can look down the throat of the torus can the regions closest to the black hole be seen. An unfavorable orientation requires the observer to be content with what can be seen outside the torus, and to get glimpses of the riches inside through indirect means, like reflection of the radiation directed along the throat, but away from the observer.

The space density of quasars increases towards earlier epochs of the universe, and available evidence seems to suggest that this increase is brought about by a general increase in the luminosity of the quasar population with redshift. This behaviour is most consistent with a picture in which the activity occurs intermittently in each of a set of galaxies. The conditions which make these galaxies the chosen ones for harboring, and displaying, a black hole, the evolution of these objects with time, and the presence or absence of black holes in galaxies which do not now seem to have an active galactic nucleus, remain to be investigated.

Afterthought

It has been a great pleasure to write this article for a volume to celebrate C. V. Vishveshwara's 60th year. He has been to me a guide, philosopher and friend. He has none of the unappealing qualities which are occasionally attributed to guides and philosophers; to friendship he brings joy and warmth, not tempered by the differences in age or persuasion. I can only hope that I continue to benefit from all these qualities over the years to come, and with equal enthusiasm, but perhaps a shaking hand, pen another story in celebration of his birthday in the distant future.

References

1. Blandford, R.D. (1990). In Blandford, R.D., Netzer, H. and Woltjer, L. (eds.), *Active Galactic Nuclei* (Springer-Verlag, Berlin), p161.
2. Boyle, B.J., Jones, L.R., Shanks, T., Marano, B., Zitelli, V. and Zamorani, G. (1991). In Crampton, D. (ed), *The Space Distribution of Quasars* (Astronomical Society of the Pacific, San Francisco)C91, p.191.
3. Cavaliere, A. and Padovani, P. (1989). *Astrophys. J.* **340**, L5.
4. Haenhelt, M.G. and Rees, M.J. (1993). *Mon. Not. Roy. Astr. Soc.* **263**, 168.
5. Kembhavi, A.K. and Narlikar J.V. 1998. *Quasars and Active Galactic Nuclei* (Cambridge University Press, Cambridge).
6. Köhler, T., Groote, D., Reimers, D. and Wisotzki, L. (1997). *Astro. Astrophys.* **325**, 502.
7. Osterbrock, D.E. 1989. *Astrophysics of Gaseous Nebulae and Active Galactic Nuclei* (University Science Books, Mill Valley).
8. Penrose, R. (1969). *Rivista Nuovo Cimento* **1**, 252.
9. Perry, J.J. (1994). In Bicknell, G.V., Dopita, M.A. and Quinn, P.J. (eds.), *The First Stromlo Symposium: The Physics of Active Galaxies* (Astronomical Society of the Pacific, San Francisco), p.417
10. Rees, M.J. (1984). *Ann. Rev. Astron. Astrophys.* **22**, 471.
11. Sanders, D.B., Phinney, E.S., Neugebauer, G., Soifer, B.T. and Matthews, K. (1989). *Astrophys. J.* **347**, 29.
12. Small, T. A., and Blandford, R., (1992), *Mon. Not. Roy. Astr. Soc.* **259**, 725.
13. Terlevich, R. (1992). In Fillipenko, A.V. (ed.), *Relationships Between Active Galactic Nuclei and Starburst Galaxies. Proc. of the Taipei Astrophysics Workshop.* (Astronomical Society of the Pacific, San Francisco), p133
14. Wagh, S.M. and Dadhich, N. (1989). *Phys. Rep.* **183**, 137.

In our analysis one component is taken as the parent quasar while the other visible component would be a radio source ejected at a relativistic speed in our general direction. Presumably this component would be only one out of many ejects which were emitted isotropically or equatorially by the quasar but remains invisible because the relativistic intensity shifts make them inconspicuous. This model, which presumably goes back to Rees, might be described as the 'three penny model' after the Brecht and Weill opera in which the chorus sings:

Therefore, some are in darkness;
Some are in the light, and these
You may see, but all those others
In the darkness no one sees

— BEHR, SCHUCKING, VISHVESHWARA and WALLACE

After a long hibernation in its mathematical lair, the black hole has travelled fast along the path of physics and has arrived in the realm of astronomy–sometimes dangerously poised on the brink of mythology. Immense amount of work has been carried out towards deciphering its secrets and investigating its influence on physical processes. As Churchill might have put it, never in the field of human knowledge so much has been said by so many about something that is so close to being nothing. Still, unanswered questions and unsolved problems remain.

— C. V. VISHVESHWARA

19. ENERGETIC PHOTON SPECTRA AS PROBES OF THE PROCESS OF PARTICLE ACCELERATION IN ACCRETION FLOWS AROUND BLACK HOLES

R. COWSIK
Indian Institute of Astrophysics,
Koramangala, Bangalore - 560 034, India.

1. Introduction

Most of the powerful astronomical objects discovered over the past three decades appear to be energised by the accretion of matter on to condensed massive bodies residing in their centres. X-ray binaries, active galactic nuclei, extragalactic double radio sources displaying relativistic jets and other such interesting phenomena are some examples for which an important part of the gravitational energy associated with the inflow of matter is radiated away as non-thermal particles and quanta. Excellent overviews of the relevant astronomical literature and of the physical processes related with accretion flows on to black holes are provided by Rees [1], Begelman et al.[2] and by Frank et al. [3]. In recent years the evidence for the existence of both supermassive black holes ($M_{BH} \sim 10^6 - 10^8 M_\odot$) and stellar mass black holes is accumulating rapidly [4-7].

The work on the dynamical models of accretion started more than a century ago and several important findings such as the Eddington-limit on the luminosity of the accreting object, the existence of sonic point and various branches in the solutions of spherically symmetric accretion by Bondi and the time-reversed solutions for supersonic stellar winds by Parker were noted several decades ago. In the early 1960s with the discovery of quasars and galactic X-ray sources the general ideas of accretion found direct observational support and the effort to understand fully this fascinating phenomenon began with great energy and earnestness on both observational and theoretical fronts.

Zeldovich and independently Salpeter were the first to suggest that black

B. R. Iyer and B. Bhawal (eds.), Black Holes, Gravitational Radiation and the Universe, 309–317.

holes of mass $10^6 - 10^8 M_\odot$ accreting surrounding matter may be responsible for energising the quasars. Shklovsky, Hayakawa and Matsuoka noted that accretion in a binary system formed by a compact object and a normal star may lead to high accretion rates. Following this Prendergast and Burbidge pointed out that because of the angular momentum in the binary system the accreted material will have a disc like distribution around the compact object. The so called standard or the α model of the accretion disc was worked out by Shakura and Sunyaev in 1973 which showed that the effective temperature of the emitted X-rays decreased as $T \sim M^{-1/4}$ with increasing mass of the compact object. Shapiro, Lightman and Eardley and independently Sunyaev and Titarchuk developed the two temperature disc models with the inner part swelling up into a high temperature thick disc or a corona. Stochastic Comptonization of the photons through the hot electron gas led to hard X-rays with power law energy spectra in conformity with the observations, for example of Cyg X-1 [8]. Rees pointed out that because of high masses of the black holes expected in quasars, the α models cannot explain the X-ray emission seen from these objects. These and other details are reviewed extensively in the book by Frank et al. [3], by Shapiro and Teukolsky [9] and in several of the articles in this Festschrift [24].

This article is based mainly on our earlier work [10] and in this paper we emphasise two important results: (1) It is not essential that the transport of particles and photons be through a hot and turbulent medium to generate non-thermal spectra - their diffusion through the accretion flow will suffice for generating the observed power laws. (2) The observed spectra from quasars, etc., are so hard that they can be generated only in accretion flows that go through a shock transition.

The existence of such shocks have been noted by Babul, Ostriker and Mézáros [11], M.Yokosawa [12], S.K.Chakrabarti [13] amongst others. Extensive reviews of the processes of shock acceleration by Drury [14], Völk [15], Blandford and Eichler [16], Berezhko and Kinmsky [17] and by Jones and Ellison [18] provide details of several recent developments.

2. The Mathematical Model

The accretion flow of plasma on to a black hole is described by the function V(x,t) and we assume that the transport of energetic particles and protons (jointly called particles) through such a medium may be described as spatial diffusion. Under this approximation their distribution function is nearly isotropic in the local plasma rest frame. We also describe the transport process in the test particle approximation in that these have no influence on the flow V(x,t). This is done mainly for mathematical convenience; however the results thus obtained may illustrate some crucial aspects of the phenomenon under study. The basic transport equation expressing the conservation of particles under the conditions just outlined, and the additional assumption that V^2/c^2 is small was originally derived by Parker [19] and by Gleeson and Axford [20] in the context of cosmic-ray transport through the solar wind. In terms of the omnidirectional distribution function $f(p,\underline{x},t)$ and the magnitude of the particle momentum, p, the equation takes the form

$$\frac{\partial f}{\partial t} + \nabla \cdot S + \frac{1}{3p^2}\frac{\partial}{\partial p}\left(V \cdot \nabla p^3 f\right) = Q, \tag{1a}$$

$$S = -\kappa \cdot \nabla f + V f - \frac{1}{3} V p^{-2} \frac{\partial}{\partial p}\left(p^3 f\right), \tag{1b}$$

where $4\pi p^2 S$ is the differential current density or streaming, Q is the source or injection term, and κ is the spatial diffusion tensor (for isotropic scattering $\kappa = \kappa_o \delta_{i,j}$ and $\kappa_o = \frac{1}{3}\lambda \mathrm{v}$). Eqs.(1a) and (1b) simplify to yield.

$$\frac{\partial f}{\partial t} - \nabla \cdot (\kappa_o \nabla f) + V \cdot \nabla f - \frac{1}{3} \nabla \cdot V p \frac{\partial f}{\partial p} = Q. \tag{2}$$

A few salient aspects of the above equation are worth noting. The first three terms are familiar terms of convective-diffusive transport. The last term on the left-hand side describes the evolution of f in p-space, for example, the adiabatic deceleration of particles in the expanding solar wind with $\nabla \cdot V > 0$. In the converging accretion flow with $\nabla \cdot V < 0$ this term leads to adiabatic acceleration of the particles. This term also describes the acceleration of particles at the quasi stationary shock at the base of the accretion flow because of plasma compression. Even though the compression is peaked only at the shock, the particles are coupled to it as they alternately scatter upstream and downstream of the shock fronts. In the absence of injection of particles f is continuous across the shock and integration of Eq.(1a) yields a finite value for the streaming S. The acceleration of particles in convergent supersonic accretion flow is particularly efficient because the acceleration occurs throughout the flow and at the shock; further, the particles as they diffuse are convected to the central regions of greater compressions and to the shock where $\nabla \cdot V$ is singular.

We now consider an idealised supersonic accretion flow which undergoes a quasistationary shock transition at $r = r_s$, and parameterize $V = V_o\,(r/r_s)^{-\alpha}$ and $\kappa_{rr} = \kappa_o\,(r/r_s)^p \times (p/p_o)^\gamma$ for $r > r_s$. We also assume that $Q = N(r - r_s) \cdot \delta(p - p_o)$ for mathematical convenience; the form of the solution for injection elsewhere will not be very sensitive to the specific location of injection. For $r > r_s$ the particle distribution function satisfies the equation

$$-\kappa_o \left(\frac{p}{p_o}\right)^\gamma \frac{1}{r^2}\frac{\partial}{\partial r}\left[r^2 \left(\frac{r}{r_s}\right)^p \frac{\partial f}{\partial r}\right] - V_o \left(\frac{r}{r_s}\right)^{-\alpha} \frac{\partial f}{\partial r}$$

$$+\frac{1}{3}V_o\frac{1}{r^2}\frac{\partial}{\partial r}\left[r^2\left(\frac{r}{r_s}\right)^{-\alpha}\right]p\frac{\partial f}{\partial p} \quad = 0. \tag{3}$$

It is convenient to describe the compression across the shock in terms of σ defined by the equation

$$V(r = r_s - \varepsilon) = V_o(1 - 3\sigma^{-1}), \tag{4}$$

so that for a strong shock σ=4. The boundary conditions on f at $r = r_s$ are given by

$$\kappa_o\left(\frac{p}{p_o}\right)^{\gamma}\frac{\partial f}{\partial r} + \kappa_-(r_s,p)\frac{\partial f_-}{\partial r} + \frac{V_o}{\sigma}p\frac{\partial f}{\partial p} = N(4\pi p_o r_s)^{-2}\delta(p - p_o), \tag{5a}$$

$$f(r_s,p) = f_-(r_s,p). \tag{5b}$$

3. Spectra of photons and energetic particles generated by accretion flows

In the previous section we have set-up in some detail the basic equations that describe the energising of the photons and particles in accretion flows, even though these are available in the paper by Cowsik and Lee [10]. The purpose is to provide easy access to any one who wishes to investigate various solutions to these equations with either analytical or numerical techniques. For the present purposes we reproduce few of their analytic solutions which are valid for $p >> p_o$.

3.1. SPECTRUM OF PHOTONS DIFFUSING THROUGH THE FLOW

Photons up to an energy of $\sim\ m_ec^2$ scatter off electrons with the Thompson cross section which is independent of energy; thus in terms of our parametrization in section 2, $\gamma = 0$. Further the scattering mean free path is inversely proportional to the electron density n_e; since in a stationary flow $n_e(r)V(r)r^2$ is a constant, we have

$$\alpha + \beta = 2. \tag{6}$$

From the work of Cowsik and Lee we cull out two solutions:

$$f_1 = p^{-3\beta^{-1}(1+\beta)} \quad \text{for} \quad \sigma = \infty, \tag{7}$$

$$f_2 = \left(\frac{p}{p_o}\right)^{-\sigma[1-2\beta\eta^{-1}(\frac{1}{3}\sigma-1]}, \tag{8}$$

where $\eta = r_s V_o / \kappa_o$. The solution, f_1 for $\sigma = \infty$ corresponds to the absence of any shock transition in the accreting flow and is valid for $p >> p_o$; this solution was also independently obtained by Blandford and Paynes [21]. The solution f_2 is valid for $p >> p_o$ and for $\eta >> 1$, i.e. when the effect of the shock on the distribution function is pronounced. Thus we find that power-law behavior are generic to diffusive transport in accretion flows:

$$f \sim p^{-\Gamma}. \tag{9}$$

This clearly shows that transport in turbulent systems is not essential for obtaining such nonthermal spectra. Now it is interesting to look at the likely values the exponent Γ might take and compare it with the observed spectra of quasars and other astronomical objects.

Let us first look at f_1. For an accretion flow in which the speed approaches the free fall velocity, we have $V \sim r^{-1/2}$ or $\alpha = 1/2$ and $\beta = 3/2$ leading to $\Gamma = 5$. In the extreme situation with $V \sim$ constant, or $\alpha = 0$ and $\beta = 2$ we obtain $\Gamma = 4.5$. Remembering that Γ is the exponent of the distribution in phase space the power-spectrum emerging from such flows will have an exponent $\delta = \Gamma - 3$, i.e. 2 or 1.5 in the cases just discussed. The observed spectra of quasars in hard photons (see Figures 1,2) have spectral power with exponent close to 1. Thus accretion flows without shocks can not reproduce their spectra. Alternatively, the exponent of the power law that is obtained in accretion flows with shocks is seen from Eq.(8) to be

$$\Gamma = \sigma \left[1 - 2\beta\eta^{-1} \left(\frac{1}{3}\sigma - 1 \right) \right], \tag{10}$$

which is harder than that for f_1 and indeed even harder than the spectrum that obtains in planar shocks with $\Gamma \approx \sigma$ [21,22]. In any case for large η (i.e. small κ) we obtain for high Mach number shocks $\sigma = 4$

$$\Gamma \approx 4, \tag{11}$$

which corresponds to $\delta \approx 1$ in good agreement with the observations as shown in Figure 1. As the energy of the photon approaches $E_\nu = m_e c^2$ the scattering cross-section changes from the constant Thompson value to the Klein-Nishina form yielding decreasing σ with increasing energy. This causes the diffusion mean free path increase with energy and the solutions indicate that the spectrum of photons cuts off exponentially with energy beyond $\sim m_e c^2$.

3.2. ACCELERATION OF SUPRATHERMAL CHARGED PARTICLES

As the energy of protons and atomic nuclei increases beyond about 10 MeV per nucleon their scattering cross-section on electrons and other nuclei decrease with energy so that the scattering of magnetic inhomogenities in the

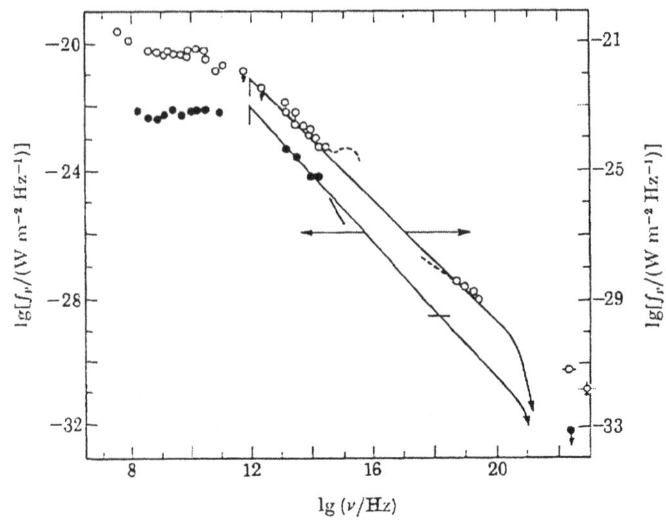

Figure 1. Power-law fits to the spectra of two quasars ($\Gamma_1 = 0.97, \eta = 30, \sigma = 4.05$; $\Gamma_2 = 1.07, \eta = 30, \sigma = 4.22$); •, 3C 120; ∘, 3C 273. The γ-rays in 3C 273 are probably produced by high energy nuclear interactions of charged particles also accelerated in the accretion flow (Figure taken from Ref.10).

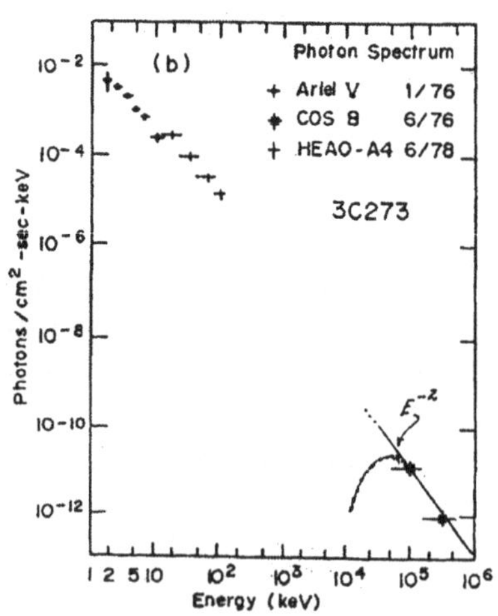

Figure 2. Number spectrum of 3C 273; Ariel-5 (Sanford). HEAO-I A4 (Primini et al.). COS-B (Swanenburg et al.) (taken from Annals of New York Academy of Sciences. Note that $F(E) \sim E^{-2}$ fits both the hard x-ray and the gamma-ray data well.

plasma dominates their transport. Thus, unlike photons discussed in the earlier section, the diffusion constant is not necessarily inversely proportional to the plasma density. It is convenient to assume that turbulence downstream of the shock is sufficiently strong to make $\kappa_-(r,p) \approx 0$ for $r < r_s$. For transport at larger distances $r > r_s$ let us consider the special case $\alpha + \beta = 1$ and $\gamma = 0$ i.e. κ is independent of p. We refer to the work of Cowsik and Lee(1982) for the full gamut of solutions but will note here that the spectra are power laws in general. For $\eta = r_s V(r_s)/\kappa(r_s)$ large, the phase space power law index p is given by

$$\Gamma = \sigma[1 + (1+\beta)\eta^{-1}] \tag{12}$$

whose leading term $\Gamma = \sigma$ is identical to the power law slope that is obtained in a planar shock discussed earlier by Axford et al. [22] and by Blandford and Ostriker [23].

Now for a high Mach-number shock $\sigma = 4$ and for large η the spectral index becomes

$$\Gamma \approx 4 \tag{13}$$

similar to protons as given in Eq.(11). As noted by us earlier these energetic particles will collide with the particles in the plasma and will generate secondary photons. Of particular interest are the high energy gamma rays generated by the decay of neutral pions, π^0, produced in nuclear interactions of the energetic particles. The spectra of these gamma rays at high energies is identical to that of the parent particle population, but as E_γ approaches $m_\pi c^2$ the spectrum flattens and has a broad peak at $m_\pi c^2/2 \sim 70$ MeV. What is remarkable is that γ-ray observations of the quasar 3C273 with COS-B satellite fit well with $\Gamma \approx 4$ which corresponds to E_γ^{-2} spectral intensities (see Figure 2). It is the same theoretical spectrum predicted for strong shocks that we used in fig.1 to the hard X-ray spectrum of the object. More recently EGRET instrument aboard the Compton Gamma-Ray Observatory observed the γ-ray spectrum of the source 2EG J1746-2852 which also fits well with E_γ^{-2} expected in the presence of a strong shock. In the absence of the shock the accretion flow will generate E_γ^{-3} or steeper spectra and will not fit the observations (see Figure 3).

4. Discussion and conclusions

The idea that powerful astronomical sources are energised by accretion flows around black holes is gaining observational support. The spectra of these objects from ~ 1 eV to about 1 GeV can be reproduced by transport of photons and suprathermal particles through accretion flows which suffer a shock transition before disappearing into the black hole. Such shocks are generic to accretion flows where radiation and magnetic fields play a

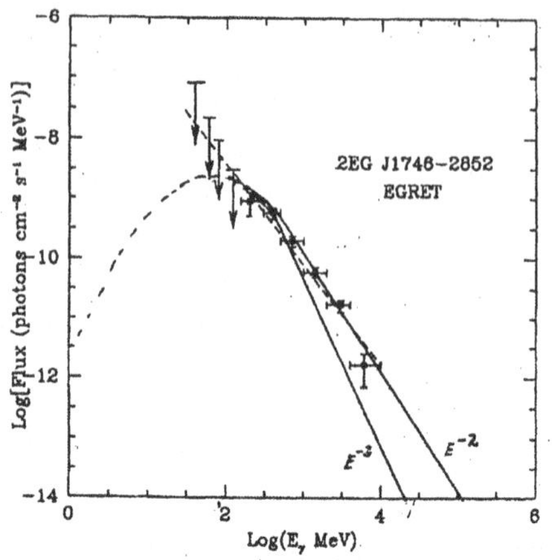

Figure 3. Note that $F(E) \sim E^{-3}$ expected for acceleration in accretion flows without shock transitron is too steep to reproduce the EGRET data on gamma-rays, but $F(E) \sim E^{-2}$ expected for acceleration in supersonic flows with a shock transitron reproduces the data well.

dynamical role and also to flows which are not perfectly spherically symmetric. High Mach number shocks with small diffusion coefficients generate spectra which are essentially parameter independent and it is remarkable that such spectra reproduce the observations extending over 9-decades in photon energy. If the shocks are absent, the accretion flow generates spectra which are considerably steeper than those emitted by these sources.

5. Dedication

Professor C.V. Vishveshwara has been a good friend and I have known his family for a very long time. It is indeed a great pleasure to wish him and his family all the very best and dedicate this article to him on his 60th birthday.

References

1. M.J.Rees, Physica Ser. **17**, 193 (1978); Ann. Rev. Astr. Astrophys., **22**, 471, (1989).
2. M.C.Begelman, R.D.Blandford and M.J.Rees, Rev. Mod. Phys., **56**, 225 (1984).
3. J.Frank, A.King and D.Raine, *Accretion Power in Astrophysics,* (Cambridge University Press, (1992).
4. M.Miyoshi et al. Nature, **373**, 127 (1995).
5. G.Fabbiano and J.Z.Juda, Ap. J. **478**, 542 (1997).
6. D.Banet, J.E.McClintok and J.E.Grindley, Ap.J. **473**, 963 (1996), and the references therein.
7. V.R.Chitnis, A.R.Rao and P.C.Agrawal, to appear in Astr. and Astrophys., (1998).
8. E.P.Liang and P.L.Nolan, Sp. Sci. Rev., **38**, 353 (1984).

9. S.L.Shapiro and S.A.Teukolsky, *Black Holes, White Dwarfs and Neutron Stars* (Wiley-Interscience, 1983).
10. R.Cowsik and M.A.Lee, Proc. Roy.Soc.Lond., **A383**, 409 (1982).
11. A.Babul, J.P.Ostriker and P.Mezaros, Ap.J. **347**, 59 (1989).
12. M.Yokosawa, PASJ, **46**, 73 (1994).
13. S.K.Chakrabarti, *Theory of shocks in accretion flows,* (World Scientific, 1992).
14. L.O'C. Drury, Rep. Prog. Phys., **46**, 973 (1983).
15. H.J.Völk, in *High Energy Astrophysics*, Proc. 19th Recontre de Moriond, (ed. Tran Than Van, Editions Fontrieres, 1984)
16. R.D.Blandford and D.Eichler, Phys. Repts., **154**, 1 (1987).
17. E.G.Berezhko and G.F.Krymsky, Usp. Fiz. Nauk., **154**, 49 (1988).
18. F.C.Jones and D.C.Ellison, Space Sci. Rev., **58**, 259 (1991).
19. E.N.Parker, Planet and Space Sci., **13**, 9 (1965).
20. L.J.Gleeson and W.I.Axford, Ap.J.Lett., **149**, (1967).
21. R.D.Blandford and D.G.Payne, M.N.R.A.S. **194**, 1033 (1980).
22. W.I.Axford, Proc. IAU Symp. on the Origin of Cosmic Rays, (Eds. G.Setti et al. 1980).
23. R.D.Blandford and J.P.Ostriker, Ap.J.Lett., **221**, L29 (1978).
24. See articles 16, 17 and 18 by Paul J. Wiita; K.Menou, E. Quataert & R. Narayan; and Ajit K.Kembhavi, in this volume.

The weakness of the gravitational interaction makes it exceedingly unlikely that gravitational radiation will ever be the subject of direct observation

— F.A.E. PIRANI (1962)

The theoretical predictions did not provide much encouragement. While several 'unusual' celestial objects were pinpointed as possible, or even likely, sources of x-rays, it did not look as if any of them would be strong enough to be observable with instrumentation not too far beyond the state of the art. Fortunately, we did not allow ourselves to be dissuaded. As far as I am personally concerned, I must admit that my motivation for pressing forward was a deep-seated faith in the boundless resourcefulness of nature, which so often leaves the most daring imagination of man far behind.

— B. ROSSI (1974)

The problems of the experimentalist working on gravitation differ from [those in other fields]. Here there is an elegant, well defined theory but almost no experiments. The situation is almost orthogonal to that in both high energy physics and solid state physics. Far from experimental science being a crutch on which theory leans, in the case of general relativity, theory has far outstripped experiment, and the big problem is one of finding significant experiments to perform. This situation raises serious problems for theory and our understanding of gravitation. For where there are no experiments the theory easily degenerates into purely formal mathematics. ... In other fields of physics the experimentalist is faced with the problem of choosing the most important out of a large number of possible experiments. With gravitation the problem is different. There are so few possible experiments, and their importance is such, that any and all significant experiments should be performed.

— ROBERT H. DICKE (1964)

20. BLACK HOLE PERTURBATION APPROACH TO GRAVITATIONAL RADIATION

Post-Newtonian Expansion for Inspiralling Binaries

MISAO SASAKI
Department of Earth and Space Science, Osaka University, Toyonaka 560, Japan.

1. Introduction

It is expected that the near future large-scale laser interferometers such as LIGO, VIRGO, GEO600 and TAMA300 will directly detect gravitational waves from various astrophysical sources [1]. Consequently there is a growing interest in the study of sources of gravitational waves in the universe and the most promising detectable source is the gravitational radiation from a compact binary, in particular a neutron star - neutron star (NS-NS) binary, during the final phase of its inspiralling. The orbit of a compact binary will gradually spiral in due to gravitational radiation and its frequency will eventually sweep through the detectable frequency range of those interferometers, approximately 10 Hz to 1000 Hz, before the coalescence. To extract out the information contained in thus detected gravitational waves efficiently as well as correctly, it has therefore become very important to know the detailed evolution of an inspiralling binary and to construct accurate theoretical templates for the waveforms. Since the orbital velocity v will be relativistic ($v \gtrsim 0.1c$ for a NS-NS binary) when the gravitational waves become detectable, a simple estimate based on the Newtonian quadrupole formula is insufficient; instead we must take into account the relativistic corrections seriously.

The standard method to calculate inspiralling waveforms from an inspiralling compact binary is the post-Newtonian (PN) expansion of the Einstein equations, in which the orbital velocity is assumed to be small ($v \ll c$) [2]. However, toward the end of the inspiralling stage, the orbital velocity can become as large as $v \sim 0.3c$. Hence it may be necessary to carry out the PN expansion to an extremely high order in v/c in order to construct sufficiently accurate template waveforms. The PN calculations

B. R. Iyer and B. Bhawal (eds.), Black Holes, Gravitational Radiation and the Universe, 319–334.

however become increasingly difficult as the order increases. As of October 1997, the highest order achieved so far is 2.5PN order or $O((v/c)^5)$ beyond the Newtonian approximation [3], and perhaps the 3.5PN order will be the highest achievable order in near future.

Then there arises an issue whether the 3.5PN approximation is accurate enough for our purpose. It is thus necessary to explore another reliable method to calculate the higher order PN corrections or to evaluate the convergence of the PN expansion. To this end, the black hole (BH) perturbation approach has been found to be very useful. Using the BH perturbation theory, we can calculate gravitational waves from a particle of mass μ orbiting a black hole of mass M, assuming $\mu \ll M$. Although it is restricted to the case $\mu \ll M$, one can develop a PN expansion scheme with a relatively simple algorithm to calculate very high order PN corrections to gravitational waves. Moreover, numerical calculations which include fully relativistic effects can be performed to test the convergence of the PN expansion.

Gravitational perturbations of a black hole were first studied by Regge and Wheeler [4] about 40 years ago. Since then numerous studies have been done to understand the properties of a black hole and to apply the results to astrophysical situations. Among others, we would like to mention that Professor Vishveshwara made pioneering contributions to the understanding of black holes [5]. Then a master equation for perturbations of a Kerr BH was derived by Teukolsky [6] and its equivalence to the Regge-Wheeler equation was shown by Chandrasekhar [7] in the case of a Schwarzschild BH.

In this contribution, we focus on a recently developed PN expansion method to solve the gravitational perturbation induced by an orbiting particle. This approach was initiated by Poisson [8] who considered the case of a circular orbit around a Schwarzschild BH and obtained the waveforms and luminosity to 1.5PN order. It should be noted that the so-called tail effect (curvature scattering of waves) appears at 1.5PN order and Poisson succeeded in obtaining the tail term in a way much easier than the standard PN expansion method. In the case of a Schwarzschild BH, Sasaki [9] improved Poisson's method and Tagoshi and Sasaki [10] obtained the gravitational waveforms and luminosity for a circular orbit to 4PN order. The calculation was extended to 5.5PN order by Tanaka, Tagoshi, and Sasaki [11]. In the case of a particle orbiting a Kerr BH, a method has been developed in parallel to the Schwarzschild case. We shall not cite all the papers here but refer to Tagoshi, Shibata, Tanaka and Sasaki [12] in which the 4PN calculation has been done, and references therein. We also mention that Mano, Suzuki and Takasugi have found an elegant way to construct the solution of the homogeneous Teukolsky equation [13]. Their method is more powerful when dealing with the Kerr case.

Below we first review the Teukolsky formalism for gravitational perturbations and outline a post-Newtonian expansion method adapted to the case of a particle orbiting a Schwarzschild BH. We discuss only the Schwarzschild case, mainly because an extension to the Kerr case is straightforward but much more complicated and partly because the higher order PN corrections in the Kerr case contain terms that are relevant only to a rapidly rotating Kerr BH while our main concern here is not limited to such a case. Then we present the result of the 5.5PN luminosity formula for a particle in a circular orbit around a Schwarzschild BH. Finally, based on this result, we discuss convergence of the PN expansion for some typical compact binaries.

In the rest of this contribution, we use the units $c = G = 1$.

2. Regge-Wheeler-Teukolsky Formalism

2.1. TEUKOLSKY EQUATION

We consider a particle of mass μ orbiting a Schwarzschild BH of mass $M \gg \mu$. The gravitational perturbations induced by a particle is conveniently described by a Newman-Penrose quantity ψ_4. On the Schwarzschild background, it is decomposed as

$$\psi_4 = \frac{1}{r^4} \sum_{\ell m} \int d\omega e^{-i\omega t} \, {}_{-2}Y_{\ell m}(\theta, \varphi) R_{\ell m \omega}(r), \tag{1}$$

where ${}_{-2}Y_{\ell m}$ are the spherical harmonics of spin weight $s = -2$. The radial function $R_{\ell m \omega}$ satisfies the inhomogeneous Teukolsky equation [6],

$$\Delta^2 \frac{d}{dr}\left(\frac{1}{\Delta}\frac{dR_{\ell m \omega}}{dr}\right) - V(r) R_{\ell m \omega} = T_{\ell m \omega}, \tag{2}$$

where

$$V(r) = \frac{r^2}{\Delta}\left[\omega^2 r^2 - 4i\omega(r - 3M)\right] - \lambda, \quad \Delta = r(r - 2M), \tag{3}$$

and $\lambda = (\ell - 1)(\ell + 2)$ is the eigenvalue of the $s = -2$ spherical harmonics. The source term $T_{\ell m \omega}$ is given in terms of the energy momentum tensor of the particle.

To solve Eq. (2), we construct the Green function with an appropriate boundary condition. The physical boundary condition is that waves are purely outgoing at infinity and purely ingoing at horizon. Therefore we consider the following two independent solutions of the homogeneous Teukolsky equation:

$$R^{\rm in} \to \begin{cases} B^{\rm trans} \Delta^2 e^{-i\omega r^*} & \text{for } r \to 2M, \\ r^3 B^{\rm ref} e^{i\omega r^*} + r^{-1} B^{\rm inc} e^{-i\omega r^*} & \text{for } r \to +\infty, \end{cases} \tag{4}$$

$$R^{\rm up} \to \begin{cases} C^{\rm up} e^{i\omega r^*} + \Delta^2 C^{\rm ref} e^{-i\omega r^*} & \text{for } r \to 2M, \\ C^{\rm trans} r^3 e^{i\omega r^*} & \text{for } r \to +\infty, \end{cases} \tag{5}$$

where $r^* = r + 2M \ln(r/2M - 1)$ is the tortoise coordinate. Then the solution is expressed as

$$R_{\ell m\omega} = \frac{1}{W} \left\{ R^{\rm up} \int_{2M}^{r} dr' \frac{R^{\rm in} T_{\ell m\omega}}{\Delta} + R^{\rm in} \int_{r}^{\infty} dr' \frac{R^{\rm up} T_{\ell m\omega}}{\Delta} \right\}, \tag{6}$$

where

$$W = \frac{1}{\Delta} \left[\left(\frac{d}{dr} R^{\rm up} \right) R^{\rm in} - R^{\rm up} \left(\frac{d}{dr} R^{\rm in} \right) \right] = 2i\omega C^{\rm trans} B^{\rm inc}. \tag{7}$$

Taking $r \to \infty$, the asymptotic behavior of the solution at infinity is found to be

$$\begin{aligned} R_{\ell m\omega}(r \to \infty) \quad &\to \frac{r^3 e^{i\omega r^*}}{2i\omega B^{\rm inc}} \int_{2M}^{\infty} dr' \frac{T_{\ell m\omega}(r') R^{\rm in}(r')}{\Delta^2(r')} \\ &\equiv \tilde{Z}_{\ell m\omega} r^3 e^{i\omega r^*}. \end{aligned} \tag{8}$$

The two independent polarizations of gravitational waves h_+ and $h_\times$ at infinity are related to ψ_4 by

$$\psi_4 = \frac{1}{2}(\ddot{h}_+ - i\ddot{h}_\times). \tag{9}$$

Thus we have

$$h_+ - ih_\times = -\frac{2}{r} \sum_{\ell m} \int \frac{d\omega}{\omega^2} \tilde{Z}_{\ell m\omega} \; {}_{-2}Y_{\ell m}(\theta, \varphi) e^{-i\omega(t - r^*)}. \tag{10}$$

In particular, for a circular orbit at radius $r = r_0$, the source term $T_{\ell m\omega}$ takes the form,

$$T_{\ell m\omega} \sim \left[a_0 \delta(r - r_0) + a_1 \delta'(r - r_0) + a_2 \delta''(r - r_0) \right] \delta(\omega - m\Omega), \tag{11}$$

where $\Omega = M/r_0^{3/2}$ is the orbital angular frequency $d\varphi/dt$ and we have assumed the orbit to lie on the equatorial plane ($\theta = \pi/2$) without loss of generality. The explicit form of $T_{\ell m\omega}$ will be given in section 3. It then follows that $\tilde{Z}_{\ell m\omega}$ takes the form,

$$\tilde{Z}_{\ell m\omega} = Z_{\ell m} \delta(\omega - m\Omega). \tag{12}$$

Further, because of the symmetry of the orbit and that of the spin-weighted spherical harmonics, we have $\overline{Z_{\ell m}} = (-1)^\ell Z_{\ell -m}$. In this case Eq. (10) reduces to

$$h_+ - ih_\times = -\frac{2}{r}\sum_{\ell m}\frac{Z_{\ell m}}{\omega_m^2}\,{}_{-2}Y_{\ell m}(\theta,\varphi)e^{-im\omega_m(t-r^*)}, \tag{13}$$

where $\omega_m = m\Omega$, and the time-averaged energy flux at infinity is given by

$$\left\langle\frac{dE}{dt}\right\rangle = \sum_{\ell=2}^{\infty}\sum_{m=1}^{\ell}\frac{|Z_{\ell m}|^2}{2\pi\omega_m^2}. \tag{14}$$

Thus, turning back to Eq. (8), given the source term $T_{\ell m\omega}$ which is readily calculated by solving the geodesic equation, our task is to find the ingoing-wave solution $R^{\rm in}(r)$ for r where $T_{\ell m\omega}$ is non-vanishing and the incident amplitude $B^{\rm inc}$ at infinity. For a circular orbit, the radius of the orbit and the angular velocity are related as $M/r_0 = v^2 = (r_0\Omega)^2$. The PN expansion assumes that the orbital velocity is much smaller than the speed of light; $v \ll 1$. On the other hand, the frequency of the gravitational waves is given by $\omega = m\Omega$. Hence we have

$$\omega r = O(v), \quad M\omega = O(v^3). \tag{15}$$

This implies that what we need to do is (a) to calculate $R^{\rm in}$ at $r\omega = O(v) \ll 1$ and (b) to evaluate $B^{\rm in}$ to a required order of $M\omega = O(v^3)$. By rescaling the Teukolsky equation, it can be expressed in terms of a non-dimensional variable $z = \omega r$ and a non-dimensional parameter $\epsilon = 2M\omega$. Then, recalling that $M = GM$ if we recover the gravitational coupling constant G, we see that (a) corresponds to the PN expansion of $R^{\rm in}$ in the near zone (yet far from the horizon) and (b) to a post-Minkowskian (PM) expansion. As we shall see below, it is straightforward to obtain the PN expansion of $R^{\rm in}$ while it is rather involved to obtain $B^{\rm inc}$. Hence our main task is to develop a PM expansion scheme to solve the Teukolsky equation.

2.2. PM EXPANSION OF THE INGOING-WAVE SOLUTION

It is possible to develop a PM expansion scheme for the Teukolsky equation itself [13]. However, since the Teukolsky equation is known to be equivalent to the Regge-Wheeler equation [7], and the latter equation has a familiar form in the limit $M \to 0$, we consider a PM expansion scheme for the Regge-Wheeler equation.

The homogeneous Regge-Wheeler equation takes the form [4],

$$\left[\frac{d^2}{dr^{*2}} + \omega^2 - V_{RW}(r)\right]X_{\ell\omega}(r) = 0, \tag{16}$$

where

$$V_{RW}(r) = \left(1 - \frac{2M}{r}\right)\left(\frac{\ell(\ell+1)}{r^2} - \frac{6M}{r^3}\right). \qquad (17)$$

The homogeneous Teukolsky equation is transformed to this equation by the transformation [7],

$$R_{\ell\omega} = \frac{\Delta}{c_0}\left(\frac{d}{dr^*} + i\omega\right)\frac{r^2}{\Delta}\left(\frac{d}{dr^*} + i\omega\right) r X_{\ell\omega}, \qquad (18)$$

where $c_0 = (\ell-1)\ell(\ell+1)(\ell+2) - 12iM\omega$. The inverse transformation is

$$X_{\ell\omega} = \frac{r^5}{\Delta}\left(\frac{d}{dr^*} - i\omega\right)\frac{r^2}{\Delta}\left(\frac{d}{dr^*} - i\omega\right)\frac{R_{\ell\omega}}{r^2}. \qquad (19)$$

Then the asymptotic forms of $X^{\rm in}$ are given by

$$X^{\rm in} \to \begin{cases} A^{\rm ref} e^{i\omega r^*} + A^{\rm inc} e^{-i\omega r^*} & \text{for } r \to \infty \\ A^{\rm trans} e^{-ikr^*} & \text{for } r \to 2M, \end{cases} \qquad (20)$$

where the coefficients $A^{\rm inc}$, $A^{\rm ref}$ and $A^{\rm trans}$ are related to $B^{\rm inc}$, $B^{\rm ref}$ and $B^{\rm trans}$ in Eq. (4), respectively, by

$$B^{\rm inc} = -\frac{1}{4\omega^2} A^{\rm in}, \quad B^{\rm ref} = -\frac{4\omega^2}{c_0} A^{\rm ref}, \quad B^{\rm trans} = \frac{1}{d_\omega} A^{\rm trans}, \qquad (21)$$

where $d_\omega = 16M^3(1 - 2iM\omega)(1 - 4iM\omega)$.

Using the variable $z = r\omega$ and the parameter $\epsilon = 2M\omega$, the Regge-Wheeler equation is rewritten as

$$\left[\frac{d^2}{dz^{*2}} + 1 - \left(1 - \frac{\epsilon}{z}\right)\left(\frac{\ell(\ell+1)}{z^2} - \frac{3\epsilon}{z^3}\right)\right] X_\ell^{\rm in} = 0, \qquad (22)$$

where $z^* = z + \epsilon \ln(z - \epsilon) = r^*\omega + \epsilon \ln \epsilon$. The PM expansion corresponds to solving this equation with the boundary condition (20) by expanding it in terms of ϵ. Poisson did this to $\mathcal{O}(\epsilon)$ [8]. However, a naive expansion of Eq. (22) would give a very complicated set of iterative equations if we go beyond the first order. Further, it would be difficult to impose the correct boundary condition that $X_\ell^{\rm in} \propto e^{-iz^*}$ at horizon ($z \to \epsilon$), since z^* involves ϵ in itself.

A way to circumvent these difficulties is to take a different choice of the dependent variable [9]. Setting

$$X_\ell^{\rm in} = e^{-i\epsilon \ln(z-\epsilon)} z \xi_\ell(z), \qquad (23)$$

we find that Eq. (22) becomes

$$\left[\frac{d}{dz^2}+\frac{2}{z}\frac{d}{dz}+\left(1-\frac{\ell(\ell+1)}{z^2}\right)\right]\xi_\ell=\epsilon e^{-iz}\frac{d}{dz}\left[\frac{1}{z^3}\frac{d}{dz}\left(e^{iz}z^2\xi_\ell(z)\right)\right]. \quad (24)$$

Since ϵ appears only as an overall factor on the r.h.s., this form of the equation is most suited for an iterative treatment. Thus expanding ξ_ℓ with respect to ϵ as

$$\xi_\ell(z)=\sum_{n=0}^{\infty}\epsilon^n\xi_\ell^{(n)}(z), \quad (25)$$

we obtain the recursive equations,

$$\left[\frac{d}{dz^2}+\frac{2}{z}\frac{d}{dz}+\left(1-\frac{\ell(\ell+1)}{z^2}\right)\right]\xi_\ell^{(n)}(z)=S^{(n)}, \quad (26)$$

where

$$S_\ell^{(n)}=e^{-iz}\frac{d}{dz}\left[\frac{1}{z^3}\frac{d}{dz}\left(e^{iz}z^2\xi_\ell^{(n-1)}(z)\right)\right].$$

Noting that the operator on the l.h.s. of Eq. (26) is just the one for the spherical Bessel functions, it can be cast into the integral form,

$$\xi_\ell^{(n)}=n_\ell\int dz\,z^2 j_\ell\,S_\ell^{(n)}-j_\ell\int dz\,z^2 n_\ell\,S_\ell^{(n)}, \quad (27)$$

where j_ℓ and n_ℓ are the spherical Bessel functions. Our strategy is to perform the above indefinite integral iteratively and choose the integration constants appropriately so as to satisfy the boundary condition at horizon.

Since $z^* \to \epsilon+\epsilon\ln(z-\epsilon)$ as $z\to\epsilon$, the boundary condition at horizon implies that ξ_ℓ is regular at $z\to\epsilon$. We have to implement this boundary condition to Eq. (27). However, since the iterative equations (26) have been obtained by expanding Eq. (24) in powers of ϵ by regarding $z=\omega r$ as the independent variable, it is implicitly assumed that $\epsilon \ll z$. Consequently, we cannot apply the above expansion at the horizon $z=\epsilon$ where the ingoing-wave boundary condition is to be imposed. To implement the boundary condition correctly, we have to consider a series solution of ξ_ℓ which is valid near the horizon as well as in the range $\epsilon \ll z$ and match it to the series solution of the form (25). This was explicitly done to $\mathcal{O}(\epsilon)$ by Poisson and Sasaki [14].

Let us denote the two independent solutions expanded in terms of z by $\xi_{j\ell}$ and $\xi_{n\ell}$ where $\lim_{\epsilon\to 0}\xi_{j\ell}=j_\ell$ and $\lim_{\epsilon\to 0}\xi_{n\ell}=n_\ell$. To consider the boundary condition at horizon, we change the independent variable to $x := r/2M = z/\epsilon$ and expand ξ_ℓ around $x = 1$. Let us express this expansion as

$$\xi_\ell^{\rm H}(x) = \sum_{n=0}^{\infty} \epsilon^n \xi_\ell^{\{n\}}(x), \tag{28}$$

where all the $\xi_\ell^{\{n\}}$ are regular at $x = 1$. It was found in [14] that $\xi_\ell^{\{0\}}$ behaves as $x^\ell = \epsilon^{-\ell} z^\ell \sim \epsilon^{-\ell} j_\ell$ for $x \to \infty$ ($\leftrightarrow \epsilon \ll z$) while $\epsilon\xi_\ell^{\{1\}}$ contains a term that behaves as $\epsilon x^{-\ell-1} = \epsilon^{\ell+2} z^{-\ell-1} \sim \epsilon^{\ell+2} n_\ell$. This means that we have

$$\xi_\ell(z) = \xi_{j\ell}(z) + c_\ell\, \epsilon^{2\ell+2} \xi_{n\ell}(z)\,, \tag{29}$$

where $c_\ell = \mathcal{O}(1)$ and we have normalized the solution so that $\lim_{\epsilon\to 0} \xi_\ell = j_\ell$. Therefore, in the sense of the PM expansion, we can solve for ξ_ℓ iteratively by imposing the condition that it be less singular than $z^{-\ell-1}$ for $z \to 0$ up to $n = 2\ell+1$. In other words, although we are dealing with perturbations of a BH, we will obtain the same answer for ξ_ℓ even if we replace the central body with any compact object like a neutron star until the $(2\ell+1)$-th PM order inclusive. However, in the PN sense, since we evaluate ξ_ℓ in the near zone, i.e., for $z = \mathcal{O}(v)$, the contribution from $\xi_{n\ell}$ becomes $\mathcal{O}(v^{4\ell+5})$ relative to $\xi_{j\ell}$. Since $\ell \geq 2$, we find that the existence of the horizon affects the homogeneous solution at and beyond $\mathcal{O}(v^{13})$ in the near zone. Here we should correct a mistake in Ref.[9]. There it was erroneously argued that the outgoing gravitational waves are unaffected by the inner boundary condition until we reach $\mathcal{O}(\epsilon^6) = \mathcal{O}(v^{18})$. As shown above, this is true only in the PM sense but not in the PN sense.

Since $j_\ell = \mathcal{O}(z^\ell)$ as $z \to 0$, we have $X_\ell^{in} \to \mathcal{O}(\epsilon^{\ell+1}) e^{-iz^*}$, or $A_\ell^{\rm trans} = \mathcal{O}(\epsilon^{\ell+1})$. On the other hand, from the asymptotic behavior of j_ℓ at $z = \infty$, we find $A_\ell^{\rm inc}$ and $A_\ell^{\rm ref}$ are of order unity. Then using the Wronskian argument, we obtain

$$|A_\ell^{\rm inc}| - |A_\ell^{\rm ref}| = \frac{|A_\ell^{\rm trans}|^2}{|A_\ell^{\rm inc}| + |A_\ell^{\rm ref}|} = \mathcal{O}(\epsilon^{2\ell+2}). \tag{30}$$

Thus $|A_\ell^{\rm inc}| = |A_\ell^{\rm ref}|$ up to $\mathcal{O}(\epsilon^{2\ell+1})$. This fact implies that we can make $A_\ell^{\rm inc}$ and $A_\ell^{\rm ref}$ to be complex conjugate to each other to $\mathcal{O}(\epsilon^{2\ell+1})$. Hence the imaginary part of $X_\ell^{\rm in}$, which reflects the boundary condition at horizon, appears at $\mathcal{O}(\epsilon^{2\ell+2})$ because the Regge-Wheeler equation is real. This is consistent with the argument given in the previous paragraph. Provided we choose the phase of $X_\ell^{\rm in}$ in this way, $\mathrm{Im}\left(\xi_\ell^{(n)}\right)$ for a given $n \leq 2\ell+1$ is completely determined in terms of $\mathrm{Re}\left(\xi_\ell^{(r)}\right)$ for $r \leq n-1$. Explicitly, setting

$$\xi_\ell^{(n)} = f_\ell^{(n)} + i g_\ell^{(n)}\,, \tag{31}$$

we find

$$g_\ell^{(1)} = j_\ell \ln z\,, \quad g_\ell^{(2)} = -\frac{1}{z} j_\ell + f_\ell^{(1)} \ln z\,,$$
$$g_\ell^{(3)} = \left(\frac{1}{3}(\ln z)^3 - \frac{1}{2z^2}\right) j_\ell - \frac{1}{z} f_\ell^{(1)} + \ln z\, f_\ell^{(2)}\,, \quad \cdots. \tag{32}$$

With the above result in mind, we perform the integrals in Eq. (27) iteratively. Since the ℓ-th multipole contribution to the luminosity is smaller by an order $v^{2(\ell-2)}$ relative to the quadrupole, the required PM order of $A^{\rm inc}$ and hence of ξ_ℓ, for the evaluation of the gravitational waves to 5.5PN order is $n = [5 - 2\ell/3]$ where $[x]$ is the maximum integer not exceeding x.

To express the results in a concise way, we introduce some new functions. First we define

$$B_{jj} := \int_0^z z j_0 j_0 dz = -\frac{1}{2} C, \quad B_{nj} := \int_0^z z n_0 j_0 dz = -\frac{1}{2} S, \tag{33}$$

where

$$S = \int_0^{2z} dy \frac{\sin y}{y}, \quad C = \int_0^{2z} dy \frac{\cos y - 1}{y}.$$

As an extension of these integral sinusoidal functions, we introduce the following functions recursively:

$$B_{jJ} := \int_0^z dz z j_0 D_0^J, \quad B_{nJ} := \int_0^z dz z n_0 D_0^J, \tag{34}$$

where J stands for a sequence of j and n, say, $J = jnnj$, and we have also introduced an extension of the spherical Bessel functions by

$$D_\ell^j := j_\ell, \quad D_\ell^n := n_\ell\,, \tag{35}$$

and

$$D_\ell^{nJ} := n_\ell B_{jJ} - j_\ell B_{nJ}\,, \quad D_\ell^{jJ} := j_\ell B_{jJ} + n_\ell B_{nJ}\,. \tag{36}$$

For $n = 1$, we find a formula which is valid for all ℓ,

$$f_{\ell m}^{(1)} = \frac{(\ell-1)(\ell+3)}{2(\ell+1)(2\ell+1)} j_{\ell+1} - \left(\frac{\ell^2-4}{2\ell(2\ell+1)} + \frac{2\ell-1}{\ell(\ell-1)}\right) j_{\ell-1}$$
$$+ R_{\ell,0} j_0 + \sum_{m=1}^{\ell-2} \left(\frac{1}{m} + \frac{1}{m+1}\right) R_{\ell,m} j_m - 2D_\ell^{nj}\,. \tag{37}$$

where $R_{m,k} = z^2(n_m j_k - j_m n_k)$. For $n = 2$, we obtain

$$
\begin{aligned}
f_2^{(2)} &= -\frac{389}{70z^2}j_0 - \frac{113}{420z}j_1 + \frac{1}{7z}j_3 + 4D_2^{nnj} - \frac{5}{3z}D_2^{nj} + \frac{10}{3}D_1^{nj} \\
&\quad + \frac{6}{z}D_0^{nj} + \frac{107}{105}D_{-3}^{nj} - \frac{107}{210}j_2 \ln z - \frac{1}{2}j_2(\ln z)^2 , \\
f_3^{(2)} &= \frac{1}{4z}j_4 + \frac{323}{49z}j_2 - \frac{5065}{294z^2}j_1 - \left(\frac{1031}{588z} + \frac{445}{14z^3}\right) j_0 + \frac{65}{6z^2}n_0 - \frac{65}{6z}n_1 \\
&\quad - \frac{3}{z}D_3^{nj} + \frac{13}{3}D_2^{nj} + \frac{9}{z}D_1^{nj} + \frac{30}{z^2}D_0^{nj} - \frac{13}{21}D_{-4}^{nj} \\
&\quad + 4D_3^{nnj} - \frac{13}{42}j_3 \ln z - \frac{1}{2}j_3(\ln z)^2 , \\
f_4^{(2)} &= \frac{56}{165\,z}j_5 + \left(-\frac{5036}{33\,z^4} + \frac{30334}{1155\,z^2}\right) j_4 + \left(\frac{35252}{33\,z^5} - \frac{30334}{165\,z^3}\right. \\
&\quad \left. + \frac{14401}{3465\,z}\right) j_3 - \left(\frac{5036}{11\,z^5} + \frac{45461}{693\,z^3} + \frac{36287}{9240\,z}\right) j_1 + \left(\frac{140}{z^3} - \frac{5}{18\,z}\right) n_0 \\
&\quad - \frac{49}{6\,z}n_2 - \frac{21}{5\,z}D_4^{nj} + \frac{149}{30}D_3^{nj} - \frac{10}{3\,z}D_2^{nj} + \frac{105}{z^2}D_1^{nj} + \frac{210}{z^3}D_0^{nj} \\
&\quad - \frac{20}{z}D_0^{nj} + \frac{1571}{3465}D_{-5}^{nj} + 4D_4^{nnj} - \frac{1571}{6930}\, j_4 \ln z - \frac{1}{2}\, j_4 (\ln z)^2 . \qquad (38)
\end{aligned}
$$

For $n = 3$, the results are

$$
\begin{aligned}
f_2^{(3)} &= \frac{214\, F_{2,0}[z(\ln z) j_0]}{105} - \frac{107\, D_{-4}^{nj}}{630} - \frac{457\, D_{-2}^{nj}}{70} - \frac{2629\, D_0^{nj}}{630} \\
&\quad + \frac{16949\, D_2^{nj}}{4410} - \frac{107\,(\ln z)\, D_2^{nj}}{105} + (\ln z)^2\, D_2^{nj} - \frac{2\, D_4^{nj}}{49} - 12\, D_{-1}^{nnj} \\
&\quad - 18\, D_1^{nnj} + \frac{2\, D_3^{nnj}}{3} - 8\, D_2^{nnnj} - \frac{197\, j_{-3}}{126} + \frac{2539\, j_{-1}}{3780} \\
&\quad + \left(\frac{107}{70} + \frac{3 \ln z}{2}\right) (\ln z) j_{-1} + \frac{21 j_1}{100} + \left(\frac{349}{140} + \frac{9 \ln z}{4}\right) (\ln z) j_1 \\
&\quad - \frac{457\, j_3}{1050} + \left(\frac{29}{252} - \frac{\ln z}{12}\right) (\ln z) j_3 + \frac{j_5}{504} , \\
f_3^{(3)} &= \frac{26\, F_{3,0}[z(\log z) j_0]}{21} + \frac{13\, D_{-5}^{nj}}{98} + \frac{14075\, D_{-3}^{nj}}{882} - \frac{1424\, D_{-1}^{nj}}{63} \\
&\quad - \frac{2511\, D_1^{nj}}{70} + \frac{269\, D_3^{nj}}{70} - \frac{13\,(\log z)\, D_3^{nj}}{21} + (\log z)^2\, D_3^{nj} - \frac{D_5^{nj}}{18} \\
&\quad + 60\, D_{-2}^{nnj} + 34\, D_0^{nnj} - \frac{710\, D_2^{nnj}}{21} + \frac{6\, D_4^{nnj}}{7} - 8\, D_3^{nnnj} + \frac{75\, j_{-4}}{28} \\
&\quad - \frac{19\, j_{-2}}{56} - \left(\frac{65}{14} + \frac{15 \ln z}{2}\right) (\ln z) j_{-2} - \frac{2789\, j_0}{3780}
\end{aligned}
$$

$$-\left(\frac{221}{84}+\frac{17\ln z}{4}\right)(\ln z)j_0-\frac{7495\,j_2}{5292}+\left(\frac{4867}{1764}+\frac{355\ln z}{84}\right)(\ln z)j_2$$
$$-\frac{10963\,j_4}{32340}+\left(\frac{15}{196}-\frac{3\ln z}{28}\right)(\ln z)j_4+\frac{4\,j_6}{1485}\,, \tag{39}$$

where

$$F_{\ell,0}[X]=n_\ell\int_0^z dz j_0 X - j_\ell\int_0^z dz n_0 X.$$

Taking $z\to\infty$ and recalling the definition of ξ_ℓ, Eq. (23), the amplitudes $A^{\rm inc}$ are calculated as

$$A_2^{\rm inc}=-\frac{1}{2}ie^{-i\epsilon(\ln 2\epsilon+\gamma)}\exp\left[i\left\{\epsilon\frac{5}{3}-\epsilon^2\frac{107}{420}\pi+\epsilon^3\left(\frac{29}{648}-\frac{107}{1260}\pi^2\right.\right.\right.$$
$$\left.\left.\left.+\frac{\zeta(3)}{3}\right)\right\}\right]\times\left[1-\epsilon\frac{\pi}{2}+\epsilon^2\left(\frac{25}{18}+\frac{5}{24}\pi^2+\frac{107}{210}(\gamma+\ln 2)\right)\right.$$
$$\left.+\epsilon^3\left(-\frac{25}{36}\pi-\frac{107}{420}(\gamma+\ln 2)\pi-\frac{\pi^3}{16}\right)+\cdots\right],$$
$$A_3^{\rm inc}=\frac{1}{2}e^{-i\epsilon(\ln 2\epsilon+\gamma)}\exp\left[i\left\{\epsilon\frac{13}{6}-\epsilon^2\frac{13}{84}\pi+\epsilon^3\left(-\frac{29}{810}-\frac{13}{252}\pi^2\right.\right.\right.$$
$$\left.\left.\left.+\frac{\zeta(3)}{3}\right)\right\}\right]\times\left[1-\epsilon\frac{\pi}{2}+\epsilon^2\left(\frac{169}{72}+\frac{5}{24}\pi^2+\frac{13}{42}(\gamma+\ln 2)\right)\right.$$
$$\left.+\epsilon^3\left(-\frac{169}{144}\pi-\frac{13}{84}(\gamma+\ln 2)\pi-\frac{\pi^3}{16}\right)+\cdots\right],$$
$$A_4^{\rm inc}=\frac{1}{2}ie^{-i\epsilon(\ln 2\epsilon+\gamma)}\exp\left[i\left\{\epsilon\frac{149}{60}-i\epsilon^2\frac{1571}{13860}\pi\right\}\right]$$
$$\times\left[1-\epsilon\frac{\pi}{2}+\epsilon^2\left(\frac{22201}{7200}+\frac{5}{24}\pi^2+\frac{1571}{6930}(\gamma+\ln 2)\right)+\cdots\right],$$
$$A_\ell^{\rm inc}=\frac{1}{2}i^{\ell+1}e^{-i\epsilon(\ln 2\epsilon)}\exp\left[i\frac{\epsilon}{2}\left\{\psi(\ell)+\psi(\ell+1)+\frac{(\ell-1)(\ell+3)}{\ell(\ell+1)}\right\}\right]$$
$$\times\left[1-\epsilon\frac{\pi}{2}+\cdots\right], \tag{40}$$

where $\gamma=0.5772\cdots$ is the Euler constant, $\zeta(3)=1.202\cdots$ and

$$\psi(\ell)=\sum_{k=1}^{\ell-1}\frac{1}{k}-\gamma\,.$$

The corresponding amplitudes $B^{\rm inc}$ are readily obtained from Eq. (21).

As for the near zone PN expansion of ξ_ℓ, it is straightforward to obtain all the necessary expressions by simply expanding Eq.(26) around $z = 0$ and equating the terms of the same power in z of both sides. Since it is pointless to show the explicit forms, we omit them here. Once we obtain all the necessary solutions for $X^{\rm in}$, it is straightforward to calculate $R^{\rm in}$ by using the transformation formula (18).

3. 5.5PN Luminosity Formula for Circular Orbit

We consider a circular orbit around a Schwarzschild BH. As mentioned before, we can take the orbit to lie on the equatorial plane without loss of generality. Then the energy and the z-component of the angular momentum of the particle per unit mass are given by

$$E = \frac{r_0 - 2M}{\sqrt{r_0(r_0 - 3M)}}, \quad l_z = \sqrt{\frac{Mr_0^2}{r_0 - 3M}}. \tag{41}$$

The source term $T_{\ell m\omega}$ for the Teukolsky equation is then calculated to be

$$\begin{aligned} T_{\ell m\omega} = \mu\pi\Big(& -2\,{}_0b_{\ell m}(r_0 - 2M)^2\delta(r - r_0) \\ & -2ir_0(r_0 - 2M)\,{}_{-1}b_{\ell m}\Big[(r_0 - 2M)\delta'(r - r_0) - (2 - i\omega r_0)\delta(r - r_0)\Big] \\ & + {}_{-2}b_{\ell m}\Big[r_0^2(r_0 - 2M)^2\delta''(r - r_0) \\ & \quad +2r_0\{i\omega r_0^2(r_0 - 2M) - 3r_0^2 + 8Mr_0 - 4M^2\}\delta'(r - r_0) \\ & \quad +\{4r_0^2 - 8M^2 - \omega^2 r_0^4 - 6i\omega r_0^2(r_0 - 2M)\}\delta(r - r_0)\Big]\Big) \\ & \quad \times\delta(\omega - m\Omega), \end{aligned} \tag{42}$$

where

$$\begin{aligned} {}_0b_{\ell m} &= \frac{1}{2}\left[(\ell-1)\ell(\ell+1)(\ell+2)\right]^{1/2}\,{}_0Y_{\ell m}\left(\frac{\pi}{2},0\right)\frac{Er_0}{r_0 - 2M}, \\ {}_{-1}b_{\ell m} &= \left[(\ell-1)(\ell+2)\right]^{1/2}\,{}_{-1}Y_{\ell m}\left(\frac{\pi}{2},0\right)\frac{l_z}{r_0}, \\ {}_{-2}b_{\ell m} &= {}_{-2}Y_{\ell m}\left(\frac{\pi}{2},0\right)l_z\Omega_\varphi. \end{aligned} \tag{43}$$

Inserting the above into Eq. (8), we find the amplitude $Z_{\ell m}$ defined in Eq. (12) as

$$Z_{\ell m} = \frac{\pi}{i\omega r_0^2 B^{in}_{\ell\omega}}\Bigg\{\Bigg[-{}_0b_{\ell m} - 2i\,{}_{-1}b_{\ell m}\left(1 + \frac{i}{2}\omega r_0^2/(r_0 - 2M)\right)$$

$$+i\,{}_{-2}b_{\ell m}\omega r_0(1-2M/r_0)^{-2}\left(1-M/r_0+\frac{1}{2}i\omega r_0\right)\Big] R^{in}_{\ell m}$$
$$+\left[i\,{}_{-1}b_{\ell m} - {}_{-2}b_{\ell m}\left(1+i\omega r_0^2/(r_0-2M)\right)\right] r_0 R^{in\,\prime}_{\ell\omega}(r_0)$$
$$+\frac{1}{2}\,{}_{-2}b_{\ell m} r_0^2 R^{in\,\prime\prime}_{\ell\omega}(r_0)\Big\}. \tag{44}$$

Finally, inserting the above to Eq. (14), we obtain the luminosity to $\mathcal{O}(v^{11})$ as

$$\left\langle\frac{dE}{dt}\right\rangle = \left(\frac{dE}{dt}\right)_N \Bigg[1-\frac{1247}{336}v^2+4\pi v^3-\frac{44711}{9072}v^4-\frac{8191\pi}{672}v^5$$
$$+\left(\frac{6643739519}{69854400}-\frac{1712\gamma}{105}+\frac{16\pi^2}{3}-\frac{3424\ln 2}{105}-\frac{1712\ln v}{105}\right)v^6$$
$$-\frac{16285\pi}{504}v^7$$
$$+\left(-\frac{323105549467}{3178375200}+\frac{232597\gamma}{4410}-\frac{1369\pi^2}{126}\right.$$
$$\left.+\frac{39931\ln 2}{294}-\frac{47385\ln 3}{1568}+\frac{232597\ln v}{4410}\right)v^8$$
$$+\left(\frac{265978667519\pi}{745113600}-\frac{6848\gamma\pi}{105}-\frac{13696\pi\ln 2}{105}-\frac{6848\pi\ln v}{105}\right)v^9$$
$$+\left(-\frac{2500861660823683}{2831932303200}+\frac{916628467\gamma}{7858620}-\frac{424223\pi^2}{6804}\right.$$
$$\left.-\frac{83217611\ln 2}{1122660}+\frac{47385\ln 3}{196}+\frac{916628467\ln v}{7858620}\right)v^{10}$$
$$+\left(\frac{8399309750401\pi}{101708006400}+\frac{177293\gamma\pi}{1176}\right.$$
$$\left.+\frac{8521283\pi\ln 2}{17640}-\frac{142155\pi\ln 3}{784}+\frac{177293\pi\ln v}{1176}\right)v^{11}\Bigg], \tag{45}$$

where $v=(M/r_0)^{1/2}$ and $(dE/dt)_N$ is the Newtonian quadrupole luminosity given by

$$\left(\frac{dE}{dt}\right)_{\rm N} = \frac{32\mu^2M^3}{5r_0^5} = \frac{32}{5}\left(\frac{\mu}{M}\right)^2 v^{10}. \tag{46}$$

It should be noted that since the emitted gravitational waves are dominated by the quadrupole, an observationally relevant quantity is the quadrupole frequency $\omega_2 = 2\Omega$. Hence the orbital angular frequency Ω is observable.

In the present case, the PN expansion parameter v is directly related to Ω as $v = (M\Omega)^{1/3}$. We find our result is in complete agreement with the standard PN result up to $\mathcal{O}(x^5)$ [3] in the limit $\mu/M \ll 1$.

4. Convergence Property of PN expansion

Using the formula (45) and the 'exact' numerical result, it is possible to estimate the effect of PN approximations to the orbital evolution of inspiralling compact binaries [11]. We ignore the finite mass effect and interpret M as the total mass and μ as the reduced mass of the system. In Fig. 1,

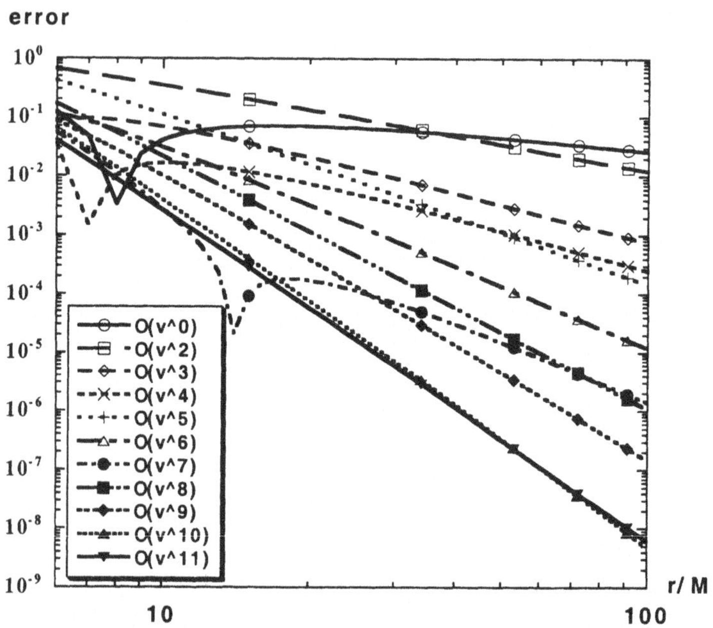

Figure 1. The error of the PN formulas as functions of the Schwarzschild radial coordinate r for $6 \le r/M \le 100$.

we show the error in the PN formula as a function of the orbital radius r. The error is defined by

$$\text{error} = \left| 1 - \left(\frac{dE}{dt}\right)_n \Big/ \left(\frac{dE}{dt}\right) \right|, \tag{47}$$

where $(dE/dt)_n$ and (dE/dt) denote the $(n/2)$PN formula and the numerical result, respectively. We see that the Newtonian and the 2PN formulas

TABLE 1. The relative error of cycle $\Delta N^{(n)} = |N^{(n)} - N^{(n-1)}|$ for typical inspiralling compact binaries. The first row shows the total cycle calculated by the Newtonian quadrupole formula.

n	$(1.4M_\odot,1.4M_\odot)$	$(10M_\odot,10M_\odot)$	$(1.4M_\odot,10M_\odot)$	$(1.4M_\odot,70M_\odot)$
0	16000	600	3578	898
2	356	54	216	212
3	228	60	208	296
4	11	5	15	31
5	12	7	20	53
6	11	8	22	75
7	1.2	1.0	2.6	10
8	0.12	0.14	0.3	2.2
9	0.82	0.80	1.9	8.9
10	0.09	0.08	0.20	0.87
11	0.03	0.03	0.07	0.40

become accidentally accurate within the radius $r \lesssim 10M$ and the convergence of the PN expansion is very slow in that region. On the other hand, for large radius $r \gtrsim 50M$, the convergence is almost uniform, indicating the validity of the PN expansion.

It has been suggested that whether the error in the total cycle of gravitational waves is less than unity or not gives a useful guideline to examine the accuracy of the PN formulas as templates [15, 16]. Here we calculate the total cycle $N^{(n)}$ of gravitational waves from an inspiralling binary by using the $(n/2)$PN luminosity formula, $(dE/dt)_n$, and the orbital energy formula $(dE/dv)_n$ which is truncated at $\mathcal{O}(v^n)$:

$$N^{(n)} = \int_{v_f}^{v_i} dv \, \frac{\Omega}{\pi} \frac{(dE/dv)_n}{|(dE/dt)_n|}, \tag{48}$$

where $v_i = (M/r_i)^{1/2}$, $v_f = (M/r_f)^{1/2}$, and r_i and r_f are the initial and final orbital separations of the binary. We define the relative difference of cycle $\Delta N^{(n)}$ as $\Delta N^{(n)} \equiv |N^{(n)} - N^{(n-1)}|$. We adopt $r_f = 6M$ and r_i is the radius at which the frequency of wave is 10Hz, i.e., $r_i/M \sim 347(M_\odot/M)^{2/3}$. The results for typical binary systems are given in Table 1. This table suggests that the $3 \sim 4$PN formula will give sufficiently accurate waveforms for NS-NS or BH-BH binaries whose total mass are less than $20M_\odot$. Although the required PN order has not been achieved yet in the standard PN analysis, our result indicates that the PN expansion is applicable to the inspiral phase of these compact binaries, which is a good news. On the other hand, the convergence for the case of BH-NS binaries with a large BH mass ($M \gtrsim 10M_\odot$) is very slow. This is because r_i/M become smaller for a larger mass BH and relativistic corrections become more important. However, for this case, we have $\mu \ll M$ and the BH perturbation approach should be reliable enough that we do not have to perform the PN expansion.

I would like to thank Bala Iyer and Biplab Bhawal for inviting me to make a contribution to this volume. It is my great honor. This contribution is based on work in collaboration with Yasushi Mino, Masaru Shibata, Hideyuki Tagoshi and Takahiro Tanaka [17].

References

1. See, e.g., K.S. Thorne, in *Relativistic Cosmology: Proceedings of the 8th Nishinomiya-Yukawa Memorial Symposium*, ed. M. Sasaki (University Academy Press, Tokyo, 1994), p.67.
2. See, e.g., C.M. Will, in *Relativistic Cosmology: Proceedings of the 8th Nishinomiya-Yukawa Memorial Symposium*, ed. M. Sasaki (University Academy Press, Tokyo, 1994), p.83.
3. L. Blanchet, T. Damour, B.R. Iyer, C.M. Will and A.G. Wiseman, *Phys. Rev. Lett.*, **74**, L3515 (1995); L. Blanchet, Phys. Rev. **D54**, 1417 (1996).
4. T. Regge and J.A. Wheeler, *Phys. Rev.*, **108**, 1063 (1957).
5. C.V. Vishveshwara, *J. Math. Phys.*, **9**, 1319 (1968); *Phys. Rev.*, **D1**, 2870 (1970); Nature **227**, 936 (1970).
6. S. A. Teukolsky, *Astrophys. J.*, **185**, 635 (1973).
7. S. Chandrasekhar, Proc. R. Soc. London A **343**, 289 (1975); *The Mathematical Theory of Black Holes*, (Oxford University Press, Oxford, 1983).
8. E. Poisson, *Phys. Rev.*, **D47**, 1497 (1993).
9. M. Sasaki, *Prog. Theor. Phys.*, **92**, 17 (1994).
10. H. Tagoshi and M.Sasaki, *Prog. Theor. Phys.*, **92**, 745 (1994).
11. T. Tanaka, H. Tagoshi, and M. Sasaki, *Prog. Theor. Phys.*, **96**, 1087 (1996).
12. H. Tagoshi, M. Shibata, T. Tanaka, and M. Sasaki, *Phy. Rev.*, **D54**, 1439 (1996).
13. S. Mano, H. Suzuki and E. Takasugi, *Prog. Theor. Phys.*, **95**, 1079 (1996); *Prog. Theor. Phys.*, **96**, 549 (1996); S. Mano and E. Takasugi, *Prog. Theor. Phys.*, **97**, 213 (1997).
14. E.Poisson and M.Sasaki, *Phys. Rev.*, **D51**, 5753 (1995).
15. C. Cutler et al., *Phys. Rev. Lett.*, **70**, 2984 (1993).
16. E. Poisson, *Phys. Rev.*, **D52**, 5719 (1995).
17. Y. Mino, M. Sasaki, M. Shibata, H. Tagoshi and T. Tanaka, in *Perturbative and Numerical Approaches to Gravitational Radiation*, ed. T. Nakamura, *Prog. Theor. Phys. Suppl.*, No. 128 (1997), in press.

21. MORE QUASI THAN NORMAL!

Can Oscillations of Relativistic Stars and Black Holes be Observable with Future Gravitational - Wave Detectors

NILS ANDERSSON
Institute für Theoretische Astrophysik, Universität Tübingen, Auf der Morgenstelle 10, D-72076 Tübingen, Germany.

It is a great pleasure for me to contribute to this Festschrift in Vishu's honour. I met him in 1987 as I started my postgraduate studies in Uppsala, Sweden. Or more precisely, that was when I first encountered his name. As I took my first stumbling steps along the road of research I came upon a paper in Nature, where Vishu described his discovery of what we now call the quasi-normal modes of a black hole. I still remember digesting every single word of that paper, hoping that somehow I could unlock its secrets by staring at the printed words. Sadly, this approach failed miserably.

I did not actually meet Vishu until several years later when I travelled to India for the Silver Jubilee meeting of the IAGRG. During this trip I met not only Vishu the scientist, I was also introduced to his private persona. My impression of this as well as our later encounters, can be summed up in a simple way: Vishu is a very funny guy! As his colleagues, or students, we are fortunate. Not only does Vishu unveil scientific mysteries to us, he also makes us laugh.

1. Asteroseismology and gravitational waves

The periodically changing luminosity of distant Cepheids, RR Lyrae and Mira variables has long been interpreted as evidence of stellar oscillation. As a star expands and contracts its luminosity increases and decreases, almost as if it was breathing. It is believed that most stars oscillate both radially – as in the above cases – and nonradially. To detect nonradial pulsations is difficult, but the evidence from observations of O, B, β Cephei, 53 Persei, ζ Ophiuchi, Ap Stars, white dwarfs etcetera is strong. The most impressive example of a nonradially oscillating star is – perhaps not surprisingly – the Sun. Many thousands of the Sun's nonradial pulsation modes have by now been detected. Naturally, these observations have prompted a wealth of theoretical studies of the pulsation properties of stars. One would like to

B. R. Iyer and B. Bhawal (eds.), Black Holes, Gravitational Radiation and the Universe, 335–350.

be able to use the detected pulsation modes to probe the interior properties of stars. This approach – known as Asteroseismology, or Helioseismology in the specific case of the Sun – has been quite successful.

With this in mind it is natural to wonder whether the new generation of gravitational-wave detectors (the multi-kilometer interferometers LIGO, VIRGO and GEO and the cryogenic bar-detectors – maybe spherical ones) that should come on-line in the near future, will provide further observations of stellar pulsation. If so, the strongest sources ought to be the most compact objects in the universe: neutron stars and black holes. Observations of pulsating neutron stars and/or black holes through the emitted gravitational waves would be very good news, since these objects are difficult to observe directly in any other way. And the truth is that we still know precious little about them. Despite strong evidence for their existence we have no direct observations of black holes. Similarly, present observations of neutron stars, eg. as radio pulsars or X-ray sources, rule out very few of the proposed equations of state for supranuclear matter.

In this chapter I will discuss the oscillations of neutron stars and black holes in the framework of general relativity. The main focus will be on the pulsation modes I expect to be the most relevant as far as gravitational-wave detection is concerned. That is, I will not mention radial oscillations or modes associated with the electromagnetic field of a neutron star etcetera. For details on such topics the reader will have to look elsewhere. In fact, I recommend the reader to look elsewhere – a good starting point would be the cited papers – also for details about what I do discuss. The present decription is not at all exhaustive.

2. Fluid sloshing about

A detailed study of the pulsation modes of a realistic neutron star model – with some preferred equation of state, a strong magnetic field, rapid rotation, and a partly superfluid interior – would be hopelessly complicated. In fact, we know far too little about most of these features to even be able to create such a stellar model. Moreover, even if we could model a rapidly rotating star with all the desirable features appropriately, the pulsation problem would still be too hard. Basically, one would expect each dynamical feature of the star, such as the fluid motion, wobbling magnetic fields, temperature variations etcetera, to be associated with more or less unique sets of pulsation modes. To entangle the complicated web of, most likely coupled, pulsation modes in a reasonably realistic stellar model is an awesome task.

If we want to study the properties of a pulsating star, and understand the origin of the pulsation modes, a more profitable approach would be to strip off all features we expect to be irrelevant for a particular mode from the star. Later on in the game – once the details of the simple model are understood – these features can be put back in and we will find out if our simplifying assumptions were justified or not.

In this spirit I begin this survey with the simplest possible model of a neutron star: A non-rotating ball of zero temperature, uniform density, fluid [1]. While this model may not be terribly realistic, it actually does yield relevant results. One should remember that the proposed neutron star equations of state are "almost constant density" (in the sense that the density changes only slightly in the central parts of the star and then falls off dramatically close to the surface). Furthermore, older neutron stars are expected to be quite cold. After some 10^5 years the temperature in the star should have fallen from the initial 10^{10} K to 10^5 K or so. The pulsation properties of the constant density star are also fairly simple: In Newtonian theory there exists only one mode for each multipole l [I assume that the angular behaviour of the mode-functions is expressed in terms of spherical harmonics $Y_{lm}(\theta, \varphi)$]. This is the so-called f-mode ("f" for fundamental) and it was first calculated to be (see [2])

$$\omega_f^2 = \frac{2l(l-1)}{2l+1} \left(\frac{M}{R^3} \right) , \tag{1}$$

by Lord Kelvin more than a hundred years ago. Here M and R are the mass and radius of the star, respectively. I have given the mode-frequency that follows if one assumes a time-dependence $e^{-i\omega t}$, and have used geometrized units ($c = G = 1$). In more human units, this implies that the quadrupole f-mode of a generic neutron star, with $M = 1.4 M_\odot$ and $R = 10$ km, will have a frequency $f \approx 2$ kHz. This should be compared to calculations for realistic equations of state that yield f-mode frequencies in the range 1.5-4 kHz [3, 4].

As a suitable refinement we can consider compressible stellar models, say polytropes $p \sim \rho^\Gamma$. For such models the f-mode is still present, but there will also be an infinite (for each l) set of modes associated with acoustic waves in the fluid. These modes are called p-modes, because pressure is the main restoring force acting on them. For compressible, homogeneous spheres one can show that [2]

$$\omega_p^2 = [\Delta_m + \sqrt{\Delta_m^2 + l(l+1)}] \left(\frac{M}{R^3} \right) , \tag{2}$$

where

$$2\Delta_m = [2l + 3 + m(2m + 2l + 5)]\Gamma - 4 , \quad m = 0, 1, 2, ... \tag{3}$$

and Γ is the polytropic index. This means that for an $\Gamma = 2$ polytrope the first quadrupole p-mode for a generic neutron star would have a frequency of 7 kHz. Again, this number is reasonable: Calculations for a set of proposed equations of state lead to p-mode frequencies in the range 5-9 kHz [4, 5, 6].

Several dissipation mechanisms will affect each pulsation mode. For example, the fluid motion will be damped by viscous heating. A mode that leads to large displacements close to the surface may shake the magnetic field, which would generate electromagnetic radiation. As the fluid sloshes about it will also generate gravitational waves. These do, of course, not come into play in Newtonian theory but must be accounted for in a relativistic investigation. In fact, gravitational-wave emission is expected to be the most efficient damping mechanism for both f- and p-modes. Although important it is a tiny effect. The frequency of each Newtonian mode is hardly altered at all. The eigenfrequencies simply pick up small imaginary parts to account for the damping. The fact that the frequencies become complex - valued led to the relativistic modes being referred to as "quasi-normal" modes, rather than "normal" modes (as in, for example, the case of a vibrating string). To make a distinction is useful for many reasons, a mathematical one being that the relativistic modes are not generally expected to form a complete set that can be used to expand any given function. On the other hand, the modes cannot be considered completely "quasi". It seems natural that the pulsations of a star are damped with time. Anyway, fully fledged calculations for several equations of state confirm that the fluid pulsation modes are slowly damped; the f-mode typically lives a few tens of a second while the first p-mode can last several seconds [3, 4] .

We have now arrived at the picture of relativistic stellar pulsation theory that followed with the seminal paper of Thorne and Campolattaro in 1967 [7]. This picture would remain virtually unchanged until 1986, but before describing the way that our understanding of relativistic stellar pulsation has changed in recent years I will take a detour via black holes.

3. Wobbling black holes

In the late 1960s it was becoming clear that a black hole is a realistic end product of stellar evolution. Hence, there was a surge of interest in the astrophysical manifestations of such completely collapsed objects. This is where Vishu first enters this story. As he himself describes it [8]: "Misner proposed the following problem for my PhD thesis. Take two ... black holes. Revolving around each other, they come closer as energy is radiated away in the form of gravitational waves. They coalesce ... still rotating and radiating. Study the whole process, computing all the characteristics of the emitted gravitational radiation. Fine, I said, thy will be done! At that time

I did not realize the magnitude of this problem. Had I pursued it, I might have entered the Guinness book of records as the oldest graduate student alive..."

Indeed, considering that the inspiral/coalescence problem is still unsolved thirty years later, and that the effort to crack it is now spearheaded by a so-called Grand Challenge collaboration (comprising most of the American numerical relativity groups with access to the most powerful computers of today) Vishu is probably right. Had the solution of the general binary black-hole problem been a requirement for him to obtain a Ph.D, he would not yet have graduated. So Vishu ended up focussing on a somewhat more modest problem: The stability of the Schwarzschild spacetime. This was still a problem of some importance. If one could establish that a black hole was dynamically unstable, i.e. fell apart if "an ant sneezed in its vicinity", it could not be considered an astrophysically relevant object.

At this time the dynamical properties of black holes were not at all well understood. Regge and Wheeler had broken the ground by deriving the equations governing small perturbations of a nonrotating black hole in 1957 [9], but their published equations contained some errors. As a first step Vishu, together with Edelstein, rederived and corrected these equations [10]. Having done this, Vishu proceeded with the stability analysis. A black-hole perturbation can be represented by a function with time-dependence $e^{-i\omega t}$, and the stability analysis essentially comprises of proving that no acceptable solutions are associated with unstable frequencies (such that $\mathrm{Im}\,\omega > 0$). After some struggle – the fact that it had not yet been shown that the equations for the so-called even parity perturbations could be written as a single wave-type equation (this was later shown by Zerilli [11]) complicated the analysis considerably – Vishu managed to show [12] that the black-hole equations permitted no such growing solutions, and his thesis was "in the bag".

Inspired by a debate at his thesis defence, Vishu turned to another interesting problem. As he describes it [8]: The question was "how do you observe a solitary black hole? To me the answer seemed obvious. It had to be through scattering of radiation, provided the black hole left its fingerprint on the scattered wave. ... So, I started pelting the black hole with Gaussian wave packets. If the wave packet was spatially wide, the scattered one was affected very little. It was like a big wave washing over a small pebble. But when the Gaussian became sharper, maxima and minima started emerging, finally levelling off to a set pattern when the width of the Gaussian became comparable to or less than the size of the black hole. The final outcome was a very characteristic decaying mode, to be christened later as the quasi-normal mode. The whole experiment was extraordinarily exciting."

It is easy to appreciate Vishu's excitement. He had discovered a phe-

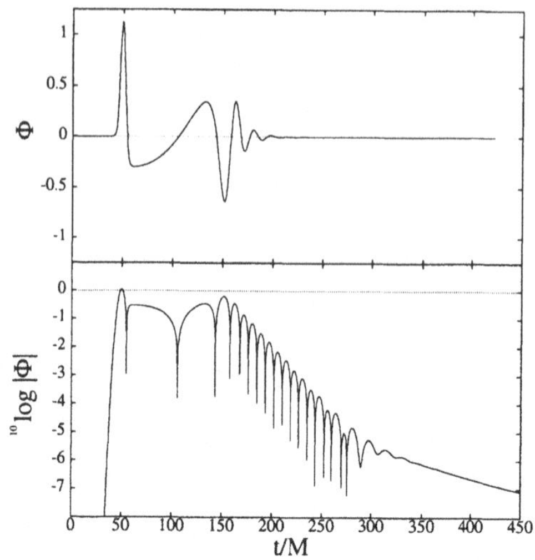

Figure 1. A recreation of Vishu's scattering experiment: The response of a Schwarzschild black hole as a Gaussian wavepacket of scalar waves impinges upon it. The first bump (at $t = 50M$) is the initial Gaussian passing by the observer on its way towards the black hole. Quasi-normal-mode ringing clearly dominates the signal after $t \approx 150M$. At very late times (after $t \approx 300M$) the signal is dominated by a power-law fall-off with time.

nomenon that may eventually enable a direct identification of a black hole [13]. Subsequent work has shown that each black hole has its own characteristic modes, the frequencies of which depend only on the mass, the angular momentum and the electric charge of the hole [14]. Of course, none expects a macroscopic black hole to have a considerable net charge. The mass and the rotation rate should be far more important, and the quasi-normal modes associated with the Kerr solution [15] are likely to be the astrophysically relevant ones.

The fact that the quasi-normal modes depend only on the black hole itself, not on the perturbing agent, is worth emphasizing. It is also notable, that the quasi-normal modes dominate the radiation from most dynamical processes that involve a black hole. Numerical simulations ranging from the formation of a black hole in a gravitational collapse [16] to the collision of two equal mass black holes [17], provide clear cut evidence: No matter how you kick a black hole, its response will be dominated by the quasi-normal modes. Only at very late times will the modes fail to describe the resultant radiation. As can be seen in Figure 1, the late-time behaviour is governed by a power-law tail (see [18] for further discussion).

The black-hole perturbation problem is essentially (at least for nonrotating holes) one of scattering off a single potential barrier [14]. The perturbation equations can be written as wave equations with slightly different effective potentials. Typical examples are: the Regge-Wheeler equation for axial (odd parity) perturbations of a Schwarzschild black hole [9] and the Zerilli equation for the polar (even parity) case [11]. In both cases, the effective potential represents a potential barrier that peaks close to $r = 3M$ (the unstable photon orbit).

It was the analogy with quantum mechanics that inspired Vishu's scattering experiment. We can also use it to infer the mode-frequencies. It is commonly accepted that scattering resonances (the quantum analogues of quasi-normal modes) arise for energies close to the top of a potential barrier. In the black-hole case, this immediately leads to the approximation [19]

$$\text{Re } \omega_0 \approx \frac{1}{3\sqrt{3}M}\left(l + \frac{1}{2}\right) . \tag{4}$$

This is an accurate approximation of the fundamental Schwarzschild quasi-normal mode for large l. It is less accurate for the first few l (the true answer for $l = 2$ is Re $\omega_0 M \approx 0.3736$, so the error of the approximation is something like 30 percent in that case) but it is still a useful estimate. For the imaginary part of the frequency – in quantum language: the lifetime of the resonance – the curvature of the potential at the peak contains the relevant information. Schutz and Will [19] used the WKB approximation to infer that

$$\text{Im } \omega_0 \approx -\frac{\sqrt{3}}{18M} , \tag{5}$$

which is accurate to within 10 percent for the fundamental mode.

Let us translate the results for the fundamental gravitational-wave quasi-normal mode of a nonrotating black hole into more human units. We then get a frequency

$$f \approx 12\text{kHz}\left(\frac{M_\odot}{M}\right) , \tag{6}$$

while the associated e-folding time is

$$\tau \approx 0.05\text{ms}\left(\frac{M}{M_\odot}\right) . \tag{7}$$

The quasi-normal modes of a black hole are clearly very shortlived. In fact, we can compare a black hole to other resonant systems in nature by defining a quality factor

$$Q \approx \frac{1}{2}\left|\frac{\text{Re } \omega_n}{\text{Im } \omega_n}\right| . \tag{8}$$

Our quasi-normal-mode approximations then lead to $Q \approx \ell$. This should be compared to the result for the f-mode of a neutron star: $Q \sim 1000$, or

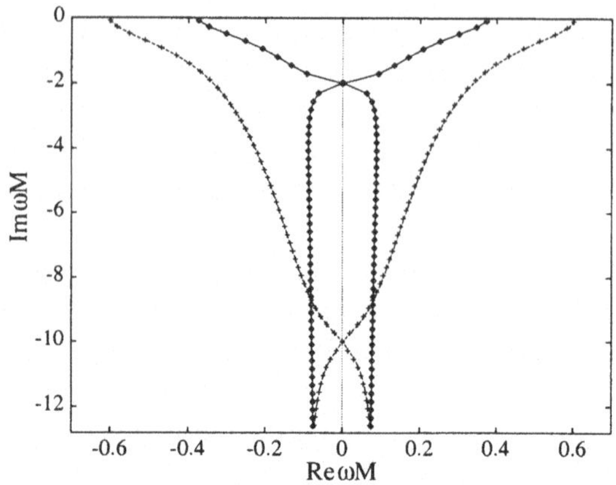

Figure 2. The spectrum of quasi-normal modes for a Schwarzschild black hole. The first fifty modes for $\ell = 2$ (diamonds) and 3 (crosses) are shown.

the typical value for an atom: $Q \sim 10^6$. The Schwarzschild black hole is clearly a very poor oscillator in comparison.

This gives us some kind of idea about the fundamental mode of the black hole, that tends to dominate the emerging radiation. But there are other ones, i.e. some kind of overtones. In fact, there exist an infinite number of quasi-normal modes for each value of l. They have ever increasing imaginary parts and thus become shorter lived as one proceeds along the spectrum [15, 20, 21, 22]. As Nollert [23] has shown, the asymptotic behaviour is (for gravitational perturbations)

$$\omega_n M \approx 0.0437 - \frac{i}{4}\left(n + \frac{1}{2}\right) + \mathcal{O}\left(\frac{1}{\sqrt{n+1}}\right) , \qquad (9)$$

when the mode-index $n(\geq 0)$ is very large. The result is independent of the value of ℓ to leading order, cf. Figure 2.

Although it is hard to imagine a physically realistic situation where the net electric charge of a black hole is significant, investigations of the Reissner-Nordström solution do not lack interest. On the contrary, since it provides a more general framework than the Schwarzschild geometry, this case contributes to our understanding of more complicated spacetimes. One interesting feature of charged black holes is that a variation in the electromagnetic field will inevitably induce gravitational radiation and vice versa. A quasi-normal mode of a Reissner-Nordström black hole therefore generally corresponds to emission of both gravitational and electromagnetic radiation [14].

A first approximation of the effect of an increasing electric charge on the quasi-normal modes can be obtained through the WKB approximation

[24]. One finds that

$$\text{Re}\,\omega_0 \approx \left(l+\frac{1}{2}\right)\left[\frac{M}{r_0^3}-\frac{Q^2}{r_0^4}\right]^{1/2}, \tag{10}$$

$$\text{Im}\,\omega_0 \approx -\frac{1}{2}\left[\frac{M}{r_0^3}-\frac{Q^2}{r_0^4}\right]^{1/2}\left[\frac{3M}{r_0}-\frac{4Q^2}{r_0^2}\right]^{1/2}, \tag{11}$$

in the limit $l \gg 1$. Here $Q \leq M$ is the charge of the black hole, and we have defined r_0 as the position where the black-hole potential attains its maximum value ($2r_0 = 3M+\sqrt{9M^2-8Q^2}$). These approximations indicate the general behaviour of mode as the charge of the black hole increases. The oscillation frequency increases, and the damping rate reaches a maximum value after which it decreases rapidly. These features have been verified by more accurate numerical work [25, 26].

A peculiar recent result is worth mentioning in this context. As the charge of the black hole increases towards the extreme value $Q = M$ the quasi-normal-mode frequencies trace out spirals in the complex frequency plane [27]. These spirals are more pronounced for the higher overtones of the black hole. The reason for the peculiar behaviour of the modes of a near extreme black hole is not well understood at present.

An extremely charged black hole is, of course, a much rarer sight in the universe than a rapidly rotating one. Conservation of angular momentum during the gravitational collapse indicates that most newly born black holes spin very fast. Thus the modes of such black holes are of great astrophysical interest. Before discussing what happens to the modes when we spin up the black hole, we recall that in the non-rotating case the modes occur in complex-frequency pairs ω_n and $-\bar{\omega}_n$ (the bar denotes complex conjugation). This is apparent in Figure 2. As the black hole spins up, each Schwarzschild mode splits into a multiplet of $2l+1$ distinct ones (in analogy with the Zeeman splitting in quantum mechanics). These modes are associated with the various values of m, where $-l \leq m \leq l$, which determine the modal dependence on the azimuthal angle through $e^{im\varphi}$ [28]. As is straightforward to deduce, modes for which Re ω_n and m have the same sign are co-rotating with the black hole. Similarly, modes such that Re ω_n and m have opposite signs are counter-rotating. The effect that rotation has on the mode-frequencies can, to some extent, be deduced from this fact.

Let us first consider the counter-rotating modes: These will appear to be slowed down by inertial frame dragging close to the black hole. Hence, their oscillation frequencies will tend to decrease as $a \to M$. At the same time, numerical calculations show that the damping rate stays almost constant. For the co-rotating modes, the effect is the opposite. Frame-dragging tends to increase the frequencies. Additionally, the modes become much longer lived, cf. [28]. The available numerical results for counter-rotating modes are well approximated by [29]

$$\text{Re } \omega_0 \approx \frac{1}{M}\left[1 - \frac{63}{100}(1 - a/M)^{3/10}\right] , \tag{12}$$

and

$$\text{Im } \omega_0 = \frac{(1 - a/M)^{9/10}}{4M}\left[1 - \frac{63}{100}(1 - a/M)^{3/10}\right] . \tag{13}$$

From this we can see that the mode becomes undamped in the limit $a = M$. One can actually show (since the case $a = M$ is amenable to analytic methods [30, 31]) that there exists an infinite sequence of real resonant frequencies with a common limiting point in that case [28]. The limiting frequency is the same as the upper limit for so called super-radiance, e.g. $\omega = m/2M$. That the modes become undamped can be understood from the fact that the angular frequency of an extreme Kerr black hole is $1/2M$. As $a \to M$ the long lived quasi-normal modes rotate uniformly with the black hole.

That the modes of a black hole rotating close to the extreme limit can be very long-lived would – greatly improve the chances for detection with future gravitational-wave detectors – provided that these modes will actually be excited in a realistic process (which is not at all clear at present)

4. Trapped gravitational waves

Let us now return to the problem of relativistic stellar pulsation. Does the existence of black-hole quasi-normal modes tell us that our previous description was somewhat incomplete? Indeed it does! Apart from providing a possible way of taking the fingerprints of the black holes that may be scattered around the universe, the black-hole quasi-normal modes indicate that the dynamical properties of spacetime itself may be relevant. In our discussion of the stellar problem we did not allow the spacetime any dynamical features of its own. We assumed that the inclusion of relativistic effects would only add damping to each stellar pulsation mode. We did not query whether there would be modes associated with the spacetime itself.

That spacetime might play a more active role was first recognized by Kokkotas and Schutz in 1986 [32]. To assess the problem they devised a simple toy model of two strings coupled by means of a spring. A finite string represented the stellar fluid, a semi-infinite string represented the dynamical spacetime, and the spring coupled the two systems together. By solving the resultant equations for solutions corresponding to purely outgoing waves on the semi-infinite string – basically an undergraduate problem in mathematical physics – they showed that two distinct classes of modes exist. First, the normal modes of the "fluid string" become slowly damped because of the coupling to the semi-infinite string. Second, a new class of – very short lived – modes can be associated with the "spacetime string". This is clear evidence that the relativistic perturbation equations ought to allow not only the fluid modes, but also rapidly damped modes (in some ways similar to the quasi-normal modes of black holes). Indeed, such modes were later calculated [5, 33, 34] and named w-modes because of their close association with gravitational waves. For a typical neutron star the w-modes have frequencies in the range 8-12 kHz and damp out in something like 0.02-0.1 ms [4].

Another important development followed a few years later when Chandrasekhar and Ferrari reformulated the stellar perturbation problem [35]. They showed that the perturbation equations can be written in such a way that they do not explicitly contain any fluid functions (such as the variation in the density or the fluid displacement). This reformulation was motivated by a desire to consider the problem as one of gravitational-wave scattering, in complete analogy with the black-hole problem that Chandrasekhar had long been interested in [14].

In one of their papers on relativistic stars, Chandrasekhar and Ferrari [36] revived the interest in the axial perturbations of stars. The equations governing axial perturbations of a relativistic star can be cast in the form of a single wave-equation (identical to the Regge-Wheeler equation in the exterior vacuum). This was first shown by Thorne and Campolattaro in 1967 [7], but the axial equation was dismissed in all subsequent mode-investigations because it could not lead to oscillations in the stellar fluid. Typically, an axial perturbation induces rotation in the fluid and, unless the star is already rotating or the fluid has a non-zero shear modulus, there will be no oscillations. Furthermore, all the Newtonian fluid modes for a nonrotating star belong to the polar class of perturbations.

The seemingly obvious idea that the axial wave equation could lead to gravitational-wave modes was overlooked until the work of Chandrasekhar and Ferrari [36]. They realized that, if the star was made so compact that the surface was inside $r = 3M$, the peak of the black-hole potential barrier would be unveiled and a potential well would form inside the star. Clearly, there ought to be resonant states associated with this potential well, and they calculated a few such rather long-lived modes for different ultracompact stellar models.

It is relevant – especially since this is written as a tribute to Vishu – to note that the idea of modes trapped inside ultracompact stars was not new. It had already been proposed by Vishu and his colleagues in the context of neutrinos and stars. The neutrino problem is remarkably similar to that for axial gravitational waves. One of the potentials for neutrinos has a potential well inside $r = 3M$ and, together with Kembhavi, Vishu had calculated the complex-frequency neutrino modes in 1980 [37]. Having done this, Vishu and his colleagues investigated whether neutron stars described by "realistic" equations of state could ever be ultracompact [38, 39]. Although they found no such objects for the then available high-density equations of state, they managed to show – using the so-called core-envelope model – that stable ultracompact objects with causal cores can exist. The conclusion may be that the features of ultracompact stars are of little astrophysical relevance, but it is nevertheless clear that Vishu was far ahead of the rest of us also in this case.

Already from the first investigation it was clear that the dynamical spacetime can play an interesting role for compact stars. In fact, we now know that the spectrum of gravitational-wave modes of a star is remarkably rich. Recall that a uniform density star has only one Newtonian mode for each l (the f-mode). As calculations by myself, Kojima and Kokkotas show [40], there are two infinite sets of gravitational-wave mode for the same star. The first one belong to the polar perturbations and correspond immediately to the modes predicted by Kokkotas and Schutz [32]. The second set is associated with axial perturbations. One can show that these axial gravitational-wave modes become increasingly long-lived as the star is made more compact, and when $r < 3M$ they coincide with the modes found by Chandrasekhar and Ferrari [36].

Our present understanding is that axial and polar perturbations lead to similar sets of w-modes [40], cf. Figure 3. This is perhaps not very surprising since the fluid plays an irrelevant role for these modes. And one can recall that the axial and polar quasi-normal modes of nonrotating black holes are identical! So an explanation of the physics behind the w-modes cannot rely on the fluid motion in any way. Rather, these modes must be understood in terms of the interaction between gravitational waves and the bowl of curvature provided by the stellar matter. A naive argument that we should have some kind of standing wave solutions inside the star, perhaps of form $\sin(\omega r)$, leads to frequencies [40]

$$\omega_n R \sim n\pi \ , \quad n = 1, 2, 3... \tag{14}$$

Although not exact in any way, this result agrees quite well with detailed calculations. It shows that there exists an infinite number of w-modes, whose frequencies depend on the size of the star.

5. Prepare the wobblescopes

In the introduction, I stated that a strong motivation for studying the detailed pulsations of relativistic stars and black holes was the hope that

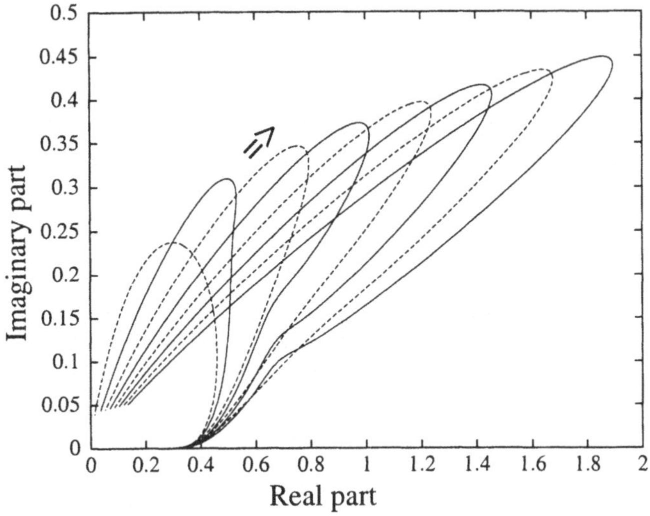

Figure 3. Gravitational-wave modes for a sequence of increasingly compact uniform density stars. The figure shows Im ωM vs Re ωM for the polar (solid) and axial (dashed) w-modes. Only the first four axial and polar modes are shown, but it seems likely that an infinite number of such modes exist. Note the remarkable qualitative agreement between the two sets of modes. The arrow indicates the direction of increasing stellar compactness.

the associated gravitational waves can be detected by future detectors, such as LIGO, VIRGO or GEO. Now that we have some understanding of the character of the various pulsation modes that may be relevant it is appropriate to address the issue of detection [41]. Should we really hope to detect these modes? The answer to that question is unclear, and may not be known until the modes are actually observed (or not as the case may be). What is clear is that we need more accurate modelling of, for example, the gravitational collapse that follows a supernova. At present we do not know how much energy will be radiated through the oscillation modes of a nascent neutron star or black hole. But the released energy could be considerable. One generally expects a newly formed neutron star (or black hole) to pulsate wildly during the first few seconds following the collapse. This pulsation will be damped mainly through gravitational waves, that carry the signature of the collapsed object. Since the energy stored in the pulsation can potentially be of the same order as the kinetic energy of the collapse the amplitude of the emerging gravitational waves can be considerable. In fact, a significant part of the mass energy of the newly formed object can be radiated away in this way. But even so, it is not clear that these gravitational waves are detectable on Earth.

However, it is straightforward to arrive at simple order of magnitude estimates for the gravitational waves from pulsating compact objects. These will tell us whether we are being overly optimistic if we expect to observe the evidence of pulsating black holes and neutron stars in the near future. (As for the distant future, who knows?) Let us use the standard formula for the flux of gravitational waves in a weak field;

$$F = \frac{c^3}{16\pi G}|\dot{h}|^2 = \frac{1}{4\pi r^2}\frac{dE}{dt} \, . \tag{15}$$

This relation should hold far away from the radiating object. Combining this with i) $dE/dt = E/2\tau$ where $\tau = 1/\mathrm{Im}\,\omega$ is the e-folding time of the pulsation and E is the available energy ii) the assumption that the signal is monochromatic (with frequency f) and iii) the knowledge that the effective amplitude achievable after matched filtering scales as the square-root of the number of observed cycles: $h_{\mathrm{eff}} = h\sqrt{n} = h\sqrt{f\tau}$, we get the estimates [41]

$$h_{\mathrm{eff}} \sim 2.2 \times 10^{-21} \left(\frac{E}{10^{-6}M_\odot c^2}\right)^{1/2} \left(\frac{2\mathrm{kHz}}{f}\right)^{1/2} \left(\frac{50\mathrm{kpc}}{r}\right) , \tag{16}$$

for the f-mode, and [41]

$$h_{\mathrm{eff}} \sim 9.7 \times 10^{-22} \left(\frac{E}{10^{-6}M_\odot c^2}\right)^{1/2} \left(\frac{10\mathrm{kHz}}{f}\right)^{1/2} \left(\frac{50\mathrm{kpc}}{r}\right) , \tag{17}$$

for the fundamental w-mode. For the slowest damped quadrupole mode of a nonrotating $10M_\odot$ black hole we would get

$$h_{\mathrm{eff}} \sim 2.8 \times 10^{-21} \left(\frac{E}{10^{-6}M_\odot c^2}\right)^{1/2} \left(\frac{1.2\mathrm{kHz}}{f}\right)^{1/2} \left(\frac{50\mathrm{kpc}}{r}\right) . \tag{18}$$

I have used typical parameters for the pulsation modes, and the distance scale used is that to SN1987A. In this volume of space one would not expect to see more than one event per ten years or so. However, the assumption that the energy released through gravitational waves in a supernova is of the order of $10^{-6}M_\odot c^2$ is probably conservative. If a substantial fraction of the binding energy of a neutron star were released through the pulsation modes they could potentially be observable all the way out to the Virgo cluster. Then we could hope to see several events per year. Suppose we want to know how much energy must go into each mode to achieve an effective gravitational-wave amplitude $h_{\mathrm{eff}} \sim 10^{-21}$ when the source is at a distance of 15 Mpc. We can invert the above relations, and find that at least 1% of a solar mass must be radiated through these modes if they are to be detectable at the Virgo distance. The specific numbers are $E \approx 0.019M_\odot c^2$ for the f-mode, $E \approx 0.096M_\odot c^2$ for the w-mode and $E \approx 0.011M_\odot c^2$ for the black-hole quasi-normal mode. To assume that this amount of energy actually goes into these modes seems optimistic. But then again, at the present time we don't know!

The rough estimates above can be interpreted in several different ways. An incurable optimist would take them as clear evidence that the pulsation modes of neutron stars and black holes are detectable, and as soon as his new detector comes on-line he will start observing black holes and newly formed stars. A total pessimist would obviously say that this can never be done, and actually give up building his machine. In my view, the appropriate interpretation is the middle of the road opinion: These are interesting numbers... It is clear that we have to push the proposed technology to the limit, but we cannot rule out a future mode-detection. Besides, the pay-off of such a detection would be great: Undisputable evidence that black holes exist. Detailed information about the interior structure of neutrons stars. These are, undoubtedly, beautiful thoughts...

References

1. B.F. Schutz, *A first course in general relativity*, Cambridge Univ. Press (1985).
2. J.L. Tassoul, *Theory of rotating stars*, Princeton Univ. Press (1978).
3. L. Lindblom and S. Detweiler, *Ap. J. Suppl.*, **53**, 73 (1983).
4. N. Andersson and K.D. Kokkotas, *in preparation*, (1997).
5. N. Andersson, K.D. Kokkotas and B.F. Schutz, *MNRAS*, **274**, 1039 (1995).
6. N. Andersson and K.D. Kokkotas, *Pulsation modes for increasingly relativistic polytropes* submitted to *MNRAS*, (1997).
7. K.S. Thorne and A. Campolattaro, *Ap. J.*, **149**, 591 (1967).
8. C.V. Vishveshwara, *On the black-hole trail...: A personal Journey*, *Curr. Sci.*, **71**, 824 (1996).
9. T. Regge and J.A. Wheeler, *Phys. Rev,*, **108**, 1063 (1957).
10. L.A. Edelstein and C.V. Vishveshwara, *Phys. Rev. D.*, **1**, 3514 (1970).
11. F.J. Zerilli, *Phys. Rev. D.*, **2**, 2141 (1970).
12. C.V. Vishveshwara, *Phys. Rev. D.*, **1**, 2870 (1970).
13. C.V. Vishveshwara, *Nature*, **227**, 936 (1970).
14. S. Chandrasekhar, *The Mathematical Theory of Black Holes* Oxford Univ. Press (1983).
15. E.W. Leaver, *Proc. Roy. Soc. London A*, **402**, 285 (1985).
16. C.T. Cunningham, R.H. Price and V. Moncrief, *Ap. J.*, **224**, 643 (1978); **230**, 870 (1979).
17. P. Anninos, D. Hobill, E. Seidel, L. Smarr and W-M. Suen, *Phys. Rev. Lett.*, **71**, 2851 (1993).
18. N. Andersson, *Phys. Rev. D.*, **55**, 468 (1997).
19. B.F. Schutz and C.M. Will, *Ap. J.*, **291**, L33 (1985).
20. H.P. Nollert and B.G. Schmidt, *Phys. Rev. D.*, **45**, 2617 (1992).
21. N. Andersson, *Proc. R. Soc. London A*, **439**, 47 (1992).
22. N. Andersson and S. Linnæus, *Phys. Rev. D.*, **46**, 4179 (1992).
23. H.P. Nollert, *Phys. Rev. D.*, **47**, 5253 (1993).
24. K.D. Kokkotas and B.F. Schutz, *Phys. Rev. D.*, **37**, 3378 (1988).
25. E.W. Leaver, *Phys. Rev. D.*, **41**, 2986 (1990).
26. N. Andersson, *Proc. R. Soc. London A*, **442**, 427 (1993).
27. N. Andersson and H. Onozawa, *Phys. Rev. D.*, **54**, 7470 (1996).

28. S. Detweiler, *Ap. J.*, **239**, 292 (1980).
29. L.S. Finn, *Phys. Rev. D.*, **46**, 5236 (1992).
30. S.A. Teukolsky and W.H. Press, *Ap. J.*, **193**, 443 (1974).
31. M. Sasaki and T. Nakamura, *Gen. Rel. Grav.*, **22**, 1351 (1990).
32. K.D. Kokkotas and B.F. Schutz, *Gen. Rel. and Grav.*, **18**, 913 (1986).
33. K.D. Kokkotas and B.F. Schutz, *MNRAS.*, **255**, 119 (1992).
34. M. Leins, H.P. Nollert and M.H. Soffel, *Phys. Rev. D.*, **48**, 3467 (1993).
35. S. Chandrasekhar and V. Ferrari, *Proc. R. Soc. London A*, **432**, 247 (1991).
36. S. Chandrasekhar and V. Ferrari, *Proc. R. Soc. London A*, **434**, 449 (1991).
37. A.K. Kembhavi and C.V. Vishveshwara, *Phys. Rev D.*, **22**, 2349 (1980).
38. B.R. Iyer and C.V. Vishveshwara, pp. 109–127 in *A random walk in relativity and cosmology*, Ed: N. Dadhich, J, Krishna Rao, J.V. Narlikar and C.V. Vishveshwara, Wiley Eastern Ltd, New Delhi (1985).
39. B.R. Iyer, C.V. Vishveshwara and S.V. Dhurandhar, *Class. Quantum Grav.*, *2*, 219–228 (1985).
40. N. Andersson, Y. Kojima and K.D. Kokkotas, *Ap. J.*, **462**, 855 (1996).
41. N Andersson and K D Kokkotas, *Phys. Rev. Lett.*, **77**, 4134 (1996).

22. THE TWO BLACK HOLE PROBLEM

Beyond linear perturbations

R. H. PRICE
Department of Physics, University of Utah,
Salt Lake City, UT 84112 USA.

1. Introduction

The decade of the seventies started with several important papers on black hole perturbations, paper that signalled what would be the "hot" topic in relativity for the next seven or eight years. In one of those decade-starting papers [1] Vishu computed the scattering of gravitational waves off a Schwarzschild hole. I remember looking at that scattered waveform and wondering "why do black holes ring like that?" The answers, of course, have been supplied over the years.

It is interesting to realize that as Vishu turns sixty, this area of study that he helped to start – black hole (BH) perturbations – is again important and fashionable, though the context has changed. In the seventies we were trying to understand issues of principle. With some gravitational wave detectors now operational, and more elaborate detectors scheduled to be operational in a few years, we can hope to "see," with resonant masses, or laser interferometers, that ringing of black holes that Vishu first observed with equations and a computer.

The advent of the era of detectors has shifted computations from principles to details, and one might ask what role there is for perturbation theory in such work. To be observable, after all, radiation generating processes should be strong, not perturbative. The most interesting source to theorists is surely the collision of two BHs, but the merging of two horizons to form a single horizon is clearly a strong field event, not perturbative. It would seem that only numerical relativity can give the needed information. But the application of numerical relativity to BH collisions has turned out to be an enormously difficult undertaking, and further advances are needed in both computer speed and in the numerical formulation of the problem [2]. At the same time, new ways have been found to wring useful information about BH collisions from perturbation theory. It has turned out, that

B. R. Iyer and B. Bhawal (eds.), Black Holes, Gravitational Radiation and the Universe, 351–366.

what could be wrung was considerably more than might have been guessed. As a result, perturbation theory is playing an important role in giving us early answers, and in aiding the development of numerical relativity, which will be needed for our ultimate understanding of collision processes.

The new era of BH perturbations has required extending previous ideas. It has led to a sharper understanding of venerable methods, and has raised some interesting new questions. I want to use this article as an opportunity to take a step back from the details that fill recent papers, and that can obscure underlying ideas. I would like to discuss the broader issues that I have seen arise in this recent work.

2. The Origins of CLAP

The simplest BH collision is what I shall call the "Smarr problem," after Larry Smarr who pioneered the use of big computers in relativity with the study of this problem in the middle seventies [3]. The Smarr problem is the head-on collision of two equal mass nonrotating holes. The holes start from rest, more specifically from the 3-geometry of the momentarily stationary 2-BH initial value solution found by Misner [4]. Up until quite recently it was essentially only that problem for which supercomputer results were available.

It was Smarr also who suggested the applicability of perturbation theory [5] by drawing attention to the case in which the BHs start from small separation, what has subsequently been called by a few of us the "Close Limit Approximation," or "CLAP." It helps to look at a little of the details. With appropriate coordinate transformations, the Misner initial geometry can be written as

$$ds^2_{\text{Misner}} = \mathcal{F}(r,\theta)^4 \left(\frac{dr^2}{1-2M/r} + r^2 d\Omega^2 \right) \tag{1}$$

with

$$\mathcal{F} \equiv 1 + 2\,(1+\frac{M}{2R})^{-1} \sum_{\ell=2,4\ldots} \kappa_\ell(\mu_0)\,(M/R)^{\ell+1}\,P_\ell(\cos\theta)\ . \tag{2}$$

Here $R = \left(\sqrt{r}+\sqrt{r-2M}\right)^2/4$ and μ_0 is a parameter representing the initial separation of the holes. The functions $\kappa_\ell(\mu_0)$ are simple, known functions of separation that vanish in the limit $\mu_0 \to 0$. Clearly the difference $\mathcal{F}-1$ can be considered as an initial perturbation of the Schwarzschild spatial geometry, parameterized by μ_0, and that this initial perturbation can be evolved [6, 7, 8] with the techniques of perturbation theory that Vishu helped to develop.

The original motivation for a CLAP calculation was to provide a check of the numerical relativity codes. We hoped that at very small values of μ_0 perturbation theory and supercomputers would find more-or-less the same value of the radiated energy. We found that they agreed, but much better than we could have hoped. The agreement in radiated energy was good out to moderately large values of μ_0, and the waveforms at large distance – a more sensitive test – were in excellent agreement.

At first it seems strange that perturbation theory would work for moderately large values of μ_0. The Misner geometry, for any value of μ_0, has two throats, and is topologically different from the single throat Schwarzschild geometry. Near the throats, then, the Misner spacetime is in no sense a perturbation of that of Schwarzschild. For small μ_0, however, those strong differences occur only well inside the horizon, and can have no effect on the outgoing radiation. This argument does not apply when μ_0 is moderately large, say larger than 1.36 in which case the initial apparent horizon is topologically two 2-spheres. For such values the strongly distorted central region can in principle influence the generation of radiation, and perturbation theory might be expected to fail. The fact that it succeeds is instructive. It is due to the fact that the large central non-perturbative features of the central geometry get "radiated" into the final hole.

The unanticipated success of simple perturbation theory is due then to the nature of the horizon. As an ally to theorists using perturbation theory it swallows all the complexity of the strongly distorted central region. This type of perturbation analysis would not, for example, work for the collision of two neutron stars starting close together. The strongly nonperturbative information telling us that it is two stars, not one, would be able to communicate with distant regions, and would strongly influence the gravitational waves generated. There is some irony in this. The nature of the event horizon ("the inner boundary") is at present one of the major difficulties in the use of numerical relativity to study BH collisions.

3. Other Early Results

The CLAP result for the Smarr problem was followed by a nontrivial modification: the "boosted problem" in which the initial data represents holes that are initially moving. Of interest, of course, was the relationship of such a model to the late stage of a more complex astrophysical collision. (For a recent review see Flanagan and Hughes [9].) When supercomputer results for such initial data became available, the agreement between numerical and perturbative evolution was again found to be remarkable, even for initial configurations that might have seemed quite nonperturbative, and even with the addition of further approximations. (The work of Baker *et al.* [10]

used the additional assumption of small momentum, and solved only to first order in the momentum.)

Among the studies of initially boosted BHs, the paper of Abrahams and Cook [11] deserves special comment. It is somewhat outside the main plot of this story, but is too interesting to ignore. Abrahams and Cook applied perturbation theory to initial data formally representing holes that started from initial proper separation L (between the minimum area throats), and parameter P representing the inward momentum of each hole. They chose L, P pairs to correspond to an initial configuration in which the apparent horizons of the two holes were just barely touching, a configuration that was deemed close enough to ensure the validity of CLAP evolution.

In 1994, when this work was done, numerical relativity had not yet been applied to the boosted initial data problem. With no supercomputer results for direct comparison, Abrahams and Cook tried to make a rough indirect comparison with the existing supercomputer data for the Smarr problem. To do this, they supposed that their starting L, P configuration had evolved from a precursor stationary configuration in which the precursor separation L_0 of the throats could be inferred from the L, P pair by straightforward application of Newtonian physics. The supercomputer results for the Smarr problem, with initial physical separation L_0 were then compared with the Abrahams-Cook result for L, P. The method has blatant flaws, in particular that the initial L, P data has little relation to the 3-geometry and extrinsic curvature of a slice of the spacetime that will evolve from the Misner initial data. Despite the fact that the Abrahams-Cook method was "obviously wrong," it gave remarkably good agreement with the supercomputer results and raised an interesting possibility: For the purpose of computing outgoing radiation, the true initial data may not be important. The outgoing radiation may be determined only by some special features of the initial data, and these seem to be connected in a robust way with a simple physical picture of the motion of the holes.

The reason for these excellent results is still not understood. With Carlos Lousto, I have explored the dependence of outgoing radiation to the details of initial conditions in the limit of an extreme ratio of colliding BH masses [12], a limit that turns out to be more sensitive to initial details than is the equal mass case. This work confirmed the already suspected idea that for generating radiation, initial extrinsic curvature is much more important than the details of the initial 3-geometry, so it is how the holes are moving that is crucial. This is, however, just one clue in the mystery.

4. The Need for Second-Order "Error Bars"

The initial motivation for CLAP perturbation theory had been a verification of supercomputer results, but the surprisingly wide range of validity of the CLAP results suggested that perturbation theory could explore some collision scenarios on its own, while numerical relativity, and terraflop machines, continue development.

For supplying reliable answers, however, linearized perturbation theory has an awkward shortcoming. We know that it gives an accurate answer for sufficiently small perturbations from some background (such as the Schwarzschild geometry), but we have no *a priori* way of saying how "small" a small perturbation must be. Thus we cannot make a computation of the radiation strength expected from a specific collision if we do not have a justification for CLAP perturbation theory to apply to that collision. A remedy for this would be only to consider collisions in which the BHs start extremely and incontrovertibly close together. But collisions in which CLAP methods "obviously" apply are collisions in which the spacetime is very close to the unperturbed spacetime, and hence in which very little energy is radiated. What one really wants to do is not to retreat to the safety of very small perturbations, but rather push perturbation theory to the limit of its applicability, to the cases of largest perturbations and strongest radiation.

Physically based estimates are of some utility in this regard. We might, for example, decide always to trust CLAP estimates when the colliding holes are initially inside a common apparent horizon. Experience shows though that CLAP estimates can be reasonably accurate for rather larger initial separations. To choose the "common apparent horizon" criterion, or any similar rule of thumb, for CLAP validity, involves a risk on the one hand of being too conservative and arriving at astrophysically useless lower bounds on the gravitational wave energy or, on the other hand, or of being too optimistic and raising false hopes of intense radiation.

What is needed is an objective criterion for assessing the accuracy of linearized perturbation calculations and the range of applicability of such calculations. In the absence of numerical relativity results the most reliable way to establish such a criterion is with higher order perturbation theory. To understand why this approach is helpful let us remember how linearized calculations are done. The perturbation parameter, call it ϵ, simply acts as a multiplier for the perturbations. In this sense linearized perturbation calculations are scale invariant. The size of ϵ enters only in the end, in a rather trivial way. If, for example, we are computing a first order waveform, the result of our computation will have the character $\psi_1(t) = \epsilon f_1(t)$. If we multiply ϵ by 100 and divide f_1 by 100, we are describing the same physical waveform, but the numerical value of ϵ associated with this waveform is very different. Within the framework of linearized calculations there is no internal mechanism for signaling when the calculations are valid.

Now let us imagine that we have chosen some perturbation parameter ϵ, and we are performing two computations of outgoing radiation waveforms, the first computation gives $\psi_1(t)$ correct only to first order, and the second $\psi_{(1+2)}(t)$ correct up to second order. The results will have the character $\psi_1(t) = \epsilon f_1(t)$ and $\psi_{(1+2)}(t) = \epsilon f_1(t) + \epsilon^2 f_2(t)$. When we choose a value of ϵ for which $\psi_1(t)$ differs significantly from $\psi_{(1+2)}(t)$, then we know we have chosen a "large" value of ϵ, and perturbation theory is unreliable. If, on the other hand, $\psi_{(1+2)}(t)$ differs from $\psi_1(t)$ by (say) 1%, we can infer that linearized perturbation theory itself will be accurate to around a percent or so. These conclusions are not mathematically rigorous, of course. Special cases can easily be contrived in which there is excellent, or even total, agreement of first and second order results, and yet the answer is wrong by orders of magnitude. But such cases are indeed contrived, and one expects the first vs. second order criterion for validity to have the same general robustness as the rules of thumb for truncation error in numerical work.

There is a secondary motivation for second order calculations: it gives somewhat better approximations. This secondary benefit should not be overvalued. For very small ϵ the second order corrections are very small and unimportant. Outside the range of validity of perturbation theory the results, even with second order corrections, are not valid. It is only in the small range of ϵ near the limit of validity that the improved accuracy of second order calculations is worth the bother. On the other hand, one should not undervalue this benefit, since it is the results at this limit of validity that are of the greatest interest in connection with sources of gravitational radiation.

The strategy for applying higher order perturbation theory to BH collisions consists of three steps: (i) Develop the technical details of second order theory for the case of collisions. (ii) Apply the second order theory to collisions in which comparisons can be made with numerical relativity, to ensure that there is no flaw in the general idea of second order theory as a source of "error bars." (iii) Apply perturbation theory, with second order error bars, to cases still beyond the capabilities of supercomputers.

The first two steps have been carried out in a geographically broad collaboration involving Reinaldo Gleiser, Oscar Nicasio, Jorge Pullin and myself [13]. Both the eventual success and the unanticipated difficulties make an interesting story that will be sketched in the remainder of this article. The third step is well underway.

5. Fundamental Issues in Perturbation Theory

At the outset it seemed that the details of second order perturbation theory would be more a matter of complexity, than of fundamental issues. That was not the case. It turned out that the simplicity of linearized theory covers many sins. When one goes to higher order, one must repent those sins. As a particularly relevant example let us note that perturbation theory in general relativity is often introduced through the idea of a metric $g_{\alpha\beta}$ that is very "close" in some sense to a background metric $g^B_{\alpha\beta}$. One then writes

$$g_{\alpha\beta} = g^B_{\alpha\beta} + h_{\alpha\beta} \ , \tag{3}$$

thereby *defining* $h_{\alpha\beta}$ as the difference between one exact solution of Einstein's equations, $g_{\alpha\beta}$, and another $g^B_{\alpha\beta}$. One next puts the right hand side of Eq. (3) in Einstein's equations, and keeps terms only to first order in $h_{\alpha\beta}$. This procedure, seems quite reasonable, and indeed, can be justified for linearized theory, but it runs into significant problems, as we shall see, when one tries to extend it to higher orders.

I will now give another, more careful, scheme for understanding perturbation theory. For definiteness, I will concentrate on vacuum solutions of Einstein's equations. Fundamental to the concept of perturbation theory is the idea of a one parameter *family* of solutions $g_{\mu\nu}(x^\alpha;\epsilon)$ of Einstein's vacuum equations $G_{\lambda\tau}[g_{\mu\nu}] = 0$. Here $g_{\mu\nu}$ is understood to be a solution of the equations for *all* values of the real parameter ϵ in some neighborhood of zero.

It is good to give a specific example of such as scheme, an example very important to early work on collisions. The Misner solution (along with the condition of vanishing extrinsic curvature) solves the Einstein initial value equations. One can therefore evolve this initial data, for any μ_0, forward in time and find a parameterized family of solutions, the "Misner spacetimes."

The Misner geometry cannot truly be considered as a perturbation of the Schwazschild geometry since the Misner geometry, for any μ_0, has a topology (two throats connecting two asymptotically flat universes) different from the Schwarzschild topology (one throat connecting two asymptotically flat universes). But we are not really interested in the entire spatial extent of the Misner solution. Whatever is interior to the initial horizon will have no effect on the dynamics outside the horizon, the dynamics that determines what is radiated to infinity. It is intuitively clear, and is straightforward to demonstrate in detail, that as $\mu_0 \to 0$ the geometry outside the horizon approaches the Schwarzschild geometry outside the horizon. It turns out that the metric is not analytic in μ_0 near $\mu_0 = 0$, but one can introduce a parameterization $\epsilon = \epsilon(\mu_0)$ in terms of which the 3-geometry *is* analytic. In this manner one arrives, *in principle*, at a 1-parameter family of solutions of Einstein's equations, and these solutions (e.g., the metric functions in some coordinates system) are analytic functions of ϵ. We can now write an expansion of the metric functions, in the form

$$g_{\alpha\beta} = g^B_{\alpha\beta} + \epsilon h^{(1)}_{\alpha\beta} + \epsilon^2 h^{(2)}_{\alpha\beta} + \cdots \tag{4}$$

and, in principle, we can put this family of solutions into Einstein's nonlinear vacuum field equations $G_{\lambda\tau}[g_{\mu\nu}] = 0$, and can expand the equations in ϵ. The zero order part is simply Einstein's equations for the background metric, and contains no information.

The first order terms are a set of homogeneous linear partial differential equations for the unknown functions $h^{(1)}_{\alpha\beta}$. We write this set of equations as

$$\mathcal{L}_{\mu\nu}(h^{(1)}_{\alpha\beta}) = 0 \tag{5}$$

in which the differential operators $\mathcal{L}_{\mu\nu}$ contain information about the background solution. To second order in ϵ the perturbations in the metric can enter in two ways: (i) The $h^{(2)}_{\alpha\beta}$ terms enter linearly, in exactly the same manner as did the $h^{(1)}_{\alpha\beta}$ terms in the Einstein equations to first order. (ii) The $h^{(1)}_{\alpha\beta}$ terms enter quadratically due to the nonlinear character of Einstein's equations. The second order equations then have the form:

$$\mathcal{L}_{\mu\nu}(h^{(2)}_{\alpha\beta}) = Q(h^{(1)}_{\alpha\beta}) \ . \tag{6}$$

Here the $\mathcal{L}_{\mu\nu}$ operators are precisely the same operators as in Eq. (5). The terms on the right are quadratic in $h^{(1)}_{\alpha\beta}$ can be considered as known "source" terms if the equations of first order theory have already been solved for $h^{(1)}_{\alpha\beta}$. For higher order this pattern continues with slight variations. On the left hand side we have $\mathcal{L}_{\mu\nu}$ acting on the unknown metric perturbations $h^{(n)}_{\alpha\beta}$; on the right we have a "source" term constructed from the already-known lower order perturbations $h^{(k)}_{\alpha\beta}, k = 1, 2, \ldots n-1$.

We might now ask what the relationship is between this "family of spacetimes" approach, and that of Eq. (3). We could rewrite that equation as

$$g_{\alpha\beta} = g^B_{\alpha\beta} + \epsilon h_{\alpha\beta} \tag{7}$$

so that it looks similar to Eq. (4), but it would differ in an important way. The metric in Eq. (4) is a solution of Einstein's equations for any value of ϵ. By contrast, the metric in Eq. (7) is a solution only for $\epsilon = 1$. The expression in Eq. (7) does not constitute a family of solutions, and we cannot expand in ϵ, and let ϵ go to zero. As a practical matter, this does not make much difference. The equations one arrives at by starting with Eq. (3), are the correct equations for linearized perturbation theory. But if we try to go to second-order theory, there is no clear path with the formalism of Eq. (7).

The nonlinear nature of second order perturbations adds some new features to the business of multipole decompositions. In linearized work one uses the spherical symmetry of the background (and hence of the differential operators $\mathcal{L}_{\mu\nu}$), and one decomposes the metric perturbations $h^{(1)}_{\alpha\beta}$ into multipoles. Since the multipoles form a complete basis, in our second order perturbation work we expand both the first order perturbations $h^{(1)}_{\alpha\beta}$, and the second order perturbations $h^{(2)}_{\alpha\beta}$ in spherical harmonics. When we solve the familiar first-order problem, in the familiar way, due to the spherical symmetry of $\mathcal{L}_{\mu\nu}$ the first order multipoles do not mix. Thus the $\ell = 2$ part of the initial data generates the $\ell = 2$ part of the outgoing radiation, and so forth. The new wrinkle in second order work is the nonlinearity on the right hand side of Eq. (6). Because of this the second order quadrupole ($\ell = 2$) part of $h^{(2)}_{\alpha\beta}$, will in general be generated by $h^{(1)}_{\alpha\beta}$ source terms of all ℓ values. If one is interested in the quadrupole part of $h^{(2)}_{\alpha\beta}$, one needs in principle to construct the source terms in Eq. (6) with all first order multipoles, then project out the $\ell = 2$ part of the source. (It turns out that in the case of the Misner data there is a convenient minor simplification: the only first order perturbation is a quadrupole perturbation, so one needs only put quadrupole first order perturbations into the source terms in Eq. (6)).

6. Gauge Transformations to Higher Order

By far, the most interesting and potentially misleading aspect of higher order perturbation theory is how to deal with coordinate arbitrariness. We can imagine a 1 parameter family of coordinate transformations $x^{\alpha'} = x^{\alpha'}(x^\beta; \epsilon)$ that can be expanded as

$$x^{\alpha'} = x^\alpha + \epsilon\xi^\alpha + \epsilon^2\zeta^\alpha + \cdots \tag{8}$$

We are familiar with the first term in such an expansion from linearized theory. We can consider ξ^α to be a vector, and the transformation it induces on the first order perturbations $h^{(1)}_{\alpha\beta}$ can be described with a Lie derivative along the field ξ^α and we call it a gauge transformation. When we go to higher order we will continue to call the coordinate changes "gauge transformations" but the geometric interpretation is no longer directly applicable. In particular, the quantities ζ^α cannot be considered to be vector fields. It is interesting to note that soon after our initial work on second order perturbations was underway, Bruni *et al.* [14] gave an alternative scheme for dealing with gauge transformations in higher order theory. In that scheme one starts with a general gauge transformation written in a form different from that in Eq. (8), and a geometric interpretation can be applied to the transformation at any order. That geometric scheme is surely more elegant than our approach, but we have found our approach to be convenient for the sometimes lengthy computations that are needed.

In order to understand coordinate issues at second order, one needs to understand that there are at least two different types of coordinate transformations that one must distinguish. Simplest is a "second order" transformation. This is one in which there are no first order changes in the coordinates, i.e., one in which $\xi^\alpha = 0$. For such a transformation, the second order part of the transformations (i.e., that due to ζ^α) induces transformations to second order in the metric perturbations. For such transformations the first order perturbations $h^{(1)}_{\alpha\beta}$ are invariant, and the second order perturbations $h^{(2)}_{\alpha\beta}$ are changed by ζ^α in precisely the same was as $h^{(1)}_{\alpha\beta}$ would be changed by ξ^α in linearized theory. A second type of transformation is a first order transformation carried out to second order. This is one in which $\xi^\alpha \neq 0$. Due to the nonlinearity of diffeomorphisms, such a transformation induces changes in the second order perturbations $h^{(2)}_{\alpha\beta}$, changes that are quadratic in ξ^α.

In linearized perturbations of spherical spacetimes, a familiar way of dealing with coordinate arbitrariness is to choose the Regge-Wheeler (RW) gauge [15], a convenient way of fixing the gauge in each multipole, by specifying that the perturbations in certain components of the metric vanish. We have carried to second order this idea of fixing certain components to be zero. It is a useful demonstration of the two-step way in which gauge transformations must be carried out in second order work. For the Misner initial data, the axisymmmetry and lack of rotation guarantees that there will be only even parity perturbations, to all orders of perturbations. The Regge-Wheeler gauge then requires the vanishing, for each multipole, of three particular even parity coefficients of multipole perturbations denoted $h_0^{(k)}, h_1^{(k)}, G^{(k)}$. Here we are using the superscript k to denote the order of the perturbation. In our scheme we want these perturbations to vanish both to first ($k = 1$) and second ($k = 2$) order. As is familiar in linearized work, the $k = 1$ job can be done with a first order gauge transformation. This first order gauge transformation will change the second order metric perturbations, but for the moment that is of no concern to us. Our second step is to use a second order gauge transformation to put the second order perturbations into RW form. Since the transformation is second order, it does not affect the first order perturbations, and hence leaves them in RW form. It should be noticed that this two step procedure "first order transformation followed by second order transformation" cannot be reversed. There is no point in starting with a second order transformation to bring the second order metric perturbations into RW form. The subsequent first order transformation would take the second order perturbations out of RW form.

In this second order work, gauge transformations are not merely a matter of principle, but must be used explicitly at two stages of the work. First, initial data may be given in a form that is not compatible with the Regge-Wheeler gauge. An explicit transformation must then be done from the gauge of the initial data to the RW gauge. Second, after the perturbations are computed in the RW gauge, the metric must be transformed to an asymptotically flat gauge in which we can extract information about gravitational radiation from the transverse traceless perturbations.

7. Success with Second Order Computations, and New Difficulties

The second order method, as described above, was applied to the Smarr problem to see whether second order calculations did indeed serve the purpose of giving error bars for perturbation theory [13]. The results are plotted in Fig. 1, which gives radiated energy, as a fraction of total spacetime mass, as a function of the parameter μ_0 characterizing the initial separation of the throats in the Misner geometry. Plotted are the radiated energy calculated in three ways. The dashed line shows the results of first order theory; These results do not disappoint. At around $\mu_0 \approx 1.8$ the first order and first+second results start to diverge significantly. In the absence of numerical results we would have to infer that this is the limit of perturbation theory. We see, in fact, that this is correct. At around this value of μ_0, first order theory begins to be unreliable, and very soon after first +second theory becomes unreliable.

This is a rather convincing demonstration of how well this method of estimating validity works, and it is rather important to understand the implication. In retrospect one might note that $\mu_0 \approx 1.8$ is the dividing line between initial horizons that enclose both holes (for $\mu_0 < 1.8$) or do not, and one might claim that this explains why perturbation theory fails for $\mu_0 > 1.8$. But without the clarity of hindsight, one would almost certainly have thought that a configuration with disjoint initial horizons was so dramatically distorted that perturbation theory would fail badly. One would likely have guessed that the limit of validity would be perhaps around $\mu_0 \approx 1.2$. This entails a difference of more than an order of magnitude in radiation energy.

The rather remarkable agreement of the waveforms of first+second calculations [13], and those of numerical relativity has been published elsewhere, but is a side benefit. The main goal, clearly achieved in Fig. 1, is

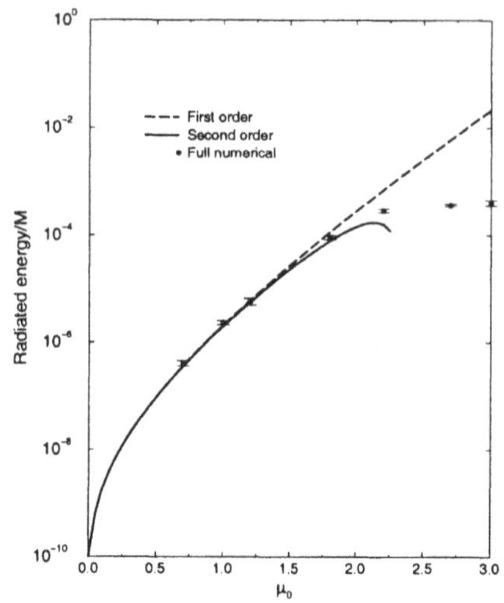

Figure 1. Energy radiated in the "Smarr problem," the collisions of two initially stationary holes starting with the Misner initial conditions. The radiated energy is given as a fraction of M, the spacetime mass. The parameter μ_0 is a measure of the initial separation of the two holes. The dashed curve shows the results of first order theory; the solid curve gives first+second order results; the isolated points are the results of fully nonlinear numerical relativity.

the test of the second order error bars.

The second application of second order work was to the axially symmetric collision of two holes with initial momentum, the "boosted collision" [10]. A satisfactory second order version of this [16] required more than a year of extra work (and puzzlement) after the second order work had been done on the Smarr problem, and after it seemed that the slippery gauge issues had been worked out. They had, but new problems were to arise.

There is no unique way of doing a computation correct to second order. Two calculations that differ in some details may both be formally second order, and will be equally accurate in the limit of small ϵ. They will in general differ, however, in higher orders of ϵ and one method might be somewhat better than another method at large ϵ. In general this divergence of the two methods will differ significantly only when perturbation theory is beginning to fail. In working with initially moving holes, however, there is an unwelcome (but interesting) complication.

In our work with initially moving holes the momentum of the holes as well as the separation was taken to be small. In principle, we considered a one parameter curve in the two dimensional momentum-separation parameter space. We were interested primarily in the quadrupole perturbations. (We knew, e.g., from supercomputer results, that the outgoing radiation was heavily quadrupole dominated.) We were certainly not interested in monopole perturbations, since monopoles do not radiate! But the interaction of low order perturbations in the monopole and low order perturbations in the quadrupole produce higher order perturbations in the radiated quadrupole. The effect though is higher order, and should not be particularly worrisome. It was, in fact, explicitly included in our earliest calculations. The change in the mass of the spacetime (from that of the background solution) due to the motion of the initial holes represented a monopole perturbation, which was calculated and included in the scheme.

What was eventually seen was that the perturbation in the monopole was much larger than that of other multipoles. For perturbations producing only small quadrupole distortions there were changes in the spacetime mass on the order of 100%. Therefore if one uses a method that is formally second order, validity is limited to a very small ϵ due to the sensitivity of the monopole to the perturbations. The remedy for this is to use a nonperturbative method of computing the change in the monopole. Rather than compute the change in the mass by perturbation theory, for any specific initial value data set the mass is computed numerically (from the full nonlinear theory). This is more "numerical" than perturbation theory usually is, but it is quite feasible, and involves none of the daunting problems of evolving the initial data. With this somewhat hybrid method one could again compare first order results with those of first+second, and could compare with numerical results. This work, only recently completed [16, 17], turned out to be just as successful as the Smarr problem in validating the method of second order error bars.

Computations are now underway applying perturbation theory to the collision of rotating holes. This is particularly exciting because the radiated energy should be much larger, and because this is a problem to which numerical relativity has not yet been applied. The relatively easy first order analysis has been done, indicating the potential of very strong radiation. Only second order error bars will tell us what to believe from these results.

8. Modified approaches, Promises and Problems

The new usefulness of perturbation theory has raised some questions about the "practice" of perturbation theory. These closely related questions are interesting in connection with the theoretical underpinnings of perturbation theory, but also have already come up as practical matters in connection with numerical relativity and gravitational wave sources.

A starting point for these issues is the question of whether we are doing perturbation calculations in the most efficient way, since at each step, we are not "updating" information about the metric as much as we could be.

To make this more specific, let us start with Eq. (3), taking it as a definition of $h_{\alpha\beta}$. In terms of this notation we can write Einstein's equations, with no approximations, as

$$L_{\mu\nu}^{(g^B_{\lambda\tau})}(h_{\alpha\beta}) = \mathcal{S}(g^B_{\lambda\tau})(h_{\alpha\beta}) \ . \tag{9}$$

The rather awkward notation means the following: On the left hand side of the equation we have the terms that are formally linear in $h_{\alpha\beta}$. The second order linear operator that acts on them, L, depends on the details of the background metric $g^B_{\lambda\tau}$, and this dependence is included in the notation. (The operator L in Eq. (9) is, in fact, identical to $\mathcal{L}$ in Eq. (5), but the change in notation will presently be important.) The right hand side of Eq. (9) represents all of the terms that are formally nonlinear in $h_{\alpha\beta}$. With Eq. (9) we can now describe "standard" perturbation theory as follows. We write $h_{\alpha\beta}$ as the series $\epsilon h^{(1)}_{\alpha\beta} + \epsilon^2 h^{(2)}_{\alpha\beta} + \cdot$. We then solve Eq. (9) iteratively. That is, we start by putting in zero on the right, and take $h_{\alpha\beta}$ to be only $\epsilon h^{(1)}_{\alpha\beta}$ on the left. We next put in $\epsilon h^{(1)}_{\alpha\beta}$ on the right, keep terms only to order ϵ^2 on the right, and take $h_{\alpha\beta}$ to be $\epsilon h^{(1)}_{\alpha\beta} + \epsilon^2 h^{(2)}_{\alpha\beta}$ on the left, and so forth.

At each step of this iteration one improves the precision with which the metric is known, but in this scheme that improved precision is not being fully exploited, since the operator L, continues to be based on the background geometry. One can remove this apparent shortcoming easily, by replacing L on the left of Eq. (9) with an operator based on the exact metric; one simply replaces the metric $g^B_{\lambda\tau}$ by $g_{\lambda\tau}$. This will, of course, generate additional nonlinear terms in $h_{\alpha\beta}$. These terms are moved to the right, and we arrive at a new formulation

$$\tilde{L}_{\mu\nu}^{(g_{\lambda\tau})}(h_{\alpha\beta}) = \tilde{\mathcal{S}}(g^B_{\lambda\tau})(h_{\alpha\beta}) \ . \tag{10}$$

To use this formulation, the first step is no different from the previous one. One replaces $g_{\lambda\tau}$ by $g^B_{\lambda\tau}$ on the left, sets the right hand side to zero, solves for $h_{\alpha\beta}$, and interprets it as $h^{(1)}_{\alpha\beta}$. In the second stage of iteration one replaces $g_{\lambda\tau}$ on the left by $g^B_{\lambda\tau} + \epsilon h^{(1)}_{\alpha\beta}$, and the differential operator $\tilde{L}$ takes a new "improved" form.

It seems intuitively persuasive that this iterative procedure will usually, if not always, give higher accuracy and/or quicker convergence than the standard procedure. (One might draw an analogy to iterative solutions of matrix equations carried out by the Jacobi and the Gauss-Seidel methods. The latter method always uses the most recently updated solution, and usually converges more quickly, and is more accurate at any stage of iteration.) These advantages in principle of the modified scheme inherent in Eq. (10), however, may be much less important than a huge disadvantage in practice. In perturbation work, the background metric is almost always a metric with symmetries. In the case of a Schwarzschild background, one has spherical symmetry and can greatly simplify the problem of solving for each new order of perturbation by decomposing it in spherical harmonics. The Schwarzshild background also has time-translational symmetry and a Fourier or Laplace transform could be used for the time dependence. In standard perturbation theory then we are using a simple linear operator and the same operator at each order. If we were to "update" the operator at each order this enormous simplification would be lost, and the advantage of improved efficiency or accuracy would not seem to justify the change.

There may, however, be reasons more important than efficiency that motivate a change. Since the (hyperbolic) linear differential operator on the left doesn't change in the standard approach, the characteristics along which the solution propagates are unchanged. At first this seems to be a serious flaw, since the causal structure will change dramatically during a collision of holes, but these dramatic changes take place inside the event horizon, and are causally disconnected from the region of interest. But even in the relatively tranquil exterior, there is reason to worry. If we are to apply perturbation theory to cases in which (unlike the Smarr problem) there is significant radiation – say 5% of the spacetime mass – then the mass and angular momentum of the final hole, will be different from those of the initial (and presumably "background") hole.

To understand why this is something to worry about, consider a rather specific question: the shape of the waveforms resulting from an inline collision of nonrotating holes. We know that such a waveform will be dominated by quasi-normal ringing, a damped oscillation with a wavelength determined by the mass M and angular momentum J of the spacetime in which the radiation is being propagated. But which M and J will appear in these waveforms, the initial mass of the spacetime, or the mass after the emission of gravitational wave energy? In the standard approach, all propagation occurs in the spacetime of the background metric which has information only about the original M and J. One therefore expects the waveforms to show quasi-normal ringing characteristic of these parameters. If the resulting waveforms are accurate they *must* show the decrease of wavelength corresponding to the loss of mass of the spacetime in which the waves are propagating. How can this happen if the characteristics are always those of the initial M and J? Does this apparent contradiction mean that standard perturbation calculations cannot converge if radiation is strong?

In addition to a disturbing fundamental issue, this issue creates a computational dilemma. If we update the differential operator, we enormously increase the difficulty of solving the higher order perturbation problem. If we do not, we are more-or-less guaranteed to make serious errors in the waveforms of outgoing radiation. A practical solution to this dilemma, has been suggested by Abrahams [18]: One can update the mass and angular momentum, which occur only as parameters in the differential operators, and not update the higher multipole moments. This technique has already been used, to first order, in the perturbative radiation zone modules used in conjunction with numerical relativity [19]. As numerical relativity and analytic techniques advance toward the goal of computing BH collisions, it will be interesting to see what we learn about this and other ways of improving perturbation theory.

References

1. C.V. Vishveshwara, *Nature*, **227**, 936 (1970).
2. L. S. Finn in *Proceedings of GR14*, edited by M. Francaviglia *et al.* (World Scientific, Singapore, 1997) p. 147.
3. L. Smarr, in *Sources of gravitational radiation*, L. Smarr, editor, Cambridge University Press, Cambridge, England, p. 245 (1979).
4. C. Misner, *Phys. Rev.* **D118**, 1110 (1960).
5. L. Smarr, private communication.
6. R. Price, J. Pullin, *Phys. Rev. Lett.* **72**, 3297 (1994).
7. A. Abrahams, R. Price, *Phys. Rev.* **D53**, 1963 (1996).
8. P. Anninos, E. Seidel, R. Price, J. Pullin, W.-M. Suen, *Phys. Rev.* **D52**, 4462 (1995).
9. E. E. Flanagan and S.A. Hughes, preprint gr-qc/9701039.
10. J. Baker, A. Abrahams, P. Anninos, S. Brandt, R. Price, J. Pullin, E. Seidel, *Phys. Rev.* **D55**, 829 (1997).
11. A. M. Abrahams and G. B. Cook 1994, *Phys. Rev.* **D50**, R2364.
12. C. O. Lousto and R. H. Price, *Phys. Rev.* **D55**, 2124 (1997); *Phys. Rev.* **D56**, 6439 (1997); *Phys. Rev.* to appear, preprint gr-qc/9708022.
13. R. Gleiser, O. Nicasio, R. Price, J. Pullin, *Class. Quan. Grav.* **13**, L117 (1996); *Phys. Rev. Lett.* **77**, 4483 (1996); J. Pullin, *Fields Inst. Commun.* **15**, 117 (1997).
14. M. Bruni, S. Matarrese, S. Mollerach, and S. Sonego, *Class. Quan. Grav.* **14**, 2585 (1997); S. Sonego and M. Bruni, *Communications of Mathematical Physics* in press, preprint gr-qc/9708068; S. Matarrese, S. Mollerach, and M. Bruni, preprint SISSA 83/97/A, submitted to *Phys. Rev. D.* astro-ph/9707278.
15. T. Regge and J. Wheeler, *Phys. Rev.* **108**, 1063 (1957).
16. R. Gleiser, O. Nicasio, R. Price, and J. Pullin, paper in preparation.
17. This problem arises also in the analysis of radiation emitted by a single spinning holes as it settles down to its final Kerr form. See R. Gleiser, O. Nicasio, R. Price, J. Pullin, to appear in *Phys. Rev. D.*, preprint gr-qc/9710096.
18. A. M. Abrahams, private communication.
19. A. M. Abrahams, C. R. Evans, *Phys. Rev.* **D49**, 3998 (1994).

23. THE SYNERGY BETWEEN NUMERICAL AND PERTURBATIVE APPROACHES TO BLACK HOLES

EDWARD SEIDEL
Max-Planck-Institut für Gravitationsphysik,
Schlaatzweg 1, 14473 Potsdam, Germany.
AND
University of Illinois, Urbana, IL 61801.

1. Introduction

When I started graduate school, I began a thesis project in perturbation theory of spherical spacetimes. I still remember well how my advisor, Vincent Moncrief, an expert in perturbation theory, advised me to study a paper by Vishveshwara on black hole perturbations [1]. "That's the best place to find the perturbation formalism", he told me. At that time, of course, I had no idea how important this subject would continue to be years later. I was very lucky to become grounded in this subject at an "early age", and I knew it would provide important insight into problems that were intractable in numerical relativity at that time. However I did not appreciate that even as numerical relativity would become more and more mature, harnessing hundreds of processors in parallel to solve ever larger problems, perturbation theory would continue to play such an important role in so many ways. In fact, its role in numerical relativity has become even more important in recent years, as I describe below. Vishu's work in this area influenced me in ways that I appreciate even more as my own research moves into large scale numerical simulation.

2. Numerical Evolutions of Black Holes

The numerical evolution of black holes is very difficult, as one must simultaneously deal with singularities inside them, follow the highly non-linear regime near the horizons, and also calculate the linear regime in the radiation zone where the waves represent a very small perturbation on the

B. R. Iyer and B. Bhawal (eds.), Black Holes, Gravitational Radiation and the Universe, 367–382.

background spacetime metric. In axisymmetry this has been achieved, for example, for stellar collapse [2], rotating collisionless matter [3], distorted vacuum black holes with rotation [4] and without [5], and for equal mass colliding black holes [6, 7], but with difficulty. These 2D evolutions can be carried out to roughly $t = 100M$, where M is the ADM mass of the spacetime, but beyond this point large gradients related to singularity avoiding slicings usually cause the codes to become very inaccurate and crash. This is one of the fundamental problems associated with black hole evolutions: if one uses the gauge freedom in the Einstein equations to bend time slices up and around the singularities, one ends up with pathological behavior in metric functions describing the warped slices that eventually leads to numerical instabilities.

In 3D the problems are even more severe with this traditional, singularity avoiding time slicing approach. To simulate the coalescence of two black holes in 3D, evolutions of time scales $t \approx 10^2 - 10^3 M$ will be required. Traditional approaches can presently carry evolutions only to about $t = 50M$. However, in spite of these difficulties, great progress is being made on several fronts. Alternative approaches to standard numerical evolution of black holes, such as apparent horizon boundary conditions and characteristic evolution, promise much longer evolutions. Apparent horizon conditions cut away the causally disconnected region interior to the black hole horizon, allowing better behaved slicings. These have been well developed in 1D, spherically symmetric studies [8, 9, 10, 11, 12, 13] and full 3D evolutions. Characteristic evolution with ingoing null slices have very recently been successful in evolving 3D single black holes for essentially unlimited times, even with distortions away from spherical or axisymmetry [14]. These alternate approaches look promising, but will take time to develop into general approaches to the two black hole coalescence problem, and I will not have space to cover them here. Instead, I focus on how traditional approaches, aided by perturbative studies, are providing insight into the dynamics of distorted and colliding black holes. When the alternative techniques for black hole evolutions mature, they too will be aided by perturbative studies to both verify and interpret numerical simulations.

3. Interplay of Perturbative and Numerical Black Hole Studies

While numerical relativity is making steady progress in its ability to simulate the dynamics of black holes, an equally important and exciting development of the last few years has been the application of perturbation theory to aid in the verification and interpretation of numerical simulations. As I discuss below, there are various ways that perturbation theory is used in conjunction with numerical relativity, from the extraction of waveforms from fully non-linear numerical simulations, to full evolutions of black hole initial data sets, treated as perturbations about the Schwarzschild black hole background.

In this section I present a series of ideas and calculations that lead to what I call a "Ladder of Credibility". Problems need to be studied in sequence from easier to more complicated, leading ultimately to the 3D spiraling coalescence problem. I will discuss the foundations of black hole perturbation theory, and show how it can be used both to test and to interpret results of full numerical simulations, beginning with axisymmetric distorted black holes, moving on to axisymmetric black hole collisions, and building to full 3D simulations.

3.1. PERTURBATION THEORY

In this section I give a very brief overview of the theory of perturbations of the Schwarzschild spacetime. This is a topic which is very rich and has a long history, and to which Vishu has contributed immensely. A more detailed discussion can be found in [15]. In the next section, I will discuss how we apply the theory to study black hole data sets.

One begins the analysis by writing the full metric as a sum of the Schwarzschild metric $\overset{o}{g}_{\alpha\beta}$ and a small perturbation $h_{\alpha\beta}$:

$$g_{\alpha\beta} = \overset{o}{g}_{\alpha\beta} + h_{\alpha\beta}. \qquad (1)$$

One then plugs this expression into the vacuum Einstein equations to get equations for the perturbation tensor $h_{\alpha\beta}$. One can separate off the angular part of the solution by expanding $h_{\alpha\beta}$ in spherical tensor harmonics, as was originally done by Regge and Wheeler [16]. For each ℓ, m mode, one gets separate equations for the perturbed metric functions, which we now denote by $h_{\alpha\beta}^{(\ell m)}$.

There are two independent expansions: one, known as even parity, which does not introduce any rotational motion in the hole, and one, known as odd parity, which does. We will concentrate here on even parity perturbations, as these are the ones which will be relevant for studying the data-sets discussed here. The odd parity perturbations produce equations which are very similar, and both can be considered in the general case.

We also note here that this treatment is presently restricted to perturbations of Schwarzschild black holes. For the more general rotating case, one would like to use the Teukolsky formalism describing perturbations of Kerr. This is much more complicated, and has not yet been applied to numerical black hole simulations of the kind discussed in this paper. This is an important research topic that needs attention soon!

I have discussed here only the linearized theory of black hole perturbations, but this can also be extended to higher order. This has recently been accomplished by Gleiser, Nicasio, Price, and Pullin [17], who worked out equations describing perturbations of Schwarzschild to second order in an expansion parameter. Hence the metric is written as

$$g_{\alpha\beta} = \overset{o}{g}_{\alpha\beta} + \epsilon h^{(1)}_{\alpha\beta} + \epsilon^2 h^{(2)}_{\alpha\beta} \tag{2}$$

and the Einstein equations are expanded to second order in ϵ. As we will see below, this second order formalism is very useful in black hole studies [18].

When dealing with perturbations in relativity, one must be careful about interpreting the various metric components $h_{\alpha\beta}$ in terms of physics. Under a coordinate transformation of the form $x^\mu \to x^\mu + \delta x^\mu$, the metric coefficients will transform as well. One can use this gauge freedom to eliminate certain metric functions to simplify the corresponding equations for the perturbations. Another, more powerful approach, developed first by Moncrief, is to consider linear combinations of the $h_{\alpha\beta}$ and their derivatives that are actually *invariant* under the gauge transformation above. In either case, the analysis leads to a single wave equation for the perturbations of the black hole:

$$\frac{\partial^2 \psi^{(\ell m)}}{\partial r^{*2}} - \frac{\partial^2 \psi^{(\ell m)}}{\partial t^2} + V^{(\ell)}(r)\, \psi^{(\ell m)} = 0, \tag{3}$$

where the potential function $V^{(\ell)}(r)$ is given by

$$\begin{aligned} V^{(\ell)}(r) &= \left(1 - \frac{2M}{r}\right)\left\{\frac{1}{\Lambda^2}\left[\frac{72M^3}{r^5} - \frac{12M}{r^3}(\ell-1)(\ell+2)\left(1 - \frac{3M}{r}\right)\right]\right. \\ &+ \left.\frac{\ell(\ell-1)(\ell+2)(\ell+1)}{r^2\Lambda}\right\}, \end{aligned} \tag{4}$$

where r is the standard Schwarzschild radial coordinate, and r^* is the so-called tortoise coordinate, given by $r^* = r + 2M\ln(r/2M - 1)$, and Λ is a function of r described below. For all ℓ, the potential function is positive and has a peak near $r = 3M$. This equation was first derived by Zerilli [19]. Regge and Wheeler found an analogous equation for the odd parity perturbations, which are much simpler than the even parity perturbations, 13 years earlier in 1957 [16]. Note that the potential depends on ℓ, but is independent of m. This remarkable equation (along with its odd-parity counterpart, known as the Regge-Wheeler equation) completely describes gravitational perturbations of a Schwarzschild black hole. It continues to be studied by many researchers now 40 years after this perturbation program was begun in 1957.

When carried out to second order in the perturbation parameter, the first order equations are of course the same, but they are accompanied by an equation representing the second order correction. Remarkably, the second order equation has *exactly the same form* as the first order equation, with source terms which are non-linear in the first order perturbation quantities. As I mention below, the second order perturbation theory has been used in application to the head-on collision of two black holes to extend the range of validity of the first order treatment. But perhaps its most important use is in providing a check on the validity of the first order perturbation treatment. When second order corrections are small compared to first order results, one expects the first order results to be reliable.

Using the first order equation, it has been shown that the Schwarzschild matric is stable to perturbations, and has characteristic oscillation frequencies known as quasinormal modes [20]. These modes are solutions to the Zerilli equation as given in Eq. (3) which are completely ingoing at the horizon ($r^* = -\infty$) and completely outgoing at infinity ($r^* = \infty$). For each ℓ-mode, independent of m, there is a fundamental frequency, and overtones. These are very important results! One expects that a black hole, when perturbed in an arbitrary way, will oscillate at these quasinormal frequencies. This will give definite signals to look for with gravitational wave observatories.

These quasinormal frequencies are complex, meaning they have an oscillatory and a damping part (not growing—black holes are stable!), so the oscillations die away as the waves carry energy away from the system. The frequencies depend only on the mass, spin, and charge of the black hole.

There are numerous ways in which this perturbation theory has become essential in numerical black hole simulations, and the rest of this paper will concentrate on this subject. First of all, the fact that perturbation theory reveals that black holes have quasinormal mode oscillations raises expectations about the evolution of distorted black holes: they should, at least in the linear regime, oscillate at these frequencies which should be seen in fully non-linear numerical simulations. But are they still seen in highly non-linear interactions, e.g., in the collision of two black holes? Secondly, as we will see, this perturbation theory provides a method by which to separate out the Schwarzschild background from the wave degrees of freedom, which can be used to find waves in numerical simulations. Finally, as the perturbations are governed by their own evolution equation, this equation should be useful to actually evolve some classes of black hole initial data, as long as they represent slightly perturbed black holes, and this can be used as an important check of fully non-linear numerical codes. Certainly, during the late stages of black hole coalescence, the system will settle down to a slightly perturbed black hole, and numerical codes had better be able to accurately compute waves from such systems if they are to be used to help researchers find signals in actual data collected by gravitational wave observatories.

3.2. WAVEFORM EXTRACTION

In this section I show how to take this perturbation theory and apply it in a practical way to numerical black hole simulations. One considers now the numerically generated metric $g_{\alpha\beta,\mathrm{num}}$ to be the sum of a spherically symmetric part and a perturbation: $g_{\alpha\beta,\mathrm{num}} = \overset{o}{g}_{\alpha\beta} + h_{\alpha\beta}$, where the perturbation $h_{\alpha\beta}$ is expanded in tensor spherical harmonics as before. To compute the elements of $h_{\alpha\beta}$ in a numerical simulation, one integrates the numerically evolved metric components $g_{\alpha\beta,\mathrm{num}}$ against appropriate spherical harmonics over a coordinate 2–sphere surrounding the black hole. The orthogonality of the $Y_{\ell m}$'s allows one to "project" the contributions of the general wave signal into individual modes, as explained below. The resulting functions can then be combined in a gauge-invariant way, following the prescription given by Moncrief [21], leading directly to the Zerilli function. This procedure was originally developed by Abrahams [22] and developed further by various groups.

As mentioned above, we assume the general metric can be decomposed into its spherical and non-spherical parts. We assume here that the background is Schwarzschild, written in Schwarzschild coordinates, but the treatment of a more general spherical background in an arbitrary coordinate system is possible [23]. The nonspherical perturbation tensor $h_{\mu\nu}$ for even-parity perturbations can be written

$$
\begin{aligned}
h_{tt} &= -SH_0^{(\ell m)} Y_{\ell m}, && (5)\\
h_{tr} &= H_1^{(\ell m)} Y_{\ell m}, && (6)\\
h_{t\theta} &= h_0^{(\ell m)} Y_{\ell m,\theta}, && (7)\\
h_{t\phi} &= h_0^{(\ell m)} Y_{\ell m,\phi}, && (8)\\
h_{rr} &= S^{-1} H_2^{(\ell m)} Y_{\ell m}, && (9)\\
h_{r\theta} &= h_1^{(\ell m)} Y_{\ell m,\theta}, && (10)\\
h_{r\phi} &= h_1^{(\ell m)} Y_{\ell m,\phi}, && (11)\\
h_{\theta\theta} &= r^2 K^{(\ell m)} Y_{\ell m} + r^2 G^{(\ell m)} Y_{\ell m,\theta\theta}, && (12)\\
h_{\theta\phi} &= r^2 G^{(\ell m)} \left(Y_{\ell m,\theta\phi} - \cot\theta Y_{\ell m,\phi}\right), && (13)\\
h_{\phi\phi} &= r^2 K^{(\ell m)} \sin^2\theta Y_{\ell m} + r^2 G^{(\ell m)} \left(Y_{\ell m,\phi\phi} + \sin\theta\cos\theta Y_{\ell m,\theta}\right), && (14)
\end{aligned}
$$

where $S = (1 - 2M/r)$ comes from the Schwarzschild background, and r is the standard Schwarzschild radial coordinate. I have taken these expressions directly from Vishu's 1970 paper [1] where I first learned them.

Each ℓm-mode of $h_{\mu\nu}$ can then be obtained numerically by projecting the full metric against the appropriate $Y_{\ell m}$. So, for example,

$$H_2^{(\ell m)} = (1 - 2M/r) \int g_{rr,\mathrm{num}} Y_{\ell m} d\Omega. \tag{15}$$

More complex expressions can be developed for all metric functions, and on a more general background, as detailed in [24, 25]. These functions can be computed numerically, given a numerically generated spacetime, as described below.

Once the perturbation functions $H_2, K, G, etc.$ are obtained, one still needs to create the special combination that obeys the Zerilli equation. Moncrief showed that the Zerilli function that obeys this wave equation is gauge invariant in the sense discussed above, and can be constructed from the Regge-Wheeler variables as follows:

$$\psi^{(\ell m)} = \sqrt{\frac{2(\ell-1)(\ell+2)}{\ell(\ell+1)}} \frac{4rS^2 k_2^{(\ell m)} + \ell(\ell+1) r k_1^{(\ell m)}}{\Lambda}, \tag{16}$$

where

$$\Lambda \equiv \ell(\ell+1) - 2 + \frac{6M}{r}, \tag{17}$$

$$k_1^{(\ell m)} \equiv K^{(\ell m)} + SrG_{,r}^{(\ell m)} - 2\frac{S}{r} h_1^{(\ell m)}, \tag{18}$$

$$k_2^{(\ell m)} \equiv \frac{H_2^{(\ell m)}}{2S} - \frac{1}{2\sqrt{S}} \frac{\partial}{\partial r} \left(\frac{rK^{(\ell m)}}{\sqrt{S}} \right), \tag{19}$$

$$S \equiv 1 - \frac{2M}{r}. \tag{20}$$

In order to compute the Regge-Wheeler perturbation functions h_1, H_2, G, and K, one needs the spherical metric functions on some 2-sphere. We get these in 3D by interpolating the Cartesian metric functions onto a surface of constant coordinate radius, and computing the spherical metric functions from these using the standard transformation. These are straightforward but messy calculations which are covered in detail in Refs. [24, 25].

In summary, in this section I have outlined a practical approach to the use of perturbation theory as a tool to construct a gauge invariant measure of the gravitational radiation in a numerically generated black hole spacetime. There are various ways in which this information can be used, which is the subject of the following sections.

3.3. APPLICATIONS

In the sections above, I showed how to extract the gauge invariant Zerilli function at a given radius on some time slice of the numerical spacetime. In the following sections I show several ways in which this information can be used, including (a) *evolving* a numerically (or analytically) generated initial data set with perturbation theory and (b) extracting waveforms from a fully non-linear evolution, possibly from the same data set. This perturbative approach has also recently been successful in providing outer boundary conditions for a 3D numerical simulation [26].

3.3.1. *Axisymmetric Distorted Black Holes*

We begin with the study of axisymmetric single black holes that have been distorted by the presence of an adjustable torus of non-linear gravitational waves which surround it. The amplitude and shape of the initial wave can be specified by hand, as described below, and can create very highly distorted black holes. Such initial data sets, and their evolutions in axisymmetry, have been studied extensively, as described in Refs. [5, 27, 28]. For our purposes, we consider them as convenient initial data that create a distorted black hole that mimics the merger, just after coalescence, of two black holes colliding in axisymmetry [7].

Following [27], we write the 3–metric in the form originally used by Brill [29]:

$$d\ell^2 = \tilde{\psi}^4 \left(e^{2q}\left(d\eta^2 + d\theta^2\right) + \sin^2\theta d\phi^2\right), \tag{21}$$

where η is a radial coordinate related to the standard Schwarzschild isotropic radius $\bar{r}$ by $\bar{r} = e^\eta$. (We have set the scale parameter m in Ref.[27] to be 2 in this paper.) We choose our initial slice to be time symmetric, so that the extrinsic curvature vanishes. Thus, given a choice for the "Brill wave" function q, the Hamiltonian constraint leads to an elliptic equation for the conformal factor $\tilde{\psi}$. The function q represents the gravitational wave surrounding the black hole, and is chosen to be

$$q(\eta,\theta,\phi) = a\sin^n\theta\left(e^{-\left(\frac{\eta+b}{w}\right)^2} + e^{-\left(\frac{\eta-b}{w}\right)^2}\right)\left(1 + c\cos^2\phi\right). \tag{22}$$

Thus, an initial data set is characterized by the parameters (a, b, w, n, c), where, roughly speaking, a is the amplitude of the Brill wave, b is its radial location, w its width, and n and c control its angular structure. A study of full 3D initial data and their evolutions are discussed elsewhere [30, 31, 32]. If the amplitude a vanishes, the undistorted Schwarzschild solution results, leading to

$$\tilde{\psi} = 2\cosh\left(\frac{\eta}{2}\right), \tag{23}$$

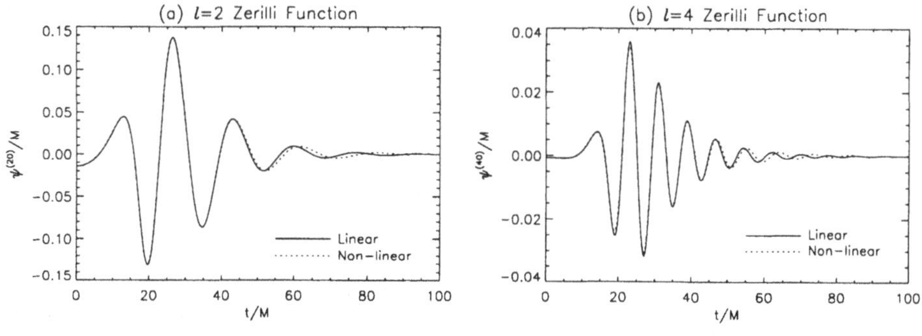

Figure 1. We show the (a)$\ell = 2$ and (b)$\ell = 4$ Zerilli functions as a function of time, extracted during linear and 2D non-linear evolutions of the data set $(a, b, w, n, c) = (0.05, 1, 1, 4, 0)$. The data were extracted at a radius of $r = 15M$.

which puts the metric (21) in the standard Schwarzschild isotropic form.

Linear Evolution: We now consider the evolution of these distorted black holes. If the amplitude is low enough, this should represent a small perturbation on a Schwarzschild black hole, and hence the system should be amenable to a perturbative treatment. The idea is to actually compute $\psi(r, t = 0)$, by extracting the Zerilli function at every radial grid point on the initial data slice, and use the Zerilli evolution equation to actually evolve the system as a perturbation. This will allow us to compare the waveform at some radius $\psi_{\text{lin}}(r_0, t)$ obtained in this way, to that obtained with a well-tested 2D axisymmetric code that performs full non-linear evolutions. This will serve as a test of the initial Zerilli function being given to the linear code, and of the linear evolution code itself. It will also help us determine for which Brill wave amplitudes this procedure breaks down. Beyond a certain point, perturbation theory will fail and non-linear effects will become important.

Let us first consider the data set $(a, b, w, n, c) = (0.05, 1, 1, 4, 0)$, in the notation above. In this case the Brill wave is initially far from the black hole, and will propagate in, hitting it and exciting the normal mode oscillations. In Figure 1 we show the $\ell = 2$ and $\ell = 4$ Zerilli functions as a function of time, at a radius of $r = 15M$. Data are shown from both the linear and 2D non-linear codes. We see that for both functions, the linear and non-linear results line up nicely until about $t = 50M$, when a phase shift starts to be noticeable. This phase shift and widening of the wave at late times is known from previous studies of numerical simulations of distorted black hole spacetimes in axisymmetry [5].

As second example, let us look at $(a, b, w, n, c) = (0.05, 0, 1, 4, 0)$. In this

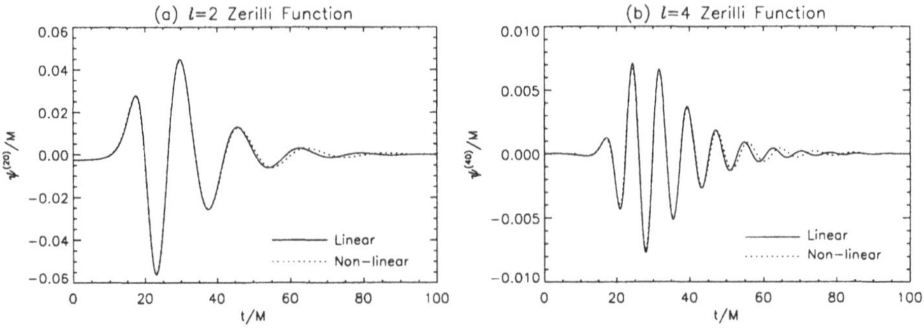

Figure 2. We show the (a)$\ell = 2$ and (b)$\ell = 4$ Zerilli functions as a function of time, extracted during linear and 2D non-linear evolutions of the data set $(a, b, w, n, c) = (0.05, 0, 1, 4, 0)$. The data were extracted at a radius of $r = 15M$.

case the Brill wave is initially right on the throat. In Figure 2 we show the $\ell = 2$ and the $\ell = 4$ waveforms as a function of time extracted at a radius of $r = 15M$. Again, data from both the linear and 2D non-linear codes are shown. The data line up well until about $t = 50M$, when phase errors again show up.

This is the first step in our ladder of credibility. We have shown that these data sets provide an important testbed for a numerical black hole evolution. First, they confirm that these data sets can be treated as linear perturbations on a Schwarzschild background, since both linear and fully non-linear evolutions agree. Second, they remarkably confirm the results of a complex, non-linear evolution code which evolves a black hole (in maximal slicing). This gives great confidence in the ability of this code to treat black holes and extract waveforms, even in the more highly distorted cases where perturbation theory breaks down (but waveform extraction will not necessarily break down, at least far from the hole). We will use this technique in various ways below.

3.3.2. *Axisymmetric Black Hole Collisions*

We now turn to another application of this basic idea of evolving dynamic black hole spacetimes with perturbation theory, but this time we consider two black holes colliding head on. It might seem to be impossible to treat colliding black holes perturbatively, but there are two limits in which perturbation theory has been shown to be incredibly successful. First, if the two holes are so close together initially that they have actually already merged into one, they might be considered as a single perturbed Schwarzschild hole (the so-called "close limit"). Using similar ideas to those discussed above, Price and Pullin and others [33, 34, 35, 36, 37] used this technique to produce waveforms for colliding black holes in the Misner [38] and Brill and

Lindquist [39] black hole initial data. These initial data sets for multiple black holes are actually known analytically. The original paper of Price and Pullin [33] is what spurred on so much interest in these many applications of perturbation theory as a check on numerical relativity. Second, when the holes are very far apart, one can consider one black hole as a test particle falling into the other. Then one rescales the answer obtained by formally allowing the "test particle" to be a black hole with the same mass as the one it is falling into [6, 7, 34].

The details of this success has provided insights into the nature of collisions of holes, and should also apply to many systems of dynamical black holes. The waveforms and energies agree remarkably well with numerical simulations. Moreover, second order perturbation theory [37] spectacularly improved the agreement between the close limit and full numerical results for even larger distances between the holes, although ultimately beyond a certain limit the approximation is simply inappropriate and breaks down.

The success of these techniques suggests, among other things, that these are very powerful methods that can be used hand-in-hand with fully nonlinear numerical evolutions, and can be applied in a variety of black hole spacetimes where one might naively think they would not work. For these reasons, many researchers are continuing to apply these techniques to the axisymmetric case with more and more complicated black hole spacetimes (e.g., the collision of boosted black holes [36], or counter-rotating, spinning black holes, colloquially known as the "cosmic screw"). Furthermore, these techniques will become even more essential in 3D, where we cannot achieve resolution as high as we can with 2D codes.

Finally, this is yet another rung on the "ladder of credibility": we now have not only slightly perturbed Schwarzschild spacetimes to consider, but also a series of highly nontrivial colliding black hole spacetimes that are now well understood in axisymmetry due to the nice interplay between perturbative and fully numerical treatments of the same problems. These then provide excellent testbeds for 3D simulations, which we turn to next.

3.3.3. *3D Testbeds*

Armed with robust and well understood axisymmetric black hole codes, we now consider the 3D evolution of axisymmetric distorted black hole initial data. These same axisymmetric initial data sets can be ported into a 3D code in cartesian coordinates, evolved in 3D, and the results can be compared to those obtained with the 2D, axisymmetric code discussed above. The 3D code used to evolve these black hole data sets is described in Refs. [10, 40], and the simulations described here are major simulations on very large supercomputers: they require about 12 GBytes of memory and take more than 24 hours on a 128 processor SGI Origin 2000 computer.

In the first of these simulations, we study the evolution of the distorted single black hole initial data set $(a, b, w, n, c) = (0.5, 0, 1, 2, 0)$. As the azimuthal parameter c is zero, this is axisymmetric and can also be evolved in 2D. In Figure 3a we show the result of the 3D evolution, focusing on the $\ell = 2$ Zerilli function extracted at a radius $r = 8.7M$ as a function of time. Superimposed on this plot is the same function computed during the evolution of the same initial data set with a 2D code, based on the one described in detail in [5, 28]. The agreement of the two plots over the first peak is a strong affirmation of the 3D evolution code and extraction routine. It is important to note that the 2D results were computed with a different slicing (maximal), different coordinate system, and a *different spatial gauge.* Yet the physical results obtained by these two different numerical codes, as measured by the waveforms, are remarkably similar (as one would hope). A full evolution with the 2D code to $t = 100M$, by which time the hole has settled down to Schwarzschild, shows that the energy emitted in this mode at that time is about $4 \times 10^{-3} M$. This result shows that now it is possible in full 3D numerical relativity, in cartesian coordinates, to study the evolution and waveforms emitted from highly distorted black holes, even when the final waves leaving the system carry a small amount of energy.

In Fig. 3b we show the $\ell = 4$ Zerilli function extracted at the same radius, computed during evolutions with 2D and 3D codes. This waveform is more difficult to extract, because it has a higher frequency in both its angular and radial dependence, and it has a much lower amplitude: the energy emitted in this mode is about three orders of magnitude smaller than the energy emitted in the $\ell = 2$ mode, yet it can still be accurately evolved and extracted. This is quite a remarkable result, and bodes well for the ability of numerical relativity codes ultimately to compute accurate waveforms that will be of great use in interpreting data collected by gravitational wave detectors. (However, as I point out below, there is a quite a long way to go before the general 3D coalescence can be studied!)

These results have been reported in much more detail in Refs.[30, 40].

3.3.4. *True 3D Distorted Black Holes*

We now turn to radiation extraction in true 3D black hole evolutions. This is of major importance for the connection between numerical relativity and gravitational wave astronomy. Gravitational wave detectors such as LIGO, VIRGO, and GEO will measure these waves directly, and may depend on numerical relativity to provide templates to both extract the signals from the experimental data and to interpret the results.

In the sections above, I showed by comparison to 2D results that a 3D code is able to accurately simulate distorted black holes. Armed with these tests, we now consider evolutions of initial data sets which are non-

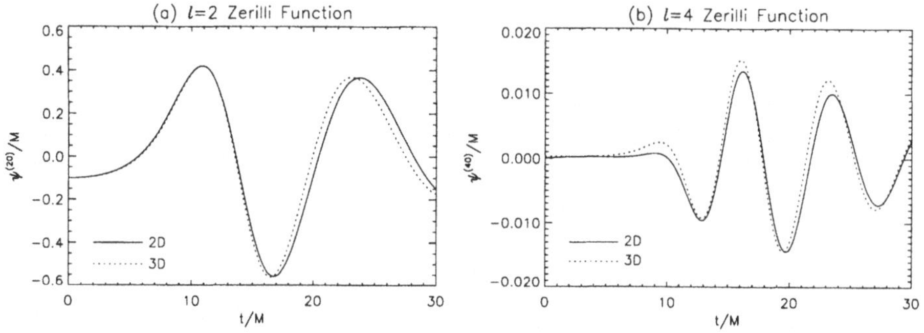

Figure 3. We show the (a) $\ell = 2$ and (b) $\ell = 4$ Zerilli functions vs. time, extracted during 2D and 3D evolutions of the data set $(a, b, w, n, c) = (0.5, 0, 1, 2, 0)$. The functions were extracted at a radius of $8.7M$. The 2D data were obtained with 202×54 grid points, giving a resolution of $\Delta\eta = \Delta\theta = 0.03$. The 3D data were obtained using 300^3 grid points and a resolution of $\Delta x = 0.0816M$.

axisymmetric distorted black holes, *i.e.*, data sets which have non-vanishing azimuthal parameter c. We also consider the evolution of the same distorted black hole data sets via linearized theory, as we did with the 2D results presented above. The techniques are the same, although the details in a full 3D treatment are more complicated. Please refer to Refs. [24, 25] for more details.

The initial data set $(a = -0.1, b = 0, c = 0.5, w = 1, n = 4)$ was evolved with the 3D numerical relativity Cartesian code described above, with 300^3 grid zones points in each coordinate direction. In Figure 4 we show the $\ell = m = 2$ Zerilli function computed during the evolution, comparing both the full non-linear theory with the linearized treatment. This is the first time a 3D non-linear numerical relativity code has been used to compute waveforms from fully 3D distorted black hole (but also see recent results using a characteristic formulation [14]). From the figure it is clear that the 3D results agree well with the perturbative treatment, even though the energy carried by these waves is very small.

Nonaxisymmetric modes are extracted not just because they *can* be, but because they should be quite *important* for gravitational wave observatories. It turns out that the $\ell = 2, m = 2$ is considered to be one of the most promising black hole modes to be seen by gravitational wave detector. In realistic black hole coalescence, the final hole is expected to have a large amount of angular momentum, possibly near the Kerr limit $a = 1$. This particular mode is one of the least damped (much less damped than for Schwarzschild, as seen here), and is also expected to be strongly excited [41]. Therefore, it is important to begin exhaustively testing the code's ability to generate and cleanly extract such nonaxisymmetric modes, even

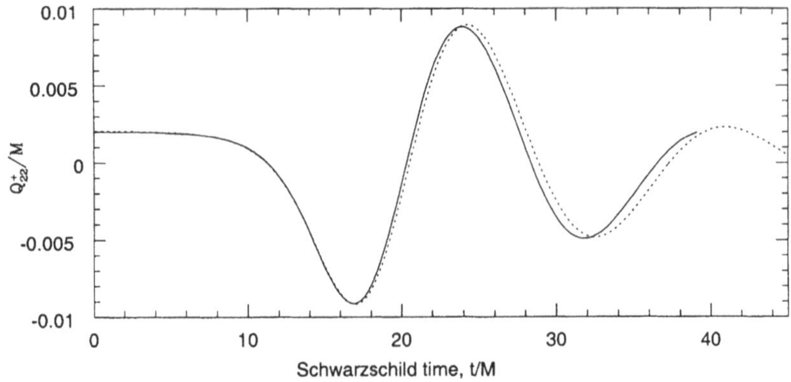

Figure 4. We show waveforms for the $\ell = m = 2$ Zerilli function extracted from the linear and non-linear evolution codes for a fully 3D, nonaxisymmetric distorted black hole. The dotted line shows the linear evolution, evolving only the Zerilli equation, and the solid line shows the non-linear evolution in full 3D cartesian coordinates, using a massively parallel supercomputer.

in the case studied here without rotation.

Many more modes can be extracted, including $\ell = 4, m$ modes, and details and analysis can be found in Refs.[24, 25, 30, 42]. Comparisons between the perturbative and non-linear results reveal that waveforms can be accurately extracted in the linear regime, and that even at very low distortion amplitudes non-linear effects appear. The comparisons with perturbation theory are essential in understanding these effects, and will continue to be for some years to come.

4. Summary

I have given a brief overview of work on evolutions of distorted black holes and black hole collisions over the last decade, from 2D axisymmetric studies to recent 3D studies. At each stage along the way, perturbation theory has turned out to be an essential ingredient in the program. Our understanding of the 2D collision of two black holes has been aided immensely from perturbation theory, and in the last year our ability to simulate true 3D distorted black holes, which model the late stages of 3D binary black hole coalescence, has matured considerably.

Unfortunately, we still have a very long way to go! Although one can now do 3D evolutions of distorted black holes, and accurately extract very low amplitude waves, the calculations one can presently do are actually very limited. With present techniques, the evolutions can only be carried out for a fraction of the time required to simulate the 3D orbiting coalescence. Most of what has been described here has been with certain symmetries, or with a single black hole in full 3D. At the present time, I am only aware of one attempt to study the collision of two black holes in 3D without any symmetries, which was recently reported by Brügmann [43]. However, this calculation is treated as a feasibility study, without detailed waveform extraction at this point, and again the evolution times are quite limited (less than those reported here).

Many new techniques are under development to extend the simulations to the time scales required for true binary black hole coalescence, such as apparent horizon boundary conditions [8, 13], hyperbolic systems [44, 45], and characteristic evolution [14]. Many of these (or perhaps all!) may be needed to handle the general, long term evolution of coalescing black holes. Each of these techniques may introduce numerical artifacts, even if at very low amplitude, to which the waveforms may be very sensitive. As new methods are developed and applied to numerical black hole simulations, they can now be tested on evolutions such as those presented here to ensure that the waveforms are accurately represented in the data.

In closing, I want to emphasize that the kind of work that Vishu helped to pioneer 30 years ago is still at the forefront of numerical simulations aided by the world's most powerful computers, and it seems clear that perturbation theory will continue to play an essential role in both the verification and physical understanding of large scale numerical simulations. There is still much work to do in this area, both on the theoretical and numerical areas. I expect it will continue to excite researchers for years to come!

5. Acknowledgments

This work has been supported by the Albert Einstein Institute (AEI) and NCSA. I am most thankful to Karen Camarda and Gabrielle Allen for carrying out much of the 3D work reported here. Among many colleagues who have contributed to this work on black holes, I also thank Andrew Abrahams, Pete Anninos, David Bernstein, Steve Brandt, David Hobill, Joan Massó, John Shalf, Larry Smarr, Wai-Mo Suen, John Towns, and Paul Walker.

I would like to thank K.V. Rao and the staff at NCSA for assistance with the computations. Calculations were performed at AEI and NCSA on an SGI/Cray Origin 2000 supercomputer.

References

1. C. V. Vishveshwara, Phys. Rev. D **1**, 2870 (1970).
2. R. F. Stark and T. Piran, Phys. Rev. Lett. **55**, 891 (1985).
3. A. M. Abrahams, G. B. Cook, S. L. Shapiro, and S. A. Teukolsky, Phys. Rev. D **49**, 5153 (1994).
4. S. Brandt and E. Seidel, Phys. Rev. D **52**, 870 (1995).

5. A. Abrahams, D. Bernstein, D. Hobill, E. Seidel, and L. Smarr, Phys. Rev. D **45**, 3544 (1992).
6. P. Anninos, D. Hobill, E. Seidel, L. Smarr, and W.-M. Suen, Phys. Rev. Lett. **71**, 2851 (1993).
7. P. Anninos, D. Hobill, E. Seidel, L. Smarr, and W.-M. Suen, Phys. Rev. D **52**, 2044 (1995).
8. E. Seidel and W.-M. Suen, Phys. Rev. Lett. **69**, 1845 (1992).
9. P. Anninos, G. Daues, J. Massó, E. Seidel, and W.-M. Suen, Phys. Rev. D **51**, 5562 (1995).
10. P. Anninos, K. Camarda, J. Massó, E. Seidel, W.-M. Suen, and J. Towns, Phys. Rev. D **52**, 2059 (1995).
11. M. A. Scheel, S. L. Shapiro, and S. A. Teukolsky, Phys. Rev. D **51**, 4208 (1995).
12. R. Marsa and M. Choptuik, Phys Rev D **54**, 4929 (1996).
13. G. B. Cook *et al.*, (1997), gr-qc/9711078.
14. R. Gomez, L. Lehner, R. Marsa, and J. Winicour, (1997), gr-qc/9710138.
15. S. Chandrasekhar, *The Mathematical Theory of Black Holes* (Oxford University Press, Oxford, England, 1983).
16. T. Regge and J. Wheeler, Phys. Rev. **108**, 1063 (1957).
17. R. J. Gleiser, C. O. Nicasio, R. H. Price, and J. Pullin, Class. Quant. Grav. **13**, L117 (1996).
18. Eg., Article by Price in this volume.
19. F. J. Zerilli, Phys. Rev. Lett. **24**, 737 (1970).
20. Eg., Article by Andersson in this volume.
21. V. Moncrief, Annals of Physics **88**, 323 (1974).
22. A. Abrahams, Ph.D. thesis, University of Illinois, Urbana, Illinois, 1988.
23. E. Seidel, Phys. Rev. D **42**, 1884 (1990).
24. G. Allen, K. Camarda, and E. Seidel, (1998), in preparation for Phys. Rev. D.
25. G. Allen, K. Camarda, and E. Seidel, (1998), in preparation for Phys. Rev. Lett.
26. A. M. Abrahams *et al.*, (1997), gr-qc/9709082.
27. D. Bernstein, D. Hobill, E. Seidel, and L. Smarr, Phys. Rev. D **50**, 3760 (1994).
28. D. Bernstein, D. Hobill, E. Seidel, L. Smarr, and J. Towns, Phys. Rev. D **50**, 5000 (1994).
29. D. S. Brill, Ann. Phys. **7**, 466 (1959).
30. K. Camarda, Ph.D. thesis, University of Illinois at Urbana-Champaign, Urbana, Illinois, 1997.
31. K. Camarda and E. Seidel, in preparation (unpublished).
32. S. Brandt, K. Camarda, and E. Seidel, in preparation (unpublished).
33. R. H. Price and J. Pullin, Phys. Rev. Lett. **72**, 3297 (1994).
34. P. Anninos, R. H. Price, J. Pullin, E. Seidel, and W.-M. Suen, Phys. Rev. D **52**, 4462 (1995).
35. A. Abrahams and R. Price, Phys. Rev. D **53**, 1972 (1996).
36. J. Baker, A. Abrahams, P. Anninos, S. Brandt, R. Price, J. Pullin, and E. Seidel, Phys. Rev. D **55**, 829 (1997).
37. R. J. Gleiser, C. O. Nicasio, R. H. Price, and J. Pullin, Physical Review Letters **77**, 4483 (1996).
38. C. Misner, Phys. Rev. **118**, 1110 (1960).
39. D. S. Brill and R. W. Lindquist, Phys. Rev. **131**, 471 (1963).
40. K. Camarda and E. Seidel, in preparation.
41. Éanna É. Flanagan and S. A. Hughes, gr-qc **9701039**, (1997).
42. K. Camarda and E. Seidel, in preparation (unpublished).
43. B. Brügmann, (1997), gr-qc/9708035.
44. C. Bona, J. Massó, E. Seidel, and J. Stela, Phys. Rev. D **56**, 3405 (1997).
45. A. Abrahams, A. Anderson, Y. Choquet-Bruhat, and J. York, Phys. Rev. Lett. **75**, 3377 (1995).

24. CAUCHY-CHARACTERISTIC MATCHING

NIGEL T. BISHOP
Department of Mathematics, Applied Mathematics and Astronomy, University of South Africa, P.O. Box 392, Pretoria 0003, South Africa.
ROBERTO GOMEZ, LUIS LEHNER, BELA SZILAGYI, JEFFREY WINICOUR
Department of Physics and Astronomy, University of Pittsburgh, Pittsburgh, PA 15260.
AND
RICHARD A. ISAACSON
Department of Physics and Astronomy, University of Pittsburgh, Pittsburgh, PA 15260.
and
Physics Division, National Science Foundation, 4201 Wilson Blvd., Arlington VA 22230.

1. Preamble

Throughout his career in theoretical gravitational physics, Vishu has been interested in questions of black holes and gravitational radiation. The subject of his useful early study of radiation from binary systems [1] is now at the forefront of research as the principal target for the first exciting experimental measurements of the LIGO project. His fundamental studies of the properties of black hole excitations and radiation [2] used the formidable technology of the mid 1960's, i.e. Regge-Wheeler perturbation theory, to analytically extract the crucial physical result that black holes were stable. In his continuing studies of the interaction of black holes with gravitational radiation [3], he first demonstrated the phenomenon that was later to be called normal mode excitations of black holes, and noted that the frequency of the emitted radiation carried with it key information which could be used to determine the mass of the invisible black hole. These issues are still at the heart of current research three decades later.

B. R. Iyer and B. Bhawal (eds.), Black Holes, Gravitational Radiation and the Universe, 383–408.

Today, the technology to study these questions has become even more formidable, requiring large groups of researchers to work hard at developing computer codes to run on the massively parallel supercomputers of today and tomorrow. This has caused gravitational theory to enter into the realm of "big science" already familiar to experimental physics, with large, geographically distributed collaborations of scientists engaged on work on expensive, remote, central facilities. The goal of this modern work is to understand the full details of black hole collisions, the fundamental two-body problem for this field. Recent developments in this area are very encouraging, but it may well take another three decades until all the riches of this subject are mined.

In this paper, we will present all the gory details of how the best current methods in computational gravitation can be forged into a single tool to attack this crucial problem, one which is currently beyond our grasp, but perhaps not out of our reach.

2. Introduction

Although Einstein's field equations for gravitation have been known for the past 80 years, their complexity has frustrated attempts to extract the deep intellectual content hidden beneath intractable mathematics. The only tool with potential for the study of the general dynamics of time-dependent, strongly nonlinear gravitational fields appears to be computer simulation. Over the past two decades, two alternate approaches to formulating the specification and evolution of initial data for complex physical problems have emerged. The *Cauchy* (also known as the ADM or "3 + 1") approach foliates spacetime with spacelike hypersurfaces. Alternatively, the *characteristic* approach uses a foliation of null hypersurfaces.

Each scheme has its own different and complementary strengths and weaknesses. Cauchy evolution is more highly developed and has demonstrated good ability to handle relativistic matter and strong fields. However, it is limited to use in a finite region of spacetime, and so it introduces an outer boundary where an artificial boundary condition must be specified. Characteristic evolution allows the compactification of the entire spacetime, and the incorporation of future null infinity within a finite computational grid. However, in turn, it suffers from complications due to gravitational fields causing focusing of the light rays. The resultant caustics of the null cones lead to coordinate singularities. At present, the unification of both of these methods [4] appears to offer the best chance for attacking the fundamental two-body problem of modern theoretical gravitation: the collision of two black holes.

The basic methodology of the new computational approach called

Cauchy-characteristic matching (CCM), utilizes Cauchy evolution within a prescribed world-tube, but replaces the need for an outer boundary condition by matching onto a characteristic evolution in the exterior to this world-tube, reaching all the way out to future null infinity. The advantages of this approach are: (1) Accurate waveform and polarization properties can be computed at null infinity; (2) Elimination of the unphysical outgoing radiation condition as an outer boundary condition on the Cauchy problem, and with it all accompanying contamination from spurious back-reflections, consequently helping to clarify the Cauchy initial value problem. Instead, the matching approach incorporates exactly all physical backscattering from true nonlinearities; (3) Production of a global solution for the spacetime; (4) Computational efficiency in terms of both the grid domain and algorithm. A detailed assessment of these advantages is given in Sec. 3.

The main modules of the matching algorithm are:

- The outer boundary module which sets the grid structures.
- The extraction module whose input is Cauchy grid data in the neighborhood of the world-tube and whose output is the inner boundary data for the exterior characteristic evolution.
- The injection module which completes the interface by using the exterior characteristic evolution to supply the outer Cauchy boundary condition, so that no artificial boundary condition is necessary.

Details of the Cauchy and characteristic codes have been presented elsewhere. In this paper, we present only those features necessary to discuss the matching problem.

3. Advantages of Cauchy-characteristic matching (CCM)

There are a number of places where errors can arise in a pure Cauchy computation. The key advantage of CCM is that there is tight control over the errors, which leads to computational efficiency in the following sense. For a given target error ε, what is the amount of computation required for CCM (denoted by A_{CCM}) compared to that required for a pure Cauchy calculation (denoted by A_{WE})? It will be shown that $A_{CCM}/A_{WE} \to 0$ as $\varepsilon \to 0$, so that in the limit of high accuracy CCM is by far the most efficient method.

In CCM a "3 + 1" interior Cauchy evolution is matched to an exterior characteristic evolution at a world-tube of constant radius R. The important point is that the characteristic evolution can be rigorously compactified, so that the whole spacetime to future null infinity may be represented on a finite grid. From a numerical point of view this means that the only error made in a calculation of the gravitational radiation at infinity is that due to the finite discretization h; for second-order algorithms this error is

$\mathcal{O}(h^2)$. The value of the matching radius R is important, and it will turn out that for efficiency it should be as small as possible. The difficulty is that if R is too small then caustics may form. Note however that the smallest value of R that avoids caustics is determined by the physics of the problem, and is *not* affected by either the discretization h or the numerical method.

On the other hand, the standard approach is to make an estimate of the gravitational radiation solely from the data calculated in a pure Cauchy evolution. The simplest method would be to use the raw data, but that approach is too crude because it mixes gauge effects with the physics. Thus a substantial amount of work has gone into methods to factor out the gauge effects and to produce an estimate of the gravitational field at null infinity from its behavior within the domain of the Cauchy computation [5, 6, 7]. We will call this method *waveform extraction, or WE.* The computation is performed in a domain D, whose spatial cross-section is finite and is normally spherical or cubic. Waveform extraction is computed on a world-tube Γ, which is strictly in the interior of D, and which has a spatial cross-section that is spherical and of radius r_E. While WE is a substantial improvement on the crude approach, it has limitations. Firstly, it disregards the effect, between Γ and null infinity, of the nonlinear terms in the Einstein equations; The resulting error will be estimated below. Secondly, there is an error, that appears as spurious wave reflections, due to the inexact boundary condition that has to be imposed at ∂D. However, we do not estimate this error because it is difficult to do so for the general case; And also because it is in principle possible to avoid it by using an exact artificial boundary condition (at a significant computational cost).

The key difference between CCM and WE is in the treatment of the nonlinear terms between Γ and future null infinity. WE ignores these terms, and this is an inherent limitation of a perturbative method (even if it is possible to extend WE beyond linear order, there would necessarily be a cut-off at some finite order). Thus our strategy for comparing the computational efficiency of CCM and WE will be to find the error introduced into WE from ignoring the nonlinear terms; And then to find the amount of computation needed to control this error.

3.1. ERROR ESTIMATE IN WE

As discussed earlier, ignoring the nonlinear terms between Γ (at $r = r_E$) and null infinity introduces an error, which we estimate using characteristic methods. The Bondi-Sachs metric is

$$\begin{aligned} ds^2 &= -\left(e^{2\beta}\frac{V}{r} - r^2 h_{AB}U^A U^B\right) du^2 \\ &\quad - 2e^{2\beta} du dr - 2r^2 h_{AB} U^B du dx^A + r^2 h_{AB} dx^A dx^B, \end{aligned} \tag{1}$$

where $A, B = 2,3$ and h_{AB} is a spherical metric that is completely described by one complex function J. The initial data required on a null cone $u = constant$ is J, and the hypersurface equations $R_{1\alpha} = 0$ then form a hierarchy from which β, U^A and $W \equiv (V - r)/r^2$ are found [8]. The evolution equation $R_{AB}h^{AB} = 0$ [8] is

$$2(rJ)_{,ur} = L_J + N_J \tag{2}$$

where L_J represents the linear part and N_J the nonlinear part.

The order of magnitude of various terms can be expressed in terms of a function $c(u, x^A)$ (whose time derivative $c_{,u}$ is the news function); note that c is not a small quantity. The expressions are

$$J = \mathcal{O}(\frac{c}{r}), \beta = \mathcal{O}(\frac{c^2}{r^2}), U^A = \mathcal{O}(\frac{c}{r^2}), W = \mathcal{O}(\frac{c^2}{r^2}). \tag{3}$$

These estimates are obtained from the hypersurface equations, and assume that the background geometry is Minkowskian. Should this not be the case then constants of order unity would be added, and the effect of this would be to amend Eq.(2) by adding terms to L_J so that it represents wave propagation on a non-Minkowskian background. However, the order of magnitude of terms in N_J would not be affected. It is straight-forward to confirm that N_J involves terms of order

$$\mathcal{O}(\frac{c^2}{r^3}). \tag{4}$$

WE estimates the news at future null infinity from data at $r = r_E$, and could be made exact if N_J were zero. Thus the error introduced by ignoring N_J is

$$\varepsilon(c_{,u}) \equiv (c_{,u})_{\text{exact}} - (c_{,u})_{WE} = \int_{r_E}^{\infty} \mathcal{O}(\frac{c^2}{r^3})dr = \mathcal{O}(\frac{c^2}{r_E^2}). \tag{5}$$

This would be the unavoidable linearization error in WE were it implemented as an "exact" artificial boundary condition by using global techniques, such as the difference potential method, to eliminate back reflection at the boundary [9]. However, this is computationally expensive [10] and has not even been attempted in general relativity. The performance of WE continues to improve but the additional error due to back reflection remains [11].

3.2. COMPUTATIONAL EFFICIENCY

A numerical calculation of the emission of gravitational radiation using a CCM algorithm is expected to be second-order convergent, so that after a fixed time interval the error is:

$$\varepsilon = \mathcal{O}(h^2) \simeq k_1 h^2, \tag{6}$$

where h is the discretization length. On the other hand, the same calculation using WE must allow for the error found in Eq.(5), and therefore after the same fixed time interval there will be an error of:

$$\varepsilon = \mathcal{O}(h^2, r_E^{-2}) \simeq \max(k_2 h^2, \frac{k_3}{r_E^2}). \tag{7}$$

We now estimate the amount of computation required for a given desired accuracy. We make one important assumption:

- The computation involved in matching, and in waveform extraction, is an order of magnitude smaller than the computation involved in evolution, and is ignored.

For the sake of transparency we also make some simplifying assumptions; If not true there would be some extra constants of order unity in the formulas below, but the qualitative conclusions would not be affected.

1. The amount of computation per grid point per time-step, a, is the same for the Cauchy and characteristic algorithms.
2. The constants k_1, k_2 in the equations above are approximately equal and will be written as k.
3. In CCM the numbers of Cauchy and characteristic grid-points are the same; Thus the total number of grid points per time-step is:

$$\frac{8\pi R^3}{3h^3}. \tag{8}$$

4. In WE, the number of grid points in D is twice the number contained in Γ; thus the total number of grid points per time-step is

$$\frac{8\pi r_E^3}{3h^3}. \tag{9}$$

It follows that the total amount of computation A (i.e., number of floating-point operations) required for the two methods is:

$$A_{CCM} = \frac{8\pi R^3 a}{3h^4}, \quad A_{WE} = \frac{8\pi r_E^3 a}{3h^4}. \tag{10}$$

Thus $R > r_E$ or $R < r_E$ determines which method requires the least amount of computation. Because of the assumptions (1) to (4) this criterion is not exact but only approximate.

As stated earlier, in a given physical situation the minimum allowed value of R is determined by the physics. However, r_E is determined by the target error (Eq.(7)); and there is also a minimum value determined by the condition that the nonlinearities must be sufficiently weak for a perturbative expansion to be possible. Thus, in a loose sense, the minimum value of r_E is expected to be related to the minimum value of R. It follows that the computational efficiency of a CCM algorithm is never expected to be significantly worse than that of a WE algorithm.

If high accuracy is required, the need for computational efficiency always favors CCM. More precisely, for a given desired error ε, Eqs.(6) and (7), and assumption (2), imply:

$$h = \sqrt{\varepsilon/k}, \; r_E = \sqrt{k_3/\varepsilon}. \tag{11}$$

Thus, substituting Eq.(11) into Eq.(10),

$$A_{CCM} = \frac{8\pi R^3 a k^2}{3\varepsilon^2}, \; A_{WE} = \frac{8\pi a k^2 k_3^{3/2}}{3\varepsilon^{7/2}}, \tag{12}$$

so that

$$\frac{A_{CCM}}{A_{WE}} = \frac{R^3 \varepsilon^{3/2}}{k_3^{3/2}} \to 0 \; as \; \varepsilon \to 0. \tag{13}$$

This is the crucial result, that the computational intensity of CCM relative to that of WE goes to zero as the desired error ε goes to zero.

4. Extraction

We describe a procedure by which information about the 4-geometry on the neighborhood of a world-tube Γ, obtained during a $3+1$ simulation of Einstein's equations, is used to extract boundary data appropriate for an exterior characteristic formulation. A numerical implementation of this characteristic formulation [12] is then used to propagate the gravitational signal to null infinity, where the radiation patterns are calculated. The process we describe here is non-perturbative, and to make it as portable as possible, it assumes only that the $3+1$ simulation can provide the 3-metric, the lapse and the shift by interpolation to a set of prescribed points.

We start by describing the world-tube Γ in Sec. 4.1, we explain what information needs to be provided by the $3+1$ simulation around the world-tube in Sec. 4.2, giving details of the transformation to null coordinates in Sec. 4.3, and the expression for the Bondi metric in Sec. 4.5. This provides the boundary data needed to start up the characteristic code as described in Sec. 4.6. The characteristic code which takes this boundary information and calculates the waveforms is described in detail in Ref.[12].

4.1. PARAMETRIZATION OF THE WORLD-TUBE

The notation and formalism are based on Misner, Thorne and Wheeler [13]. Greek indices range from 1 to 4, Latin indices range from 1 to 3, and upper case Latin indices refer to coordinates on the sphere and range from 2 to 3.

The intersections S_t of the world-tube with the space-like slices Σ_t of the Cauchy foliation are topologically spherical, and they can be parametrized by labels $\tilde{y}^A$, $A = [2,3]$ on the sphere. The intersections themselves are labeled by the time coordinate of the Cauchy foliation, $x^4 = t$. Future oriented null cones emanating from the world-tube are parametrized by the labels on the sphere $\tilde{y}^A$ and an affine parameter λ along the radial direction, with $\lambda = 0$ on the world-tube. With the identifications $\tilde{y}^1 = \lambda$ and retarded time $\tilde{y}^4 \equiv u = t$, we define a null coordinate system $\tilde{y}^\alpha = (\tilde{y}^1, \tilde{y}^A, \tilde{y}^4)$. We will later introduce a second null coordinate system $y^\alpha = (y^1, y^A, y^4)$, where $y^1 = r$, r being a surface area coordinate, and $y^A = \tilde{y}^A$, $y^4 = \tilde{y}^4$.

Following Ref.[14], we cover the unit sphere with two stereographic coordinate patches, centered around the North and South poles, respectively, where the stereographic coordinate is related to the usual spherical polar coordinates (θ, ϕ) by

$$\xi_{\text{North}} = \sqrt{\frac{1-\cos\theta}{1+\cos\theta}}\, e^{i\phi}, \quad \xi_{\text{South}} = \sqrt{\frac{1+\cos\theta}{1-\cos\theta}}\, e^{-i\phi}. \tag{14}$$

Let $\tilde{y}^A \equiv (q, p)$, for $A = [2,3]$, label the points of a stereographic coordinate patch. We adopt as these labels the real and imaginary part of the stereographic coordinate, i.e. for $\xi = \tilde{y}^2 + i\tilde{y}^3 = q + ip$.

For the computational implementation of the procedure we are describing, we introduce a discrete representation of these coordinate patches

$$\tilde{y}^2_{\mathtt{i}} = -1 + (\mathtt{i} - 3)\Delta, \quad \tilde{y}^3_{\mathtt{j}} = -1 + (\mathtt{j} - 3)\Delta, \tag{15}$$

where the computational spatial grid indices $\mathtt{i,j,k}$ run from 1 to N and $\Delta = 2/(N-5)$, N is the stereographic grid size and Δ the stereographic grid spacing.

On each patch we introduce complex null vectors on the sphere $q^A = (P/2)(\delta^A_2 + i\delta^A_3)$, where $P = 1 + \xi\bar{\xi}$. The vectors q^A, $\bar{q}^A$ define a metric on the unit sphere

$$q_{AB} = \frac{1}{2}(q_A \bar{q}_B + \bar{q}_A q_B) = \frac{4}{P^2} \begin{bmatrix} 1 & 0 \\ 0 & 1 \end{bmatrix} \tag{16}$$

with determinant $\det(q_{AB}) = 16/P^4$. The co-vector $\bar{q}_A$ satisfies the orthogonality condition $\bar{q}_A q^A = 2$, and has components $q_A = (2/P)\,(\delta^2_A + i\delta^3_A)$.

Given Cartesian coordinates $x^i = (x, y, z)$ on a space-like slice Σ_t, the intersection S_t of the world-tube Γ with Σ_t is described parametrically by three functions of the $\tilde{y}^A$, $x^{(0)i} = f^i(\tilde{y}^A)$. In the following we will fix the location of the world-tube by choosing these functions (in stereographic coordinates) as:

$$\begin{aligned} f^x(\tilde{y}^A) &= 2R\left(\frac{\Re(\xi)}{1+\xi\bar{\xi}}\right), \\ f^y(\tilde{y}^A) &= \pm 2R\left(\frac{\Im(\xi)}{1+\xi\bar{\xi}}\right), \\ f^z(\tilde{y}^A) &= \pm R\left(\frac{1-\xi\bar{\xi}}{1+\xi\bar{\xi}}\right), \end{aligned} \qquad (17)$$

where the positive (negative) sign corresponds to the north (south) patch. This defines a canonical spherical section S_t of radius R in Minkowski-space. Note that this provides a prescription for locating the world-tube which is *time-independent.*

Given two 3-dimensional quantities v^i and w^i, we introduce their Euclidean inner product $(v \cdot w) = \delta_{ij} v^i w^j$. The stereographic coordinates of points on the surface of the 3-sphere with Cartesian coordinates $x^i = (x, y, z)$ can then be represented by $y^A(x^i) = (q, p)$, with

$$q(x^i) = \frac{x}{\hat{r} \pm z}, \quad p(x^i) = \pm \frac{y}{\hat{r} \pm z} \qquad (18)$$

on the north $(+)$ and south $(-)$ patches, where

$$\hat{r}^2 = (x \cdot x). \qquad (19)$$

Note that the stereographic coordinates are invariant under a change of scale, $y^A(x^i) = y^A(cx^i)$.

The Euclidean radius of the world-tube with Cartesian coordinates $x^{(0)a}$ is given by:

$$R^2 = (x^{(0)} \cdot x^{(0)}). \qquad (20)$$

This equation provides a particularly simple alternative definition of the choice of world-tube using Cartesian coordinates. This definition also holds for each instant of Cartesian time.

4.2. 4-*D* GEOMETRY AROUND THE WORLD-TUBE

The 4-*d* geometry around the world-tube is fully specified by the 4-*d* metric and its derivatives (alternatively, by the metric and the metric connection). They determine the unit normal n^α to the $t = constant$ slices, and given the parametrization of the world-tube, the (outward pointing) normal s^α to the world-tube. They also determine the generator of the outgoing null radial geodesics through the world-tube, which in turn completes the specification of the coordinate transformation $x^\alpha \to \tilde{y}^\alpha$ in a neighborhood of the world-tube.

In practice, the necessary information is not available at the discrete set of points on the world-tube specified by Eqs. (14) and (17) where $\xi_{\mathtt{ij}} = \tilde{y}^2_{\mathtt{i}} + i\tilde{y}^3_{\mathtt{j}}$ as given in Eq. (15). However, the required variables are known on the points of the computational grid used in the simulation, a Cartesian grid $(x_{\mathtt{i}}, y_{\mathtt{j}}, z_{\mathtt{k}})$, from which we interpolate them to the world-tube points to second order accuracy.

For a standard $3+1$ simulation [15], the variables that we need to interpolate are the 3-d metric g_{ij}, the lapse α and the shift β^i. Their spatial derivatives are also interpolated. Their values at the world-tube points are stored for a number of time levels, and the time derivatives of the 3-metric, lapse and shift at the world-tube are computed by finite-differencing between these time levels.

Using all these values we can compute the 4-metric $g_{\mu\nu}$, and its first derivatives $g_{\mu\nu,\sigma}$, using the following relations:

$$
\begin{aligned}
g_{it} &= g_{ij}\beta^j \,, \\
g_{tt} &= -\alpha^2 + g_{it}\beta^i \,, \\
g_{it,\mu} &= g_{ij,\mu}\beta^j + g_{ij}\beta^j_{\ ,\mu} \,, \\
g_{tt,\mu} &= -2\alpha\alpha_{,\mu} + g_{ij,\mu}\beta^i\beta^j + 2g_{ij}\beta^i\beta^j_{\ ,\mu} \,.
\end{aligned} \tag{21}
$$

The unit normal n^μ to the hypersurface Σ_t is determined from the lapse and shift

$$
n^\mu = \frac{1}{\alpha}\left(-\beta^i, 1\right). \tag{22}
$$

Let $s^\alpha = (s^i, 0)$ be the outward-pointing unit normal to the section S_t of the world-tube at time t^n. By construction, s^i lies in the slice Σ_t, and it is known given the two vectors $\partial_{\tilde{y}^2}$, $\partial_{\tilde{y}^3}$ in S_t, defined by the parametrization of the world-tube $x^i(\tilde{y}^A)$

$$
q^i = \frac{\partial x^i}{\partial \tilde{y}^2}, \quad p^i = \frac{\partial x^i}{\partial \tilde{y}^3}. \tag{23}
$$

These may be obtained analytically from Eq. (17), the equation for the world tube. Antisymmetrizing q^i and p^i, we obtain the spatial components of the normal 1-form σ_i and its norm σ

$$
\sigma_i = \epsilon_{ijk}\, q^j p^k, \quad \sigma = \sqrt{g^{ij}\sigma_i\sigma_j} \tag{24}
$$

from which s^i is obtained by raising σ_i with the contravariant 3-metric g^{ij} on the slice Σ_t and dividing by the norm σ, yielding

$$s^i = g^{ij}\frac{\sigma_j}{\sigma}. \tag{25}$$

The generators ℓ^α of the outgoing null cone C_t through S_t are given on the world-tube by

$$\ell^\alpha = \frac{n^\alpha + s^\alpha}{\alpha - g_{ij}\beta^i s^j} \tag{26}$$

which is normalized so that $\ell^\alpha t_\alpha = -1$, where $t^\alpha = \alpha n^\alpha + \beta^\alpha$ is the Cauchy evolution vector.

The equations in this section show explicitly how to use the output data from the Cauchy simulation to completely reconstruct the full 4-geometry of the spacetime, as well as other important geometrical objects of interest, in the neighborhood of the world-tube Γ. This is all described in this section within the Cartesian coordinate system used by the 3+1 computation, and holds to the second-order accuracy assumed for this computation. In the next three sections, we will demonstrate how to use this information to redescribe the same geometry in another coordinate system, the Bondi coordinates needed for characteristic simulations. There is nothing in these sections that goes beyond the elementary concepts of defining coordinate systems and transforming tensors under a change of coordinates. Nevertheless, it is quite instructive to see just how much work lies hidden beneath the clever notation used by theorists.

4.3. COORDINATE TRANSFORMATION

In this section we build the coordinate transformation between the 3 + 1 Cartesian coordinates x^α and the (null) affine coordinates $\tilde{y}^\alpha$. We will need this in the *neighborhood* of the world-tube, not just at a *point* on the tube, in order to easily pass information back and forth between the Cauchy and characteristic computer codes in the overlap region near the world-tube. In Sec. 4.4, we will transform the metric to affine coordinates. Finally, in Sec. 4.5, we will complete the transformation from affine to Bondi coordinates.

The motivation for this indirect route to the Bondi coordinate frame deserves some comment. Both affine coordinates $\tilde{y}^\alpha$ and Bondi coordinates y^α utilize null hypersurfaces for foliations. Calculating these directly would require the numerical solution of a nonlinear partial differential equation (the eikonal equation). A much simpler, but physically equivalent, approach is to instead solve the null geodesic equation in Cartesian coordinates, in order to find the rays $x^\mu(\lambda)$ generating the required null hypersurfaces.

Secondly, we must introduce the intermediary of the affine radial coordinate λ rather than the Bondi surface area coordinate r, since the latter is actually unknown until the angular coordinates are defined. As shown below, it is only after the null rays have been found that we are able to proceed with the orderly introduction of angular coordinates $\tilde{y}^A$ in the exterior of the world-tube. Finally, null geodesics can be analytically constructed trivially in terms of λ, whereas even when we have defined the surface area coordinate r, solution of the null geodesics equation requires the numerical solution of an ordinary differential equation. With this rationale complete, we will now proceed to carry out the coordinate definitions and transform the metric.

By inspection, $x^\alpha(\lambda)$, the solution to the geodesic equation relating x^α to $\tilde{y}^\alpha$ off the world-tube is:

$$x^\alpha = x^{(0)\alpha} + \ell^{(0)\alpha}\lambda + \mathcal{O}(\lambda^2). \tag{27}$$

This expression determines $x^\alpha(\lambda)$ to $\mathcal{O}(\lambda^2)$, given the coefficients

$$x^{(0)\alpha} = x^\alpha_{|\Gamma} \quad \text{and} \quad \ell^{(0)\alpha} = x^\alpha_{,\lambda\,|\Gamma}\,, \tag{28}$$

that is, given the coordinates of the points and the generators of the null cone through the world-tube section S_t. For completeness, we repeat the remaining coordinate relations defined in Sec. 4.1. Along each outgoing null geodesic emerging from S_t, angular and time coordinates are defined by setting their values to be constant along the rays, and equal to their values on the world-tube

$$\tilde{y}^A = y^A_{|\Gamma} \quad \text{and} \quad \tilde{y}^4 \equiv \tilde{u} = t. \tag{29}$$

Given the coordinate transformation $x^\mu = x^\mu(\tilde{y}^\alpha)$, we obtain the metric in null affine coordinates $\tilde{\eta}_{\alpha\beta}$ by

$$\tilde{\eta}_{\tilde{\alpha}\tilde{\beta}} = \frac{\partial x^\mu}{\partial \tilde{y}^\alpha}\frac{\partial x^\nu}{\partial \tilde{y}^\beta} g_{\mu\nu}. \tag{30}$$

The Jacobian of the coordinate transformation is viewed as a series expansion in the affine parameter λ for each point on the world-tube. Furthermore, we may omit the λ derivatives of the x^μ, i.e. the $x^\mu_{,\lambda}$ are not needed because the radial coordinate λ is an affine parameter of the null geodesics, hence the $\tilde{\eta}_{\lambda\tilde{\mu}}$ components of the null metric are fixed:

$$\tilde{\eta}_{\lambda\lambda} = \tilde{\eta}_{\lambda\tilde{A}} = 0, \quad \tilde{\eta}_{\lambda\tilde{u}} = -1, \tag{31}$$

(This follows from the conditions $s^\alpha n_\alpha = 0$, $\ell^\alpha \ell_\alpha = 0$, $s^\alpha s_\alpha = 1$, $n^\alpha n_\alpha = -1$ and $t^\alpha \ell_\alpha = -1$). In other words, $\tilde{\eta}_{\tilde{\alpha}\tilde{\beta}}$ contains six independent metric functions. The relevant part of the coordinate transformation is then

$$x^{\mu}_{,\tilde{\alpha}} \equiv \frac{\partial x^{\mu}}{\partial \tilde{y}^{\alpha}} = x^{(0)\mu}_{,\tilde{\alpha}} + x^{(1)\mu}_{,\tilde{\alpha}}\lambda + O(\lambda^2), \quad x^{(1)\mu}_{,\tilde{\alpha}} \equiv \ell^{(0)\mu}_{,\tilde{\alpha}}, \quad \text{for} \quad \tilde{\alpha} = (\tilde{A}, \tilde{u}). \tag{32}$$

Because the specification of the world-tube is time-independent, and the slices S_t of the world-tube are by construction at $t = constant$, only the angular derivatives of the $x^{(0)i}$ for $i = 1, 2, 3$ survive, i.e. the $O(\lambda^0)$ part of the Jacobian is given by the condition $\partial t / \partial \tilde{u}_{|\Gamma} = 1$, and by the relations

$$x^{(0)i}_{,\tilde{A}} = \frac{\partial f^i(\tilde{y}^B)}{\partial \tilde{y}^A}, \tag{33}$$

which can be evaluated from the analytic expressions Eq. (17).

To evaluate the $\mathcal{O}(\lambda)$ part of the Jacobian, we note that

$$x^{\mu}_{,\lambda\tilde{A}} = \ell^{\mu}_{,\tilde{A}}, \quad x^{\mu}_{,\lambda\tilde{u}} = \ell^{\mu}_{,\tilde{u}}\,. \tag{34}$$

We will spend the remainder of this section evaluating these terms. To proceed, we see from Eq. (26) that this will require derivatives of n^{μ} and s^i. For the simple case of geodesic slicing, $\alpha = 1$, $\beta^i = 0$, the derivatives of the normal n^{μ} vanish, and the transformation is contained purely in the derivatives of s^{μ}. In general, the angular derivatives of n^{μ} at $\lambda = constant$ can be computed in terms of spatial 3+1 derivatives of the lapse and shift, transformed with the $\mathcal{O}(\lambda^0)$ Jacobian, i.e.

$$n^{\mu}_{,\tilde{A}} = n^{\mu}_{,j} x^{j}_{,\tilde{A}}\,. \tag{35}$$

The 3+1 derivatives are given by

$$\begin{aligned} n^i_{,j} &= \frac{1}{\alpha^2}\left(\alpha_{,j}\beta^i - \alpha\beta^i_{,j}\right), \\ n^t_{,j} &= -\frac{1}{\alpha^2}\alpha_{,j}. \end{aligned} \tag{36}$$

The retarded time derivative $\partial_{\tilde{u}}$ at $\lambda = constant$ is simply the $3+1$ time derivative ∂_t, therefore $n^{\mu}_{,\tilde{u}} = n^{\mu}_{,t}$ where

$$\begin{aligned} n^i_{,t} &= \frac{1}{\alpha^2}\left(\alpha_{,t}\beta^i - \alpha\beta^i_{,t}\right), \\ n^t_{,t} &= -\frac{1}{\alpha^2}\alpha_{,t}. \end{aligned} \tag{37}$$

From Eq. (25), and since the σ_k are time-independent, the time derivative of s^i is given by

$$\begin{aligned} s^i_{,t} &= g^{ik}_{,t}\frac{\sigma_k}{\sigma} - g^{ik}\frac{\sigma_k\sigma_{,t}}{\sigma^2} = -g^{im}g^{kn}g_{mn,t}\frac{\sigma_k}{\sigma} - s^i\frac{\sigma_{,t}}{\sigma} \\ &= -g^{im}g_{mn,t}s^n - s^i\frac{\sigma_{,t}}{\sigma}, \end{aligned} \tag{38}$$

where the time derivative of σ can be calculated from

$$2\sigma\sigma_{,t} = \left(\sigma^2\right)_{,t} = g^{kl}_{,t}\sigma_k\sigma_l = -g^{km}g^{ln}g_{mn,t}\sigma_k\sigma_l = -s^m s^n g_{mn,t}\sigma^2, \quad (39)$$

with the resulting expression

$$s^i_{,t} = \left(-g^{im} + s^i \frac{1}{2} s^m\right) g_{mn,t} s^n. \quad (40)$$

Similarly, it follows from Eq.(25) that

$$\begin{aligned} s^i_{,\tilde{A}} &= g^{ik}_{,j} x^j_{,\tilde{A}} \frac{\sigma_k}{\sigma} + g^{ik}\frac{\sigma_{k,\tilde{A}}}{\sigma} - g^{ik}\frac{\sigma_k \sigma_{,\tilde{A}}}{\sigma^2} \\ &= -g^{in}g^{km}g_{mn,j}\, x^j_{,\tilde{A}} \frac{\sigma_k}{\sigma} + g^{ik}\frac{\sigma_{k,\tilde{A}}}{\sigma} - s^i \frac{\sigma_{,\tilde{A}}}{\sigma}, \end{aligned} \quad (41)$$

where the $\sigma_{k,\tilde{A}}$ are obtained from the analytic expressions Eq. (17), and $\sigma_{,\tilde{A}}$ from

$$\begin{aligned} 2\sigma\sigma_{,\tilde{A}} &= \left(\sigma^2\right)_{,\tilde{A}} = \left(g^{kl}\sigma_k\sigma_l\right)_{,\tilde{A}} = g^{kl}_{,j} x^j_{,\tilde{A}} \sigma_k \sigma_l + 2\, g^{kl}\sigma_l \sigma_{k,\tilde{A}} \\ &= -g^{km}g^{ln}g_{mn,j}x^j_{,}\sigma_k\sigma_l + 2\, g^{kl}\sigma_l\sigma_{k,\tilde{A}} \\ &= -s^m s^n g_{mn,j} x^j_{,\tilde{A}} \sigma^2 + 2 s^k \sigma \sigma_{k,\tilde{A}}\,. \end{aligned} \quad (42)$$

Collecting Eqs. (41) and (42), we arrive at the angular derivatives of the normal to the world-tube

$$\begin{aligned} s^i_{,\tilde{A}} &= -g^{in}s^m g_{mn,j}\, x^j_{,\tilde{A}} + g^{ik}\frac{\sigma_{k,\tilde{A}}}{\sigma} + s^i\left(\frac{1}{2}s^m s^n g_{mn,j} x^j_{,\tilde{A}} - s^k \frac{\sigma_{k,\tilde{A}}}{\sigma}\right) \\ &= \left(g^{in} - s^i s^n\right)\frac{\sigma_{n,\tilde{A}}}{\sigma} + \left(-g^{in} + \frac{1}{2}s^i s^n\right) s^m g_{mn,j} x^j_{.} \end{aligned} \quad (43)$$

4.4. NULL METRIC $\tilde{\eta}_{\alpha\beta}$

Given the 4-metric and its Cartesian derivatives at the world-tube, we can calculate its derivative with respect to the affine parameter λ according to

$$g_{\alpha\beta,\lambda|\Gamma} = g^{(0)}_{\alpha\beta,\mu}\ell^{(0)\mu}\,. \quad (44)$$

Having obtained the relevant parts of the coordinate transformation $\tilde{y}^\alpha \to x^\alpha$, Eqs.(34), (35), (36), (37), (40), and (43), and given the metric and its λ derivative as per Eq.(44), we can expand the null metric as follows

$$\tilde{\eta}_{\tilde{\alpha}\tilde{\beta}} = \tilde{\eta}^{(0)}_{\tilde{\alpha}\tilde{\beta}} + \tilde{\eta}_{\tilde{\alpha}\tilde{\beta},\lambda}\lambda + \mathcal{O}(\lambda^2), \quad (45)$$

where the coefficients are given by

$$
\begin{aligned}
\tilde{\eta}^{(0)}_{\tilde{u}\tilde{u}} &= g_{tt|\Gamma} \, , \\
\tilde{\eta}^{(0)}_{\tilde{u}\tilde{A}} &= x^{i}{}_{,\tilde{A}} g_{it|\Gamma} \, , \\
\tilde{\eta}^{(0)}_{\tilde{A}\tilde{B}} &= x^{i}{}_{,\tilde{A}} x^{j}{}_{,\tilde{B}} g_{ij|\Gamma} \, ,
\end{aligned} \tag{46}
$$

and, for the λ derivative

$$
\begin{aligned}
\tilde{\eta}_{\tilde{u}\tilde{u},\lambda} &= \left[g_{tt,\lambda} + 2\, \ell^{\mu}{}_{,\tilde{u}} g_{\mu t} \right]_{|\Gamma} + \mathcal{O}(\lambda) \, , \\
\tilde{\eta}_{\tilde{u}\tilde{A},\lambda} &= \left[x^{k}{}_{,\tilde{A}} \left(\ell^{\mu}{}_{,\tilde{u}} g_{k\mu} + g_{kt,\lambda} \right) + \ell^{k}{}_{,\tilde{A}} g_{kt} + \ell^{t}{}_{,\tilde{A}} g_{tt} \right]_{|\Gamma} + \mathcal{O}(\lambda) \, , \\
\tilde{\eta}_{\tilde{A}\tilde{B},\lambda} &= \left[x^{k}{}_{,\tilde{A}} x^{l}{}_{,\tilde{B}} g_{kl,\lambda} + \left(\ell^{\mu}{}_{,\tilde{A}} x^{l}{}_{,\tilde{B}} + \ell^{\mu}{}_{,\tilde{B}} x^{l}{}_{,\tilde{A}} \right) g_{\mu l} \right]_{|\Gamma} + \mathcal{O}(\lambda) \, .
\end{aligned} \tag{47}
$$

The remaining components are fixed by Eq. (31).

It is worthwhile to discuss a subtlety in the rationale underlying the computational strategy used here. The purpose of carrying out an expansion in λ is to enable us to give the metric variables in a small region off the world-tube, *i.e.* at points of the grid used to discretize the null equations. This method is an alternative to the more obvious and straightforward strategy of interpolation to determine metric values at needed points on the null computational grid. At first glance, it might appear a more cumbersome way to proceed, due to the necessity of analytic calculation of the derivative of the metric on the world tube to determine coefficients needed for the metric expansion in powers of λ. On the contrary, however, *this trick allows a major simplification in computer implementation.* It allows us to use a null code which does not require any special implementation at the irregular boundary defined by the world-tube, and it automatically ensures continuity of the metric and the extrinsic curvature at the world-tube. The obvious alternative of a brute-force interpolation approach would require a full 4-dimensional evaluation with quite complicated logic, since it would lie at the edge of both the Cauchy and null computational grids. Instead, the λ expansion reduces the complexity by one dimension, allowing for much easier numerical implementation.

We also need to compute the contravariant null metric, $\tilde{\eta}^{\alpha\beta}$, which we similarly consider as an expansion in powers of λ,

$$
\tilde{\eta}^{\tilde{\mu}\tilde{\nu}} = \tilde{\eta}^{(0)\tilde{\mu}\tilde{\nu}} + \tilde{\eta}^{\tilde{\mu}\tilde{\nu}}_{,\lambda} \lambda + \mathcal{O}(\lambda^2), \tag{48}
$$

with coefficients given by

$$
\tilde{\eta}^{(0)\tilde{\mu}\tilde{\alpha}} \tilde{\eta}^{(0)}_{\tilde{\alpha}\tilde{\nu}} = \delta^{\tilde{\mu}}_{\tilde{\nu}}, \quad \tilde{\eta}^{\tilde{\mu}\tilde{\nu}}_{,\lambda} = -\tilde{\eta}^{\tilde{\mu}\tilde{\alpha}}\, \tilde{\eta}^{\tilde{\beta}\tilde{\nu}}\, \tilde{\eta}_{\tilde{\alpha}\tilde{\beta},\lambda}. \tag{49}
$$

It follows from Eq. (31) that the following components of the contravariant null metric in the $\tilde{y}^\alpha$ coordinates are fixed

$$\tilde{\eta}^{\lambda\tilde{u}} = -1 \quad \tilde{\eta}^{\tilde{u}\tilde{A}} = \tilde{\eta}^{\tilde{u}\tilde{u}} = 0\,. \tag{50}$$

Therefore the contravariant null metric can be computed by

$$\begin{aligned}
\tilde{\eta}^{\tilde{A}\tilde{B}}\tilde{\eta}_{\tilde{B}\tilde{C}} &= \delta^{\tilde{A}}_{\tilde{C}}\,,\\
\tilde{\eta}^{\lambda\tilde{A}} &= \tilde{\eta}^{\tilde{A}\tilde{B}}\tilde{\eta}_{\tilde{B}\tilde{u}}\,,\\
\tilde{\eta}^{\lambda\lambda} &= -\tilde{\eta}_{\tilde{u}\tilde{u}} + \tilde{\eta}^{\lambda\tilde{A}}\tilde{\eta}_{\tilde{A}\tilde{u}}\,,
\end{aligned} \tag{51}$$

and similarly for its λ derivative

$$\begin{aligned}
\tilde{\eta}^{\tilde{A}\tilde{B}}_{,\lambda} &= -\tilde{\eta}^{\tilde{A}\tilde{C}}\tilde{\eta}^{\tilde{B}\tilde{D}}\tilde{\eta}_{\tilde{C}\tilde{D},\lambda}\,,\\
\tilde{\eta}^{\lambda\tilde{A}}_{,\lambda} &= \tilde{\eta}^{\tilde{A}\tilde{B}}\left(\tilde{\eta}_{\tilde{u}\tilde{B},\lambda} - \tilde{\eta}^{\lambda\tilde{C}}\tilde{\eta}_{\tilde{C}\tilde{B},\lambda}\right)\,,\\
\tilde{\eta}^{\lambda\lambda}_{,\lambda} &= -\tilde{\eta}_{\tilde{u}\tilde{u},\lambda} + 2\,\tilde{\eta}^{\lambda\tilde{A}}\tilde{\eta}_{\tilde{u}\tilde{A},\lambda} - \tilde{\eta}^{\lambda\tilde{A}}\tilde{\eta}^{\lambda\tilde{B}}\tilde{\eta}_{\tilde{A}\tilde{B},\lambda}\,.
\end{aligned} \tag{52}$$

4.5. METRIC IN BONDI COORDINATES

The surface area coordinate $r(u, \lambda, \tilde{y}^A)$ is defined by

$$r = \left(\frac{\det(\tilde{\eta}_{\tilde{A}\tilde{B}})}{\det(q_{AB})}\right)^{\frac{1}{4}}, \tag{53}$$

where, for our choice of stereographic coordinates $\xi = q + ip$, we use $\tilde{y}^A = (q, p)$ and $\det(q_{AB}) = 16/(1 + q^2 + p^2)^4$. To carry out the coordinate transformation $\tilde{y}^\alpha \to y^\alpha$ on the null metric, where $y^\alpha = (r, y^A, u)$, we need to know $r_{,\lambda}$, $r_{,\tilde{A}}$ and $r_{,\tilde{u}}$. From Eq.(53) it follows

$$r_{,\lambda} = \frac{r}{4}\tilde{\eta}^{\tilde{A}\tilde{B}}\tilde{\eta}_{\tilde{A}\tilde{B},\lambda}. \tag{54}$$

Similarly,

$$r_{,\tilde{C}} = \frac{r}{4}\left(\tilde{\eta}^{\tilde{A}\tilde{B}}\tilde{\eta}_{\tilde{A}\tilde{B},\tilde{C}} - \frac{\det(q_{\tilde{A}\tilde{B}})_{,\tilde{C}}}{\det(q_{\tilde{A}\tilde{B}})}\right), \tag{55}$$

where

$$\begin{aligned}
\frac{\det(q_{\tilde{A}\tilde{B}})_{,\tilde{C}}}{\det(q_{\tilde{A}\tilde{B}})} &= -\frac{8}{1+q^2+p^2}\tilde{y}^{\tilde{C}}\\
\tilde{\eta}_{\tilde{A}\tilde{B},\tilde{C}} &= \left(x^i_{,\tilde{A}\tilde{C}}\,x^j_{,\tilde{B}} + x^i_{,\tilde{A}}\,x^j_{,\tilde{B}\tilde{C}}\right) g_{ij} + x^i_{,\tilde{A}}\,x^j_{,\tilde{B}}\,x^k_{,\tilde{C}}\,g_{ij,k}
\end{aligned} \tag{56}$$

with the ${x^i}_{,\tilde{A}\tilde{C}}$ given functions of (q,p). From Eqs.(53) and (46)

$$r_{,\tilde{u}} = \frac{r}{4}\tilde{\eta}^{\tilde{A}\tilde{B}}\tilde{\eta}_{\tilde{A}\tilde{B},\tilde{u}} , \tag{57}$$

where

$$\tilde{\eta}_{\tilde{A}\tilde{B},\tilde{u}} = \left[{x^i}_{,\tilde{A}}\,{x^j}_{,\tilde{B}}\,g_{ij,t}\right]_{|\Gamma} + \mathcal{O}(\lambda) . \tag{58}$$

The null metric $\eta^{\alpha\beta}$ in Bondi coordinates is defined on the world-tube Γ by

$$\eta^{\alpha\beta} = \frac{\partial y^\alpha}{\partial \tilde{y}^\mu}\frac{\partial y^\beta}{\partial \tilde{y}^\nu}\tilde{\eta}^{\tilde{\mu}\tilde{\nu}} . \tag{59}$$

Note that the metric of the sphere is unchanged by this coordinate transformation, i.e. $\eta^{AB} = \tilde{\eta}^{\tilde{A}\tilde{B}}$, so we need to compute only the elements η^{rr}, η^{rA} and η^{ru} on Γ, or equivalently the metric functions β, U^A and V. From Eq.(50)

$$\begin{aligned}
\eta^{rr} &= r_{,\tilde{\alpha}}\,r_{,\tilde{\beta}}\,\tilde{\eta}^{\tilde{\alpha}\tilde{\beta}} = (r_{,\lambda})^2\,\tilde{\eta}^{\lambda\lambda} + 2\,r_{,\lambda}\left(r_{,\tilde{A}}\,\tilde{\eta}^{\lambda\tilde{A}} - r_{,\tilde{u}}\right) + r_{,\tilde{A}}\,r_{,\tilde{B}}\,\tilde{\eta}^{\tilde{A}\tilde{B}} , \\
\eta^{rA} &= r_{,\tilde{\alpha}}\,\tilde{\eta}^{\tilde{\alpha}\tilde{A}} = r_{,\lambda}\,\tilde{\eta}^{\lambda\tilde{A}} + r_{,\tilde{B}}\,\tilde{\eta}^{\tilde{A}\tilde{B}} , \\
\eta^{ru} &= r_{,\tilde{\alpha}}\,\tilde{\eta}^{\tilde{\alpha}\tilde{u}} = -r_{,\lambda} .
\end{aligned} \tag{60}$$

The contravariant Bondi metric can be written in the form

$$\eta^{\alpha\beta} = \begin{bmatrix} e^{-2\beta}\dfrac{V}{r} & -e^{-2\beta}U^2 & -e^{-2\beta}U^3 & -e^{-2\beta} \\ -e^{-2\beta}U^2 & r^{-2}\,h^{22} & r^{-2}\,h^{23} & 0 \\ -e^{-2\beta}U^3 & r^{-2}\,h^{32} & r^{-2}\,h^{33} & 0 \\ -e^{-2\beta} & 0 & 0 & 0 \end{bmatrix} , \tag{61}$$

where h_{AB} is a metric on the sphere of surface area 4π, such that $h_{AB}h^{BC} = \delta_A^C$ and $\det(h_{AB}) = \det(q_{AB})$, for q_{AB} a unit sphere metric.

4.6. BONDI VARIABLES FOR STARTING UP THE NULL CODE AT THE WORLD-TUBE

The natural variables of the null formalism are certain combinations of the null metric functions, which we will give in this section. They will be expressed as an expansion in λ, as was done in the previous sections, to enable us to give these variables in a small region off the world-tube, *i.e.* at points of the grid used to discretize the null equations. *These expressions are the main results of the extraction module.*

4.6.1. *The metric of the sphere J*
Given r and $r_{,\lambda}$, and noting that

$$\eta_{AB} = \tilde{\eta}_{AB} \equiv r^2 h_{AB}, \tag{62}$$

we can write

$$\begin{aligned} h_{AB} &= \frac{1}{r^2}\eta_{AB} , \\ h_{AB,\lambda} &= \frac{1}{r^2}\left(\eta_{AB,\lambda} - \frac{2\,r_{,\lambda}}{r}\eta_{AB}\right) . \end{aligned} \tag{63}$$

In terms of q^A, $\bar{q}^A$ and the metric on the unit sphere h_{AB}, we define the metric functions

$$J \equiv \frac{1}{2}q^A q^B h_{AB}, \quad K \equiv \frac{1}{2}q^A \bar{q}^B h_{AB}, \quad K^2 = 1 + J\bar{J}. \tag{64}$$

It suffices to evaluate J, since the last relation holds for a Bondi metric. We give expressions for J and $J_{,\lambda}$

$$\begin{aligned} J &= \frac{1}{2\,r^2}q^A q^B \eta_{AB} , \\ J_{,\lambda} &= \frac{1}{2\,r^2}q^A q^B \eta_{AB,\lambda} - 2\,\frac{r_{,\lambda}}{r}J . \end{aligned} \tag{65}$$

Then, in the neighborhood of Γ, the metric of the sphere is explicitly given in Bondi coordinates to second-order accuracy by these last two expressions as

$$J(y^\alpha) = J + J_{,\lambda}\lambda + \mathcal{O}(\lambda^2). \tag{66}$$

4.6.2. *The "expansion factor" β*
From the last of Eq. (60) we obtain the metric function β

$$\beta = -\frac{1}{2}\log(r_{,\lambda}). \tag{67}$$

We want to know also its λ-derivative, but instead of calculating $\eta^{\lambda u}_{,\lambda}$ directly (which would involve $r_{,\lambda\lambda}$)

$$\beta_{,\lambda} = -\frac{\eta^{ru}_{,\lambda}}{2\,\eta^{ru}} = -\frac{r_{,\lambda\lambda}}{2\,r_{,\lambda}} , \tag{68}$$

we obtain $\beta_{,\lambda}$ from the characteristic equation (see Ref.[8], Eq.(A1))

$$\beta_{,r} = \frac{r}{8}\left(J_{,r}\bar{J}_{,r} - (K_{,r})^2\right) . \tag{69}$$

At constant angles (q,p), the relation $\partial_\lambda = r_{,\lambda}\partial_r$ holds, and we know from Eq. (65) J and $J_{,\lambda}$ for each outgoing radial null geodesic through the world-tube. Thus we can write

$$\beta_{,\lambda} = \frac{r}{8\,r_{,\lambda}}\left(J_{,\lambda}\bar{J}_{,\lambda} - (K_{,\lambda})^2\right) \tag{70}$$

and from Eq.(64), it follows that

$$K_{,\lambda} = \frac{1}{K}\Re\left(\bar{J}J_{,\lambda}\right)\ , \tag{71}$$

$$\beta_{,\lambda} = \frac{r}{8\,r_{,\lambda}}\left(J_{,\lambda}\bar{J}_{,\lambda} - \frac{1}{1+J\bar{J}}\left[\Re\left(\bar{J}J_{,\lambda}\right)\right]^2\right). \tag{72}$$

Then, β is found to second-order accuracy by:

$$\beta(y^\alpha) = \beta + \beta_{,\lambda}\lambda + \mathcal{O}(\lambda^2). \tag{73}$$

4.6.3. *The "shift" U*

The metric function U is related to the Bondi metric, Eq. (61) by

$$U \equiv U^A q_A = \frac{\eta^{rA}}{\eta^{ru}} q_A = -\left(\tilde{\eta}^{\lambda\tilde{A}} + \frac{r_{,\tilde{B}}}{r_{,\lambda}}\tilde{\eta}^{\tilde{A}\tilde{B}}\right) q_{\tilde{A}}, \tag{74}$$

where we have made use of Eq. (60). We also want the λ derivative of U

$$\begin{aligned} U_{,\lambda} &= -\left[\tilde{\eta}^{\lambda\tilde{A}}_{,\lambda} + \left(\frac{r_{,\lambda\tilde{B}}}{r_{,\lambda}} - \frac{r_{,\tilde{B}}\,r_{,\lambda\lambda}}{r^2_{,\lambda}}\right)\tilde{\eta}^{\tilde{A}\tilde{B}} + \frac{r_{,\tilde{B}}}{r_{,\lambda}}\tilde{\eta}^{\tilde{A}\tilde{B}}_{,\lambda}\right] q_{\tilde{A}}, \\ &= -\left(\tilde{\eta}^{\lambda\tilde{A}}_{,\lambda} + \frac{r_{,\lambda\tilde{B}}}{r_{,\lambda}}\tilde{\eta}^{\tilde{A}\tilde{B}} + \frac{r_{,\tilde{B}}}{r_{,\lambda}}\tilde{\eta}^{\tilde{A}\tilde{B}}_{,\lambda}\right) q_{\tilde{A}} + 2\,\beta_{,\lambda}\left(U + \tilde{\eta}^{\lambda\tilde{A}} q_{\tilde{A}}\right), \end{aligned} \tag{75}$$

where in the last line we have used Eq. (68) to eliminate $r_{,\lambda\lambda}$.

Then, U is found to second-order accuracy by:

$$U(y^\alpha) = U + U_{,\lambda}\lambda + \mathcal{O}(\lambda^2). \tag{76}$$

4.6.4. *The "mass aspect" W*

The metric function V is given in terms of η^{rr} and η^{ru} by

$$V \equiv -r\,\eta^{rr}/\eta^{ru}\ . \tag{77}$$

For Minkowski-space, $V = r$ at infinity. Thus, we introduce the auxiliary metric function $W \equiv (V-r)/r^2$, which is regular at null infinity. In terms of the contravariant null metric (with the affine parameter λ as the radial coordinate)

$$W = \frac{1}{r}\left(\frac{\eta^{rr}}{r_{,\lambda}} - 1\right) = \frac{1}{r}\left(r_{,\lambda}\tilde{\eta}^{\lambda\lambda} + 2\left(r_{,\tilde{A}}\,\tilde{\eta}^{\lambda\tilde{A}} - r_{,u}\right) + \frac{r_{,\tilde{A}}\,r_{,\tilde{B}}}{r_{,\lambda}}\,\tilde{\eta}^{\tilde{A}\tilde{B}} - 1\right). \tag{78}$$

The λ derivative of W is given by

$$\begin{aligned} W_{,\lambda} &= -\frac{r_{,\lambda}}{r}W + \frac{1}{r}\left(r_{,\lambda}\tilde{\eta}^{\lambda\lambda} + 2\left(r_{,\tilde{A}}\,\tilde{\eta}^{\lambda\tilde{A}} - r_{,u}\right) + \frac{r_{,\tilde{A}}\,r_{,\tilde{B}}}{r_{,\lambda}}\,\tilde{\eta}^{\tilde{A}\tilde{B}} - 1\right)_{,\lambda} \\ &= -\frac{r_{,\lambda}}{r}\left(\left(\frac{r_{,\lambda}}{r} + 2\beta_{,\lambda}\right)\tilde{\eta}^{\lambda\lambda} - \tilde{\eta}^{\lambda\lambda}_{,\lambda} - \frac{1}{r}\right) + \frac{2}{r}\left(\frac{r_{,\lambda}r_{,u}}{r} - r_{,\lambda u}\right) \\ &\quad + \frac{2}{r}\left(r_{,\lambda\tilde{A}} - \frac{r_{,\lambda}r_{,\tilde{A}}}{r}\right)\tilde{\eta}^{\lambda\tilde{A}} + 2\frac{r_{,\tilde{A}}}{r}\,\tilde{\eta}^{\lambda\tilde{A}}_{,\lambda} \\ &\quad + \frac{r_{,\tilde{B}}}{r\,r_{,\lambda}}\left(2\,r_{,\lambda\tilde{A}}\tilde{\eta}^{\tilde{A}\tilde{B}} + 2\,\beta_{,\lambda}r_{,\tilde{A}} + r_{,\tilde{A}}\tilde{\eta}^{\tilde{A}\tilde{B}}_{,\lambda}\right) - \frac{r_{,\tilde{A}}\,r_{,\tilde{B}}}{r^2}\,\tilde{\eta}^{\tilde{A}\tilde{B}}. \end{aligned} \tag{79}$$

Then, W is found to second-order accuracy by:

$$W(y^\alpha) = W + W_{,\lambda}\lambda + \mathcal{O}(\lambda^2). \tag{80}$$

5. Injection

We have now reached the halfway point in the problem of relating data between the Null and Cauchy computational approaches. We turn to injection, the final part of the problem. Injection is the inverse problem to extraction, and allows us to use data obtained by characteristic evolution in the exterior of the world-tube in order to provide the exact boundary conditions for Cauchy evolution in its interior. That is, given the metric in Bondi coordinates in a region around the world-tube Γ, two steps must be carried out. The first ingredient is to relate the Cartesian coordinate system to the Bondi null coordinates. Secondly, the Cartesian metric components must be calculated in the neighborhood of Γ in the usual way, from the Bondi metric components and the Jacobian of the coordinate transformation.

5.1. LOCATING THE CARTESIAN GRID WORLDLINES

Given a worldline of a Cartesian grid point, we need to locate, in null coordinates, its intersection with a given null hypersurface. Mathematically, this means: Given x^i and u we want to find t and the null coordinates r, as well as $y^A = y^A(x^{(0)i}) = (q, p)$ (where, at this stage, $x^{(0)i}$ is unknown).

The starting point is the geodesic equation (27)

$$x^\alpha = x^{(0)\alpha} + \lambda \ell^{(0)\alpha} + \mathcal{O}(\lambda^2), \tag{81}$$

Here $\ell^{(0)\alpha} = \ell^\alpha(u, y^A(x^{(0)i})$ is the value of the null vector given by Eq. (26) at the point on the world-tube with null coordinates $u = t^{(0)}$, $\lambda = 0$ and $y^A(x^{(0)i})$.

We set $L^\alpha \equiv \ell^\alpha(u, y^A(x^i))$, so that

$$\ell^{(0)\alpha} \equiv \ell^\alpha(x^{(0)\beta})) = L^\alpha + \mathcal{O}(\lambda), \tag{82}$$

since $y^A(x^i) = y^A(x^{(0)i}) + \mathcal{O}(\lambda)$. ($L^\alpha$ can now be interpolated onto the stereographic coordinate patch to fourth order accuracy in terms of known quantities near the world tube.)

To the same approximation as (81), we have

$$x^\alpha = x^{(0)\alpha} + \lambda L^\alpha + \mathcal{O}(\lambda^2) \tag{83}$$

so that we have four equations for the five unknowns ($x^{(0)i}, t$, and λ):

$$t = u + \lambda L^t + \mathcal{O}(\lambda^2) \tag{84}$$

$$x^i = x^{(0)i} + \lambda L^i + \mathcal{O}(\lambda^2). \tag{85}$$

To find the fifth unknown, we need one additional equation introducing new information. This is conveniently supplied by the equation defining the world-tube Eq.(20). Even though we have not yet found the values for $x^{(0)i}$, we may use Eq.(20) to eliminate these unknowns by taking the Euclidean norm of Eq.(85), whence using Eq.(19) we find

$$R^2 = \hat{r}^2 + \lambda^2(L \cdot L) - 2\lambda(L \cdot x) + \mathcal{O}(\lambda^2). \tag{86}$$

where the value of $R^2 = (x^{(0)} \cdot x^{(0)})$ is known from the beginning by the definition of the world tube, and $\hat{r}^2 = (x \cdot x)$.

This is a quadratic in λ, but we are consistently only keeping terms to linear order. This leads to

$$\Lambda = (\hat{r}^2 - R^2)/\left[2(L \cdot x)\right] \tag{87}$$

Consequently, $\Lambda = \lambda + \mathcal{O}(\lambda^2)$ and can be calculated in terms of known quantities. This leaves four unknown quantities remaining. To proceed to evaluate these in terms of known data, we set $T = u + \Lambda L^t, X^{(0)i} = (X^{(0)}, Y^{(0)}, Z^{(0)}) = x^i - \Lambda L^i$, and $Y^A = (Q, P) = y^A(X^{(0)i})$. Then, using Eq. (18)

$$Q = \frac{X^{(0)}}{R \pm Z^{(0)}}, \quad P = \pm \frac{Y^{(0)}}{R \pm Z^{(0)}}, \tag{88}$$

on the north (+) and south (−) patches. T, Λ and Y^A are accurate to $\mathcal{O}(\lambda^2)$ since $t = T + \mathcal{O}(\lambda^2)$, $\lambda = \Lambda + O(\lambda^2)$ and $y^A = Y^A + \mathcal{O}(\lambda^2)$.

Finally, the value of r is obtained from

$$r = r^{(0)} + r^{(0)}_{,\lambda}\Lambda, \tag{89}$$

and the values of $r_{,\lambda}$, $r_{,A}$ and $r_{,u}$ (used in computing the $\tilde{\eta}^{\tilde{\mu}\tilde{\nu}}$ metric) from

$$r_{,\lambda} = r^{(0)}_{,\lambda} + r^{(0)}_{,\lambda\lambda}\Lambda, \quad r_{,A} = r^{(0)}_{,A} + r^{(0)}_{,\lambda A}\Lambda, \quad r_{,u} = r^{(0)}_{,u} + r^{(0)}_{,\lambda u}\Lambda, \tag{90}$$

all to $\mathcal{O}(\lambda^2)$ accuracy.

We calculate $r^{(0)}$, the derivatives $r^{(0)}_{,\lambda}$, $r^{(0)}_{,A}$ and $r^{(0)}_{,u}$ and the mixed derivatives $r^{(0)}_{,\lambda\lambda}$, $r^{(0)}_{,\lambda A}$ and $r^{(0)}_{,\lambda u}$ at the points (Q, P) on the world-tube by fourth order interpolation.

The null metric variables are then obtained by fourth order interpolation at the Bondi coordinate points (r, Q, P, u), corresponding to the Cartesian coordinate points (x^i, T)

5.2. REVERSING THE COORDINATE TRANSFORMATION

The following describes the procedure to obtain the *contravariant* null metric $\tilde{\eta}^{\tilde{\alpha}\tilde{\beta}}$ in affine coordinates $\tilde{y}^\alpha = (u, q, p, \lambda)$ on points on the characteristic grid. This is 90% of the work in obtaining the Cartesian metric at those points. What remains is to contract this null metric with the Jacobian of the coordinate transformation $x^\mu = x^\mu(\tilde{y}^\alpha)$. The Jacobian is given in Eq. (32) as a series expansion on the parameter λ. We will use this expression, and consequently we need to work with the contravariant (rather than the covariant) metric components, since they will transform with the already known Jacobian as:

$$g^{\mu\nu} = \frac{\partial x^\mu}{\partial \tilde{y}^\alpha}\frac{\partial x^\nu}{\partial \tilde{y}^\beta}\tilde{\eta}^{\tilde{\alpha}\tilde{\beta}}. \tag{91}$$

5.2.1. *The metric on the sphere*

Recall that on each stereographic patch we introduce complex null vectors on the sphere $q^A = (P/2)(\delta^A_2 + i\delta^A_3)$ where $P = 1 + \xi\bar{\xi}$ and the q^A satisfy the orthogonality condition $\bar{q}_A q^A = 2$. The q^A define a metric on the sphere $q_{AB} = (4/P^2)\delta_{AB}$ and $q_A = (2/P)(\delta^2_A + i\delta^3_A) = q_{AB}q^B$.

By construction, the null metric restricted to the sphere, η_{AB}, is the same in Bondi coordinates y^α and affine coordinates $y^{\tilde{\alpha}}$. It can be expanded as:

$$\tilde{\eta}_{\tilde{A}\tilde{B}} \equiv \eta_{AB} = \alpha q_A q_B + \bar{\alpha}\bar{q}_A\bar{q}_B + \delta q_A\bar{q}_B + \bar{\delta}\bar{q}_A q_B. \tag{92}$$

The values of α and δ follow from the orthogonality condition

$$4\,\alpha = \eta_{AB}\bar{q}^A\bar{q}^B, \quad 4\,\delta = \eta_{AB}\bar{q}^A q^B = 4\,\bar{\delta}. \tag{93}$$

From the definition of J and K

$$J = \frac{1}{2} q^A q^B h_{AB}, \quad K = \frac{1}{2} q^A \bar{q}^B h_{AB}, \tag{94}$$

and $\eta_{AB} = r^2 h_{AB}$, it follows that

$$\alpha = \frac{1}{2} r^2 \bar{J}, \quad \delta = \frac{1}{2} r^2 K, \tag{95}$$

so we can write the metric on the sphere in terms of J and K as

$$\eta_{AB} = \frac{1}{2} r^2 \left(J\bar{q}_A\bar{q}_B + \bar{J} q_A q_B + K q_A \bar{q}_B + K \bar{q}_A q_B \right). \tag{96}$$

From the expressions for the q^A, the components of the sphere metric are given by:

$$\eta_{qq} = \frac{4\,r^2}{P^2}\left(K + \Re(J)\right), \quad \eta_{pp} = \frac{4\,r^2}{P^2}\left(K - \Re(J)\right), \quad \eta_{qp} = \frac{4\,r^2}{P^2}\Im(J). \tag{97}$$

Notice that given the Bondi coordinates of a point, the determinant is known

$$\det(\eta_{AB}) = r^4 \det(q_{AB}) = 16\,r^4/P^4, \tag{98}$$

therefore the inverse sphere metric η^{AB} such that $\eta^{AB}\eta_{BC} = \delta^A_C$ has coefficients

$$\begin{aligned}
\eta^{qq} &= \tilde{\eta}^{\tilde{q}\tilde{q}} = \frac{P^2}{4\,r^2}\left(K - \Re(J)\right), \\
\eta^{pp} &= \tilde{\eta}^{\tilde{p}\tilde{p}} = \frac{P^2}{4\,r^2}\left(K + \Re(J)\right), \\
\eta^{qp} &= \tilde{\eta}^{\tilde{q}\tilde{p}} = -\frac{P^2}{4\,r^2}\Im(J).
\end{aligned} \tag{99}$$

5.2.2. *The radial-angular metric coefficients*

We can write the metric coefficients in terms of a single coefficient γ

$$\tilde{\eta}^{\lambda\tilde{A}} = \bar{\gamma} q^{\tilde{A}} + \gamma \bar{q}^{\tilde{A}}. \tag{100}$$

In components

$$\tilde{\eta}^{\lambda q} = P\Re(\gamma), \quad \tilde{\eta}^{\lambda p} = P\Im(\gamma). \tag{101}$$

The value of γ follows from the expression for the metric function U

$$U = U^A q_A = \frac{\eta^{rA}}{\eta^{ru}} q_A = -\left(\tilde{\eta}^{\lambda\tilde{A}} + \frac{r_{,\tilde{B}}}{r_{,\lambda}} \tilde{\eta}^{\tilde{A}\tilde{B}}\right) q_{\tilde{A}}, \tag{102}$$

$$2\gamma = -U - \frac{r_{,\tilde{B}}}{r_{,\lambda}} \tilde{\eta}^{\tilde{A}\tilde{B}} q_{\tilde{A}}. \tag{103}$$

Recall that the derivatives $r_{,A}$, $r_{,u}$ are at constant λ and $r_{,\lambda}$ can be read off directly from the Bondi metric function β

$$r_{,\lambda} = e^{-2\beta}. \tag{104}$$

From the expressions for $\tilde{\eta}^{\tilde{A}\tilde{B}}$ and $q_{\tilde{A}}$,

$$\begin{aligned} r_{,\tilde{B}} \tilde{\eta}^{\tilde{A}\tilde{B}} q_{\tilde{A}} &= \frac{2}{P} \left[r_{,2} \left(\tilde{\eta}^{qq} + i\tilde{\eta}^{qp} \right) + r_{,3} \left(\tilde{\eta}^{qp} + i\tilde{\eta}^{pp} \right) \right] \\ &= \frac{P}{2r^2} \left[r_{,2} \left(K - J \right) + i r_{,3} \left(K + J \right) \right]. \end{aligned} \tag{105}$$

Thus the complex coefficient sought is given by

$$\gamma = -\frac{1}{2} U - e^{2\beta} \frac{P}{4r^2} \left[r_{,2} \left(K - J \right) + i r_{,3} \left(K + J \right) \right]. \tag{106}$$

The $r_{,A}$ can be obtained at the point where the coordinate transformation is being performed to $\mathcal{O}(\lambda^2)$, by using the angular derivatives of $r_{,\lambda}$. To this end, note that for fixed (q,p):

$$r_{,A} = \overset{(0)}{r_{,A}} + r_{,\lambda A} \lambda, \tag{107}$$

where the derivatives on the right hand side are computed on the world-tube.

5.2.3. *The $\tilde{\eta}^{\lambda\lambda}$ metric function*

By reversing the procedure to obtain W, i.e. from

$$W = \frac{1}{r} \left(r_{,\lambda} \tilde{\eta}^{\lambda\lambda} + 2\left(r_{,\tilde{A}} \tilde{\eta}^{\lambda\tilde{A}} - r_{,u} \right) + \frac{r_{,\tilde{A}} r_{,\tilde{B}}}{r_{,\lambda}} \tilde{\eta}^{\tilde{A}\tilde{B}} - 1 \right) \tag{108}$$

we obtain the remaining component of the null metric

$$\tilde{\eta}^{\lambda\lambda} = e^{2\beta} \left(rW + 1 - 2\left(r_{,\tilde{A}} \tilde{\eta}^{\lambda\tilde{A}} - r_{,u} \right) - e^{2\beta} r_{,\tilde{A}} r_{,\tilde{B}} \tilde{\eta}^{\tilde{A}\tilde{B}} \right). \tag{109}$$

We have made use of the relation $r_{,\lambda} = e^{-2\beta}$. The $r_{,u}$ and $r_{,\tilde{A}}$ are at constant λ, the former can be evaluated to $\mathcal{O}(\lambda)$ accuracy by using the value at the world-tube. To get $r_{,u}$ to $\mathcal{O}(\lambda^2)$, we need to calculate $r_{,\lambda u} = -2\,e^{-2\beta}\beta_{,u}$ on the world-tube, and use $r_{,u} = r_{,u}^{(0)} + r_{,\lambda u}\lambda$.

This completes the calculation of the final contravariant metric component, and so the injection process may be carried out at last, after the final multiplication by the Jacobian described in Eq. (91).

5.2.4. *Status of the algorithm*

The CCM algorithm described in the previous sections is quite new. It has not yet been translated into a stable numerical scheme for black hole collisions and interactions in 3-D. However, there has already been extensive testing of these techniques in simpler situations. Recently, significant codes have been built which implement these ideas for scalar waves without gravity but with full 3-D Cartesian and null spherical coordinate grid patches in flat space [17, 18]. Codes have also been written for 1-D problems with gravitation, both for spherical self-gravitating scalar waves collapsing and forming black holes [19, 20] and for cylindrically symmetric vacuum geometry [21, 22]. What has not yet been accomplished is the development of stable codes with both 3-D geometry and non-trivial strong gravitation, although codes currently under development demonstrate the stable evolution of a single moving black hole. Thus, we may expect exciting progress in the near future, and the mysterious details of collision between black holes may soon be unveiled.

5.3. ACKNOWLEDGEMENTS

This work has been supported by the Binary black Hole Grand Challenge Alliance, NSF PHY/ASC 9318152 (ARPA supplemented) and by NSF PHY 9510895 and NSF INT 9515257 to the University of Pittsburgh. N.T.B. thanks the Foundation for Research Development, South Africa, for financial support. N.T.B. and R.A.I. thank the University of Pittsburgh for hospitality. J.W. and R.A.I. thank the Universities of South Africa and of Durban-Westville for their hospitality.

References

1. C.V. Vishveshwara, *Gravitational Radiation from Binary Systems*, University of Maryland Technical Report (1965).
2. C.V. Vishveshwara, *Phys. Rev. D.*, **1**, 2870 (1970).
3. C.V. Vishveshwara, *Nature*, **227**, 936 (1970).
4. N. T. Bishop, *Class. Quant. Grav.*, **10**, 333-341 (1993).
5. A.M. Abrahams and C.R. Evans, *Phys. Rev. D.*, **42** 2585 (1990).
6. A.M. Abrahams, S.L. Shapiro and S.A. Teukolsky, *Phys. Rev. D.*, **51** 4295 (1995).
7. A.M. Abrahams and R.H. Price, *Phys. Rev. D.*, **53**, 1963 (1996).
8. N.T. Bishop, R. Gómez, L. Lehner, M. Maharaj and J. Winicour, *Phys. Rev. D.*, **56**, 6298 (1997).

9. S. V. Tsynkov, *Artificial boundary conditions based upon the difference potential method*, NASA Technical Memorandum No. 110265, Langley Research Center (1996).
10. B. Engquist and A. Madja, *Math Comput.*, **31**, 629 (1977).
11. The Binary Black Hole Grand Challenge Aliance, *Schwarzschild-perturbative gravitational wave extraction and outer boundary conditions*, (submitted for publication).
12. N. T. Bishop, R. Gómez, L. Lehner and J. Winicour, *Phys. Rev. D,* **54**, 6153 (1996).
13. C. W. Misner, K. S. Thorne and J. A. Wheeler, *Gravitation*, W. H. Freeman, San Francisco, (1973).
14. R. Gómez, L. Lehner, P. Papadopoulos and J. Winicour, *Class. Quant. Grav.* **14** (4), 977-990 (1997)
15. J.W. York, in *Sources of Gravitational Radiation*, ed. by L. Smarr, Cambridge University Press, (1979).
16. For the ADM code, see http://www.npac.syr.edu/projects/bh/ADM/adm.html.
17. N. T. Bishop, R. Gómez, P. R. Holvorcem, R. A. Matzner, P. Papadopoulos and J. Winicour, *Phys. Rev. Letters*, **76**, 4303 (1996).
18. N. T. Bishop, R. Gómez, P. R. Holvorcem, R. A. Matzner, P. Papadopoulos and J. Winicour, *J. Comp. Phys.*, **136**, 140 (1997).
19. R. Gómez, P. Laguna, P. Papadopoulos and J. Winicour, *Phys. Rev. D*, **54**, 4719 (1996).
20. R. Gómez, R. Marsa and J. Winicour, *Phys. Rev. D*, **56**, 6310 (1997).
21. C. Clarke, R. d'Inverno, and J. Vickers, *Phys. Rev. D.*, **52**, 6863 (1995).
22. M. Dubal, R. d'Inverno, and C. Clarke, *Phys. Rev. D*, **52**, 6868 (1995).

25. ASTROPHYSICAL SOURCES OF GRAVITATIONAL WAVES

B.S. SATHYAPRAKASH
Department of Physics and Astronomy
Cardiff University of Wales
P.O. Box 913, Cardiff, CF2 3YB, U.K.

Prolog When I was a student of the National College during the years 74–79 every year I eagerly awaited the month of November during which the *Bangalore Science Forum* celebrated the Science Festival by organising lectures on a variety of topics from liquid crystals to superconductivity and neurology to chromatography. It was during the 1975 Science Festival, on a gloomy evening that I attended your lecture on black holes. Though the evening was gloomy the lecture was excellent and has remained clearly in my memory. It was that lecture which focussed my otherwise topsy-turvy mind into making the crucial decision of what I wanted to do next. During your career you have motivated a large number of students and I am among those who is indebted to you for having shown the first rays of light. I wish you all the success with your untiring effort in popularising science and motivating thousands of young students. I will consider myself to be most fortunate if some day I can come and tell you that we have seen the quasi-normal modes to conclude your quest *on the trail of a black hole* and to begin a new era of gravitational wave astronomy.

1. Introduction

The twentieth century has witnessed a great revolution in the field of astronomy, starting from the construction of large optical telescopes to the building of radio antennas, infrared cameras, X-ray and gamma-ray satellites, etc. Each of these windows of observation has brought with it a new and complementary picture of the Universe and more often than not has had surprises and puzzles never ever imagined in the wildest tales of science fiction. The expansion of the Universe, microwave background, millisecond pulsars, quasars and radio galaxies, gamma-ray bursts, X-ray binaries, and, more recently, very compelling evidence for the existence of supermassive

B. R. Iyer and B. Bhawal (eds.), Black Holes, Gravitational Radiation and the Universe, 409–436.

(10^6–10^9 $M_\odot$) as well as stellar mass black holes, are but a small fraction of the novel discoveries they have brought to bare.

In the new millennium, several long-baseline interferometric gravitational wave detectors will become operational, complementing the existing resonant bar antennas, opening yet another window for observing the Universe. Gravitational wave antennas are essentially omni-directional with their response better than 50% of the average over 75% of the sky. Once built, these detectors will operate as a network taking data continuously and will initially be able to survey a volume of 10^4 Mpc^3 for inspiralling compact binary stars and supernovae — the most promising sources for these detectors — and will enhance that volume by 3 orders of magnitude in about five years after they are built, thus increasing the number of potential events a thousand-fold. It is this second phase of operation that will be most interesting from the astrophysical point of view bringing us a wealth of information from far corners of the Universe. However, the initial phase may very well see some of the strongest sources of gravitational waves such as coalescences of binaries consisting of neutron stars and/or black holes, and, with long integration, sources of continuous radiation, such as spinning neutron stars in our Galaxy, and may well reveal unexpected sources. The initial detections, though not frequent, are important from the fundamental physics point of view and will enable us to directly test the long-standing and much awaited predictions of general relativity.

In this article, I will discuss promising sources of gravitational radiation for both ground- and space-based detectors. In addition to those discussed here there are a number of sources that are rather speculative, yet very interesting—naked singularities, soliton stars, quark stars, etc.—but I won't touch upon them due to the lack of space. Though speculative, all efforts will be made in searching for these exotic events/signals in the outputs of our detectors; after all the purpose of building new detectors is, to a large extent, to test predictions and to confirm or to rule them out and to learn equally from failures and successes. For recent reviews on gravitational waves see Schutz (1997) and Thorne (1995; 1997). Lasota & Marck (1997) contains many articles devoted to gravitational waves and their detection.

2. Gravitational wave detectors

Currently, there are a number of bar detectors, both room-temperature and cryogenic, in operation. Bar detectors are resonant detectors and have a maximum sensitivity of $h \sim 10^{-20}$ Hz$^{-1/2}$, in a narrow band of 10–20 Hz. They are sensitive to supernova bursts in our Galaxy as well as to continuous wave sources whose frequency lies in their sensitivity band. Astone, Lobo & Schutz (1994) have shown that bar detectors will be very useful for observation of stochastic backgrounds in coincidence with interferometers. There are plans to build resonant spherical antennas which are truly omni-directional and just one of these can resolve the polarisation of the wave.

Interferometric detectors currently under construction will greatly increase our chance to directly detect gravitational waves and once improved and enhanced they will allow us to do gravitational wave astronomy. The Japanese are building a 300 m TAMA detector in Tokyo (Tsubono *et al.*, 1995), the British-German collaboration is constructing a 600 m interferometer called GEO600 in Hannover (Danzmann *et al.*, 1995), the French-Italian VIRGO collaboration is building a 3 km detector near Pisa (Bradaschia *et al.*, 1990) and the American LIGO project is building two 4 km arm length detectors, one in Louisiana and the other in Hanford (Abramovici *et al.*, 1992). These detectors will start taking data between 1999 and 2002. In addition, the Australians have plans to build a km-baseline detector. There is a proposal submitted to ESA to build a Laser Interferometer Space Antenna (LISA) (Bender *et al.*, 1996) which, if it comes through, may fly around 2015, or earlier. (See the article by Biplab Bhawal in this book for details on detectors).

The noise in interferometric detectors is characterised by the noise power spectral density (psd) $S_h(f)$ which is the spectral power of noise in a frequency bin. On coherently integrating a continuous wave signal for a time T, the signal strength builds up as T, while the *amplitude* noise, *viz* $\sqrt{S_h(f)}$, increases in proportion to the $T^{1/2}$. The net effect is that relative to a signal the noise amplitude goes down by $T^{-1/2}$. In Fig. 2 we plot $\sqrt{S_h(f)/T}$, which is now dimensionless, for $T = 1$ yr (left panel). The dimensionless gravitational wave amplitude (cf. Sec. 3) of continuous waves should be compared with this graph. On the other hand, effective source strengths of bursts (cf. Sec. 3), such as supernovae or coalescing binaries, should be compared with the *amplitude* noise per logarithmic frequency bin $h_{\rm SB} = \sqrt{f S_h(f)}$, which again is dimensionless. This is plotted in Fig. 2 (right panel), together with some source strengths to be discussed below. In about five years after they are built , LIGO and VIRGO will be improved to achive enhanced sensitivity. These enhanced interferometers will have a sensitivity about ten times better than their initial sensitivity thus increasing the volume of space they can survey by a thousand-fold.

The Laser Interferometer Space Antenna (LISA) is a proposal submitted to ESA for a space-based detector. It consists of three drag-free satellites, forming an equilateral triangle of size 5 Mkm, in heliocentric orbit, lagging behind the Earth's orbit by 20 deg. The plane of the triangle at an inclination of 60 deg to the Earth's orbital plane. The three arms together constitute two independent detectors from the view point of polarisation

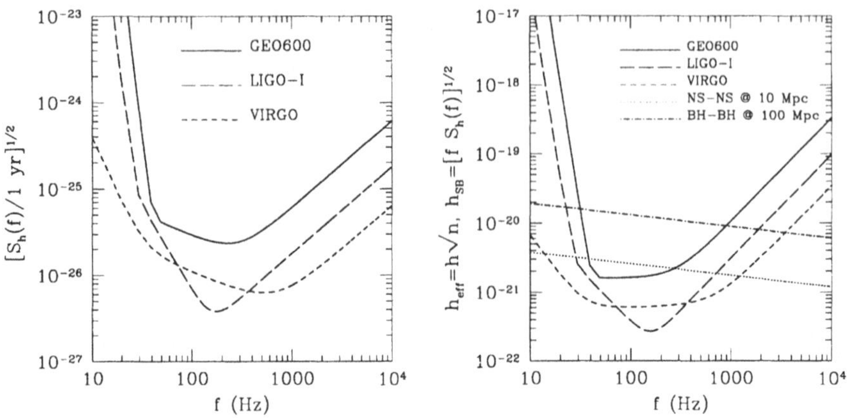

Figure 1. The rms noise after one year's of integration relative to signal strength (left panel) and sensitivity to bursts, such as binary inspiral waves and supernovae, of interferometric detectors (right panel).

measurements, but their noise will not be completely independent. The sensitivity of LISA for one year observation is shown in Fig. 3.

3. Gravitational waves: overview

Astronomical sources of gravitational radiation are extremely luminous; indeed some of the sources at even as great a distance as the Virgo supercluster of galaxies can be as luminous as Moon, though for a short time. Even so, gravitation being the weakest of all known interactions, the waves associated with it are not easily detectable. Thus, though gravitational waves are produced by almost all bodies in motion, in most cases the signal strengths are far too small to be sensed with the current detectors or those that are being built. For example, our only evidence for the emission of gravitational waves comes from the observation of the decay in the orbital period of the Hulse-Taylor binary caused by radiation back-reaction (Taylor, 1994). Yet the wave amplitude (and the gravitational wave frequency) is far too small to be sensed by any antenna, existing, under construction or planned. It is only in a few catastrophic astronomical events that the wave amplitude and frequency can be sufficiently large so as to cause our detectors to respond.

One may wonder, therefore, how is it possible to infer the presence of a body by the waves that it emits rather than by its gravitational interaction? In general relativity the effects of both the gravitational interaction and gravitational radiation are described by the tidal forces they produce on free

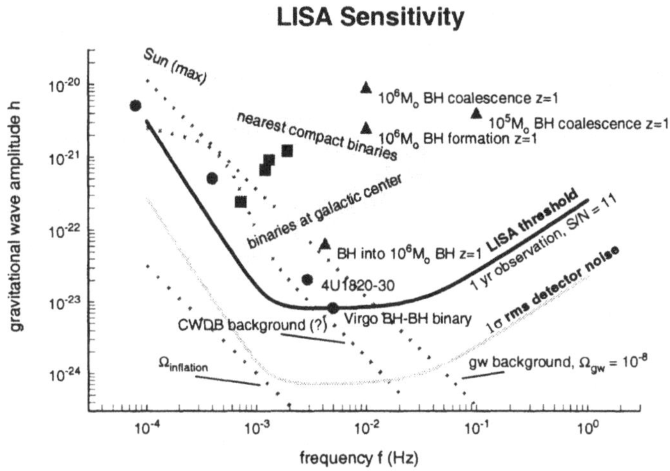

Figure 2. Sensitivity of the LISA detector corresponding to a year's of integration. The rms noise (grey line) is shown together with 11 times the rms (solid line) which on an average corresponds to a 5-σ detection for a source located in an arbitrary direction (a factor of $5^{1/2}$, see text). Many sources which LISA may see are also shown.

test masses; the Newtonian inverse-square force field simply determines the background geometry in which free particles move along geodesics. While the tidal force falls off as r^{-3}, the force due to gravitational waves that it emits decreases as r^{-1}. Therefore, the gravitational interaction is the dominant effect close to the gravitating body and the effect of the waves is pretty much negligible. However, far away from the body the effect of the waves are more readily sensed.

There is a fundamental difference in the way gravitational wave sources are observed as opposed to electromagnetic wave sources: Optical, infrared, radio and other telescopes are essentially sensitive to the intensity of radiation and build up the signal-to-noise ratio by incoherent superposition of waves emitted by a large number of microscopic sources. One therefore only obtains a gross picture of the region emitting the radiation rather than information about individual sources emitting the waves. The intensity of radiation falls off with distance as r^{-2} so that the number of sources of a given intensity increases as $r^{3/2}$, for a homogeneous distribution of sources in a flat Universe. In contrast, all gravitational wave signals, with the exception of stochastic backgrounds and signals of unknown shape, are tracked in phase and the signal-to-noise ratio is built up by coherent superposition of many cycles emitted by a source. In this case it is possible to deduce the detailed dynamics of the bodies emitting radiation and a precise picture of the source can be reconstructed. Since one is tracking the phase of the signal, the signal strength is proportional to the amplitude and only falls off as r^{-1} and therefore the number of sources of a given intensity increases as r^3. Gravitational wave signals result from coherent accelerated motion of massive objects. Unlike electromagnetism, there is no dipole radiation in Einstein's gravity. The dominant channel for emission is quadrupolar, which makes gravitational wave emission an even more an inefficient process than electromagnetic phenomena. The very nature of gravitational dynamics and radiation back reaction makes the emission of radiation either weak, or short lived, or both.

Gravitational radiation can be represented by a second rank, symmetric trace-free tensor. In a general coordinate system, and in an arbitrary gauge, this tensor has nine independent components. However, as in the electromagnetic case, gravitational radiation has only two independent states of polarisation in Einstein's theory: the plus-polarisation and the cross-polarisation (the names being derived from the shape of the equivalent force fields they produce). However, unlike electromagnetic radiation the angle between the two polarisation states is $\pi/4$ rather than $\pi/2$. Measuring the polarisation states is of fundamental importance since there are theories of gravity other than general relativity in which the number of polarisation states is more than two; in some theories even dipolar and scalar waves exist (Will, 1981). Polarisation measurements would be possible with a network of detectors, which would be an independent test to confirm or rule out alternative theories of gravity. We will see below that measuring the polarisation of the waves has astrophysical implications too. For the moment let us note that the waves can be represented by two dimensionless amplitudes, h_+ and $h_\times$.

Our detectors cannot directly measure the two independent polarisations of the waves but rather a certain linear combination of the two. The complex response R of a detector is (Schutz & Tinto, 1987; Dhurandhar & Tinto, 1988):

$$R = h_+F_+ + ih_\times F_\times, \quad F_+ = \Re(F), \quad F_\times = \Im(F), \tag{1}$$

where F_+ and $F_\times$ are the real and imaginary parts of the complex antenna pattern, which is a function of the direction (θ, ϕ) to the source and the polarisation angle ψ of the wave, in a coordinate system 'attached' to the detector. Responses of three widely separated detectors to an incident gravitational wave, together with two independent differences in arrival times among them, are, in principle, sufficient to fully reconstruct the gravitational wave. We shall use the symbol h to mean the rms response or one of the two polarisations. If the initial distance between two free masses is ℓ, the maximum change in the proper distance caused by a passing gravitational wave of amplitude h is $h\ell/2$.

Gravitational waves carry energy and momentum from the system. By demanding that the energy lost in the form of gravitational radiation is precisely balanced by the decrease of energy in the system, one can derive a simple expression for the apparent luminosity of radiation $\mathcal{F}$, at great distances from the source, in terms of the variation of the gravitational wave amplitude (Schutz, 1985): $\mathcal{F} = |\dot{h}|^2/(16\pi)$. The above relation can be used to make an order-of-magnitude estimate of the gravitational wave amplitude from a knowledge of the rate at which energy is emitted by a source in the form of gravitational waves. If a source at a distance r radiates away energy E, in a time T, predominantly at a frequency f, then the amplitude at a distance r, writing $\dot{h} = 2\pi f h$, is

$$h^2 \sim E/\left(T\pi^2 f^2 r^2\right). \tag{2}$$

The amplitude for a fiducial source is:

$$h \sim 4 \times 10^{-21} \left(\frac{E}{10^{-7} M_\odot}\right)^{1/2} \left(\frac{5 \text{ ms}}{T}\right)^{1/2} \left(\frac{200 \text{ Hz}}{f}\right) \left(\frac{40 \text{ kpc}}{r}\right). \tag{3}$$

Long baseline interferometers, such as LIGO and VIRGO, aim towards achieving a broad-band noise amplitude $\sqrt{f S_h(f)}$ of 3×10^{-22} at 200 Hz during their first stage of operation. Thus, an event at 40 kpc, a distance covering our own Galaxy, depositing energy equivalent to $10^{-7} M_\odot$ at 200 Hz in 5 ms will appear in these detectors with a maximum signal-to-noise ratio (SNR) of 13. However, averaging over all sky directions and polarisation angles reduces the expected rms response by a factor $\sqrt{5}$, which means that an event, on the average, produces an SNR a factor $\sqrt{5}$ lower than the maximum, or about 6 in the present case. Enhanced interferometers will have about ten times better sensitivity and will therefore be able to survey a distance ten times larger for such events with a similar SNR.

The numbers we have used for energy, duration and frequency are typical of what is expected for a strong type II supernova. Coalescence of massive black holes ($M \sim 10^6 \ M_\odot$) at cosmological distances ($r \sim$ Gpc) would give rise to a thousand times larger amplitude, but the waves would come off at milli-Hertz and the total energy emitted would be thousands of solar masses over thousands of seconds. Ground-based detectors wouldn't see such a source but in a space antenna, such as LISA, it would be a strong signal.

When the time development of a signal is known one can filter the detector output with a copy of the signal. This leads to an enhancement in the SNR, as compared to its narrow-band value, by the square-root of the number of cycles the signal spends in the detector band. A signal of frequency f lasting for a time T would produce $n = fT$ cycles. Using this expression and substituting for T in Eq. (3) we can define the effective amplitude $h_{\rm eff} \equiv \sqrt{n} \times h = \sqrt{E/f}/(\pi r)$:

$$h_{\rm eff} = 4 \times 10^{-21} \left(\frac{E}{0.05 M_\odot}\right)^{1/2} \left(\frac{200{\rm Hz}}{f}\right)^{1/2} \left(\frac{30\ {\rm Mpc}}{r}\right). \tag{4}$$

Thus, a well modelled signal, such as that of an inspiralling binary whose parameters have been used in the above formula, can be detected even if it is very weak, provided we have sufficiently large number of cycles. Typical number of cycles can vary from 10–100 (for supernovae and other impulsive events) to 10^4–10^{10} (for inspiral events at one end and rapidly spinning neutron stars in a year's observation period at the other) requiring a narrow-band amplitude of 3 to 10^5 lower. However, integrating a broad band signal would require an accurate model of the signal's phase evolution, which in turn places stringent demands on the required computing power to process the data on-line. From Eq. (4) we see that the effective signal amplitude depends on the total energy emitted and frequency and not on the duration. Therefore, a short-lived (1 ms) supernova can be as strong as a relatively long-lived (few to tens of seconds) inspiral signal from a binary. On the other hand, a very weak signal lasting a large number of cycles in the detector band would be detectable with as good a signal-to-noise ratio as a strong supernova burst.

4. Transients

In this Section we will discuss transient sources such as supernovae and inspiralling compact stars in binary systems. These sources last for a duration in the range of 10^{-3}–10^4 s.

4.1. SUPERNOVAE

Supernovae (SN) are classified into classes: In type I SN explosions (SNe) hydrogen lines are absent and in type II Sne hydrogen lines are present. Type I SNe are believed to be thermonuclear explosions of degenerate white dwarfs while type II SN are believed to occur as a result of the core collapse of an evolved massive ($> 9M_\odot$) star. It is possible that not all type I SN are thermonuclear explosions of white dwarfs; some of them (type Ib, Ic) may be the core collapse of an evolved massive star, as in the case of a type II SN, which has lost hydrogen either due to strong stellar winds or due to mass transfer as would happen in a binary. It is unlikely that type I SN would emit significant amount of gravitational waves. Type II SN, on the other hand, are associated with violent mass ejection with velocities of order 0.03 and formation of a compact remnant — neutron star or a black hole. The event has at its disposal the gravitational binding energy of the newly formed compact star which is:

$$E \sim 3 \times 10^{53} \left(\frac{M}{M_\odot}\right)^2 \left(\frac{R}{10\ \text{km}}\right)^{-1}\ \text{erg}. \tag{5}$$

99% of this energy is carried away by neutrinos, about 1% is transferred as kinetic energy of eject, a fraction of 10^{-4} of the total energy is emitted in the form of electromagnetic radiation. Depending on how asymmetric the collapse is, one might expect up to 10^{-7} of the total energy deposited into gravitational waves; spherically symmetric collapse, of course, cannot emit any radiation. Significant amount of gravitational waves may be emitted during essentially two stages of the collapse: rotational core collapse, bounce and oscillations, rotation-induced bars and convective instabilities set up in the core of the newly born neutron star.

Rapid rotation flattens the collapsing core inducing large quadrupole moment thus generating gravitational waves. Zweger and Müller (1996) have studied a wide range of rotational core collapse models and they conclude that the largest signals are produced by models which are (i) initially slowly rotating and have a stiff equation of state or (ii) initially rapidly rotating and have a soft equation of state. In the first case bounce occurs at densities above nuclear matter density, with a fast deceleration of the collapsing core resulting in the emission of GW signals. In the second case, the quadrupole moment is large due to rapid rotation which facilitates emission of gravitational waves. The first scenario leads to emission of type II gravitational waves that are characterised by several pronounced peaks which arise if the core bounces repeatedly, and the second leads to type I gravitational waves which are bursts followed by a ring-down phase caused by the bounce of the in-falling material followed by damped volume and surface oscillations of the inner core. In either case, unfortunately, the signals are not strong enough to be interesting sources for the first generation detectors. The maximum amount of energy emitted is of the order of $10^{-7}M_\odot$ in the frequency range 500–1000 Hz (see Müller (1997) for a review).

When the core's rotation is high enough it may cause the core to flatten before it reaches nuclear density leading to an instability that transforms the flattened core into a bar-like configuration which spins about its transverse axis. Some of these instabilities could also fragment the core into two or more pieces which then rotate about each other. Both are efficient ways of losing energy in the form of gravitational waves. It is estimated (Bonnell & Pringle, 1995) that the waves could carry up to $10^{-3}M_{\odot}$ in a few ms. On using these values in Eq. (3) we see that LIGO and VIRGO detectors can see such an event up to 50 Mpc or about 5–10 Mpc if the waves come off at 1 kHz. GEO600 will also be able to see these events provided the signal comes off around 200 Hz and comes from a reasonable direction in the sky.

Finally, convective instabilities in the core of the newly born neutron star, which last for about a second after the collapse, are likely to produce gravitational radiation due to anisotropic mass distribution and motion. Müller and Janka (1997) find that the gravitational wave amplitude at 100 kpc is $h \sim 10^{-23}$ and that the waves would come off at about 100 Hz. Since there would be about 100 cycles, one can enhance the amplitude to about 10^{-22} provided we know the development of the signal but this is still far too weak to be detected beyond about 10 kpc with high confidence.

The supernova that occurred in the Large Magellanic Cloud in 1987 (SN 1987a) would most likely have caused bar detectors to respond if they were operational at that time. In any case, coincidences among bar detectors and with other (neutrino/electromagnetic) observations would have yielded very positive results. Unfortunately, none of the bars was up and running and we possibly missed a once-in-a-century opportunity to gravitationally observe a brilliant event in our neighbourhood. Even today, supernovae are a favourite as they are expected to produce a characteristic burst of radiation and be seen in our instruments without the aid of sophisticated data processing and observed in coincidence with neutrino detectors and electromagnetic wave detectors. A supernova may not be visible optically if there is an optically thick medium between the supernova and Earth. Nevertheless, a supernova will be gravitationally visible if the collapse is sufficiently asymmetric.

It is expected that the SN event rate is about 0.03–0.01 per year in a galaxy. Enhanced interferometers can see a supernova event as far as the Virgo supercluster which contains about 200 bright galaxies and at least twice as many low-surface brightness galaxies. In addition, there are a few other smaller clusters within that distance as also a large number of field galaxies. Therefore, the supernova event rate for these instruments could be as large as tens per year. Such observations would undoubtedly be of great interest and would shed light on hitherto un-understood processes that occur when a star collapses to form a compact object.

4.2. COMPACT BINARIES

Compact binaries consisting of neutron stars (NS) and black holes (BH) are the most promising sources of gravitational radiation for both ground-

and space-based interferometric detectors. (For the sake of quoting numbers we shall assume below the mass of a neutron star to be 1.4 $M_\odot$ and that of a black hole to be $10 M_\odot$.) As mentioned in the introduction, they are the only systems in which we have seen the direct influence caused by the emission of gravitational radiation. The observation of the orbital period decay in the binary pulsar PSR1913+16 is in excellent agreement with the predictions of general relativity. The agreement is indeed better than one part in a thousand (Taylor, 1994). The Hulse-Taylor binary will coalesce over a time scale of 10^9 years, a bit too long for us to observe during our life times! However, there must be other binaries in our Galaxy with coalescence time scales much shorter than that. Indeed, based on the statistical analysis of current surveys of binary pulsars the conservative estimates are, 3 NS-NS coalescences per year out to a distance of 200 Mpc (Phinney, 1991; Narayan, *et al.*, 1991). More recently, there have been simulations of compact binary formation and evolution by many authors (Lipunov, Postnov, Prokhorov, 1997a; Tutkov & Yungelson, 1993; Yungelson & Zwart, 1998; Zwart & Yungelson, 1997). There is a large uncertainty in these calculations, due to lack of knowledge about the formation of compact binaries in our Galaxy, and the event rate could be a factor 50 smaller or a factor 1000 larger. This means the nearest/farthest distance within which the event rate could be several per year is 25/1000 Mpc. The initial LIGO/VIRGO interferometers are expected to have a span of about 30–40 Mpc, i.e. three times further away than the Virgo cluster, and have, therefore, a fair chance to see an inspiral event during their 4/5 years of operation. Enhanced interferometers, as remarked earlier, will have a span ten times larger, that is up to 300–400 Mpc and will begin to see several/many hundred events each year.

We have no idea about the coalescence rates of NS-BH or BH-BH binaries; no one has seen these objects yet. Even if the absolute coalescence rates of these systems is smaller than the NS-NS coalescences, since the SNR for a binary inspiral signal is proportional to $m^{5/6}$, m being the total mass, more massive binaries can be seen to a greater distance and hence these systems may well be the first to be observed by the interferometer network (Lipunov, Postnov, Prokhorov 1997a, 1997b, 1997c) or it may very well be that NS-BH systems are too scarce and BH-BH binaries may never form (Yungelson & Zwart, 1998; Zwart & Yungelson, 1997). However, assuming rates of coalescence similar to NS-NS rates, one finds that the NS-BH coalescence events could be a factor 30 larger and BH-BH coalescences a factor 130 larger. A thorough investigation of the problem is in order before we can draw concrete conclusions.

There are three distinct phases in the evolution of a binary during which gravitational waves of different signatures are emitted. The first is the phase during which the two stars slowly spiral-in due to radiation reaction emitting waves of greater amplitude and frequency until the two stars begin to tidally interact with each other and merge. At that stage the inspiral phase ends and the complicated dynamics governing the merger of the two stars will, it is expected, lead to the emission of gravitational waves depositing a large amount of energy in the so-called coalescence waves. At the end of coalescence either the system will form a black hole, provided the mass remaining after collision and merger and bounce is be large enough, or a massive neutron star. In the case of a newly born neutron star one can expect certain kinds of instabilities to facilitate the emission of radiation lasting for some time as remarked in the last Section. The formation of a black hole will be accompanied by the emission of radiation, called quasi-normal modes, that is characteristic of the mass and angular momentum of the hole and no other detail. These waves will carry the unique signature of the event and it is undoubtedly the only way in which we will ever directly witness the formation of a black hole. We will discuss the nature of each of the aforementioned phases in the following Sections.

Inspiral waves The inspiral waves are computed using post-Newtonian and post-Minkowskian approximation schemes (Blanchet *et al.*, 1995). The phasing of the wave-forms is presently known to a high order in post-Newtonian theory, that is to order v^7 (v being the velocity in the system) beyond the dominant quadrupole order (Blanchet *et al.*, 1998). However, the post-Newtonian scheme, being an asymptotic expansion in the parameter v, converges rather poorly. Cutler *et al.* (1993) pointed out the importance of knowing the phasing of an inspiral wave-form accurately in order to extract the signal and this was confirmed by Poisson (1995) who concluded that even corrections of order v^8 are not good enough. Indeed, based on computations of the signal emitted by a test particle in orbit around a Schwarzschild black hole (Tanaka *et al.*, 1996) we know that a wave-form at order v^{11} does not agree well with the exact wave-form (Damour, Iyer & Sathyaprakash, 1998). The problem has basically to do with the poor convergence of the post-Newtonian series, which is an asymptotic series and hence suffers from poor analytic properties. Fortunately, recent work (Damour, Iyer & Sathyaprakash, 1998) has shown that the relevant physical quantities, gravitational binding energy of the system and gravitational luminosity, entering the phasing formula can be suitably re-defined and the new quantities are then amenable to the use of Padé approximation, which helps in the acceleration of the convergence. It is envisaged that these improved wave-forms — called *P-approximants* — will be used in the design of search templates to detect inspiral waves.

A binary source will become visible in a detector when the frequency of radiation it is emitting enters the detector band. In most cases the detectors will also be able to track the source until the coalescence phase. However, whether or not the inspiral phase will be visible in a detector depends on the mass of the source. This is because the inspiral phase ends when the binary reaches its last stable orbit (lso) at which time the two stars are roughly at a distance $R = 6m$ from each other, where m is the total mass of the system. In terms of the frequency of gravitational radiation this is given by

$$f_{\rm lso} = \frac{1}{6^{3/2}\pi m} = 1.5 \left(\frac{m}{M_\odot}\right)^{-1} \text{ kHz.} \tag{6}$$

Binaries that terminate before reaching a detector's sensitivity band will be unobservable by that detector. For most ground-based interferometers the band where most of the signal power is extracted starts around 50 Hz, which means the heaviest binaries they will observe is around $100M_\odot$.

A typical NS-NS binary entering a detector of lower frequency cutoff of 100 Hz will last for a duration of 3 sec and that duration rises in proportion to $f_{\rm s}^{-8/3}$ in a detector whose lower cutoff $f_{\rm s}$ is lower, increasing to about 15 mins, for the same system, when $f_{\rm s} = 10$ Hz. More precisely, the *chirp time*—the time spent by a binary of total mass m and mass ratio η in a detector of lower frequency cutoff $f_{\rm s}$—is given by

$$\tau_0 = \frac{5}{256 \eta m^{5/3} (\pi f_{\rm s})^{8/3}} \simeq 24.2 \left(\frac{0.25}{\eta}\right) \left(\frac{2.8M_\odot}{m}\right)^{5/3} \left(\frac{40 \text{ Hz}}{f_{\rm s}}\right)^{8/3} \text{ s.} \tag{7}$$

During this time the signal will sweep as many as

$$N_{\rm cyc} = \frac{1}{32\pi\eta\,(\pi m f_{\rm s})^{5/3}} \simeq 1600 \left(\frac{2.8M_\odot}{m}\right)^{5/3} \left(\frac{40 \text{ Hz}}{f_{\rm s}}\right)^{5/3}, \tag{8}$$

cycles. The above expressions clearly show the importance of having as small a lower cutoff $f_{\rm s}$ as possible. On the one hand we will begin to see more and more massive binaries, which can be seen to a greater distance and hence, possibly, more frequently, and on the other hand, the time spent by and the number of cycles of, a given binary, which determine the SNR, are sharp functions of the lower cutoff, increasing rapidly as $f_{\rm s}$ decreases.

The energy emitted in the process of inspiral is equal to the binding energy of the system when the inspiral waves shut off at $R = 6m$:

$$E = -\frac{m_1 m_2}{2R} = -\frac{\eta m^2}{2R} = -\frac{\eta m}{12}, \tag{9}$$

where we have substituted $R = 6m$ and $\eta = m_1 m_2/m^2$ is the symmetric mass ratio whose maximum value is $1/4$. Thus, the maximum energy emitted by a binary in its inspiral phase (practically the time it spends in the detector band if it reaches lso during observation) is about 1/50 of its total mass, which for a NS-NS binary is roughly $0.05M_\odot$ [cf. Eq. (4)]. The effective amplitude at a given frequency f can be obtained by noting that the distance between the two stars is related to gravitational wave frequency via $R = (M/\pi f)^{1/3}$, and using $|E| = \eta m^2/2R = \eta m^{5/3}(\pi f)^{2/3}/2$, in the expression for the effective amplitude, $h_{\rm eff} = \sqrt{E/f}/(\pi r)$:

$$h_{\rm eff} = \frac{\eta^{1/2} m^{5/6} f^{-1/6}}{\pi^{2/3} r}. \tag{10}$$

The above equation gives,

$$\begin{aligned} h_{\rm eff} &= 3\times 10^{-21}\left(\frac{30\ \rm Mpc}{r}\right)\left(\frac{\eta}{0.25}\right)^{1/2}\left(\frac{m}{2.8M_\odot}\right)^{5/6}\left(\frac{1\ \rm kHz}{f}\right)^{1/6} \\ &= 4\times 10^{-17}\left(\frac{1\ \rm Gpc}{r}\right)\left(\frac{\eta}{0.25}\right)^{1/2}\left(\frac{m}{10^6 M_\odot}\right)^{5/6}\left(\frac{1\ \rm mHz}{f}\right)^{1/6} \end{aligned} \tag{11}$$

Stellar mass compact binaries close to their coalescence can be observed by ground-based antennas while observation of more massive systems, especially the coalescence of binaries consisting of supermassive black holes or a supermassive black hole and a stellar mass compact object (a neutron star or a black hole), will require space-based detectors which have a good sensitivity in the low-frequency region.

Inspiral events are wonderful test beds for general relativity, for both its fundamental predictions as well as for testing the theory in the strongly nonlinear regime. Association of an inspiral event with an electromagnetic event, such as the observation of a gamma- or X-ray burst, would help to deduce the speed of gravitational waves to a phenomenal accuracy. (Even a day's delay in the arrival times of gravitational and electromagnetic radiation from a source at a distance of million light years would determine the relative speeds to better than one part in 10^8). This will require a good timing accuracy to determine the direction to the source and to signal astronomers to guide their telescopes in that direction for electromagnetic observation. One can also set a limit on the mass of the graviton by observing the inspiral waves. C.M. Will (1998) points out that if graviton has mass then it would alter the phasing of the waves and hence by tracking the inspiral waves one can bound graviton's mass to 2.5×10^{-22} eV using ground-based detectors and 2.5×10^{-26} eV using space-based detectors in future observations of inspiral events.

A network of detectors will be able to determine the polarisation of the waves. While Einstein's general relativity predicts only two independent polarisations, there are other theories of gravitation in which more states of polarisation, and even dipolar waves, are predicted. Therefore, an unambiguous determination of the polarisation of the waves will be of fundamental importance. Moreover, astronomical observations of binaries cannot yield the total mass but only the combination $m \sin i$, where i is the inclination of the binary's orbit to the line of sight. However, measurement of polarisation can determine the angle i since the polarisation state depends on the binary's inclination with the line of sight.

The post-Newtonian expansion of the inspiral wave form has unfolded the rich cache of non-linear physics that governs the dynamics of a binary. There are many interesting effects such as the tails of gravitational waves, spin-spin interaction, precessional effects, nonlinear tails, tails-of-tails, etc., which can all be deciphered from a strong inspiral signal. For instance, Blanchet & Sathyaprakash (1995) have shown that the effect of nonlinear tails on the phasing of gravitational waves can be detected in the inspiral waves emitted by BH-BH binaries.

Schutz (1986) pointed out that inspiral waves are standard candles in the sense that by measuring their luminosity we can infer the distance to source. In the quadrupole approximation, the wave's amplitude (as well as its phase) depends only on a certain combination of the masses called the *chirp mass*, $\mathcal{M} = \eta^{3/5} m$, and the distance r to the source. If we can measure the amplitude and the chirp mass independently, then we can infer the distance to the source directly. Inspiral waves are detected by using matched filtering technique which enables an accurate estimation of the chirp mass. In addition, the SNR enables us to measure the signal's amplitude as well. Thus, inspiral waves serve as standard candles. This will allow us to determine the Hubble constant very accurately by identifying the host galaxy and measuring its red-shift. It should be noted that the inspiral wave's phase is quite sensitive to post-Newtonian corrections and when they are included the mass-degeneracy is lifted. It is, therefore, possible to determine both masses of the binary by using post-Newtonian search templates.

In the recent times there is a growing evidence for the existence of super-massive black holes (10^6–10^9 $M_\odot$) at the centres of not only active galaxies but also that of ordinary galaxies. It is now believed that every galaxy hosts a super-massive black hole at its centre. High red-shift galaxy catalogues show interactions between galaxies which is clearly indicative of collisions, mergers, etc. Galaxy collisions must be accompanied by inspiral and merger of the associated black holes. Space-based detectors will be sensitive to gravitational waves from such systems even at a red-shift $z = 1$ (roughly 3 Gpc) and would obtain a SNR $\sim 10^4$.

The origin of massive black holes is not clear but they must be accreting ordinary stars, compact stars and black holes, in their vicinity, there-by increasing their mass. Accretion of ordinary stars is not likely to produce coherent emission of gravitational waves as the star would fragment much before reaching the horizon. However, spiralling-in of compact stars or black holes would definitely be producing coherent radiation. These events will be more frequent than merger of massive black holes and will be very interesting phenomena from fundamental physics point of view. We do not yet fully understand trajectories of test masses around Kerr black holes; observation of such events will greatly help in testing strong field predictions of general relativity and understanding black hole spacetimes including uniqueness of the Kerr solution.

Since inspiral signals are standard candles, observations of massive black hole coalescences at cosmological distances $z \sim 1$ by space-based detectors can facilitate an accurate determination of the distance to the source. Space-based detectors observe a massive binary inspiral for a whole year and they have, thus, the baseline of the Earth's orbit around the Sun to triangulate the source on the sky. They can do this to an accuracy of a degree at high SNR. At a distance of 3 Gpc this is about 10 Mpc, a scale over which no more than one virialised galaxy cluster can be found. Thus, an optical identification of the host galaxy cluster and its red-shift, would enable the measurement of the deceleration parameter q_0 and hence the density parameter Ω of the Universe. A single source at $z \gtrsim 1$ is enough to measure both H_0 and Ω to an accuracy of better than 1%. Thus, space-based detectors can potentially contribute quite a lot to further our understanding of fundamental science and cosmology.

Sathyaprakash & Dhurandhar (1991; 1994) gave the formalism to optimally set up a lattice of templates to search for binary inspiral signals. This was extended to the case of first-post-Newtonian corrected signal by Sathyaprakash (1994) wherein it was shown that the inclusion of post-Newtonian templates, which adds an extra dimension to the search parameter space, does not significantly increase the number of search templates over quadrupole wave-form. Apostolatos (1995) studied the effect of spin-orbit and spin-spin interaction on the number of templates. He concluded that even if the magnitudes of the spins are large, their effect on wave-form detectability is small if the spin vectors do not precess. Owen (1996) and Owen & Sathyaprakash (1998) have spelled out a detailed prescription for obtaining the metric on the space of wave-forms given any noise spectrum and any restricted post-Newtonian wave-form. The resulting values of $\mathcal{N}$ for different detectors and second post-Newtonian wave-forms, are shown in Table 1.

Coalescence waves The physics of the coalescence waves is not known and there is a lot of effort to solve this important problem by using semi-analytical and numerical techniques. It is unlikely that a solution will be

TABLE 1. Number of templates required for a search with $m_{\rm min} = 0.2 M_\odot$ and minimal match=0.97. Divide by 73 for a $1 M_\odot$ search.

Interferometer	1PN	1.5PN	2PN
LIGO-I	2.5×10^5	5.3×10^5	4.7×10^5
VIRGO	1.4×10^7	1.4×10^7	1.3×10^7
GEO600	4.3×10^5	8.5×10^5	7.5×10^5

in place by the time the first detectors will begin to operate. In view of this it is important to make an educated guess of what to expect and prepare our search algorithms to efficiently dig into detector noise to look for these signals. The fact that the coalescence waves will be preceeded by inspiral waves makes the search easier, though it is not inconceivable that while the former may be observable with the aid of accurate search templates the latter may not be. Flanagan & Hughes (1998) estimate that for binary systems of total mass in excess of 25 $M_\odot$, the coalescence waves are likely to be significantly stronger than the inspiral waves. The span of a detector is larger for heavier binaries and therefore, these authors conclude, it is likely that the first gravitational wave events will be coalescences of massive neutron stars rather than inspiral waves.

A prediction about coalescence waves which can say what is the total energy that will be deposited in the form of gravitational waves during this phase, regardless of the details of the spectral content and the duration over which the waves are emitted, will suffice to conclude whether or not these waves will be observable. Of course, to extract the full signal it will eventually be necessary to know the exact phasing of the waves. But to judge whether merger waves are detectable it is enough to know the total energy emitted in the detector bandwidth. As we have seen in Sec. 3 the signal-to-noise ratio obtainable for any event simply depends on the total energy emitted, not on the luminosity, provided the structure of the wave-form is known. It is therefore easy to see why the work by Flanagan and Hughes predicts that coalescence wave-form will be stronger. It is related to the fact that they assume that 10% of the binding energy will be emitted in the form of gravitational waves during the coalescence phase.

MACHOs The results of the recent microlensing experiments (Alcock *et al.*, 1996) have revealed massive compact halo objects (MACHOs) of mass $0.5^{+0.3}_{-0.2} M_\odot$. Nakamura *et al.* (1997) argue that if MACHOs are black holes then they must have formed in the early Universe and they estimate that our Galaxy must contain about 10^8 black hole binaries with inspiral time scales less than the Hubble time. If this is the case then the rate of MACHO coalescences in our Galaxy is $\sim 5 \times 10^{-2}\ \mathrm{yr}^{-1}$, implying an event rate of few per year within 15 Mpc. The gravitational waves from such black hole binaries can be detected by the first generation of interferometers in the LIGO/VIRGO/TAMA/GEO600 network.

Binary merger and quasi-normal modes Vishveshwara (1970) showed that gravitational waves scattered off a black hole have a characteristic waveform when the incident wave has frequencies beyond a certain value depending on the size of the black hole. Press (1971) showed that perturbed black holes have quasi-normal modes of vibration and in that process emit gravitational radiation whose frequency and damping time are characteristic of the hole's mass and angular momentum. Flanagan & Hughes (1998) estimate that during the quasi-normal mode ringing by the black hole formed as a result of the coalescence of two neutron stars, the energy emitted is equivalent to 3% of the system's total mass. By matched filtering the ringdown waves, it would be possible to accurately determine the black hole mass and angular momentum.

5. Continuous waves

A source of continuous radiation is of interest for the primary reason that such sources can be observed over and over again and therefore single interferometers are sufficient to confirm their existence. But, as mentioned in the Introduction, we cannot expect a source to be continuous as well as strong. Normally, the source of gravitational waves is the gravitational binding energy or rotational kinetic energy. These could both be of the same order for a relativistic system. Suppose we have a neutron star of mass $M = 1M_\odot$, radius $R = 10$ km, spinning at a frequency $f = 100$ Hz. It has a rotational energy equal to $E = I\omega^2/2$, where $I = MR^2$ is the moment of inertia, $\omega = 2\pi f$. If all this rotational energy is radiated away as gravitational waves over, say, about a million years, then the amplitude of the waves using Eq. (3) is:

$$h = 3.2 \times 10^{-26} \left(\frac{10\ \mathrm{kpc}}{r}\right) \left(\frac{E}{2 \times 10^{-4} M_\odot}\right)^{1/2} \left(\frac{T}{10^{14}\ \mathrm{s}}\right)^{1/2} \left(\frac{f}{200\ \mathrm{Hz}}\right). \tag{12}$$

We see, therefore, that the amplitude of continuous waves cannot be too large. Now, what regulates the emission so that it is spread over millions of years? The answer is quite easy to see: Neutron stars have very high self-gravity $\varphi = M/R \sim 0.15$ and therefore it is very hard to produce any deformation which, together with rotation, is needed in emitting gravitational waves. Below we will discuss some mechanisms that may produce ellipticity in a neutron star but it is always too low to extract gravitational waves efficiently.

Our Galaxy is expected to have at least 10^8 spinning neutron stars that form roughly at a rate of one every 100 years. It is expected that a population of neutron stars will be in binaries. There are a number of ways in which a spinning neutron star could radiate away gravitational waves (if the neutron star is axisymmetric, of course, then there will be no gravitational wave emission): (i) Neutron stars normally spin at high rates (several to 500 Hz) and this must induce some equatorial bulge and flattening of the poles. Additionally, the presence of a magnetic field may cause the star to spin about an axis that is different from the symmetry axis leading to a time-varying quadrupole moment. (ii) The star may have some density inhomogeneities in the core/crust set up during its formation and/or subsequent convectively unstable motions of the core. (iii) The presence of an accretion disc, with its angular momentum not necessarily aligned with that of the neutron star, can potentially alter axisymmetry. That and electromagnetic radiation reaction torques can induce and sustain wobble. (iv) The normal modes of the neutron star fluid (radial and other oscillations) can extract rotational energy and re-emit in the form of gravitational waves. (v) There are certain classical and relativistic instabilities in the neutron star fluid which may cause the star to radiate away energy in the form of gravitational radiation.

Lumpy neutron stars If I_{zz} is the moment of inertia about the spin axis of a neutron star of frequency f then the gravitational amplitude at a distance r is:

$$h = 3 \times 10^{-27} \left(\frac{10 \text{ kpc}}{r}\right) \left(\frac{I_{zz}}{10^{45}\ g\ cm^2}\right) \left(\frac{f}{200 \text{ Hz}}\right)^2 \left(\frac{\epsilon}{10^{-6}}\right), \tag{13}$$

where ϵ is the ellipticity of the pulsar. It is an unknown but one can obtain an upper limit on it by attributing the observed spin-down of pulsars to gravitational radiation back reaction, namely that the change in the rotational energy $E = I\omega^2/2$ is equal to gravitational wave luminosity. By using a simple model of an equatorial plane of elliptical cross section of semi-major axis a_1 and semi-minor axis a_2, the ellipticity $\epsilon \equiv 1 - a_2/a_1$ is related to the spin-down rate of a pulsar via

$$\epsilon = 5.7 \times 10^{-6} \left(\frac{P}{10^{-2} \text{ s}}\right)^{3/2} \left(\frac{\dot{P}}{10^{-15}}\right)^{1/2}. \tag{14}$$

Since one knows the observed values of P and $\dot{P}$ one can obtain an upper limit on ϵ using the above equation. Following this method Schutz has found that for the Crab pulsar $\epsilon \leq 7\times10^{-4}$ and a gravitational amplitude of 10^{-24}. Indeed, he has obtained in this way the upper limit on the amplitude of all pulsars for which period decay has been measured. He concludes that Crab will have sufficiently large amplitude to be observable with the aid of the GEO600 detector by one year of continuous observing. There is one other pulsar, J0437-4715, which has an upper limit just above GEO600's broad band noise. But experimenters can greatly improve noise in a small band of about 10-20 Hz at the cost of worsened sensitivity in the rest of the band by using a technique called signal recycling. Such a technique would allow one to dig out a system such as J0437-4715. But pulsars are known to have other mechanisms by which they are spinning down; accelerated charged particles escaping from the system will carry away energy and so do the radio waves. It is therefore unlikely that neutron stars will be observed in this way. There are other mechanisms which may make a neutron star emit radiation over shorter time scales and in the following we will consider some of these.

The CFS instability and r-modes Chandrasekhar (1970) and Friedman & Schutz (1978) discovered an instability in the fundamental or f-mode of a neutron star fluid that goes unstable at high spin frequencies of the star, which is now called the CFS instability. Imagine exciting one of these modes in a non-spinning star. After a while the energy in the mode will be radiated away and the mode will decay. Now consider a spinning neutron star in which one of these modes is excited. These fluid modes have a certain pattern speed on the surface of the star which is fixed relative to the star. For every mode moving in the same sense as the spin of the star there will be another moving in the opposite direction. For low spin rates both these modes will decay in course of time by emitting gravitational waves. But as the neutron star is spun up above a critical spin rate, an external inertial observer sees both modes to be travelling in the same sense as the spin of the star. Thus, while the mode co-rotating with the star will decay, the mode counter-rotating with the star will grow in amplitude and emit more and more radiation. This will go on until the mode has sucked out enough angular momentum to make the mode appear to be counter-rotating with respect to an inertial observer. It is suspected that the CFS instability will not work in the presence of viscosity and hence it may be unimportant in old neutron stars. However, newly born neutron stars will be very hot and viscous forces may be unimportant in them. The CFS instability remained only a theoretically interesting possibility since one did not know how to spin up a neutron star to the CFS instability point, until Wagoner came up with a brilliant way of achieving this (see below).

Recently, Andersson (1998) has discovered another class of modes in the context of neutron stars called r-modes which are unstable at all spin frequencies of the star. It is proposed that r-modes are responsible for the limit on the spin frequencies of newly born neutron stars (Lindblom, Owen & Morsink, 1998). Owen *et al.* (1998) have computed the efficiency with which these modes extract energy out of the system and the expected gravitational wave amplitude from isolated neutron stars and from the ensemble of all sources up to cosmological distances. They conclude that advanced LIGO/VIRGO interferometers will be able to unambiguously detect such a background by coincident observation with a nearby bar detector. Much needs to be done in this area to assess the efficiency of these modes to extract energy from the system and up-convert it to gravitational waves when neutron stars are formed by gravitational collapse.

Wagoner radiation In general, neutron stars will not be rotating fast enough for the CFS instability to operate. However, Wagoner (1984) suggested that accretion discs can spin up a neutron star to the point where the CFS instability can operate. From then on, all accreted angular momentum will be radiated away in the form of gravitational waves. There are two types of X-ray binaries: the low- and high-mass X-ray binaries. These X-ray binaries are believed to consist of neutron stars with massive main-sequence companion entering its red-giant phase. To explain the observed X-ray luminosity one must suppose that the neutron star is accreting mass at the rate $\dot{M} \sim 10^{-10} M_\odot \text{ yr}^{-1}$. It is thought that this phase will only last for about 10^4 years and hence the mass gained by the neutron star is not appreciable. However, together with mass, there is also accretion of angular momentum which spins up the star and provides the rapid rotation that is needed in order for the CFS instability to operate. Once the CFS instability becomes operative the angular momentum accreted will not spin up the neutron star any further, it is rather radiated away in the form of gravitational waves. Since the angular momentum accretion rate is proportional to mass accretion rate, the gravitational wave amplitude of such a system will be proportional to the X-ray luminosity. One can therefore compute gravitational wave amplitude without needing to know any other details of the system including the distance to the source:

$$h = 6 \times 10^{-26} \left(\frac{\dot{M}}{10^{-10} M_\odot \text{ yr}^{-1}} \right)^{1/2} \left(\frac{100 \text{ Hz}}{f} \right) \left(\frac{1 \text{ kpc}}{r} \right). \quad (15)$$

The frequency of gravitational waves depends on the mode excited, which is unknown, and not the neutron star spin frequency. In course of time, the neutron star and its companion come closer to each other and the neutron star will be essentially buried in the atmosphere of the giant and the system will not appear to be an X-ray emitting binary as the X-rays will be absorbed. In such systems, called Thorne-Zytkov objects (Thorne & Zytkov, 1977), the mass accretion rate could be as high as $\dot{M} \sim 10^{-8} M_\odot \text{ yr}^{-1}$ and therefore the system could be an efficient radiator of gravitational waves. We note that the Wagoner mechanism not only provides a way of generating gravitational waves but it also predicts that neutron stars at the centres of accretion discs will have a limiting spin frequency.

Wobble radiation If a body has a substantial centrifugal bulge and during course of time its spin axis is misaligned with the symmetry axis then there is a precessional or wobble motion induced in the star. This wobble motion can cause the star to radiate away gravitational waves. If the wobble angle is θ and the ellipticity is ϵ then the gravitational wave amplitude is:

$$h \sim 6 \times 10^{-25} \left(\frac{1 \text{ kpc}}{r}\right) \left(\frac{f}{500 \text{ Hz}}\right)^2 \left(\frac{\theta\epsilon}{10^{-6}}\right)^2. \tag{16}$$

The frequency of radiation is at the precessional sideband of the spin frequency, viz, $f = f_{\text{spin}} + f_{\text{prec}}$. Normally, $f_{\text{prec}} \ll f_{\text{spin}}$ so that the frequency of the gravitational waves is nearly equal to the spin frequency of the star itself and NOT twice that value unlike in the case of a lumpy neutron star. Left to itself the wobble angle decays due to radiation reaction very quickly and the gravitational radiation emission will be shutoff. However, there may be electromagnetic and other torques present in the star which will maintain the wobble angle for long periods of time (Jones, 1997). Much work needs to be done in this direction especially to firm up whether the radiation emitted by the Bildsten mechanism (see below) comes off at the spin frequency or at twice the spin frequency of the star.

Bildsten Mechanism In the recent years, Rossi satellite observations of the X-ray emitting binaries have shown quasi-periodic oscillations (QPO) in their X-ray power spectra. It is believed that this QPO is a result of beating of two frequencies one of which is that of the neutron star. The neutron star spins inferred in this way quite dramatically lie in a narrow range of 250–350 Hz and are all within 20 % of 300 Hz. A neutron star is born with a high spin rate (several 100 Hz) but quickly spin downs to moderate rates (several 10 Hz). In a binary system, when the companion becomes a red giant the neutron star starts accreting mass and angular momentum. Though the mass accretion rate is very low ($\dot{M} \sim 10^{-10} M_\odot \text{ yr}^{-1}$) the accretion of angular momentum can spin up a neutron star substantially. Indeed, the millisecond pulsars are believed to be old pulsars in binaries recycled in this way. It is puzzling as to why the spin frequencies of neutron stars in X-ray binaries are all in a narrow range. Bildsten suggests that a Wagoner-type mechanism is operative in these systems which is limiting the spin frequencies of these neutron stars. However, unlike Wagoner radiation, the agent responsible for emission of gravitational waves is not the CFS instability but density inhomogeneities formed in the crust of the neutron star due to accretion-induced heating and temperature gradients set up in the crust. Bildsten suggests that absence of efficient heat transport processes make it possible to set up temperature gradients. Provided that the temperature distribution has large scale asymmetry then the temperature sensitive electron captures in the deep crust can provide the quadrupole ($\sim 10^{-7}MR^2$) needed to radiate away accreted angular momentum and limit the spin frequency. This mechanism is present only during accretion and decreases rapidly once the accretion halts. The frequency, unlike in the case of Wagoner radiation, will be known in advance. Bildsten predicts that the gravitational wave strength will be $h \sim (0.5\text{–}3) \times 10^{-26}$ for many of these sources and that the LIGO/VIRGO and signal recycled GEO600, can detect the strongest of these sources, Sco X-1, at a SNR of 5 with a few years of integration.

The search problem for continuous waves from spinning neutron stars is the most compute-intensive job in gravitational wave data analysis. Today, there is little hope that all-sky searches lasting for a year or more, can be made. It is not difficult to see why this is such a difficult problem: Firstly, the data has to be collected continuously for months together and at a good sensitivity. No one has run interferometers for periods as long as that and we do not yet know if this would be possible. Secondly, though a neutron star emits a periodic signal in its rest frame, save for the neutron star spin-down which indeed induces some modulation in the wave-form, because of Earth's acceleration relative to the source, the detector does not see a periodic wave. The wave is both frequency- and amplitude-modulated. One can, fortunately, de-modulate these effects since Earth's motion is known quite accurately, and hence recover the original periodic signal. *But* de-modulation requires a knowledge of the source's direction and its frequency, which are unknown in a blind search. The angular resolution one obtains in a year's integration is $\Delta\theta = \lambda/D$, where λ is the wave length of radiation and D is the baseline of the detector in a year's integration, namely 1 A.U. Thus, for $f = 100$ Hz we have $\Delta\theta = 10^{-5}$ rad or about two arcsec. Now, assuming that the source may be in any one of the 4 arcsec^2 patches on the sky we get the number of patches in the sky for which we will have to try out a de-modulation correction to be $4\pi/(\Delta\theta)^2 = 4\pi 10^{10}$. It is quite an impossible task to apply Doppler de-modulation to the detector output for each of these $\sim 10^{11}$ patches and compute as many Fourier transforms.

One, therefore, asks the question given a compute power what is the best possible search one can do? Is there any advantage in going from a one-step search to a two- or multi-step hierarchical search? What about directional searches? These are some of the problems for which we have some answer; but a great deal of work is needed and is currently under progress, to improve and optimise search algorithms. Brady, *et al.* (1998) have computed the number of days of data that a TFLOPS-class computer can analyse on-line (that is, analyse T-hours of data in T hours) and carry out a blind (that is, unknown direction, frequency and spin-down rate) search. Unfortunately, the longest data we can integrate on-line, for neutron stars with spin frequencies $f \leq 100$ Hz and spin-down rates less than 1000 years, is about 18 days. This yields a SNR lower by a factor of 5 as compared to a year's worth of observing. On-line searches for neutron stars with $f \leq 500$ Hz (largest observed frequencies of millisecond pulsars) and spin-down rates of 40 years (shortest observed spin-down rates), can only be made for a data set lasting for a duration of 20 hours or less. If the source's position is known in advance, but not its frequency, then one can carry out an on-line search, again with a TFLOPS-class computer, for the frequency of the source in a data set that is worth 3 months long. This is good news since there are many known pulsars and X-ray binary systems that are potential sources of radiation. In addition, the obvious targeted search locations are the centre of the Galaxy and globular clusters.

6. Stochastic backgrounds

There are myriads of sources of gravitational waves distributed in the Universe that will produce stochastic backgrounds of radiation. A given stochastic background will be characteristic of the sources that are responsible for it and it may be possible to discriminate different backgrounds provided their spectral characteristics are sufficiently different. Quantum fluctuations in the early Universe is believed to have generated a gravitational wave background (Grishchuk, 1997). Observation of this primordial background of gravitational waves is of fundamental importance giving us the unique opportunity to learn about the physical processes that operated in the first moments of creation. A stochastic background is normally measured by the spectral energy density of the waves in relation to the critical density of the Universe. More precisely, if ρ is the Fourier transform of the energy density of the waves then the spectral density $\Omega(f)$ is defined as $\Omega(f) = f\rho_c^{-1}(d\rho/df)$, where $\rho_c = 3H_0^2/8\pi$, and H_0 is the present value of the Hubble parameter. The value of the Hubble parameter is not known to a good accuracy and is believed to be in the range [50,100] km s^{-1} Mpc^{-1}. For our purposes we shall take H_0 to be 100 km s^{-1} Mpc^{-1}. In this Section we will discuss the prominent backgrounds that have been envisaged and the prospects of the various detectors to observe them.

Primordial background Just like the early Universe generated electromagnetic wave background—the cosmic microwave background (CMBR)—it is believed that the gravitational wave background was also produced at the same time as a result of quantum fluctuations in the early Universe. Primordial background radiation would freely travel to us from almost the very moment of creation since, unlike electromagnetic radiation which is strongly coupled to matter and hence interacted with it until the decoupling era at $z \sim 3000$, gravitational radiation is very weakly coupled to matter. Hence, its detection would help us get a picture of the first moments after the big bang. The observation of anisotropies in the CMBR by the COBE satellite experiment, has set limits on the strength of the gravitational wave background. The CMBR anisotropies arise as a result of both density inhomogeneities as well as the gravitational wave background. There is, at the moment, no way of ascertaining the relative contributions of these and hence we can only set upper limits on both. On the scale of COBE observations (*i.e.*, $f_{\rm gw} \sim 10^{-16}$ Hz) $\Omega_{\rm gw} < 7 \times 10^{-11}$ (see, e.g., Allen (1997)). By using pulsar timings Kaspi *et al.* (1994) have constrained the background at 10^{-8} Hz to $\Omega_{\rm gw} < 10^{-8}$. It is possible that gravitational waves were generated at a later stage due to physical processes such as up-conversion of false vacuum energy density during inflation, collision of cosmic strings or domain walls, etc. The waves generated during inflation are constrained by the COBE data and they are far too weak to be detected by any planned ground-based or space-based detectors. On the other hand, advanced LIGO detectors may observe the background generated by the collisions of a cosmic string network. This model predicts a fairly flat spectrum for gravitational wave background as there is no inherent scale in the problem and therefore the background may be visible in many different frequency windows, including advanced LIGO, provided the sensitivity is good enough. (For a recent review of primordial backgrounds see Allen (1997) and references therein.)

Supernovae background Blair & Ju (1996) have suggested that even though an individual supernova may not be detectable out to a great distance, the background produced by all supernovae within the Hubble radius might indeed be detectable. As we increase the span of our detectors, the number of supernovae increases not only because the volume of space increases, but also because it is believed that the star formation rate, and hence the rate of supernova explosion, was about two times larger at about a red-shft $z = 2$. Working with this reasonable hypothesis Blair & Ju (1996) find that the supernova background could be as large as $\Omega_{SN} \sim 10^{-6}$, in the frequency range of advanced bars and interferometers. They further suggest that coincidence between an advanced interferometer and a resonant bar within 50 km of the interferometer, will enable the detection of this background. Such a detection/non-detection, can shed light on the history of star formation rate, a subject of vigorous debate amongst astrophysicists.

Galactic binary background The Hulse-Taylor binary is orbiting much too slowly for it to be observable in the GEO600/LIGO/VIRGO or LISA detectors. However, there must be other compact binaries in our Galaxy whose orbital periods are a lot shorter than the Hulse-Taylor system but still quite large for them to last millions of years. As mentioned earlier, some of these binaries with orbital periods $P \sim 10^{-4}$–10^{-2} s will be observable in space-based detectors. However, there are such a large number of them in our Galaxy which are so distant, and hence their individual amplitudes will be so feeble, that they will not be identifiable separately. In addition to binaries of compact stars, there are also other binaries consisting of white dwarfs, cataclysmic variables, etc., which will also contribute to the background radiation produced by compact binaries. The net effect is that these sources will appear as a background noise in space interferometers as shown in Fig. 3. By studying the nature of this background we will learn a lot about binary population in our Galaxy.

Galactic pulsar background Giazotto, Gourgoulhon & Bonazzola (1997) have suggested that the continuous radiation from pulsars could also produce a background of radiation which will limit the sensitivity, this time, of our ground-based interferometers. As mentioned earlier, there are about 10^9 neutron stars in our Galaxy of which, Giazotto et.al. conclude, about 2×10^5 will contribute to the background radiation. This background radiation will be prominent, and observable in the LIGO/VIRGO detectors, in the frequency range of 5–10 Hz, at an rms amplitude of $h_{rms} \sim 2 \times 10^{-26}$, where the rms is computed over 10^5 sources. At frequency of 10 Hz, wavelength of gravitational waves will be around 30,000 km. Hence, it would be possible to cross-correlate data from two distant detectors, such as the two LIGOs, or LIGO and VIRGO, and discriminate the background against other sources of noise.

7. Summary

In this article we have discussed astrophysical sources of gravitational radiation that are expected to be seen in ground- and space-based gravitational wave antennas. With the new millennium a new generation of gravitational wave detectors will go on-line with potentialities to bring to fruition the long and hard search for gravitational radiation pioneered by Joseph Weber. The initial interferometers, such as GEO600, have a fair chance of seeing some of the strongest periodic sources in X-ray emitting binaries, but will have to wait for km sized LIGO/VIRGO to come up for confirmation. If continuous waves turn out to be weak after all, then the GEO600/LIGO/VIRGO network should together make first detections by observing black hole-black hole or neutron star-black hole binaries. Continuous upgrade of LIGO/VIRGO should make it possible for the network to detect several interesting and possibly new sources within the next decade. Through the gravitational window to the Universe we will undoubtedly resolve many of the present mysteries but it will surely open a whole new world of puzzles.

References

ABRAMOVICI, A., *et al.* (1992) *Science,* **256**, 325.

ALCOCK, C. *et al.* (1996) *astro-ph/9606165.*

ALLEN, B. (1997) in Relativistic Gravitation and Gravitational Radiation, J.-P. Lasota and J.-A. Marck (eds.) Cambridge Univ. Press, Cambridge.

ANDERSSON, N. (1998) *gr-qc/9706075.*

APOSTOLATOS, T. A. (1995) *Phys. Rev. D,* **52**, 605.

ASTONE, P., LOBO, J.A., & SCHUTZ, B.F. (1994) *Class. Quant. Grav.,* **11,** 2093.

BENDER, P. *et al.* (1996) *LISA: Pre-Phase A Report,* MPQ 208 (Max-Planck-Institut für Quantenoptik, Garching, Germany).

BLANCHET, L., DAMOUR, T., IYER, B.R., WILL, C.M. & WISEMAN, A.G. (1995) *Phys. Rev. Lett.,* **74**, 3515.

BLANCHET, L., IYER, B.R. & JOGUET, B. (1998) paper in preparation.

BLANCHET, L. & SATHYAPRAKASH, B.S. (1995) *Phys. Rev. Lett.,* **74,** 1067.

BLAIR, D. & JU, L. (1996) *Mon. Not. Roy. Astron. Soc.,* **283,** 648.

BONNELL, I.A. & PRINGLE, J.E. (1995) *Mon. Not. Roy. Astron. Soc.,* **273,** L12.

BRADASCHIA, C., *et al.* (1990) *Nucl. Inst. A.,* **289**, 518.

BRADY, P., CREIGHTON, T., CUTLER, C.W. & SCHUTZ, B.F. (1998) *Phys. Rev. D,* **57,** 2101.

CHANDRASEKHAR, S. (1970) *Phys. Rev. Lett.,* **24,** 611.

CUTLER, C. *et al* (1993) *Phys. Rev. Lett.,* **70**, 2984.

DAMOUR, T., IYER, B.R. & SATHYAPRAKASH, B.S. (1998) *Phys. Rev. D,* **57,** 885.

DANZMANN, K., *et al.* (1995) *in Gravitational Wave Experiments, Proceedings of the Edoardo Amaldi Conference,* World Scientific, p. 86.

DHURANDHAR, S.V. & SATHYAPRAKASH, B.S. (1994) *Phys. Rev. D,* **49**, 1707.

DHURANDHAR, S.V. & TINTO, M (1988) *Mon. Not. R. Astron. Soc.,* **234**, 663.

FLANAGAN, É. É. & HUGHES, S. (1998) *Phys. Rev. D,* **57,** 4535.

FRIEDMAN, J.L. & SCHUTZ, B.F. (1978) *Astrophys. J.,* **222,** 281.

GIAZOTTO, A., BONAZZOLA & GOURGOULHON, E. (1997) *Phys. Rev. D,* **55,** 2014.

GRISHCHUK, L.P. (1997) *Class. Quant. Grav.,* **14,** 1445.

JONES, I. (1997) *private communication.*

KASPI, V., TAYLOR, J., RYBA, M. (1994) *Astrophys. J.,* **428,** 713.

LASOTA, J.-P. & MARCK, J.-A. (1997) Relativistic Gravitation and Gravitational Radiation, Cambridge Univ. Press, Cambridge.

LINDBLOM, L., OWEN, B.J. & MORSINK, S.M. (1998) *gr-qc/9803053.*

LIPUNOV, V.M., POSTNOV, K.A. & PROKHOROV, M.E. (1997a) *Astron. Lett.,* **23,** 492.

LIPUNOV, V.M., POSTNOV, K.A. & PROKHOROV, M.E. (1997b) *Mon. Not. R. Astron. Soc.* **288,** 245.

LIPUNOV, V.M., POSTNOV, K.A. & PROKHOROV, M.E. (1997c) *New Astronomy* **2,** 43.

MÜLLER, E. 1997) *in Relativistic Gravitation and Gravitational Radiation,* J.-P. Lasota and J.-A. Marck (eds.) Cambridge Univ. Press, Cambridge (1997).

MÜLLER, E. & JANKA, H.-T. (1997) *Astron. Astrophys.,* **317,** 140.

NAKAMURA, T., SASAKI, M., TANAKA, T. & THORNE, K.S. (1997) *Astrophys. J.,* **487,** L139.

NARAYAN, R., PIRAN, T. & SHEMI, A. (1991) *Astrophys. J.,* **379,** L17.

OWEN, B.J. (1996) *Phys. Rev. D.,* **53**, 6749.

OWEN, B.J., LINDBLOM, L., CUTLER, C., SCHUTZ, B.F., VECCHIO, A. & ANDERSSON, N. (1998) *gr-qc/9804044.*

OWEN, B.J. & SATHYAPRAKASH, B.S. (1998) *Phys. Rev. D.,* submitted.

PHINNEY E S (1991) *Astrophys. J.,* **380,** L17.

POISSON, E. (1995) *Phys. Rev. D,* **52**, 5719.

PRESS, W.H. (1971) *Astrophys. J. Lett.,* **170**, L105.

SATHYAPRAKASH, B.S. & DHURANDHAR, S.V. (1991) *Phys. Rev. D,* **44**, 3819.

SATHYAPRAKASH, B.S. (1994) *Phys. Rev. D,* **50**, R7111.

SCHUTZ, B.F. (1985) *A First course in general relativity,* Cambridge Univ. Press, Cambridge.

SCHUTZ, B.F. (1986) *Nature,* **323,** 310.

SCHUTZ, B.F. (1997) *in Relativistic Gravitation and Gravitational Radiation,* J.-P. Lasota and J.-A. Marck (eds.) Cambridge Univ. Press, Cambridge.

SCHUTZ, B.F. AND TINTO, M. (1987) *Mon. Not. R. Astron. Soc.,* **224**, 131.

TANAKA, T., TAGOSHI, H. & SASAKI, M. (1996) *Prog. Theor. Phys.,* **96,** 1087.

TAYLOR, J.H., 1994, *Rev. Mod. Phys.,* **66,** 711.

THORNE, K.S. (1995) *in Proceedings of the Snowmass 95 Summer Study on Particle and Nuclear Astrophysics and Cosmology,* E. W. Kolb and R. Peccei (eds) World Scientific, pp. 398.

THORNE, K.S. (1997) *in Critical Problems in Physics: Proceedings of Princeton University's 250th Anniversary Conference,* M. Dementi, V. Fitch, and D. Marlow (eds), Princeton University Press, in press (gr-qc/9704042).

THORNE, K.S. & ZYTKOV, A. (1977) *Astrophys. J.,* **212,** 832.

TSUBONO, K., *et al.* (1995) *in Gravitational Wave Experiments, Proceedings of the Edoardo Amaldi Conference,* World Scientific, p. 112.

TUTKOV, A.V. & YUNGELSON, L.R. (1993) *Mon. Not. Roy. Astron. Soc.,* **260,** 675.

VISHVESHWARA, C.V. (1970) *Nature,* **227,** 936.

WAGONER, R.V. (1984) *Astrophys. J.,* **278,** 345.

WILL, C.M. (1981) *Theory and Experiment in Gravitational Physics,* Cambridge Univ. Press, Cambridge.

WILL, C.M. (1998) *Phys. Rev. D,* **57,** 2061.

YUNGELSON, L. & ZWART, S.F.P. (1998) *gr-qc/9801127* to appear in the Proceedings of the GWDAW-2, Orsay, Nov. 1997.

ZWART, S.F.P. & YUNGELSON, L. (1997) *gr-qc/9710347* to appear in A & A.

ZWEGER, T & MÜLLER, E. (1996) *Report MPA976, Max-Planck-Institut für Astrophysik, Garching.*

26. GRAVITATIONAL RADIATION FROM INSPIRALLING COMPACT BINARIES

MOTION, GENERATION and RADIATION REACTION

BALA R. IYER
Raman Research Institute,
Bangalore 560 080, India.

In 1978 when I met Ajit Kembhavi he was excited about someone called Vishveshwara who had come back from USA to the Raman Research Institute in Bangalore and was the head of a relativity group consisting of D. M. Chitre and N. D. Haridass. Ajit had worked on a project with him on compact objects and Sanjeev Dhurandhar was soon to also join him as a post doc. I was then a student with Arvind Kumar at the Bombay university working on problems of quantum field theories (QFT) in curved spacetimes(CST). The first relativity meeting we went to was the Einstein Centenary symposium at Physical Research Laboratory, Ahmedabad. Though I have many wonderful memories of the symposium the most memorable one was Vishu's lecture entitled 'Black Holes for Bedtime' [1]. To me it was a magical experience; an exotic cocktail of science, art, humour and caricature. Equations of the kind I was struggling with in my thesis were not necessarily abstract and unspeakable. They could as well be translated in the best literary tradition.

In the Einstein meeting all the well-known speakers like Vishu seemed to be inaccesible stars and hence I was elated when he accepted me as a postdoc at RRI in 1980. QFT in CST was then past its peak and Vishu encouraged me to explore problems in classical general relativity with possible astrophysical implications. Over the years we worked on problems related to existence and stability of ultracompact objects, accretion in presense of magnetic fields, exact solutions, separability properties of the Dirac equation, black holes in higher dimensions and Gauss Bonnet theories, and Frenet Serret methods for black holes and gyrosopic precession.

It has always been a pleasure working with Vishu. There is no pressure, no generation gap, a natural possibility to grow and contribute your best, an easy personal rapport, a refreshing sense of humour, an unassuming erudition and most importantly a warm and wonderful human being.

B. R. Iyer and B. Bhawal (eds.), Black Holes, Gravitational Radiation and the Universe, 437–459.

Over the years we have tread many trails together: research, teaching, editing, reading, movies, schools, organisation, science education, friendship, discussions, dreams, beer! There is a trail we want to intensely explore: To write a book together. I hope that we can do it in the coming years.

1. Introduction

The Binary pulsars 1913+16 and 1534+12 establish the reality of gravitational radiation [2]. They also provide proof of the validity of Einstein's general relativity in the strong field regime [3]. More importantly they are prototypes of inspiralling compact binaries which are strong sources of gravitational waves for ground based laser interferometric detectors like LIGO and VIRGO

With an orbital period of eight hours, the frequency of gravitational waves from the binary pulsar 1913+16 today is very low: (10^{-4} Hz). However in about three hundred million years the two stars will inspiral and the gravitational waves will sweep upward in frequency to about 10 Hz. In the following fifteen minutes before the neutron stars collide and coalesce the frequency will rise to about 1000 Hz with increasing amplitude producing a characteristic chirp waveform containing about 16000 cycles. Though the gravitational wave signal is extremely weak and buried deep in the detector noise the large number of precisely predictable cycles in the detector bandwidth brings the characterstic signal strength to the realm of the measurable. This enables one to use the technique of matched filtering initially for detection and later for estimation of parameters of the inspiralling binary [4]. The information content in these events is of excellent quality. If they are detected with a suitably high signal to noise ratio they should allow one to do astronomy. For instance, it could (i) provide precise measurements of the masses of the objects, possibly of their spins and probably, in the case of neutron stars, of their radii; (ii) allow one to measure cosmological distance directly and provide a cleaner determination of H_0 and q_0; (iii) test nonlinear structure of radiative gravitation; (iv) perform new tests of the existence of a scalar component to gravitation (v) probe black hole physics [5]. Estimates of the rate of such coalescence events are about a few per year upto 200 Mpc. Advanced LIGO which would look upto cosmological distances [5] would get to numbers of hundreds per year.

The phenomenal success of the high-precision radio wave observation of the binary pulsar makes crucial use of an accurate relativistic 'Pulsar timing formula' [6, 7].

$$\phi_n^{\mathrm{PSR}} = F[t_n; p_i] , \tag{1}$$

linking the rotational phase of the spinning pulsar (stroboscopically observed when $\phi_n^{\rm PSR} = 2\pi n$ with $n \in \mathcal{N}$) to the time of arrival t_n on Earth of an electromagnetic pulse, and to some parameters p_i. Similarly precise gravitational-wave observation of inspiralling compact binaries would require an equivalent accurate 'Phasing formula' i.e. an accurate mathematical model of the continuous evolution of the gravitational wave phase

$$\phi^{\rm GW} = 2\Phi = F[t; p_i] , \tag{2}$$

involving a set of parameters $\{p_i\}$ carrying information about the emitting binary system (such as the two masses m_1 and m_2). Since the equations of motion is not yet available at such higher orders the conventional approach heuristically relies on a standard energy-balance argument. From this it follows that the time evolution of the orbital phase Φ is determined by two functions: an energy function $E(v)$, and a flux function $F(v)$. The argument v is defined by $v = (\pi\, m\, f^{\rm GW})^{1/3}$, which can be rewritten in terms of the instantaneous *orbital* angular frequency Ω, $v \equiv (m\Omega)^{1/3} \equiv x^{1/2}$ ($m \equiv m_1 + m_2$ denotes the total mass of the binary). The (dimensionless) energy function E is defined by

$$E_{\rm tot} = m(1 + E) \tag{3}$$

where $E_{\rm tot}$ denotes the total relativistic energy (Bondi mass) of the binary system. The flux function $F(v)$ denotes the gravitational luminosity of the system. The three quantities v, E and F are invariantly defined (as global quantities in the instantaneous center of mass frame), so that the two functions $E(v)$, $F(v)$ are coordinate-independent constructs. Denoting the symmetric mass ratio by $\eta \equiv m_1 m_2/(m_1 + m_2)^2$, the energy balance equation $dE_{\rm tot}/dt = -F$ gives the following parametric representation of the phasing formula Eq. (2) (for the orbital phase)

$$t(v) = t_c + m \int_v^{v_{\rm lso}} dv \, \frac{E'(v)}{F(v)} , \tag{4}$$

$$\Phi(v) = \Phi_c + \int_v^{v_{\rm lso}} dv v^3 \, \frac{E'(v)}{F(v)} , \tag{5}$$

where t_c and Φ_c are integration constants and $v_{\rm lso}$ is the velocity corresponding to the 'last stable orbit'. Note that $E'(v) < 0$, $F(v) > 0$ so that both t and Φ increase with v.

The accurate mathematical modelling of gravitational wave signals from inspiralling compact binaries requires solutions to two different but related problems referred to respectively as the "wave generation problem" and the "radiation reaction problem" [8], adequate for treating compact objects. The wave generation problem deals with the computation of the gravitational waveforms generated by the binary when the orbital phase and frequency of the binary take some given values ϕ and ω taking into account propagation and nonlinear effects. This problem involves computing the (tensorial) amplitude of each harmonic of the wave corresponding to frequencies which are multiples of the orbital frequency, with the predominant harmonic being at twice the orbital frequency. The radiation reaction problem consists of determining the evolution of the orbital phase $\phi(t)$ itself as a function of time, from which one deduces the orbital frequency $\omega(t) = d\phi(t)/dt$. The actual time variation of $\phi(t)$ is nonlinear because the orbit evolves under the effects of gravitational radiation reaction forces. In principle it should be determined from the knowledge of the radiation reaction forces acting locally on the orbit. However these forces are at present not known with sufficient accuracy, so in practice the phase evolution is determined by equating post-Newtonian energy flux in the waves or energy loss (averaged over one orbit) and the decrease of the correspondingly accurate binding energy of the binary. In order not to suffer a very severe reduction in signal-to-noise ratio, one will have to monitor the phase evolution with an accuracy of one tenth of a cycle over the tens of thousands of cycles during the entire passage through the frequency bandwidth of the detector. Consequently, the radiation-reaction part of the problem which determines the time evolution of the phase of the gravitational wave signal needs more crucial attention for successful detection. Determination of the 'Energy function' on the other hand requires the solution to the problem of 'motion'. The problems of motion, generation and radiation reaction are the constitutive elements of this analysis and the algebraic complexity introduced by nonlinearity makes it mandatory to make as many independent checks and counter checks as is possible.

None of the above problems can be solved exactly. They are treated by a combination of approximation methods like Post-Newtonian approximation, Post-Minkowskian approximation and Perturbations about a Curved Background. Since the details of the last approach are reviewed in Sasaki's article in this volume we only discuss the first two here and list the main features of these two schemes in the following section:

2. Post-Newtonian versus Post-Minkowskian Approximation

The two main approximation schemes that feature in the analytical studies of inspiralling binaries are:

1. Post-Newtonian Approximation (PNA): It is based on the assumption of an everywhere weak gravitational field together with that of slow motions. It is an expansion in $\beta \sim v/c \sim L/\lambda \sim (L/c)/P$ where v, L, λ and P are the characteristic velocity, size, wavelength and period of the

system respectively. It uses newtonian concepts like absolute space with an Euclidean metric and absolute time. It uses newtonian techniques and in this viewpoint Einstein theory provides small numerical corrections to Newtonian theory. The equations in this scheme are a hierarchy of Poisson equations which are solved by instantaneous potentials. In the PNA one looks for solutions of the field equations which are formal expansions in $1/c$ and hence it is also called the slow motion expansion. At higher orders PNA lead to to *divergent* integrals since they are based on instantaneous potentials – such potentials lead to divergent integrals at some order since they correspond to expansion in powers of r/c which grow like positive powers of r causing *infrared divergent* integrals. However, this may be avoided by assuming its validity in the near-zone only. This implies that the source is well-within the near-zone where retardation effects are small and also that time derivatives of the field are smaller than space-derivatives. More seriously, the PNA breaks down at 4PN since hereditary effects arise at this level and instantaneous potentials can no longer suffice to approximate the situation. A word about nomenclature: Each power of $(v/c)^2$ corresponds to one post-Newtonian order (1PN) and thus 4PN above refers to corrections including $(v/c)^8$. All the corrections(orders) are relative to the lowest order. This, 1PN in the equations of motion refers to terms including $1/c^2$ while 1PN accurate energy flux refers to terms including $1/c^{12}$.
2. Post-Minkowskian Approximation (PMA): It is based on the assumption of the weakness of the gravitational field and hence an expansion in $\gamma_i = GM/c^2R$ where M is the characteristic mass and R the characteristic size of the compact object. It makes crucial use of the conceptual framework of Minkowski geometry and its causality properties. The equations in this scheme reduce to a hierarchy of wave equations on Minkowski background which are solved by retarded potentials. In the PMA one looks for solutions to Einstein's equations which are formal expansions in powers of G. It is also called nonlinearity, weak field, or fast motion approximation. The basic complication is in the nonlinear iteration. The PMA shows no signs of internal inconsistency. It applies all over the weak field zone. In vacuum the PM expansion is reliable i.e., any solution of the perturbation equations comes from the Taylor expansion, when $G \to 0$, of a family of exact solutions [9].

3. Problem of Motion

It may be worth mentioning that unlike linear EM, non-linear GR has the feature that its field equations contain the equations of motion. In their original form the Einstein equations(EE) $G_{\mu\nu} = 8\pi G T_{\mu\nu}/c^4$ are mathematically not very convenient since they do not form a partial differential system of well defined type. To alleviate this difficulty one 'relaxes' the field equations by use of special coordinate conditions. One of the most common choice is the harmonic (De Donder or Lorentz) coordinate condition

$$\mathcal{G}^{\mu\nu}_{,\nu} = 0,$$

where the 'Gothic' contravariant metric is defined by

$$\mathcal{G}^{\mu\nu} = \sqrt{g} g^{\mu\nu} \quad ; \quad g = -\det(g_{\mu\nu}).$$

The EE then become

$$\mathcal{G}^{\alpha\beta}\mathcal{G}^{\mu\nu}_{,\alpha\beta} + Q^{\mu\nu}(\mathcal{G}, \partial\mathcal{G}) = \frac{16\pi G}{c^4} g T^{\mu\nu}.$$

where $Q^{\mu\nu}$ is a complicated quadratic form in the first derivatives of $\mathcal{G}$. In terms of deviation from the flat metric $h^{\mu\nu} = \mathcal{G}^{\mu\nu} - \eta^{\mu\nu}$ the above equations become

$$\partial_\nu h^{\mu\nu} = 0,$$
$$\Box h^{\mu\nu} = \frac{16\pi G}{c^4}|g|T^{\mu\nu} + \Lambda^{\mu\nu}(h),$$

where $\Lambda^{\mu\nu}$ includes all the nonlinearities of the field equations and is at least quadratic in h and its first and second derivatives: $\Lambda^{\mu\nu} = N^{\mu\nu}(h, h) + O(h^3)$. The harmonic coordinates are well adapted to Lorentz-covariant formulations. Here all the ten metric coefficients satisfy hyperbolic equations.

Within the PMA one then assumes a formal asymptotic expansion

$$\mathcal{G}^{\alpha\beta} \equiv \sqrt{g}\, g^{\alpha\beta} = \eta^{\alpha\beta} + \gamma_i h_1^{\alpha\beta} + \gamma_i^2 h_2^{\alpha\beta} + \cdots \gamma_i^n\, h_n^{\alpha\beta} + \cdots \tag{6}$$

One is thus led to a formal hierarchy of inhomogeneous wave equations for the $h_n^{\mu\nu}$ of the form:

$$\Box_f\, h_n^{\alpha\beta} = T^{\alpha\beta} + N_n^{\alpha\beta}\ (h_1, h_2 .... h_{n-1}) \equiv S_n^{\alpha\beta}. \tag{7}$$

The equations must be supplemented by a prescription to pick the physically correct solution. The Fock conditions consisting of appropriate fall-off together with the 'no incoming-radiation' condition at past infinity implies that the solution is determined by the flat spacetime retarded Green function as:

$$h_n^{\mu\nu} = \int d^4x G_{\rm ret}^{(f)}(x - x') S_n^{\mu\nu}(x'). \tag{8}$$

The N-body problem as in newtonian gravity is decomposed into an external problem and an internal problem. The former refers to the problem of defining and determining the motion of the center of mass and the latter to motion of each body around the center of mass. The effacement of internal structure in the external problem and effacement of external structure on the internal problem involves subtle issues in the problem of motion and we cannot do better than refer the reader to the beautiful review by Damour [10].

The topic of EOM for compact binary systems received careful scrutiny in the years following the discovery of the binary pulsar. There have been three different approaches to the complete kinematical description of a two body system upto the level where radiation damping first occurs (2.5PN). Damour's method explicitly discusses the external motion of two condensed bodies without ambiguities. The method employs the best techniques to treat various subproblems. (a) A PMA to obtain the gravitational field outside the bodies incorporating a natural 'no incoming-radiation condition' whose validity is not restricted to only the near-zone. (b) A matched asymptotic expansion scheme to prove effacement and uniquely determine the gravitational field exterior to the condensed bodies. (c) An Einstein Infeld Hoffmann Kerr(EIHK) type approach to compute equations of orbital motion from knowledge of the external field only. The nth approximate EOM is obtained from the integrability condition on the $(n+1)^{th}$ approximated vacuum field equations. (d) Use of Riesz's analytic continuation technique to evaluate surface integrals. The final EOM at 2.5PN level are expressed only in terms of instantaneous positions, velocities and spins in a given harmonic coordinate system and given explicitly in Ref.[10]. The two mass parameters in these formulas are the Schwarzschild masses of the two condensed bodies.

The conservative part of the EOM upto 2PN (excluding the secular 2.5PN terms) are not deducible from an conventional Lagrangian (function of positions and velocities) in harmonic coordinates, but only from a generalised Lagrangian (depending on accelerations). This is consistent with the result in classical field theory that in Lorentz-covariant field theories there exists no (ordinary) Lagrangian description at $O(c^{-4})$ [11]. This Lagrangian is invariant under the Poincare group and thus allows one to construct ten Noetherian quantities that would be conserved during the motion. These include the 'Energy', 'Angular Momentum', 'Center of Mass' and thus a solution to the problem of 'motion' provides the first element $E(v)$ that enters into the phasing formula. For inspiralling compact binaries the 2.5PN accurate binding energy reads

$$E(v) = -\frac{c^2}{2} m\eta x \left[1 - \frac{1}{12}(9+\eta)x - \frac{1}{8}\left(27 - 19\eta + \frac{\eta^2}{3}\right)x^2 + \mathcal{O}(x^3)\right], \tag{9}$$

where m, η, v, and x are defined as before. The EOM to 2PN accuracy in this case is given by:

$$\mathbf{a} = -\frac{Gm}{r^3}\left[1-(3-\eta)\gamma+\left(6+\frac{41}{4}\eta+\eta^2\right)\gamma^2\right]\mathbf{x} \tag{10}$$

where $\gamma = Gm/c^2r$. The EOM for the general case is given in [10] and crucially used in the following studies of generation [12, 13] and radiation reaction [14].

Schafer's [15] approach on the other hand is based on the Hamiltonian approach to the interaction of spinless point particles with the gravitational wave field. The Hamiltonian formulation is best done in the Arnowitt-Deser-Misner (ADM) coordinates in which two metric coefficients satisfy hyperbolic equations (evolution) while the remaining eight are of elliptic type(constraints). It uses a different gauge that allows an elegant separation of conservative and damping effects. One recovers the damping force acting on the Hamiltonian subsystem of instantaneously interacting particles coming from its interaction with the dynamical degrees of freedom of the gravitational field. In this approach point masses are used as sources and regularisation uses Hadamard's 'partie finie' based on Laurent's series expansion regularisation.

The last approach due to Grischuk and Kopejkin [16] on the other hand is based on (a) PNA scheme (b) assumption that bodies are non-rotating 'spherically-symmetric' fluid balls. The symmetry is in the coordinate sense. The EOM of the center of mass of each body are obtained by integration of the local PN EOM. These are explicitly calculated retaining all higher derivatives that appear. One then reduces the higher derivatives by EOM and obtains the final results. Formally collecting the various relativistic corrections into a 'effective mass', one can have a PN proof of effacement of internal structure and provide a plausibility argument for validity of 'weak field formulas' for compact objects.

The fact that three independent methods: PM + EIHK, PN + Perfect fluid and PN + delta functions give formally identical equations of motion at 2PN order is a strong confirmation of the validity of the numerical coefficients in the EOM. This work provides the basis for the timing formula mentioned earlier. The damping terms can be considered as perturbation to a Lagrangian system which is multiperiodic – a radial period and a angular period corresponding to periastron precession – and leads to the observed secular acceleration effect in the binary pulsar. No balance argument is involved at any stage.

The work on 3PN generation crucially requires the EOM at 3PN accuracy and the situation is now under investigation at the 3PN level. The equations of motion are the geodesic equations in the space-time generated by the two particles. The gravitational field is computed using the standard post-Newtonian theory from the stress-energy tensor appropriate for point particles (i.e. involving delta functions). Because the metric coefficients are to be evaluated at the location of the particles, the geodesic equations must be computed using a process of regularization of the infinite self-field of point particles. Work is in progress to obtain the 3PN contributions by different techniques. These include the MPM method supplemented by Hadamard 'partie-finie' [17], the Epstein Wagoner Will Wiseman method [18] as also the Hamiltonian formalism [19]. As mentioned above upto 2.5PN three distinct computational techniques led to a unique EOM. Though, preliminary investigations have even raised questions about whether this sort of uniqueness will persist at 3PN [20], the work of Blanchet, Faye and Ponsot [17] indicate that unique results will obtain.

4. The Multipolar Post-Minkowskian Formalism

The use of multipole expansion methods in combination with the PMA scheme offers one of the most powerful techniques in gravitational radiation studies [21]. After decomposing the gravitational field generated by an isolated source in the exterior in terms of a formal post-Minkowskian series one decomposes each coefficient of the PM expansion in terms of its multipole moments. The use of a multipolar decomposition for the field simplifies the resolution of the field equations. Since the gravitational field is a second rank tensor, symmetric trace-free tensors are a very convenient representation to implement multipole decomposition rather than tensor spherical harmonics. They lead to a more transparent definition of multipole moments. Thus we have,

$$h_n^{\alpha\beta} = \sum_{\ell\geq 0} h_{nL}^{\alpha\beta}\, \hat{n}^L\,(\theta,\phi), \tag{11}$$

$$\hat{n}^L \equiv n^{<i_1 i_2 \cdots i_l>} \equiv STF_{i_1 i_2 \cdots i_l} n^{i_1 i_2 \cdots i_l}. \tag{12}$$

Choosing harmonic coordinates such that $\partial_\beta\, h_n^{\alpha\beta} = 0$ we look for MPM metrics which are stationary in the past and Minkowskian at spatial infinity then. This allows for a well-defined iteration at any order incorporating Fock's no incoming radiation condition. In combination with a PNA in the inner region this leads to a powerful method to deal with not only generation of gravitational waves by relativistic sources but also reaction of gravitational waves on the source.

The source-free scalar wave equation has as its most elementary retarded solution $\phi = F(t-r/c)/r$, where F is an arbitrary function of retarded time. From this one can construct the most general retarded solution by repeated spatial differentiation: $\phi_{i_1 i_2 \cdots i_l} \equiv \phi_L = \partial_{i_1}\partial_{i_2}\cdots\partial_{i_l}\phi \equiv \partial_L\phi$. Similarly in the tensor case relevant here the most general solution for linearised gravity can always be written – modulo an infinitesimal gauge transformation that preserves the harmonic gauge condition – in terms of two arbitrary functions M_L and S_L which can be chosen to be symmetric and trace-free(STF) with respect to all the l indices L.

To proceed further we note that the source term for h_2 that involves h_1 are in the form of multipole expansions. Consequently one cannot use the usual retarded integral to solve the wave equation since the multipole expansion is valid only in D_e. To deal with this problem one considers a fictitious source constructed by multiplying the actual source by with r^B where B is a complex number. The retarded integral of this source admits after analytic continuation a Laurent expansion near $B \to 0$ whose finite part (i.e. coefficient of zeroth power of B in the expansion) is the solution we are looking for. This solution is also in the form of a multipole expansion; however it does not by itself satisfy the harmonic gauge condition. Thus we supplement our solution by a solution of the homogeneous wave equation to satisfy the harmonic gauge condition. This treatment can be extended to higher orders and thus we have

$$h_n^{\mu\nu} = p_n^{\mu\nu} + q_n^{\mu\nu}, \tag{13}$$

$$p_n^{\mu\nu} = FP_{B=0}\Box_R^{-1}(r^B \Lambda_n^{\mu\nu}), \tag{14}$$

$$\partial_\nu q_n^{\mu\nu} = -\partial_\nu p_n^{\mu\nu}. \tag{15}$$

The above solution is proved to be the most general and thus the general radiative field outside an isolated system depends on only two sets of time-dependent multipole moments. In the exterior near zone the field admits a PN expansion in terms of functions of the form $(\ln c)^p/c^k$ where $p, k \in N$.

5. Parametric Representation of Motion

At the newtonian level the motion of a binary system can be expressed in the parametric form:

$$\frac{2\pi}{P}t = u - e\sin u, \tag{16}$$

$$r = a(1 - e\cos u), \tag{17}$$

$$\phi = 2\arctan\left(\sqrt{\frac{1+e}{1-e}}\tan(\frac{1}{2}u)\right). \tag{18}$$

In the above P is the orbital period, e the ecentricity and a the semimajor axis of the orbit. All these parameters are functions of the the energy E and angular momentum J of the orbit. The parameter u is called the eccentric anomaly.

Damour and Deruelle [6] devised a very useful parametrisation structurally similar to the above and valid at 1PN level. This quasi-Keplerian representation uses, instead of an eccentricity e, three different eccentricities e_r, e_ϕ and e_t. The above construction has been generalised to the 2PN order by Damour, Schafer and Wex [22, 23]. This generalised quasi-Keplerian representation introduces in addition another parameter v and in this case we have:

$$2\pi\frac{t-t_0}{P} = u - e_t \sin u + \frac{f_p}{c^4}\sin v + \frac{g_p}{c^4}(v-u), \tag{19}$$

$$r = a(1 - e_r \cos u), \tag{20}$$

$$2\pi\frac{\phi-\phi_0}{\Phi} = v + \frac{f_\phi}{c^4}\sin(2v) + \frac{g_\phi}{c^4}\sin(3v), \tag{21}$$

$$v = 2\arctan\left(\sqrt{\frac{1+e_\phi}{1-e_\phi}}\tan(\frac{1}{2}u)\right). \tag{22}$$

All the parameters P, Φ, a, e_t, e_r, e_ϕ, f_p, g_p, f_ϕ, and g_ϕ are functions of the 2PN conserved E and J characterising the orbit. The above representation is one of the inputs to derive the timing formula to 2PN accuracy [24, 25]. It has also been used to discuss the evolution of the orbital elements to 2PN accuracy under 2PN radiation reaction [13] and obtain a useful form for the 2PN accurate polarisations in the case of quasi-elliptic orbits [26].

6. The Wave Generation Formalism

Einstein's far field quadrupole equation is the solution of the generation problem to the lowest order (hence referred to as Newtonian order) but applies only to objects held together by non-gravitational forces. Fock and Landau-Lifshiftz provided two very different methods to generalise to weakly self-gravitating systems. The above two approaches are the starting points for the two methods available today to calculate gravitational wave generation to higher orders: The Blanchet-Damour-Iyer(BDI) [27] approach and the Epstein-Wagoner-Thorne-Will-Wiseman (EWTWW) [28, 29, 30] approach. The relativistic corrections are called post-Newtonian corrections according to the PN order of the EOM of the source needed to reduce all accelerations with consistent accuracy. Thus the 1PN formalism retains all terms in the radiation field and reaction computed consistently using 1PN EOM. An interesting and different approach to the quadrupole formula is due to Haridass and Soni [31] who obtained the one graviton transition operator and thus the classical energy loss formula for gravitational radiation from Feynman graphs of helicity ± 2 theories of gravitation.

6.1. BDI APPROACH

Blanchet, Damour and Iyer build on a Fock type derivation using the double-expansion method of Bonnor [32]. This approach makes a clean separation of the near-zone and the wave zone effects. It is mathematically well defined, algorithmic and provides corrections to the quadrupolar formalism in the form of compact support integrals or more generally well defined analytically continued integrals. The BDI scheme has a modular structure: the final results are obtained by combining an 'external zone module' with a 'radiative zone module' and a 'near zone module'. For dealing with strongly self-gravitating material sources like neutron stars or black holes one needs to use a 'compact body module' together with an 'equation of motion module'. It correctly takes into account all the nonlinear effects.

It should be noted that, in generation problems, as one goes to higher orders of approximation two independent complications arise. Though algebraically involved in principle the first is simpler: contributions from higher multipoles. The second complication is not only algebraically tedious but technically more involved: contributions from higher nonlinearities e.g for 2PN generation cubic nonlinearities need to be handled.

The general approach to solve the generation problem may be broken up into the following steps:

1. Integrate the Einstein field equations in the vacuum exterior region D_e by means of a Multipolar Post-Minkowskian series. The exterior solution is parametrised by moments M_L and S_L called the algorithmic moments. The mass monopole and dipole moments as well as the current dipole moment are necessarily constant to satisfy the harmonic gauge condition.

2. In the far wave zone rewrite the solution in suitable coordinates to find the observable moments of the radiative field that a detector would measure. This involves going over from the harmonic coordinates to the radiative or Bondi coordinates to correct for the logarithmic deviation of the true light cones from the flat line cones in the wave zone D_w. In these coordinates we have,

$$h_{ij}^{\mathrm{TT}}(\mathbf{X},T) = \frac{4G}{c^2R}\mathcal{P}_{ijkm}(\mathbf{N})\sum_{\ell=2}^{\infty}\frac{1}{c^\ell \ell!}\Big\{N_{L-2}U_{kmL-2}(T-R/c) - \frac{2\ell}{(\ell+1)c}N_{aL-2}\varepsilon_{ab(k}V_{m)bL-2}(T-R/c)\Big\} + \mathcal{O}\left(\frac{1}{R^2}\right) \quad (23)$$

where U_L and V_L are the 'mass' and 'current' type radiative moments. In terms of these radiative moments the total power or luminosity is given by,

$$\mathcal{L} = \sum_{\ell=2}^{+\infty}\frac{G}{c^{2\ell+1}}\Big\{\frac{(\ell+1)(\ell+2)}{(\ell-1)\ell\ell!(2\ell+1)!!}U_L^{(1)}U_L^{(1)}$$

$$+\frac{4\ell(\ell+2)}{(\ell-1)(\ell+1)!(2\ell+1)!!c^2}V_L^{(1)}V_L^{(1)}\Bigg\}. \qquad (24)$$

The observable or radiative moments U_L and V_L are related to the algorithmic moments and we have:

$$U_L(T_R) = M_L^{(l)}(T_R) + \sum_{n\geq 2}\frac{G^{n-1}}{c^{3(n-1)+2k}}X_{nL}(T_R) \qquad (25)$$

$$\varepsilon_{ai_li_{l-1}}V_{aL-2}(T_R) = \varepsilon_{ai_li_{l-1}}S_{aL-2}^{(l-1)}(T_R) + \sum_{n\geq 2}\frac{G^{n-1}}{c^{3(n-1)+2k}}Y_{nL}(T_R) \qquad (26)$$

where k is a positive integer representing the number of contractions between indices on the moments in that particular term and functions X_{nL} and Y_{nL} represent some nonlinear and in general nonlocal functionals of n moments M_L and S_L whose general form reads:

$$X_{nL}(T),\ Y_{nL}(T) = \sum\int_{-\infty}^{T}dv_1\cdots\int_{-\infty}^{T}dv_n\mathcal{X}(T,v_1,\cdots,v_n)M_{L_1}^{(a_1)}(v_1)\cdots S_{L_n}^{(a_n)}(v_n). \qquad (27)$$

The kernel $\mathcal{X}$ made from Kronecker and Levi- Civita symbols, has a complicated index structure and depends on variables having dimensions of time. The structure of X_{nL} and Y_{nL} embody the fact that the gravitational field in higher approximations depends on the 'history' of the source and that propagation of radiation is not only along light cones but also inside them.
3. Finally one needs to relate the field in D_e to the inner field in the source. To this end one does two things: Re-expand the external post-Minkowskian field in a post-Newtonian expansion. Integrate the *non-vacuum* field equations in the near zone D_i by means of a post-newtonian expansion using as source variables $\sigma = (T^{00}+T^{ss})/c^2$, $\sigma_i = T^{0i}/c$ and $\sigma_{ij} = T^{ij}$. This choice simplifies the 1PN solution and hence the subsequent iterations. Starting with the source terms at the lowest order one solves for the gravitational field $h^{\mu\nu}$. This solution for $h^{\mu\nu}$ is then used in the relevant nonlinear terms to generate a more accurate source term at the next order. This in turn determines a h to higher accuracy. To 2PN accuracy the solutions are determined in terms of potentials V, V_i and W_{ij} which are retarded integrals associated with sources σ, σ_i and $\sigma_{ij}+(1/4\pi G)(\partial_iV\partial_jV-(1/2)\delta_{ij}\partial_kV\partial_kV)$ respectively.
4. One finally matches the two solutions in the exterior near zone $D_i\cap D_e$ to relate the algorithmic moments to the source properties. This is most

conveniently done by relating the 'exterior potentials' in terms of which the exterior solution is expressed to multipole expansions of the corresponding 'inner potentials'. This leads one to a result that the linear piece of the external potential $h_1^{\mu\nu}$ is – modulo a linear gauge transformation – given by a multipole expansion of the form

$$Gh_1^{\mu\nu}[M_L, S_L] = -\frac{4G}{c^4}\sum_{l\geq 0}\frac{(-)^l}{l}\partial_L\left[\frac{1}{r}\mathcal{F}_L^{\mu\nu}\left(t-\frac{r}{c}\right)\right]+\mathcal{O}(\varepsilon^7)$$

where the reducible moment is given by

$$\mathcal{F}_L^{\mu\nu} = FP_{B=0}\int d^3x|\mathbf{x}|^B\hat{x}_L\int_{-1}^{1}dz\delta_l(z)\left(\overline{|g|T^{\mu\nu}}+\frac{c^4}{16\pi G}\overline{\Lambda}^{\mu\nu}(V,W)\right)$$

In the above equation the bar above any symbol is a reminder that the corresponding symbol is post-Newtonian expanded to the appropriate accuracy and not retained in its retarded form. In the final step the 'reducible' moment needs to be decomposed into its 'irreducible' parts and this technical problem is the same as discussed in the multipole analysis of linearised gravity [33]. For instance, one finally obtains for the mass-algorithmic moment to 2PN accuracy:

$$\begin{aligned}
M_L(t) &= \mathrm{FP}_{B=0}\int d^3\mathbf{x}|\mathbf{x}|^B\Big\{\hat{x}_L\left[\sigma+\frac{4}{c^4}(\sigma_{ii}U-\sigma P_{ii})\right]\\
&+\frac{|\mathbf{x}|^2\hat{x}_L}{2c^2(2\ell+3)}\partial_t^2\sigma-\frac{4(2\ell+1)\hat{x}_{iL}}{c^2(\ell+1)(2\ell+3)}\partial_t\left[\left(1+\frac{4U}{c^2}\right)\sigma_i\right.\\
&+\left.\frac{1}{\pi Gc^2}\left(\partial_k U[\partial_i U_k-\partial_k U_i]+\frac{3}{4}\partial_t U\partial_i U\right)\right]\\
&+\frac{|\mathbf{x}|^4\hat{x}_L}{8c^4(2\ell+3)(2\ell+5)}\partial_t^4\sigma-\frac{2(2\ell+1)|\mathbf{x}|^2\hat{x}_{iL}}{c^4(\ell+1)(2\ell+3)(2\ell+5)}\partial_t^3\sigma_i\\
&+\frac{2(2\ell+1)}{c^4(\ell+1)(\ell+2)(2\ell+5)}\hat{x}_{ijL}\partial_t^2\left[\sigma_{ij}+\frac{1}{4\pi G}\partial_i U\partial_j U\right]\\
&+\frac{1}{\pi Gc^4}\hat{x}_L\Big[-P_{ij}\partial_{ij}^2U-2U_i\partial_t\partial_i U+2\partial_i U_j\partial_j U_i\\
&\qquad-\frac{3}{2}(\partial_t U)^2-U\partial_t^2 U\Big]\Big\}+\mathcal{O}(\varepsilon^5)\,. \qquad (28)
\end{aligned}$$

After elimination of the mathematical intermediaries appearing in the formalism like the algorithmic moments, the basic structure of the final results of the BDI formalism is the following: The observable 'radiative moments' U_L and V_L giving the angular dependence of the asymptotic gravitational wave amplitude $h_{ij}^{TT}(T,R,\theta,\phi)$ are given in terms of the source-related potentials as a series of terms of increasing nonlinearity.

6.1.1. *A Sampler of Nonlinear effects*

The nonlinear nature of general relativity leads to interesting physical phenomena which we catalog next. The first such effect is the interaction between the time dependent mass quadrupole and the static mass monopole. This represents the back-scatter of linear waves by the space-time curvature generated by the mass energy and referred to as tails. Thus gravitational radiation not only propagates on the light cone but also inside it. Tails appear in the radiation field at 1.5PN order and contribute to the far zone flux. They appear also in the radiation reaction forces at 1.5PN order so that the 'balance equations' are correctly satisfied. In the formal structure of the theory the appearance of tails imply a dependence on the past history of the source and hence a non-locality in time. As mentioned earlier this signals the breakdown of the PNA at 4PN. It has important observational consequences in the dynamics of coalescing binaries.

The second such effect is the interaction of the quadrupole moment with itself [34]. This includes a non-local contribution which causes a permanent change in the wave amplitude before and after the burst as first pointed out by Christodoulu [35]. The physical interpretation of this effect as the re-radiation of gravitational waves by the stress-energy tensor of the linear waves was clarified by Thorne [36]. It appears at 2.5PN in the radiation field. The memory effect does not contribute to the energy loss and hence has poor observable consequences. In addition to this non-local effect the quadrupole-quadrupole interaction includes many instantaneous terms; these unlike the non-local term are transients.

The last such investigated effect involves the cubic interaction between the time varying mass quadrupole and two static mass monoples [37]. In addition to a second order scattering of the linear waves, it also includes the scattering of the tail of waves from the static mass monopole M. The latter called 'tails of tails' is of order 3PN in the radiation field and, though small, is still important for the detection of inspiralling compact binaries.

6.2. THE EPSTEIN-WAGONER FORMALISM

The Epstein and Wagoner (EW) [28] approach, also starts by rewriting the Einstein equations in a "relaxed" form. As in electromagnetism one can write down a *single* formal solution valid everywhere in spacetime based on the flat-spacetime retarded Green function. The retarded integral equation for $h^{\alpha\beta}$, can then be iterated in a slow-motion ($v/c < 1$), weak-field ($||h^{\alpha\beta}|| < 1$) approximation as shown by Thorne [29]. Unlike in the electromagnetic case, however, the non-linear field contributions make the integrand of this retarded integral non-compact. The EW formalism leads to integrals that are not well defined, or worse, are divergent. Though at the first few PN orders different arguments were given to ignore these issues they provide no justification that the divergences do not become fatal at higher orders. Consequently, the EW formalism did not appear to be a reliable route to discuss higher PN approximations. Recently, Will and Wiseman have critically examined the EW formalism and provided a solution to the problem of its divergences. The resolution involves taking literally the statement that the solution is a *retarded* integral, *i.e.* an integral over the *entire* past null cone of the field point. Unlike in the original EW treatment, only the part of the integral that extends over the intersection between the past null cone and the material source and the near zone is approximated by a slow-motion expansion involving spatial integrals of moments of the 'source'. Undefined and divergent integrals result from PN expansions if these spatial integrals are extended to infinity. Will and Wiseman restrict these integrals to the boundary of the near zone $\mathcal{R}$ chosen roughly to be a wavelength of the gravitational radiation. The slow motion expanded form is not used to evaluate the integral over the rest of the past null cone exterior to the near zone ("radiation zone"). A coordinate transformation is used to convert the integral into a convenient form for easy evaluation and it is manifestly convergent for reasonable past behavior of the source. All integrations are explicitly finite and convergent and all contributions from the near-zone spatial integrals that grow with $\mathcal{R}$ (and that would have diverged had $\mathcal{R} \to \infty$) are actually *cancelled* by corresponding terms from the radiation-zone integrals. The procedure, as expected, has no dependence on the artificially chosen boundary radius $\mathcal{R}$ of the near-zone. The new EW method proposed by Will and Wiseman can thus be carried to higher orders in a straightforward, albeit very tedious manner and the result is a manifestly finite, well-defined procedure for calculating gravitational radiation to high PN orders. Moreover, part of the tail terms at 3/2PN and 2PN order serve to guarantee that the outgoing radiation propagates along true null directions of the asymptotic curved spacetime, despite the use of flat spacetime wave equations in the solution.

6.3. SUMMARY OF RESULTS

The end result of the computations of the previous subsection are expressions for the radiative mass and current multipole moments characterising the source distribution. Once they are on hand one can proceed to compute the associated gravitational waveform. From the waveform, the far zone energy flux may be computed by time differentiation (this is why one needs the EOM) and integration over all directions. The energy flux can also be computed directly from the moments and this provides a simple check on the algebraic correctness of the long computations. The angular momentum flux can also be computed for non-circular orbits.

For nonspinning compact objects (mass monopoles) Wagoner and Will [38] and later Blanchet and Schafer [39] obtained the 1PN accurate energy flux and discussed the evolution of orbital period. The corresponding angular momentum flux and evolution of other orbital elements were studied by Junker and Schafer [40]. Wiseman discussed the linear momentum flux and the recoil effect in binaries [41]. Blanchet and Schafer discussed the tail effect in energy [42] while Rieth and Schafer extended it to angular momentum [43]. At 2PN the cubic nonlinearity needs to be handled and this was provided by Blanchet, Damour and Iyer [44] in the case of circular orbits. This was independently computed by Will and Wiseman [12] using their improved Epstein Wagoner formalism. They also provided the waveform and energy flux for general (non- circular) orbits. A summary of these results is presented in [45] and the associated 2PN accurate gravitational polarisations is available in ref.[46]. Recently, Gopakumar and Iyer [13] using the BDI approach obtained the waveform, energy flux and associated angular momentum flux and proved the equivalence to the Will Wiseman results. They also used the generalised quasi-Keplerian representation of Damour, Schafer and Wex to compute the evolution of the orbital elements to 2PN accuracy.

The extension of these results to 3PN accuracy is an algebraically heavy and conceptually involved exercise. The multipolar post-Minkowskian approach has been extended to compute the 3PN accurate mass quadrupole (source) moment, 2PN current quadrupole moment and 2PN mass octupole moment of a system of two point masses moving on a circular orbit [47, 48]. From the moments the total energy flux has been computed to 3.5PN order. The regularization of the equations of motion should be consistent with the computation of the multipole moments. It is shown that the arbitrary constants associated with the Hadamard partie finie drop out from the final result (providing one of the sensitive tests of the computation). The Hadamard regularization, based on the Hadamard partie finie, thus seem to provide a good method in this context. Furthermore there is agreement with the known test particle limit. Hopefully in the near future the EW formalism [18, 19] should provide a check on these results.

The extension of the above results to spinning bodies (current dipole) has also been given by Kidder, Will and Wiseman [49, 51] and Owen, Tagoshi and Ohashi [52]. The effects of rotationally induced and tidally induced quadrupole and higher moments on orbital evolution and gravitational wave generation [53, 54] have been also investigated. These are found to be negligible except in the final coalescence stage for neutron star binaries.

As a sample we quote below the 2PN accurate mass quadrupole for circular orbits [45, 46]:

$$
\begin{aligned}
I_{ij} &= \eta m STF_{ij}\left\{x_{ij}\left[1-\frac{\gamma}{42}(1+39\eta)-\frac{\gamma^2}{1512}(461+18395\eta+241\eta^2)\right]\right. \\
&\left.+\frac{r^2}{c^2}v_{ij}\left[\frac{11}{21}(1-3\eta)+\frac{\gamma}{378}(1607-1681\eta+229\eta^2)\right]\right\}. \qquad (29)
\end{aligned}
$$

The general expression for non-circular orbits may be found in [13]. The corresponding 2PN accurate energy flux is given by:

$$
\begin{aligned}
\mathcal{L} &= \frac{32c^5}{5G}\eta^2 x^5\left\{1-\left(\frac{1247}{336}+\frac{35}{12}\eta\right)x+4\pi x^{3/2}\right. \\
&+\left(-\frac{44711}{9072}+\frac{9271}{504}\eta+\frac{65}{18}\eta^2\right)x^2 \\
&\left.-\left(\frac{8191}{672}+\frac{535}{24}\eta\right)\pi x^{5/2}+\mathcal{O}(x^3)\right\}, \qquad (30)
\end{aligned}
$$

with γ and x as defined earlier. The solution to the generation problem thus provides the second input for phasing once we make the assumption of energy balance. In terms of the adimensional time variable :

$$
\Theta = \frac{c^3\eta}{5GM}(t_c - t), \qquad (31)
$$

where t_c denotes the instant of coalescence, the orbital phase is given by

$$
\begin{aligned}
\phi(t) &= \phi_0 - \frac{1}{\eta}\left\{\Theta^{5/8}+\left(\frac{3715}{8064}+\frac{55}{96}\eta\right)\Theta^{3/8}-\frac{3\pi}{4}\Theta^{1/4}\right. \\
&+\left(\frac{9275495}{14450688}+\frac{284875}{258048}\eta+\frac{1855}{2048}\eta^2\right)\Theta^{1/8} \\
&\left.-\left(\frac{38645}{172032}+\frac{15}{2048}\eta\right)\pi\ln\Theta+\mathcal{O}(\Theta^{-1/8})\right\}, \qquad (32)
\end{aligned}
$$

where ϕ_0 is a constant phase determined by initial conditions.

In the next section we discuss the possible checks we can make to verify this.

7. Radiation Reaction Problem

As in electromagnetism, radiation reaction forces arise in gravitation from the use of retarded potentials satisfying time asymmetric boundary conditions like no-incoming boundary condition at past null infinity. As in earlier cases the problem is more complicated because of the nonlinearity of general relativity.

The approach to gravitational radiation damping has been based on the balance methods, the reaction potential or a full iteration of Einstein's equation. The first computation in general relativity was by Einstein [55] who derived the loss in energy of a spinning rod by a far-zone energy flux computation. The same was derived by Eddington [56] by a direct near-zone radiation damping approach. He also pointed out that the physical mechanism causing damping was the effect discussed by Laplace [57], that if gravity was not propagated instantaneously, reactive forces could result. An useful development was the introduction of the radiation reaction potential by Burke [58] and Thorne [59] using the method of matched asymptotic expansions. In this approach, one derives the equation of motion by constructing an outgoing wave solution of Einstein's equation in some convenient gauge and then matching it to the near-zone solution. Restricting attention only to lowest order Newtonian terms and terms sensitive to the outgoing (ingoing) boundary conditions and neglecting all other terms, one obtains the required result. The first complete direct calculation à la Lorentz of the gravitational radiation reaction force was by Chandrasekhar and Esposito [60]. Chandrasekhar and collaborators [61, 62] developed a systematic post-Newtonian expansion for extended perfect fluid systems and put together correctly the necessary elements like the Landau-Lifshiftz pseudotensor, the retarded potentials and the near-zone expansion. These works established the balance equations to Newtonian order, albeit for weakly self-gravitating fluid systems. The revival of interest in these issues following the discovery of the binary pulsar and the applicability of these very equations to binary systems of compact objects follows from the works of Damour [63, 64] and Damour and Deruelle [6] discussed earlier.

Many other approaches to radiation reaction problems have emerged in the last five years. E.g., given the formulas for the far-zone energy and angular momentum fluxes to a particular PN accuracy, to what extent can one infer the radiation reaction acceleration in the (local) EOM? Given the algebraic complexity of various computations and subtle evaluations of various small coefficients, it is worthwhile to check the obvious consistency requirement on the far-zone fluxes. To this end, Iyer and Will (IW) [65, 66] proposed a refinement of the text-book [67] treatment of the energy balance method used to discuss radiation damping. This generalization uses both energy and angular momentum balance to deduce the radiation reaction force for a binary system made of nonspinning structureless particles moving on general orbits. Starting from the 1PN conserved dynamics of the two-body system, and the radiated energy and angular momentum in the gravitational waves, and taking into account the arbitrariness of the 'balance' upto total time derivatives, they determined the 2.5PN and 3.5PN terms in the equations of motion of the binary system. The part not fixed by the balance equations was identified with the freedom still residing in the choice of the coordinate system at that order. The explicit gauge transformations they correspond to has also been constructed. Blanchet [68], on the other hand, obtained the post-Newtonian corrections to the radiation reaction force from first principles using a combination of post-Minkowskian, multipolar and post-Newtonian schemes together with techniques of analytic continuation and asymptotic matching. By looking at "antisymmetric" waves – a solution of the d'Alembertian equation composed of retarded wave minus advanced wave, regular all over the source, including the origin – and matching, one obtains a radiation reaction tensor potential that generalizes the Burke-Thorne reaction potential, in terms of explicit integrals over matter fields in the source. The *validity* of the balance equations upto 1.5PN is also proved. By specializing this potential to two-body sytems, Iyer and Will [66] checked that this solution indeed corresponds to a unique and consistent choice of coordinate system. This provides a delicate and non-trivial check on the validity of the 1PN reaction potentials and the overall consistency of the direct methods based on iteration of the near-field equations and indirect methods based on energy and angular momentum balance. It should be noted that the 'balance method' by itself cannot fix the particular expression for the reactive force in a given coordinate system. In order to solve a practical problem (in which we erect a particular coordinate system), the method is in principle insufficient by itself, but it provides an extremely powerful check of other methods based on first principles. Gopakumar, Iyer and Iyer [14] have applied the refined balance method to obtain the 2PN radiation reaction – 4.5PN terms in the equation of motion. Different facets of the IW choice like the functional form of the reactive acceleration have been systematically and critically explored and a better understanding of the origin of redundant equations is provided by studying variants obtained by modifying the functional forms of the ambiguities in energy and angular momentum. These reactive solutions are general enough to treat as particular cases any reactive acceleration obtained from first principles in the future. The radiative 3.5PN ADM hamiltonian has been obtained by Jaranowski and Schafer [69].

Work on radiation reaction in the test particle case has focussed on understanding the evolution of Carter constant in Kerr geometry. Ryan [70] has investigated the effect of gravitational radiation reaction, first on circular, and later even for non-equatorial orbits around a spinning black hole. Kennefick and Ori [71] developed a computational scheme in which the radiation reaction force is determined by the 'physical retarded' radiation field rather than the radiative field. This allows them to determine the evolution of all associated constants of motion. Capon and Schutz[72] have looked at a 'local expression' for radiation reaction by evaluating its self field as an integral over the particle world line. Recently Mino, Sasaki and Tanaka [73] have derived the leading order correction to the equation of motion of a particle which presumably describes the effect of gravitational radiation reaction by two methods: One approach is analogous to the DeWitt and Brehme [74] method in the case of electromagnetic radiation, where the conservation law of the total (matter + e.m. field) stress-energy tensor is integrated across a tube surrounding the particle world-line, giving the equations of motion including radiation reaction. The other method uses on the other hand asymptotic matching. Quinn and Wald [75] have discussed an axiomatic approach to gravitational radiation reaction and their results are consistent with those of Ref. [73]. Gergely, Perjes and Vasuth [76] have included the spin effects on gravitational radiation reaction using the BDI approach and their results are in accordance with those of Refs. [51, 70]

8. Concluding remarks

Far from being an esoteric and abstruse theory driven by aesthetic considerations, we are in a situation where experiments are driving the theory of general relativity. It is interesting that in the macroscopic world the computations of small higher order corrections so reminiscent of Lamb shift corrections in quantum electrodynamics are inexpendable. We are on the threshold of opening another window to this marvellous universe and with the inauguration of the new gravitational wave astronomy more than ever before general relativity will have found its true home.

Vishu began his career working on Gravitational Radiation[77] and should be happy twice over: Glad he was not so ahead of his time that he gained a place of notoriety in the Guinness book of records [78]! Glad once again that the path he almost tread at the beginning of his research career is probably the most explored today not only the world over but also in India! Maybe he will join the black hole hunt with gravitational waves.. Life after all begins at sixty!!

References

1. C. V. Vishveshwara in *Gravitation, Quanta and the Universe*, Eds. A. R. P. Prasanna, J. V. Narlikar and C. V. Vishveshwara, Wiley Eastern, New Delhi (1980).
2. J. H. Taylor, A. Wolszczan, T. Damour, and J. M. Weisberg, Nature, **355**, 132 (1992).
3. T. Damour, *Gravitational Radiation*, Eds. N. Deruelle and T. Piran (North Holland, Amsterdam, 1983), p.59 and references therein.
4. C. Cutler, T. A. Apostolatos, L. Bildsten, L. S. Finn, É. E. Flanagan, D. Kennefick, D. M. Marković, A. Ori, E. Poisson, G. J. Sussman, and K. S. Thorne, Phys. Rev.

Lett. **70**, 2984 (1993).
5. K. S. Thorne in *300 Years of Gravitation*, Eds. S. W. Hawking and W. Israel, (Cambridge Univ. Press, Cambridge, 1987), p.330; *Proc. of IAU symposium 165, Compact Stars in Binaries*, Eds. J. Van Paradijs, E. Van den Heuvel and E. Kuulkers, (Kluwer Academic Publishers, 1995); *Proc. of the Snowmass 95 summer study on particle and nuclear astrophysics and cosmology*, Eds. E. W. Kolb and R. Peccei, (World Scientific, 1995)
6. T. Damour and N. Deruelle, C. R. Acad. Sci. Paris **293**, 537 (1981); **293**, 877 (1981); Phys. Lett., 87A, 1981, 81.
7. T. Damour, B. R. Iyer and B. S. Sathyaprakash, Phys. Rev. D **57**, 885 (1998).
8. T. Damour, *Gravitation in Astrophysics*, Eds. B. Carter and J. B. Hartle, (Plenum Press, New York and London, 1986), p.3
9. T. Damour and B. G. Schmidt, J. Math. Phys., **31**, 2441().
10. T. Damour, in *300 Years of Gravitation*, edited by S. W. Hawking and W. Israel (Cambridge University Press, London, 1987), p. 128.
11. J. Martin and J. L. Sanz, J. Math. Phys. **20**, 25 (1979).
12. C. M. Will and A. G. Wiseman, Phys. Rev. D **54**, 4813 (1996).
13. A. Gopakumar and B. R. Iyer, Phys. Rev. D **56**, 7708 (1997).
14. A. Gopakumar, B. R. Iyer and Sai Iyer, Phys. Rev. D **55**, 6030 (1997).
15. G. Schafer, Ann. Phys. (N.Y.) **B161**, 81 (1985); *Gravitational wave detection*, Ed. A. Krolak, (Banach center publications, Warzawa, 1997), p. 43.
16. L. P. Grishchuk and S. M. Kopejkin, in *Relativity in Celestial Mechanics and Astrometry*, edited by J. Kovalevsky and V. A. Brumberg (Reidel, Dordrecht, 1986), p. 19.
17. L. Blanchet, G. Faye and B. Ponsot, work in progress.
18. C. M. Will and M. E. Pati, work in progress;
19. G. Schäfer and P. Jaranowski, grqc-9712075 (1997), grqc-9802030 (1998).
20. P. Jaranowski, in *Gravitational wave detection*, Ed. A. Krolak, (Banach center publications, Warzawa, 1997), p. 55.
21. L. Blanchet and T. Damour, Phil. Trans. R. Soc. London A **320**, 379 (1986); Phys. Rev. D **37**, 1410 (1988).
22. T. Damour and G. Schäfer, Nuovo Cimento B **101**, 127 (1988).
23. G. Schäfer and N. Wex, Phys. Lett. **174 A**, 196, (1993); erratum **177**, 461.
24. N. Wex, CQG **12**, 983, (1995).
25. N. Wex and R. Rieth, in *Symposia Gaussiana*. Eds. Behera, Fritsch and Lintz (W. Gruyter and Co., Berlin, 1995).
26. A. Gopakumar and B. R. Iyer, in preparation(1998).
27. L. Blanchet, Proc. R. Soc. Lond. A **409**, 383 (1987); L. Blanchet and T. Damour, Ann. Inst. H. Poincaré (Phys. Théorique) **50**, 377 (1989); T. Damour and B. R. Iyer, Phys. Rev. D **43** 3259 (1991); Ann. Inst. H. Poincaré (Phys. Théorique) **54**, 115 (1991); CQG **11**, 1353 (1994); L. Blanchet and T. Damour, Phys. Rev. D**46**, 4304 (1992). L. Blanchet, Phys. Rev. D **51**, 2559 (1995); L. Blanchet, T. Damour, and B. R. Iyer, Phys. Rev. D **51**, 5360 (1995); L. Blanchet, Phys. Rev. D **54**, 1417 (1996).
28. R. Epstein and R. V. Wagoner, Astrophys. J. **197**, 717 (1975).
29. K. Thorne, Rev. Mod. Phys. **52**, 299 (1980).
30. C. M. Will and A. G. Wiseman, Phys. Rev. D **54**, 4813 (1996).
31. N. D. Haridass and V. Soni, J. Phys. A **15**, 473, (1982).
32. W. B. Bonnor, Philos. Trans. R. Soc. London A **251**, 233 (1959).
33. T. Damour and B. R. Iyer, Phys. Rev. D **43** 3259 (1991).
34. L. Blanchet, *Quadrupole Quadrupole gravitational waves*, CQG (1998) to appear.
35. D. Christodoulou, Phys. Rev. Lett. **67**, 1486 (1991).
36. K. Thorne, Phys. Rev. D **45**, 520 (1992).
37. L. Blanchet, *Gravitational wave tails of tails*, CQG (1998) to appear.
38. R. V. Wagoner and C. M. Will. Astrophys. J **210**, 764 (1976).

39. L. Blanchet and G. Schafer, Mon. Not. R. Astron. Soc. **239**, 845 (1989).
40. W. Junker and G. Schafer, Mon. Not. R. Astron. Soc. **254**, 146 (1992).
41. A. G. Wiseman, Phys. Rev. D **46**, 1517 (1992); Phys. Rev. D **48**, 4757 (1993).
42. L. Blanchet and G. Schäfer, CQG **10**, 2699 (1993).
43. R. Rieth and G. Schafer, CQG **14**, 2357 (1997).
44. L. Blanchet, T. Damour, and B. R. Iyer, Phys. Rev. D **51**, 5360 (1995).
45. L. Blanchet, T. Damour, B.R. Iyer, C. M. Will and A. G. Wiseman, Phys. Rev. Lett. **74**, 3515 (1995).
46. L. Blanchet, B. R. Iyer, C. M. Will and A.G. Wiseman, CQG **13**, 575, (1996).
47. L. Blanchet, B. R. Iyer and B. Joguet, paper in preparation;
48. L. Blanchet, G. Faye, B. R. Iyer, B. Joguet and B. Ponsot, work in progress;
49. L. E. Kidder, C. M. Will, and A. G. Wiseman, Phys. Rev. D **47**, R4183 (1993).
50. L. E. Kidder, Phys. Rev. D **52**, 821 (1995).
51. L. E. Kidder Phys. Rev. D **52**, 821 (1995).
52. B. Owen, H. Tagoshi and A. Ohashi, gr-qc??
53. L. E . Kidder (unpublished).
54. K. Taniguchi and M. Shibata, *Phys. Rev. D.* (submitted), Kyoto University preprint KUNS 1464, Osaka University preprint OU-TAP 68.
55. A. Einstein, Sitzber. Preuss. Akad. Wiss. (Berlin), (1918).
56. A. S. Eddington, *The mathematical theory of relativity*, (Cambridge Univ. Press, Cambridge, 1924).
57. P. S. Laplace, *Mécanique céleste*, (Courcier, Paris). Second Part: Book 10, chap. 7.
58. W. L. Burke, unpublished Ph. D. Thesis, California Institute of Technology (1969); J. Math. Phys. **12**, 401 (1971).
59. K. S. Thorne, Astrophys. J. **158**, 997 (1969).
60. S. Chandrasekhar and F. P. Esposito, Astrophys. J. **160**, 153 (1970).
61. S. Chandrasekhar, Astrophys. J. **158**, 45 (1969).
62. S. Chandrasekhar and Y. Nutku, Astrophys. J. **158**, 55 (1969).
63. For a comprehensive review and detailed references to original literature see T. Damour in *Gravitational radiation* edited by N. Deruelle and T. Piran, (North Holland, Amsterdam, 1982).
64. T. Damour, C. R. Acad. Sc. Paris, **294**, 1355 (1982).
65. B. R. Iyer and C. M. Will, Phys. Rev. Lett. **70**, 113 (1993).
66. B. R. Iyer and C.M. Will, Phys. Rev. D **52** ,6882 (1995).
67. L. D. Landau and E. M. Lifschitz, *The Classical Theory of Fields* (Pergamon, New York, 1975), 4th ed., p. 357.
68. L. Blanchet, Phys. Rev. D **47**, 4392 (1993); L. Blanchet, Phys. Rev. D **55**, 714 91997).
69. P. Jaranowski and G. Schafer, Phys. Rev. D **55**, 4712 (1997).
70. F. D. Ryan, Phys. Rev. D **52** , R3159 (1995); Phys. Rev. D **53** , 3064 (1995).
71. D. Kennefick and A. Ori, Phys. Rev. D **53**, 4319 (1996); A. Ori, Phys. Rev. D **55**, 3444 (1997).
72. R. A. Capon (1997) in *Mathematics of Gravitation: Gravitational Wave Detection*, **41**, p. 65, Ed. A. Królak, Banach Center Publications, Warsaw.
73. Y. Mino, M. Sasaki and T. Tanaka, Phys. Rev. D **55**, 6030 (1997).
74. B. S. DeWitt and R. W. Brehme, Annals of Physics, **9**, p. 220, (1960).
75. T. C. Quinn and R. M. Wald, Phys. Rev. D **56** 3381 (1997).
76. L. Gergely, Z. Perjés and M. Vasúth, Phys. Rev. D.**57**, 876, (1998); **57**, 3423 (1998).
77. C.V. Vishveshwara, *Gravitational Radiation from Binary Systems*, University of Maryland Technical Report (1965).
78. C. V. Vishveshwara *On the black hole trail...: A personal journey* Current Science, **71**, 824 (1996).

On the other hand, as emphasised recently by Eisenstaedt, the use of this post-Newtonian approximation method has had, from a conceptual point of view, the adverse side-effect of introducing implicitly a 'neo-Newtonian' interpretation of general relativity. Indeed, technically this approximation method follows the Newtonian way of tackling gravitational problems as closely as possible. But this technical reduction of Einstein's theory into the Procrustean bed of Newton's theory surreptitiously entails a corresponding conceptual reduction: the Einsteinian problem of motion being, in fact, conceived within the Newtonian framework of an absolute (coordinate) space and an absolute (coordinate) time. However, some recent developments oblige us to reconsider the problem of motion within Einstein's theory.

– THIBAULT DAMOUR

All this means that it is worth looking back in detail at the foundations of the problem of motion. The most natural way to do so is to begin by discussing the Newtonian way of tackling the problem of motion. Our principal aim in doing so will be to clarify the basic ideas which allow one to put the Newtonian problem of motion into a manageable form. In doing so we shall go into the details of the way these ideas are implemented in the Newtonian framework, some of which will seem fairly trivial to the reader; but we think it is important to go through them for the following reason: precisely because these details are technically trivial (or too well known), one is liable to miss the ideas which underlie them, so that when tackling the corresponding relativistic problem, one has the tendency to *transplant*, within Einstein's mathematical framework, exactly the *same technical developments*, instead of trying to *transmute*, within Einstein's conceptual framework, the *ideas* underlying them.

— THIBAULT DAMOUR

27. GROUND-BASED INTERFEROMETRIC DETECTORS OF GRAVITATIONAL WAVES

BIPLAB BHAWAL
TAMA project,†
National Astronomical Observatory,
2-21-1 Osawa, Mitaka, Tokyo 181, Japan

"Difficulty is an excuse history never accepts." – *Samuel Grafton*

1. Prologue

I still remember how, after looking at some jumbled-up combination of algebraic characters obtained in a hard way as a solution of a field equation in a complicated spacetime, Vishu used to whistle with some sort of happiness while simultaneously flipping the three joined-up middle fingers of his right hand and moving the same hand in air to explain to me the physics behind such a solution representing a wave!

....When, right after finishing my doctoral work, I decided to get myself seriously involved in the search for gravitational waves, perhaps from somewhere behind Vishu blew his whistle! So, here I am going to describe the ground-based interferometric detectors which will most probably be the first of all kinds of detectors to record the audio frequency gravitational waves – the trails of compact stars and blackholes – things Vishu loves. We also hope that in near future our computers would be able to convert these ripples of spacetime into audible whistles!

The aim of this article is to provide an introduction to this field as well as an updated review of the developments that happened over the last few years. It is rather a short introduction to a vast subject which has been gaining enrichment with inputs from perhaps an uncountable number of fields of study as the days are passing by. It is mainly motivated by the discussions I regularly had with my friends working in various projects and it must be obvious that I am not an expert of technical details of all these aspects. The readers are assumed to be familiar with some of the

†Address after December 8, 1997: LIGO Project, Caltech, Pasadena, CA 91125, USA.

B. R. Iyer and B. Bhawal (eds.), Black Holes, Gravitational Radiation and the Universe, 461–479.

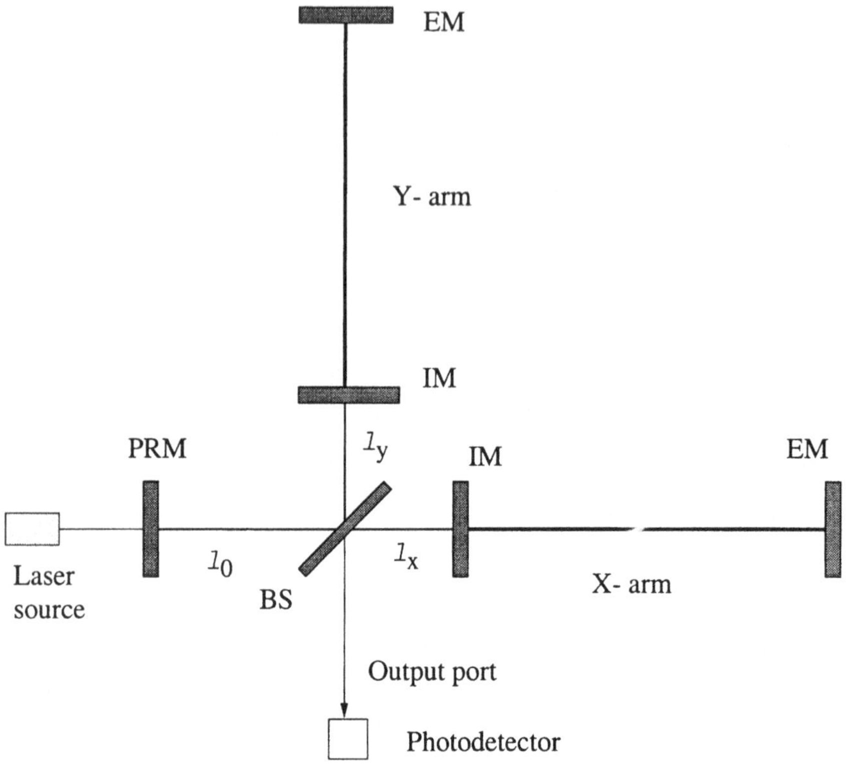

Figure 1. Configuration of a long-baseline interferometric GW detector. IM – input mirror, EM – end mirror, PRM – power recycling mirror, BS – beam-splitter

basic features of the interferometric detectors, although the next section introduces a few aspects of it.

2. Common features of interferometric detectors

The effect of a gravitational wave (GW) is to produce a transverse shear strain and this fact makes the Michelson interferometer an obvious candidate for a detector. However, a number of modifications on it were necessary to make the interferometer suitable for the detection of GWs. The following six features are essential for any kind of ground-based long-baseline interferometric detector. A schematic diagram of such a detector is shown in Fig.1. Further modifications could be made only in order to enhance the sensitivity in a particular range of frequency or to utilize some advantages for some specific purpose.

2.1. LONG BASELINE

Since the GW strains space stretching it in one direction and compressing it in the orthogonal direction, the change in length of an arm is equal to the size of the strain times the initial length of the arm. However, since this strain is a time-dependent one, we need to take into account the time-scale of this strain while deciding about the duration for which light should be allowed to acquire the changing phase cumulatively inside each arm. This duration is termed as *storage time* of light. Obviously, if this strain has just one frequency component (say, that for GW from a solitary pulsar), we would like to optimize the instrument by fixing arm-length so that the storage time equals half the time-period of the GW strain. Any other duration in between zero and the time-period of GW leads to a smaller value for the acquired cumulative phase. A value equal to the full time-period produces no change at all. Unfortunately, this simple game with sinusoidal waves leads to a conclusion that, in order to optimize an interferometer for a GW frequency of 100 Hz, we need an interferometer of arm-length 750 Km; Also, the desired length is inversely proportional to the targetted frequency.

The way to solve this problem is to have the light in each arm reflect back and forth several times from each back mirror. This can be accomplished in two ways : (i) using an optical delay line (DL), where a single beam strikes the mirror in a different spot on each traversal of the arm [1], (ii) using a Fabry-Perot (FP) cavity where partial beams from many traversals meet in a single spot on each mirror [2].

In a DL system, we know that there's always one beam and so the light has only one time-parameter. That is why it has peaks of its sensitivity around some integral multiples of the optimal frequency corresponding to this storage time. Any two peaks are separated from each other by a point characterised by a frequency where the sensitivity goes to zero. Larger the storage time, narrower are these peaks. Larger the frequency, shorter the height of the peaks. On the other hand, in a FP system, at any moment of time, the light coming out of the cavity consists of a number of partial beams each of which has been stored inside for a different length of time. So, within a finite bandwidth, it is likely that each frequency of GW will have its corresponding partial beam with a storage time exactly equal or very near to the optimum value. This allows the FP system to have a sensitivity curve which never goes to zero. If one goes on increasing the storage time, the sensitivity becomes inversely proportional to the frequency.

A theoretical analysis and comparison of these two schemes [3, 4] bring out a number of merits and demerits for both these schemes. However, there are a number of technical reasons why FP cavity turns out to be the choice for any long-baseline interferometer. These are as follows: (i) The vacuum system (as explained later in this section) needed for accomodating a DL system is much larger than that for a FP system. This leads to much higher cost of construction and maintenance. (ii) In a DL system, the beam goes back and forth quite close to one another both in position and angle. This leads to lot of noise originating from the scattered light. (iii) Since the lateral space needed to accomodate a FP cavity is less, it is possible to accomodate more than one interferometer inside the same enclosure. In fact, the design of LIGO at Hanford accomodates one full-length (4 Km) and one half-length (2 Km) interferometer. This allows certain advantages that can be gained from local coincidence experiments [5].

The only configuration where the use of a cavity (of ring-type – in the present context) might turn out to be disadvantageous is an interferometer of Sagnac topology proposed mainly by the Stanford group [6], since such an inclusion takes away two important virtues of such a topology, i.e., insensitivity to variation in laser frequency and low-frequency fluctuations of the arm lengths. However, the technical problems originating from a large vacuum system for DL still remains. Also, a thorough analysis performed recently by the Garching group [7] has shown that a Sagnac topology does not really offer advantages over Michelson topology for the purpose of GW detection.

2.2. SUSPENDED TEST MASS

In order to allow the test masses (in the present context – mirrors) to feel the presence of GW in a natural way and to follow a geodesic path in the background of flat spacetime perturbed by tiny *ripples* of the wave, we need to suspend those as pendulums. So, at a frequency range far above the resonant frequency, ω_0 ($\approx$ 1Hz), the pendular restoring forces do not find opportunity to make themselves felt before the oscillating input reverses the sign. The suspension system actually consists of a cascade of massive pendulums connected one after the other. From the behavior of a forced harmonic oscillator, we can guess that each of these stages also helps in attenuating the position noise from the suspension noise to the swinging mass - it acts like a filter attenuating a signal by a factor $K(\omega) \propto \omega_0^2/\omega^2$ above the resonant frequency of the system.

2.3. DARK FRINGE OPERATION

At the input port of a Michelson interferometer, light both enters and comes out of the interferometer. On the other hand, although light comes out of the interferometer at its output port, *nothing* enters the system through this port. However, when we intend to study the system quantum-mechanically, we cannot apply such a phrase to this context; Instead of *nothing*, there is always something which is entering the system through this port - that is the vacuum-state fluctuation of light. This fluctuation mixes up with the coherent state of the laser light [8]. As a result, from time to time we may find some extra photons (more than what we expect in a certain set-up of fixed value of phase difference between the arms) coming out of the system through this port in a random manner.

This originates due to the fact that we are trying to count a number of discrete independent events, i.e., photons coming out at the output port characterised by a mean number, $N_{\rm out}$ during a certain measurement interval, τ. In such a case of measurement, the outcome is characterised by the Poissonian distribution of probability function. This randomness in the measurement of output light is termed as *shot noise* or *photon counting noise* and can be expressed in terms of an equivalent difference in length of the arms of the interferometer, ΔL. This turns out to be inversely proportional to the square-root of the number of photons (an intrinsic feature of the Poissonian statistics) and also the cosine of $\phi/2$ where ϕ is the difference in phase set between the two arms. So, we see that if ϕ is set to zero, the noise will be minimum. This is called *dark fringe operation* since, under such a condition, light interferes destructively at the output port (the word *fringe* is not appropriate here since it is power which is measured and it is not a fringe which is looked at. However, somehow this word continued to stick to this choice of the operating condition). To reduce the noise further we need to increase the power of laser.

2.4. POWER RECYCLING

If light returning from the arms to the beam splitter is set to interfere destructively at the output port then by energy conservation, the light must interfere constructively in the direction of the laser source and get out of the interferometer through the input port. So, it was suggested [9] that this wasted light be recycled back to the interferometer by putting a mirror in front of the source. This enhances the laser power at the beam-splitter and thus reduces the shot-noise. Power recycling in Michelson interferometer with FP cavities in arms was first demonstrated at MIT[10]. The maximum power built-up is achieved when the recycling mirror transmittivity exactly matches the total of the fractional power losses inside the interferometer. This maximum enhancement factor has an approximately inverse relationship with the total loss inside the interferometer and will be about 40 for the initial LIGO and VIRGO.

2.5. TEST MASS WITH LOW LEVEL OF DISSIPATION

Thermal noise [11] is expected to be a dominant noise source in laser interferometric detectors. According to the fluctuation-dissipation theorem, for a linear system in thermal equilibrium at a temperature T, the displacement power spectrum is proportional to the real part of the admittance which represents the dissipation in the system. So, each process that contributes a mechanical loss to the free motion of the mirror also produces a noise source. The following two main sources of dissipation pose a severe limit to the sensitivity of interferometric detectors: (i) internal vibration modes of test masses [12, 13, 14] (ii) suspension modes of pendulum [15, 16]. As a result, very high quality factor material (so with less dissipation) is required for both the suspension wires and the test masses. Also, efforts are made to keep the test masses as massive as possible to reduce the thermal noise. It is also important to avoid mechanical clamping of suspension wires or membranes which also leads to lowering of pendulum mode quality factor.

2.6. VACUUM ENCLOSURE

The statistical fluctuation of the residual gas density can induce a fluctuation of the index of refraction and thus of the interferometer phase difference. It is thus necessary to enclose the arms of the interferometer in a vacuum chamber. Initial vacuum performance is required to be approximately $< 10^{-6}$ torr but pressures as low as 10^{-9} torr may be needed by future advanced interferometers. By the time this article is published, hopefully initial testings on the LIGO, Hanford vacuum system would be completed. With a length of 4+4 Km and a diameter of 1.2 m, this will be the largest (volume) man-made vacuum system of the universe.

3. Controlling the interferometer

If the interferometer has such great advantages over the bar (far larger bandwidth and far larger potential sensitivity), then why didn't Braginsky, Weber and others build interferometers instead of bars? When I asked Braginsky in the mid-1970s, he replied that bar detectors are simple, while interferometers are horrendously complex.
– Kip S. Thorne, *Black holes and time warps: Einstein's outrageous legacy* (W.W. Norton, NY, 1994)

The interferometer with FP cavities in arms and recycling mirror is a system of coupled cavities. In addition to maintaining the *dark fringe* condition, all of these cavities are to be kept on resonance with the laser source. It is thus essential to develop some scheme to monitor the length of various cavities in order to maintain those in the perfect operating condition. To understand the importance of the length control system for the detector configuration, we see that even in an ordinary Michelson interferometer operating at the dark fringe condition a displacement of about 250 nanometers (one-fourth of the wavelength of infrared laser) of either of the back mirrors will cause the intensity at the output to change to full brightness.

Controlling length is, therefore, not an easy task especially since the suspended mirrors, by getting excited by the residual seismic noise, may oscillate around their equilibrium points at a low frequency with an amplitude of some tens of wavelengths. A local damping system may reduce this amplitude but, even if it succeeds to certain extent, purely local systems are not sufficient to maintain the whole interferometer at the operating condition. The system is not only coupled but also a dynamic one and these dynamical aspects in a coupled system lead to further complications by bringing forth the nonlinear responses [17]. Thus global control of the mirror distances is a necessity for continuously maintaining the resonance conditions. All these features provide newer dimensions to the interferometry of the GW detectors.

Let us first discuss about the locking schemes developed for static differences in arm-lengths of the interferometer. To achieve the desired operating point as mentioned above, we need to control four linear combinations (modes) of four independent changes in lengths: (i) Arm-length common mode, $\delta L_+ = \delta L_x + \delta L_y$, (ii) Arm-length difference mode, $\delta L_- = \delta L_x - \delta L_y$, (iii) Michelson common mode, $\delta l_+ = \delta l_x + \delta l_y$, (iv) Michelson difference mode, $\delta l_- = \delta l_x - \delta l_y$, where δL_x and δL_y represent changes in the arm-cavity lengths and δl_x and δl_y represent changes in lengths of the recycling cavities.

The problem is to extract four independent error signals related to these four lengths and to feed them back to various mirrors. This feedback system, apart from its main task of locking, in an ideal case, should have a high gain at a low frequency (up to a few Hz) and should not introduce any extra noise in the operating frequency range of the antenna (10 Hz to 10 KHz).

In order to avoid the noise due to laser-power fluctuations at low frequencies, the measurement must be shifted to the quieter MHz region using a modulation- technique. To serve this purpose the laser light is phase modulated using an electro-optic modulator. Under the approximation that the modulation index is small, the beam after the phase modulation effectively contains three frequency components - the carrier, ω and two sidebands, $\omega \pm \omega_m$, where ω_m is the modulation frequency.

A number of length control schemes were proposed earlier [18, 19] but considering various technical problems associated with these schemes, the decision of people working in all detectors has gone in favour of the frontal (or pre- or the Schnupp after its inventor, L. Schnupp) modulation technique. In this scheme, the modulation is done in between the laser and the interferometer and no modulation is done inside or on any outgoing light of the interferometer. The beam is introduced into the interferometer and after it experiences amplitude and phase changes inside it, is detected with photo detectors at various ports.

The error signals giving the deviation of cavity lengths from resonance are obtained by beating the carrier with the modulation sidebands. Therefore, it is necessary that at the observation points carrier and sidebands are sufficiently strong. So, both the carrier and sidebands are made resonant with the recycling cavity but the sidebands, unlike the carrier, are kept almost antiresonant with the arm cavities. This makes the sidebands insensitive to δL_+ or δL_- modes.

Also, the GW signal (just behaving like one of the error signals, L_- at the frequency range of our interest) must be observed at the modulation frequency at the output port. Therefore some sideband power must be present at the output port in order to create a beat signal with the carrier, if the latter leaks out at the dark fringe due to a GW. In frontal modulation, an asymmetry is introduced between l_x and l_y as shown in Fig.1. Because of this asymmetry, the fringe does not become completely dark for the sidebands at the output port. When the carrier beam leaks to the dark port, it beats with the residual sidebands and produces an amplitude change in the photo current at the modulation frequency.

In a similar way other error signals are detected at a number of ports (depending on the scheme employed). Each output of these detectors is demodulated with mixtures in two phases: in phase, the same phase as that of the modulation and in a quadrature phase, the phase shifted from that of the modulation by $\pi/2$. Outputs of these mixtures are then passed through low pass filters in order to eliminate harmonics of modulation frequency and to obtain the signals in the form of changes in photocurrents due to phase differences in various cavities.

The scheme of length control system must be able to maintain the stability and should not spoil the sensitivity achieved by the main components of the interferometer. However, it turns out to be difficult to extract the signal corresponding to the differential change in lengths of the recycling cavities (δl_+) independently. This signal is always mixed up with other signals and and, in order to obtain it, one needs to combine the signals linearly by electronic circuits or by digital processing [20, 21, 22, 23]. Although it is not much of a problem to lock the interferometer following these techniques, proper care must be taken in designing the circuits and using suitable electronic components since the noises of the detectors also add incoherently when their signals are combined.

Another proposal deals with the use of an additional carrier, a frequency-shifted subcarrier [23, 24, 25] in order to obtain signals for the Michelson lengths, δl_x and δl_y separately from those for the arm lengths. After phase modulation the subcarrier is superimposed on the main carrier with a beam-splitter. The frequencies of the subcarrier and its sidebands are so choosen that they get resonant only with the recycling cavities and not with the arm cavities. So, the subcarrier and its sidebands remain unaffected by the phase changes in the arm cavities but carry information about δl_x and δl_y. The associated disadvantage is the introduction of obvious complications into the the optical setup.

In any case it would be an extra advantage if it's possible to obtain well-separated signals (i.e., *diagonalisation*, if you see what I mean). The design and study of the control system and the operation of the interferometer would be much simplified in that case. A very promising scheme in that direction for controlling a power-recycled interferometer seems to be the one proposed by Ando et al [26]. This will be implemented in the TAMA300 interferometer. In this configuration, the recycling gain is optimized not with respect to the laser light but with respect to sidebands. This scheme, while maintaining the simplicity of frontal modulation, can extract all four signals separately (mixing is of the order of a few in thousand) at only a little expense of the recycling gain in carrier light. However, in low loss systems of advanced interferometers, where cavity-reflectivities for the carrier and sidebands would be very near to each other, it might turn out to be difficult to implement a very precise specification for the reflectivity of the recycling mirror.

The discussion above on the control schemes are based on static differences in various arm-lengths. Also, under these schemes, signals are generated only when the interferometer is very close to its desired operating point. In a GW detector, as mentioned above, the mirrors may move thereby generating nonlinear dynamical responses of the detector [17]. It is quite important to study the dynamical lock-acquisition procedure in any of the schemes mentioned above. The LIGO team [27] has reported a real-time analysis of the optical transient response of a high finesse suspended Fabry-Pẹrot Cavity (one used in Caltech 40m interferometer) to guide its acquisition of lock as the mirrors swing through resonance. This is an important first step to the implementation of guided lock acquisition for the interferometric detectors.

So far we discussed only about the longitudinal degrees of freedom keeping aside the problems related to the misalignment effects. It is desirable to maintain the beam in its fundamental transverse electric and magnetic mode (i.e. TEM00 or fundamental Hermite-Gaussian mode) as it propagates inside the interferometer and gets reflected from or transmitted through various mirrors and cavities. The higher order modes get generated whenever any mirror gets rotated from its correct position or the beams get displaced from its correct propagation axes or the mirrors have manufacturing defects [25, 28, 29]. As a consequence the stored power in arm-cavities gets reduced and weakens the signal generated by GW and the resulting light in higher order modes leaks out of the output port, thus increasing the shot noise; Both effects affect the sensitivity of the interferometer. Control schemes are being designed[28] in order to achieve an upper limit of 10^{-8} radian-rms of the optical axis for the angular orientations of the mirrors so that the maximum achievable strain sensitivity in LIGO could be utilised.

4. Gravitational waves: Its effect and detection

As we can see, the effect of GW will be felt by the interferometer in the form of a dynamic arm-length difference mode, δL_- as mentioned in the previous section. To discuss this effect in a more concrete way, let us consider a simplified case: A laser beam polarised along the Z-direction is moving along the interferometer arm of the X-direction (the other arm is along the Y-direction). In such a case, if we assume the gauge where the scalar potential for the electromagnetic radiation is zero, then only the Z-component of the vector potential turns out to be nonzero. We may thus represent this light as a scalar field equivalent to this non-zero Z-component of the vector potential.

Let us now assume that a monochromatic GW is coming along the Z-direction (for simplicity, again). This will lead to small time-dependent changes, $h(t)$ (with opposite signs) in the xx and yy components of the otherwise flat Minkowskian metric. To understand the effect of GWs on the light beam we thus need to solve the Klein-Gordon equation of curved spacetime in the background of this perturbed flat spacetime.

The resulting solution looks just like a wave which is phase modulated at the frequency of the GW, ω_g. Like any other phase-modulated waveform, the solution can be expressed in terms of Bessel function of first order which can be expanded to show that the resulting laser light is nothing but a combination of mainly three frequency components: the dominant one is the original laser frequency, ω_L and the weaker ones (proportional to $h(t)$) with sideband frequencies $\omega_L \pm \omega_g$.

All the cavities in a power-recycled interferometer are always operated in resonance with the laser frequency, so that, while in a dark fringe operation all the laser light goes back towards the source, these newly formed GW-sidebands are thrown out-of-resonance by the interferometer and comes out through the output port. There they are added up with PM-sidebands (the phase-modulated ones described in the last section) to produce an oscillating part in the interference term in the detected power with frequencies $\omega_m \pm \omega_g$. [If there were no phase-modulation and lights leaking from the arms were added to a local oscillator of frequency, ω_L, that would have produced a fluctuation in power at audio frequency, ω_g and would have got submerged in the high level of low-frequency amplitude fluctuation noise of the laser.] This is now multiplied by another local oscillator of frequency,ω_m to get back the GW-phase-modulated signal in the output power.

If the feedback loops act only at low frequency (< 10 Hz, one may do that especially if the seismic isolation is a good level above this frequency), then the GW signal may be directly obtained by looking at the interferometer main output signal. If the small bandwidth of the length control system turns out to be not compatible with the high gain required at low freq (0.1-1Hz) then the feedback will be active also at higher frequency. In that case the GW signal might be detected by looking at the feedback signal used to control the δL_- length variation. However if the signals at various detection ports are not well separated (control system not diagonalised appropriately), the feedback system spoils the sensitivity of detection to some extent by reintroducing shot noise coming from other ports. All these considerations need to be taken into account before designing a good control/detection scheme.

5. Near future

Table I provides a list of detectors which are either under construction or under consideration by the funding authorities. In the third column 'dual' refers to both power recycling and signal recycling when the signal coming out of the output port is also recycled back again by putting another mirror at this port[30]. The operation of multiple interferometers in coincidence allows such a network to achieve immunity to nonstationary, nongaussian noise. Also, since a network of three detectors provides three amplitudes and two independent time delays, it is the minimum configuration needed to determine the five parameters of the GWs, namely, two angles representing direction, two independent polarizations, h_+ and $h_\times$ in the geocentric coordinate system and the orientation of polarization angles on the sky. Two independent relative time delays, in turn, however, yield two source directions. So, a complete solution for the GW requires observation by four laser interferometric broadband detectors. This provides another advantage over three detector network: in coincidence experiments with four detectors, the solution for a transversely polarized quadrupolar wave will be overdetermined and thus any inconsistency among the data

TABLE 1. Description of the site and configuration of the planned detectors. The orientation (last column) refers to the angle made by the bisector of the arms with the local south → north direction measured in anticlockwise convention. Budgetory proposals have been made for the *-marked detectors, JGWO and AIGO. For details visit websites given at the end of this article.

Detector	Arms	Recyc.	Latitude	Longitude	Orient.
LIGO (Hanford)	FP(4Km)	Power	$46^0 27' 18.5''$N	$119^0 24'\ 27''$W	81.8^0
LIGO (Hanford)	FP(2Km)	Power	$46^0 27' 18.5''$N	$119^0 24'\ 27''$W	81.8^0
LIGO (Lousiana)	FP(4Km)	Power	$30^0 33' 46.0''$N	$90^0 46'\ 27.3''$W	153^0
VIRGO (Pisa)	FP(3Km)	Power	43.3^0N	10.1^0E	26.5^0
GEO (Hannover)	DL(0.6Km)	Dual	$52^0 14.8'$N	$9^0 49.3'$E	339.25^0
TAMA (Mitaka)	FP(0.3Km)	Power	$35^0 40'\ 25''$N	$139^0 32' 15''$E	225^0
JGWO* (Japan)	FP(3Km)	Power	—	—	-
AIGO* (Perth)	DL(0.5Km)	Dual	31.04^0S	115.49^0E	-

would be evidence for other polarizartion states [31]. One can, therefore, test Einstein's predictions regarding GW polarization using such a network.

For the present network, if the source lies anywhere near the plane formed by the LIGO/VIRGO network, the accuracies of their direction measurements and the ability to monitor more than one waveform will suffer a lot. The coincidence experiment with a fourth detector at a site far out of this plane is very important to get a good sky coverage and for the accurate determination of source parameters. A long baseline interferometer either in Japan or in Australia will be very useful for this purpose.

An exciting possibility in the effort to detect GWs would be the launching of a space-based detector - LISA in heliocentric orbit. The most important fact is that it will be able to access the low frequency window (10^{-4} to 10^{-1} Hz) to observe some important signals: massive black-hole formation, stochastic background or coalescing binary signals which lasts for quite a long time in this low frquency regime. LISA will therefore complement the advanced ground-based detectors which will be good for observing the later stage of binary coalescence.

Although space-based detectors will be sure to detect GW, there might be some time gap (at least of 15 years from now) before such a detector gets fully operational. During these next 15 years or so, we need to depend mainly on Ground-based detectors. In order to achieve most of the detection aims of ground-based interferometers (i.e., binary coalescences with an event rate of 3 or 4 per year, supernove explosions, solitary pulsars, stochastic background etc.), a sensitivity of 10^{-22} over a bandwidth of approximately 1KHz or a sensitivity spectral density of $3 \times 10^{-24}/\sqrt{\text{Hz}}$ from about 100 Hz is likely to be required. However, the initial designs of the LIGO and VIRGO detectors are somewhat more modest than this - a few times $10^{-23}/\sqrt{\text{Hz}}$. GEO, using some special techniques, could achieve a sensitivity close to LIGO-VIRGO's initial sensivities (at least above 200 Hz and within a limited bandwidth). This will make it useful for detecting GW from solitary pulsars and for its contribution to a coincidence experiment [32] in order to detect coalescing binaries or stochastic background radiation.

Increasing the sensitivity will require progress in technology involving a number of related fields of research. In order to achieve the best possible level of performance various research and development activities will continue in different directions, some of which are discussed below:

5.1. IN SEARCH OF NEW MATERIALS

As far as improvement in sensitivity of individual detectors is concerned, finding suitable materials with very low level of dissipation is the most challenging task ahead since the thermal noise affects a number of important factors related to design-decisions.

In the initial interferometers, fused silica will be used as test masses. Experiments have shown that a quality factor of 5×10^6 can be achieved [14, 33] for the internal modes. For the pendulum mode, a quality factor of 2.5×10^7 has been measured for pendulums using ultra-pure fused silica fibres welded to fused silica masses and to the points of support [34, 35]. Similar level of losses could be found using fibres of commercial grade fused quartz [36]. However, this is not sufficient to reduce the thermal noise at frequencies below or around 100 Hz where most of the contribution to SNR for the coalescing binary chirp waveform comes. The improvement in seismic isolation or photon shot noise would lose much of its effectiveness unless the thermal noise is reduced to a reasonable extent from its current level.

A number of materials (Silicon, Diamond, Zerodur, ULE: ultra-low expansion,..) have been considered for this purpose. Although, for the time being, fused silica remains to be the best choice for test masses, it seems that sapphire [37, 38] is the material to put one's hope upon. This is despite the fact that it has some demerits associated with it but the number of advantages it offers is more than that of other materials (except fused silica – for the moment). Sapphire has the following interesting (merits or demerits) thermal and mechanical properties (this list is not exhaustive and a number of other production and implementation issues need to be considered) : (i) high Q-factor quoted to be about 43 times greater than that of silica [39, 40], (ii) high thermal conductivity (about 20 times higher than silica) which means that, in a transmissive optics system, the problem of thermal lensing [41] can be avoided to a certain extent. (iii) high value of Young's modulus and low acoustic loss which ensures that the internal resonant modes have high frequency and low thermal noise. This is a major advantage which may lead to improvement of internal thermal noise by one order of magnitude. (iv) optical absorption rate is about 3.5ppm/cm, whereas the best values are reported for fused silica ($\approx$ 1 ppm/cm). However, these values need to be confirmed again. (v) a disadvantage is the problem of birefringence [42] which originates from two reasons: imperfect crystalline structure (i.e., sapphire is a uniaxial crystal – in an imperfect piece the nonconstancy in the direction of C-axis leads to changes in polarization of light beams and corresponding losses and degradation in contrast when beams from both arms meet together) and mechanical and thermal stress in it. However, recent measurements [37] claim that the birefringence induced losses in interferometer, under certain conditions, would be comparable whether sapphire or fused silica is used. (vi) it might be difficult to obtain sapphire in a big enough piece.

5.2. INCREASING THE POWER AND FINESSE

In the higher-frequency region of detection sensitivity, photon shot noise is the major limiting factor. Improvement of shot noise depends on the development of high power laser. The present level of laser power is about 15 Watt. Also, provided the thermal noise can be reduced, it would be better to increase the finesse of the arm cavities in order to make the valley of the noise curve somewhat deeper and (consequently) narrower at around 50-70Hz (the corner frequency) where most of the SNR for the detection of coalescing binaries comes up [43]. This is due to the fact that more cycles per unit frequency for the binary *chirp* waveform are received at lower frequencies. This will further enhance the energy stored in the arms.

The circulating power inside a power-recycled interferometer can be increased by either increasing the recycling gain or by increasing the finesse of the arm cavities. The former solution has the following problems associated with it: (i) Mirrors with less losses are needed since light will now be stored longer in cavities and thus with a lossy mirror, more power will be lost from the system due to increased number of bounces. (ii) For a high level of laser power circulating inside the interferometer, the heating effects in mirrors and beam-splitters which transmit the light can become serious. If light passes through the substrate of a component, heat is produced there as well as in the reflective coatings. Heat from both regions leads to thermal lensing from changes in refractive index and also to distortion in the optical surfaces [41, 44]. The wavefront distortions in the beams returning from each of the arms are usually different. This reduces the achievable contrast at the output of the interferometer and increases losses in recycling cavity.

A high finesse arm cavity may enhance the laser power inside the arms and can also solve the thermal distortion problem to certain extent by confining the high power beam within these cavities and thus not letting them pass through the sustrate of the input mirrors or the beam-splitter. However, again the associated problems turn out to be like these: (i) Since a high finesse cavity increases the storage time of light inside arms, that leads to lower bandwidth for the detector. The bandwidth requirement thus effectively puts an upper limit on the finesse of the arm cavities. (ii) The other kind of optical noise, the radiation pressure noise which is dominant at low frequencies increases. However, this will be a headache for designers only after the thermal noise and seismic noise go down to a considerable extent and the achievable laser power reaches an optimum level (where and when the photon shot noise becomes equal to the radiation pressure noise) at a low frequency of interest.

5.3. TAKING CARE OF THERMAL DISTORTION

(i) *Resonant sideband extraction (RSE) :* [4, 45] This a promising candidate for an advanced detector configuration which allows to incorporate high finesse cavities in arms and, at the same time, is capable of solving the thermal distortion problem and adapting the frequency response of the detector to a variety of requirements. In particular, the detector bandwidth can be made broader than that of the arm cavity. The configuration is identical to that for signal recycling: in this case also a mirror is placed at the output port. This mirror with the input mirrors of the arm cavities acts like a frequency-dependent compound mirror. The purpose of RSE is to reduce the storage time for the signal, in order to allow a long storage time for the carrier without sacrificing detector bandwidth. By appropriate tuning of this compound mirror, it can be made resonant with signal frequency of interest impinging on it from the arms, but highly reflective for carrier light. Thus the arm cavities can exhibit different storage times for carrier and signal: an extremely high finesse and therefore long carrier storage time and optimal signal storage times. The signal storage times are however shorter than that in the unmodified arm cavities and so the bandwidth of the detector increases from that of the arm cavity.

In signal recycling[30], on the other hand, this cavity is tuned to anti-resonance so as to obtain a lower transmittance than that of the corner mirror alone, resulting in an increased storage time (and consequently narrower bandwidth) for signal in arms. The major benefit of the resonant sideband arrangement is that the circulating power in the arm cavities can be made very large. This reduces the need for high power to pass through the substrates of the input mirrors and the beam-splitter, the components which will exhibit the effects of thermal distortion.

(ii) *Non-transmissive Optics:*

It was proposed by Drever [46] to use new types of interferometers 'which avoid transmission of light through test masses and other key components by using holographic diffraction patterns of profile or refractive index on reflecting surfaces, to couple light into and out of optical cavities and to split and combine beams'. If the problem of thermal lensing could be reduced this way, a major extra benefit would be the ability to use those materials (like sapphire) whose other physical properties are superior. However, overall losses in a diffracting system is quite high and this is a severe problem especially for achieving a high value of recycling factor.

5.4. ISOLATION FROM SEISMIC NOISE

The seismic noise spectrum falls very steeply ($\approx f^{-24}$) and is like a *wall* near a certain cut-off frequency, f_l. So, practically for $f > f_l$, the seismic noise contribution is negligibly small as compared to thermal noise and shot noise. It can be checked numerically [32] that, unless the thermal noise is improved from their presently estimated level, the variation in the value of f_l within the range (10Hz – 80Hz) does not change the SNR for coalescing binaries much. However, a less value of Seismic noise may come to be useful even in initial interferometers for detecting the incoherent signal generated by a large ensemble of neutron stars in the galaxy as proposed by Giazotto et al [47] or for being ready to detect unknown sources.

The VIRGO seismic isolation system [48] has been designed to take the lower limit down to a single digit value in Hertz. It would perhaps need more than an extra effort in order to develop and implement a system that might push this value further down where the changing gravity gradients are the dominant noise source. Experiments on seismic isolation systems based on active components using feedback elements are going on at JILA, Colorado. Most experts of the suspension system always comment that, if possible, they would love to get rid of the wire of suspension. An interesting series of experiments on magnetically levitated mirrors at room temperature have been started by Ron Drever [49] at Caltech. Some people are also thinking of electrostatic suspension systems. However, these are just the first efforts and need to solve a number of scientific and technical problems before they are implemented in GW detectors.

5.5. KNOWLEDGE OF SPURIOUS SOURCES OF NOISE

Many of the sources of noise [50] may suddenly appear without any prior notice and may almost paralyse the operation of an interferometer. These might originate from problems ranging from some *a priori* unknown design defects to matters as mundane as a ground wire touching some part of the set-up. A very difficult task is to zero-in on such sources of noise especially when the detector has a length of 4+4 Km. Once that is done, suitable measures may be taken to avoid those in the next run of the interferometer. Seiji Kawamura, out of his experience with such spurious noise-sources appearing in Caltech 40m interferometer, enumerated those in an excellent talk – *Seiji Kawamura's Top 10 sources of noises* [51].

5.6. MORE DEMANDING CONTROL TECHNIQUES

As the design and performance of the detectors will improve over the future years, it would necessitate development of more demanding control techniques. For example, even in initial detectors, thermal lensing would be a problem; thermal distortion and other inhomogeneities in input mirrors affect the sidebands if those are kept as resonant only in recycling cavities but out-of-resonance in arm cavities as mentioned in section 3. This leads to serious problems for both control and signal extraction. A control scheme in which both carrier and sidebands are doubly-resonant in the interferometer have been investigated by McClelland and others at LIGO, Caltech [52] in order to avoid this and other problems. Also, if either dual recycling and resonant sideband extraction is incorporated in FP-detectors, one will need a number of improvements in the length and alignment control scheme and in that case perhaps it would be difficult to avoid introduction of subcarrier and its sidebands in the design.

6. Epilogue

History tells us that whenever a highly sensitive instrument was built, it, in most of the cases, led to rewarding pay-offs and, in many cases, in an unexpected way. Discoveries of microwave background radiation, pulsar, binary pulsar, gamma-ray bursters, strong x-ray sources are just a few examples of this fact. At this point of time, although we are aware of some potential sources of gravitational radiation and our instruments are being built up with specific objectives to keep the sensitivity above a certain level in this targetted frequency range, it seems to be of no use to spend our time in questions like *how many events would be detected if initial interferometers achieve a sensitivity level of so much* or *would it be able to detect anything at all.* The main goal at this point of time must be to try hard to design, build and operate interferometers with its highest achievable level of sensitivity. We can reasonably hope that the sky remains rich enough to offer something known or unknown, everytime a deserving detector is ready to receive it.

References

1. Winkler, W. (1991) in *The detection of gravitational waves*, Blair, D.G., ed., Cambridge University Press, Cambridge.
2. Drever, R.W.P. (1991) in *The detection of gravitational waves* Blair, D.G., ed., Cambridge University Press, Cambridge.
3. Vinet, J.-Y., Meers, B.J., Man, C.N., Brillet, A. (1988) *Phys. Rev.*, **D38**, 433.
4. Mizuno, J. (1995) *Comparison of optical configurations for laser-interferometric gravitational-wave detectors*, Ph.D dissertation, Max-Planck-Institute für Quantenoptik, Garching.
5. Drever, R.W.P., Hough, J., Munley, A.J., Lee, S.-A., Spero, R., Whitcomb, S.E., Pugh, J., Newton, G., Meers, B.J., Brooks III, E., Gürsel, Y. (1983) in *Quantum Optics, Experimental Gravity and Measurement Theory*, Meystre, S., Scully, M.O., eds., Plenum, New York.
6. Sun, K.-X, Fejer, M.M., Gustafson, E., Byer, R.L. (1996) *Phys. Rev. Lett.*, **76**, 3053.
7. Mizuno, J., Rüdiger, A., Schilling, R., Winkler, W., Danzmann, K. (1997), *Optics Communication*, **138**, 383.
8. Caves, C.M. (1981) *Phys. Review*, **D23**, 1693.
9. Drever, R.W.P. (1983) in *Gravitational radiation*, Deruelle, N. and Piran, T., eds., North-Holland, Amsterdam, 321.
10. Fritschel, P., Shoemaker, D., Weiss, R. (1992), *Appl. Opt.*, **31**, 1412.
11. Saulson, P.R. (1994) *Fundamentals of interferometric gravitational wave detectors*, World Scientific, Singapore.
12. Bondu, F., Vinet, J.-Y. (1995), *Phys. Lett.*, **A198**, 74.
13. Gillespie, A. (1995) *Thermal noise in the initial LIGO interferometers*, Ph.D dissertation, Caltech.
14. Gillespie, A., Raab, F. (1995) *Phys. Rev.*, **D52**, 577.
15. Gillespie, A., Raab, F. (1993) *Phys. Lett.*, **A178** 357.
16. Logan, J.E., Hough, J., Robertson, N.A. (1993) *Phy. Lett.*, **A183**, 145.
17. Bhawal, B. (1998), *Journal of Optical Society of America*, **A15**, 120.
18. Weiss, R. (1972) *Electromagnetically coupled broadband gravitational antenna*, Q. Prog. Rep. Res. Lab. Electron. MIT, **105**, 54.
19. Man, C.N., Shoemaker, D., Pham-Tu, M., Dewey, D. (1990) *Phys. Lett.*, **A148**, 8.
20. Regehr, M.W. (1995) *Signal extraction and control for an interferometric gravitational wave detector*, Ph.D dissertation, Caltech.
21. Regehr, M.W., Raab, F.J., Whitcomb, S.E., (1995), *Optics Lett.*, **20** 1507.
22. Caron, B., Dominjon, A., Flaminio, R., Marion, F., Massonnet, L., Morand, R., Mours, B., Verkindt, D., Yvert, M. (1995) *Nucl. Instr. Methods in Phys. Research*, **A360**, 375.
23. Flaminio, R., Heitmann, H. (1996) *Phys. Lett.*, **A214**, 112.
24. Giaime, J. A. (1995) *Studies of laser interferometer design and a vibration isolation system for interferometric gravitational wave detectors*, Ph.D dissertation, MIT.

25. Mavalvala, N. (1997) *Alignment issues in laser interferometric gravitational-wave detectors*, Ph.D dissertation, MIT.
26. Ando, M., Kawabe, K., Tsubono, K. (1997), *Phys. Lett.*, **A237**, 13.
27. Camp, J., Sievers, L., Bork, R., Heefner, J. (1995) *Optics Lett.*, **20**, 2463.
28. Fritschel, P.K., Gonzalez, G., Mavalvala, N., Shoemaker, D., Sigg, D., Zucker, M. (1997) *Alignment of an interferometric gravitational wave detector*, LIGO internal report no. LIGO-P970017-05-D, *submitted for publication.*
29. Hefetz, Y., Mavalvala, N., Sigg, D., *J. Opt. Soc. Am.*, **B107**, 1597.
30. Meers, B.J., *Phys. Rev.*, **D38**, 2317.
31. Gürsel, Y., Tinto, M., *Phys. Rev.*, **D40**, 3884.
32. Bhawal, B. and Dhurandhar, S.V. (1995) *Coincidence detection of broadband signals by a network of the planned interferometric detectors*, IUCAA-39/95 (*unpublished*), Inter University Centre for Astronomy and Astrophysics, Pune.
33. Traeger, S., Willke, B., Danzmann, K. (1997) *Phys. Lett.*, **A225**, 39.
34. Braginsky, V.B., Mitrofanov, V.P., Okhrimenko, O.A. (1993) *Phys. Lett.*, **A175**, 82.
35. Braginsky, V.B., Mitrofanov, V.P., Tokmakov, K.V. (1996) *Phys. Lett.*, **A218**, 164.
36. Rowan, S., Hutchins, R., McLaren, A., Robertson, N.A., Twyford, S.M., Hough, J. (1997) *Phys. Lett.*, **A227**, 153.
37. Benabid, F., Notcutt, M., Ju, L., Blair, D.G. (1998), *Phys. Lett.*, **A237**, 337.
38. Ju, L., Notcutt, M., Blair, D., Bondu, F., Zhao, C.N. (1996) *Phys. Lett.*, **A218**, 197.
39. Braginsky, V.B., Mitrofanov, V.P., Panov, O.I, (1985), *Systems with small dissipation*, Univ. of Chicago Press.
40. Braginsky, V.B., Mitrofanov, V.P., Okhrimenko, O.A. (1992) *JETP Lett.*, **55**, 432.
41. Winkler, W., Danzmann, K., Rüdiger, A., Schilling, R. (1991) *Phys. Rev.*, **A44**, 7022.
42. Winkler, W., Rüdiger, A., Schilling, R., Strain, K.A., Danzmann, K. (1994) *Optics Communications*, **112**, 245.
43. Cutler, C. and Flanagan, E., Phys. Rev., **D49**, 2658 (1994).
44. Hello, P. (1996), *Couplings in interferometric gravitational wave detectors*, Mémoire d'habilitation á diriger des recherches, Université de Paris-sud, Orsay, France.
45. Heinzel, G., Mizuno, J., Schilling, R., Winkler, W., Rüdiger, A., Danzmann, K. (1996) *Phys. Lett.*, **A217**, 305.
46. Drever, R.W.P. (1995) in *Proc. of the seventh Marcel Grossman meeting on general relativity* , Keiser, M., Jantzen, R.T., eds., World Scientific, Singapore.
47. Giazotto, A., Bonazzola, S., Gourgoulhon, E. (1997) *Phys. Rev.*, **D55** 2014.
48. Giazotto, A. *for VIRGO collaboration* (1998) in *Gravitational wave detection*, Tsubono, K., Fujimoto, M.-K., Kuroda, K., eds., Universal Academic Press, Tokyo.
49. Drever, R.W.P. (1997) in *Gravitational waves: sources and detectors* Ciufolini, I. and Fidecaro, F., eds., World Scientific, Singapore.
50. Kawamura, S., Abramovici, A., Zucker, M.E., *Rev. of Sc. Instr.* (1997) **68**, 223.
51. Kawamura, S. (1997), present address: TAMA project, Mitaka, Japan.
52. McClelland, D.E., Whitcomb, S.E., Kells, B., Camp, J. (1997) *Doubly resonant sideband control for LIGO*, LIGO internal document LIGO-T970177-00-D, LIGO, Caltech.

WEBSITES:

- AIGO: http://www.anu.edu.au/Physics/ACIGA/
- GEO: http://www.geo600.uni-hannover.de/
- LIGO: http://www.ligo.caltech.edu/
- LISA: http://www.lisa.uni-hannover.de/
- TAMA: http://tamago.mtk.nao.ac.jp/
- VIRGO: http://www.pg.infn.it/virgo/

R. Drever: ... In conclusion, I say, again, that all the different techniques should be pursued. There is a place for them all.
S. Boughn: I agree that we should pursue all possible ways of detecting gravitational radiation, but after having been at it for as long as I have, I have been impressed by the incredible difficulties of continuing in any of these many directions. To take the experimental gravitational radiation program as it now stands and continue in all directions toward workable detectors is going to be very expensive in time, labor and money. Wouldn't it therefore be better to concentrate on fewer experiments?
R. Drever: I think I am a little more optimistic, perhaps, than most experimentalists in the field. I rather think, for example, that these interferometers will begin to work in the next year or two at an interesting sensitivity and will develop pretty rapidly, perhaps a little faster than the time scale you were suggesting. None of the problems look overwhelming [to me], so I am optimistic that the progress might not be that slow.
D.H. Douglass: Steve [Boughn], I think you're wrong. Our field is not expanding in all different directions. The number of groups has remained constant in the past 3 or 4 years, and some groups are actually disappearing .... I think you will be more optimistic after you get your first cooldown. [*laughter*]
L. Smarr: ... In the past, there has been some disagreement on the best approach for detecting gravitational radiation. But now, I think, we have all stepped back and taken a long-range view of our field. This was one major goal of this workshop. We have looked down the road a long way, and we can see how we're going to get there. Previously, there may have been an illusion of chaos, but now everything looks a lot more orderly. I think there is good reason to be optimistic.
R.L. Forward: At least we are now able to draw the antenna sensitivity curves and the source [strength] curves on the same graph. Surely [*laughter and applause*] this means we have come a long way.
Unidentified: That says it all.

— BATTELLE SEATTLE WORKSHOP (1978)

28. DETECTION OF GRAVITATIONAL WAVES FROM INSPIRALING COMPACT BINARIES

S. V. DHURANDHAR
Inter-University Centre for Astronomy and Astrophysics,
Post Bag 4, Ganeshkhind, Pune -7, India.

1. Introduction

The 1993 Nobel prize for physics was awarded to Hulse and Taylor for the proof of existence of gravitational waves. This was based on the observations of the binary pulsar PSR 1913+16 which showed a decay in its orbit. This decay is correctly predicted by the general theory of relativity and provides indirect evidence for the existence of gravitational waves. Direct evidence however is lacking and may not be provided until advanced ultra high sensitivity resonant bar detectors and large scale laser interferometric detectors [1, 2] are completed. This network of detectors would not only detect gravitational waves but perform as directionally sensitive gravitational wave observatories.

The direct detection of gravitational waves is perhaps the most challenging problem in experimental physics today. Success in this field will benefit both astronomy, where new information is expected to emerge and fundamental physics, where various theories of gravitation could be tested. Starting from the pioneering experiments of Weber using a bar detector, there has been a lot of effort in building detectors of higher sensitivity. In the past few years, several groups around the globe have begun constructing long baseline laser interferometric gravitational wave detectors, prototypes of which already exist in Germany, Great Britain and USA.

An important source of gravitational waves is a system of coalescing compact binaries [3]. Due to their extragalactic origin the amplitude of these signals is, in general, expected to be too low for them to be seen directly in the time series. However, these sources are relatively clean systems, in the sense that, during the inspiral, tidal effects can be ignored and also effects such as quadrupole-quadrupole interactions. Therefore the bi-

B. R. Iyer and B. Bhawal (eds.), Black Holes, Gravitational Radiation and the Universe, 481–492.

naries can be modelled as a system of two point masses whose orbit evolves quasi-statically as the system loses energy by the emission of gravitational waves. Thus the gravitational waveform emitted during the inspiral can be predicted with a fair degree of reliability and is called a *chirp* for obvious reasons.

Since the waveform of this signal can be computed with adequate accuracy, pattern matching techniques, like matched filtering, can be used to considerably enhance the signal to noise ratio [3, 4]. Therefore it should be possible to detect such events upto a large distance and hence observe a significant event rate. A reasonable event rate may be about 3 events per year out to 200 Mpc, the earliest such estimate being obtained in estimates based on formation rates of neutron star - neutron star binaries and from considerations of gamma-ray burst events.

In matched filtering, the data is correlated with a *template* waveform that "matches" the signal waveform in some optimal sense. If the maximum of the correlation exceeds a preset threshold, a detection is announced. However, even if a signal is present in the data, the accompanying noise can mask its presence by preventing such a crossing of the threshold in some cases. Thus a signal can be detected with only a certain probability called its *detection probability*. Also, there will be a non-zero probability, called the *false alarm probability*, of a false detection due to noise alone.

The simplest application of matched filtering is the detection of signals with an unknown arrival time but an otherwise known waveform. However, for signals from coalescing binaries, there are several other parameters apart from the time of arrival, like the masses and spins of the bodies among others. For such signals, it will be required to correlate the data with several templates (each template corresponding to a separate point in the space of these other parameters) and compare the maximum over the complete set of correlations with a threshold.

Since this bank of templates will be matched to only a discrete set of signals from among the continuum of possible signals, it is natural that all signals will not get detected with the same probability. However, given the available computational power, it is possible to judiciously choose the set of templates so that all signals having a given minimum amplitude are detected with a given minimum detection probability. A formalism to choose the optimal set of templates using a criterion similar to the above was presented in Refs.[6, 7] and has been termed as the S-D formalism. Here we present an improved version of the S-D formalism [8, 9] which (i) explicitly takes into account the detection probability and (ii) uses more accurate distributions for the statistic for calculating the false alarm and detection probabilities.

The gravitational waveform from a coalescing binary can be calculated as a perturbative series in v/c. The lowest order waveform is called the Newtonian waveform and the higher order ones are termed as post-Newtonian (PN). For the purpose of detection, one can use templates corresponding to various orders. The S-D formalism was originally proposed for Newtonian templates but its extension to post-Newtonian templates (having a larger number of parameters) shows [10, 11] that the use of a single template bank would be computationally expensive. An alternative strategy could be to use several template banks in a hierarchy. The first level in such a strategy would consist of a template bank with a coarse spacing in parameter space but a lower threshold. In the next level a more finely spaced set of templates and a higher threshold would be used but only around those templates, in the previous level, which produce a crossing of the lower threshold. An appropriate choice of thresholds and spacings would reduce the number of templates employed compared to that required by a one step search and hence, save on the computational cost. The strategy is known as a *hierarchical search* [8, 9] and here only the outline and the results are presented.

Here we restrict ourselves to the family of Newtonian waveforms and the noise power spectral density expected for the initial LIGO. More accurate waveforms have been avoided to prevent the extra complications that arise due to a larger number of parameters. They would obscure an important objective of this article which is to highlight the methods. However, this formalism can be easily extended to accommodate more accurate waveforms which have larger number of parameters.

2. Maximum Likelihood Detection of Newtonian Signals

We denote the parameters characterizing a family of waveforms by Θ and the family itself by $s(t;\,\Theta)$. We denote a segment of the output of a detector as $x(t)$, where $0 \leq t \leq T$. If the signals have finite durations then T should be larger than the longest signal. A particular realization of the noise contaminating $x(t)$ will be denoted as $n(t)$. For a given $x(t)$, the problem of detection consists of discriminating between the following cases,

$$x(t) = \begin{cases} n(t) & \text{when there is no signal,} \\ n(t) + s(t;\,\Theta) & \text{when there is a signal with parameters } \Theta \ . \end{cases}$$

The range of every parameter will, in general, be finite. In the absence of any prior information, it must be assumed that all the parameter values, within their respective ranges, are equally likely to occur. In such a case, the method of *maximum likelihood detection* [12] can be used. When the noise is a stationary Gaussian random process, maximum likelihood detection reduces to (a) the computation of the *test statistic*, Λ, defined below and (b) its comparison with a threshold η. If $\Lambda > \eta$ then a detection is announced or else not. The test statistic, Λ, is obtained as,

$$\Lambda = \max_{\Theta} \left(\langle\, x(t), s(t;\, \Theta) \,\rangle - \frac{1}{2} \langle\, s(t;\, \Theta), s(t;\, \Theta) \,\rangle \right) , \tag{1}$$

where the maximization is performed over all the values of the parameters. The angular brackets enclosing two functions denote an inner product which can be written in the Fourier domain as ('~' denotes a Fourier transform),

$$\langle\, x(t), g(t) \,\rangle = \int_0^\infty df \, \frac{1}{S_n(f)} \, (\tilde{x}(f)\, \tilde{g}^*(f) + \tilde{x}^*(f)\, \tilde{g}(f)) \, , \tag{2}$$

where, the Hermitian property of the Fourier transform of a real function has been used to restrict the domain of integration to positive frequencies. $S_n(f)$ is the power spectral density of the noise which is defined by,

$$\mathrm{E}[\, \tilde{n}^*(f)\, \tilde{n}(f') \,] = S_n(f)\, \delta(f - f') \, , \tag{3}$$

where, $\mathrm{E}[z]$ stands for the ensemble average of a random variable z.

The noise in ground based laser interferometric detectors will have, in general, both a Gaussian and a non-Gaussian component. However, it should be possible to remove most of the non-Gaussian component by using environmental monitors and looking for coincidence among detectors located at widely separated sites. It is, therefore, usually assumed that the detector noise will be a Gaussian random process. Over a time scale of hours, it can also be assumed to be stationary. Thus, the method of maximum likelihood detection can be used in this case.

The power spectral density of the Gaussian noise component rises very steeply towards the low frequency end due to seismic effects. At the high frequency end, it is dominated by photon shot noise which leads to a rise towards higher frequencies. Thus the data will have to be bandpassed with a low frequency seismic cutoff, f_a, and a high frequency cutoff, f_c. We use the power spectral density expected for the initial LIGO as given in Ref.[13]. Accordingly, we choose $f_a = 40$ Hz and $f_c = 1$ kHz.

Due to the presence of noise, Λ becomes a random variable. Therefore, in the presence of any signal one can only associate a probability with the event $\Lambda > \eta$. This is known as the detection probability of a signal with parameters Θ, denoted by $Q_d(\eta;\, \Theta)$. The event $\Lambda > \eta$ can also occur in the absence of any signal due to noise alone. This then leads to an error in detection. The probability of such an event is called the false alarm probability denoted by $Q_0(\eta)$. If the probability density function of Λ, the test statistic, be $p_0(\Lambda)$ in the absence of a signal and $p_1(\Lambda;\, \Theta)$ in the presence of a signal with parameters Θ, then,

$$Q_0(\eta) \;=\; \int_\eta^\infty p_0(\Lambda)\, d\Lambda \, , \tag{4}$$

$$Q_d(\eta;\,\Theta) \;=\; \int_{\eta}^{\infty} p_1(\Lambda;\,\Theta)\,d\Lambda\,. \tag{5}$$

3. The Newtonian Waveform

The lowest order approximation to the waveform of the gravitational wave emitted by a coalescing compact binary is provided by the quadrupole formalism [14]. The response of an interferometric detector to such a waveform can be written as [7],

$$h(t;\,\mathcal{A}, t_a, \xi, \Phi) = \mathcal{A}\,a(t-t_a,\,\xi)\cos(\phi(t-t_a,\,\xi)+\Phi)\,. \tag{6}$$

We choose this waveform as the signal that is to be detected. The parameter $\mathcal{A}$ takes into account the distance to the binary as well as various geometrical factors connected with the orientation of the orbital plane of the binary relative to the plane of the sky and the orientation of the detector antenna pattern [3, 15, 16]. Due to the Earth's rotational and orbital motions, the orientation of the detector antenna pattern will change with time, making $\mathcal{A}$ time dependent. For observations lasting a few minutes, however, it is effectively a constant. The other, time dependent part of the amplitude is,

$$a(t,\,\xi) = \left[1-\frac{t}{\xi}\right]^{-1/4}\,. \tag{7}$$

The phase of the waveform, $\phi(t,\xi)$ can be expressed as

$$\phi(t,\,\xi) = 2\pi\int_0^t f(t',\xi)dt'\,. \tag{8}$$

The integrand, $f(t,\xi)$, is the instantaneous frequency of the signal which is given by,

$$f(t,\,\xi) = f_a a(t,\,\xi)^{3/2}\,. \tag{9}$$

Thus, the waveform is a *chirp* whose amplitude and instantaneous frequency increase with time.

The rate at which the instantaneous frequency increases is governed by the parameter ξ, called the *chirp time*,

$$\xi = 34.54\left[\frac{\mathcal{M}}{M_\odot}\right]^{-5/3}\left[\frac{f_a}{40\mathrm{Hz}}\right]^{-8/3}\ \mathrm{sec}\,, \tag{10}$$

where $\mathcal{M}$, the *chirp mass*, is the following combination of the reduced mass μ and the total mass M of the binary, $\mathcal{M} = (\mu^3M^2)^{1/5}$. Due to the seismic cutoff, the amplitude of the signal becomes negligible when its instantaneous frequency lies below f_a. The time at which $f(t,\xi) = f_a$ is denoted by t_a which can be taken as the *time of arrival* of the signal. The phase of the signal at t_a is denoted by Φ.

The high frequency cutoff, f_c, will also force the amplitude to a negligible value for instantaneous frequencies beyond f_c. Besides this the nature of the waveform will change when the compact bodies plunge towards each other once the last stable orbit is reached. The waveform has, therefore, an effectively finite duration given, to a very good approximation by ξ. For a binary consisting of Neutron stars with $m_1 = m_2 = 1.4M_\odot$, the value of ξ is 24.8 sec.

We can also write the waveform in Eq. (6) as

$$h(t;\, \mathcal{A}, t_a, \xi, \Phi) = \mathcal{A}\, h_0(t - t_a;\, \xi) \cos(\Phi) + \mathcal{A} h_{\frac{\pi}{2}}(t - t_a;\, \xi) \sin(\Phi)\,, \qquad (11)$$

where,

$$\begin{aligned} h_0(t;\, \xi) &= a(t,\, \xi) \cos(\phi(t,\, \xi))\,, & (12)\\ h_{\frac{\pi}{2}}(t;\, \xi) &= a(t,\, \xi) \cos(\phi(t,\, \xi) + \pi/2)\,. & (13) \end{aligned}$$

These "quadrature" components have the following properties,

$$\begin{aligned} \langle\, h_0(t;\, \xi), h_0(t;\, \xi)\,\rangle &= \langle\, h_{\frac{\pi}{2}}(t;\, \xi), h_{\frac{\pi}{2}}(t;\, \xi)\,\rangle\,, & (14)\\ \langle\, h_0(t;\, \xi), h_{\frac{\pi}{2}}(t;\, \xi)\,\rangle &\approx 0. & (15) \end{aligned}$$

4. The S-D Formalism

4.1. THE TEST STATISTIC

We now apply the strategy of maximum likelihood detection to the noise and family of signals described above. The test statistic in this case can be written as,

$$\Lambda = \max_{\Theta} \left(\langle\, x(t), h(t;\, \Theta)\,\rangle - \frac{1}{2} \langle\, h(t;\, \Theta), h(t;\, \Theta)\,\rangle \right)\,, \qquad (16)$$

where $\Theta = (\,\mathcal{A}, t_a, \xi, \Phi\,)$. The maximization over $\mathcal{A}$ and Φ can be carried out, in a straightforward manner [17], using Eqs. (11), (14) and (15). This yields an equivalent test statistic or its surrogate the correlation C which is nothing but the output of the matched filter,

$$C = \max_{(t_a,\xi)} \sqrt{C_0^2(t_a,\, \xi) + C_{\frac{\pi}{2}}^2(t_a,\, \xi)}\,, \qquad (17)$$

where,

$$\begin{aligned} C_0(t_a,\xi) &= \mathcal{N}_h \int_{f_a}^{f_c} \frac{df}{S_n(f)}\, \tilde{x}(f)\, \tilde{h}_0^*(f;\,\xi)\, e^{2\pi i f t_a} + \text{c.c.}\,, \\ &= \langle\, x(t),\, \mathcal{N}_h\, h_0(t-t_a;\,\xi)\,\rangle\,, \end{aligned} \tag{18}$$

$$\begin{aligned} C_{\frac{\pi}{2}}(t_a,\xi) &= \mathcal{N}_h \int_{f_a}^{f_c} \frac{df}{S_n(f)}\, \tilde{x}(f)\, \tilde{h}_{\frac{\pi}{2}}^*(f;\,\xi)\, e^{2\pi i f t_a} + \text{c.c.}\,, \\ &= \langle\, x(t),\, \mathcal{N}_h\, h_{\frac{\pi}{2}}(t-t_a;\,\xi)\,\rangle\,. \end{aligned} \tag{19}$$

The presence of noise turns $C_0(t_a,\xi)$ and $C_{\frac{\pi}{2}}(t_a,\xi)$ into random variables. By demanding that their variances be unity, one may fix the normalization constant $\mathcal{N}_h$.

A comparison of the integrals in Eq.(18) and (19) with the definition of a linear correlation shows that for a fixed ξ, C_0 (or $C_{\frac{\pi}{2}}$) can be computed as functions of t_a, by simply obtaining a linear correlation between $x(t)$ and the template $\bar{h}_0(t;\,\xi)$ (or $\bar{h}_{\pi/2}(t;\,\xi)$). The role of the lag, τ, in a correlation is played here by t_a. Since a correlation can be efficiently computed using a Fast Fourier Transform(FFT), maximization over the parameters $\mathcal{A}$, Φ and t_a is thus taken care off in a straightforward manner. Maximization over the remaining parameter ξ is performed over a discrete set $\{\,\xi_j\,\}$.

Thus, the overall form of the detection strategy which results, given $\{\,\xi_j\,\}$ and a threshold η, is the following. For each $\xi_m \in \{\,\xi_j\,\}$, the detector output is separately correlated with two templates $\bar{h}_0(t;\,\xi_m)$ and $\bar{h}_{\frac{\pi}{2}}(t;\,\xi_m)$ to obtain $C_0(\tau,\,\xi_m)$ and $C_{\frac{\pi}{2}}(\tau,\,\xi_m)$ respectively. Then, the quantity $X(\tau;\,\xi_m) = \sqrt{C_0^2(\tau,\xi_m) + C_{\frac{\pi}{2}}^2(\tau,\xi_m)}$ is computed. We call $X(\tau;\,\xi_m)$ the *rectified* output of a template with chirp time ξ_m. We denote the maximum, over τ, of a rectified output $X(\tau;\,\xi_m)$ by λ_m. These operations lead, therefore, to the construction of a set $\{\lambda_j\}$. The test statistic, C, is then obtained as,

$$C = \max_j\{\lambda_j\}\,. \tag{20}$$

Finally, C would be compared with η.

We denote by $d(\tau,\xi_m;\,\Theta)$, the quantity,

$$d(\tau,\xi_m;\,\Theta) = \mathcal{N}_h\left[\langle\, h(t;\,\Theta), h_0(t-\tau;\,\xi_m)\,\rangle^2 + \langle\, h(t;\,\Theta), h_{\frac{\pi}{2}}(t-\tau;\,\xi_m)\,\rangle^2\right]^{1/2}. \tag{21}$$

This is the rectified output of a template when $x(t) = h(t;\,\Theta)$ and noise is absent. We call such a rectified output as the *processed form* of the signal produced by the template with chirp time ξ_m. The processed form of a signal has the property that, a change in the signal's time of arrival simply translates it along the rectified output. It is easy to show, using

Eq. (15), that the processed form is independent of the signal parameter Φ. Henceforth, we suppress Φ whenever the signal parameters in $d(\tau, \xi_m; \Theta)$ are required to be expressed explicitly.

The marginal probability density function of a rectified output sample is a Rician $Ri(z)$ when a signal is present and a Rayleigh $R(z)$ in the absence of a signal,

$$Ri(z) = z \exp\left[-\frac{1}{2}(z^2 + d^2)\right] I_0(zd) , \tag{22}$$

$$R(z) = z \exp\left[-\frac{z^2}{2}\right] . \tag{23}$$

4.2. THE AMBIGUITY FUNCTION

The maximum value that any processed form of a signal can have is called the *strength*, S, of the signal [7]. For a signal with amplitude $\mathcal{A}$,

$$S = \frac{\mathcal{A}}{\mathcal{N}_h} . \tag{24}$$

Henceforth, we use S instead of $\mathcal{A}$ to parameterize a signal.

When the template chirp time is not the same as that of the signal, the maximum of the processed form will be reduced. We call the maximum of the processed form in such a case as the *observed strength*, $S_{\rm obs}$, of the signal in that template.

Given a signal with $S = 1$ and a chirp time ξ_s, we define the *intrinsic ambiguity* function $\mathcal{H}(\xi_t, \xi_s)$ as

$$\mathcal{H}(\xi_t, \xi_s) = \max_{\tau} d(\tau, \xi_t; \Theta) , \tag{25}$$

where ξ_t is a template chirp time. In terms of the intrinsic ambiguity function, the observed strength can be obtained as,

$$S_{\rm obs} = S\,\mathcal{H}(\xi_t, \xi_s) . \tag{26}$$

It is easy to show, using the stationary phase approximation, that $\mathcal{H}(\xi_t, \xi_s)$ depends only on $|\Delta\xi|$ and not on ξ_t and ξ_s separately. This behavior is replicated by the exact intrinsic ambiguity function also. However, for binaries with massive components the plunge cutoff frequency is small. This effectively reduces the signal bandwidth and leads to a wider intrinsic ambiguity. For simplicity, we neglect this effect and take the intrinsic ambiguity to be independent of its location. The final results are not affected much by this approximation.

The practical implementation of the above formalism is as follows:

The detector output will be sampled with a sufficiently small sampling interval, Δ, to give the time series ($x_i = x(i\Delta)$; $i = 0, \ldots, N-1$). Since the upper frequency cutoff of the bandpass filter, in section 3, was chosen to be $f_c = 1$ kHz, the Nyquist sampling rate would be 2 kHz. We take the sampling frequency to be 2048 Hz, the nearest power of two, which implies $\Delta = 1.0/2048.0$ sec. The time series should of course be longer than the duration of the longest template or equivalently, the largest chirp time, ξ_{max}. Such time series will be required for the templates, $\bar{h}_0$ and $\bar{h}_{\frac{\pi}{2}}$, also. In the time series of a template with chirp time ξ_m, samples for $i > \xi_m/\Delta$ will be zero since the template has a finite duration of ξ_m. Therefore when a correlation between the template and the detector output is taken using an FFT, only the first $N - \xi_m/\Delta$ samples will be the result of a linear correlation [4]. It is desirable to have equal lengths of correlations for every template. Hence, only the first $N_p = N - \xi_{\text{max}}/\Delta$ samples will be retained in each correlation and the rest discarded. We call N_p, the *padding* for the template bank. A useful figure for N_p is $\sim 5 \times 10^5$ corresponding to $N = 256 \times 2048$ and $\xi_{\text{max}} = 32.0$ sec. A time series corresponding to $X(\tau;\, \xi_m)$ will then be obtained whose i^{th} sample we denote as $X_i(\xi_m)$ ($i = 0, \ldots, N_p$ now),

$$X_i(\xi_m) = \sqrt{C_0^2(i\Delta,\, \xi_m) + C_{\frac{\pi}{2}}^2(i\Delta,\, \xi_m)}\ . \tag{27}$$

As before, λ_m will then be found as,

$$\lambda_m = \max_i \{X_i(\xi_m)\}\ . \tag{28}$$

C is then obtained by taking the maximum over λ_m.

4.3. THE BANK OF TEMPLATES

The bank of templates should be chosen in such a way that **(i)** every waveform, having a strength S greater than a given minimum strength S_{min}, should have a detection probability greater than a given minimum detection probability $Q_{d,\text{min}}$, and **(ii)** The false alarm should stay below a specified level, $Q_{0,\text{max}}$.

It can be shown that for our purpose it is sufficient to use a two template formula for detection probability,

$$Q_d(\eta;\, S, \xi) = 1 - \int_0^\eta Ri(x, S_m)dx \int_0^\eta Ri(x, S_{m+1})dx\ . \tag{29}$$

The detection probability can be expected to be the smallest for signals having a strength S_{min} and chirp time $\xi = (\xi_m + \xi_{m+1})/2$ for $\xi_m \in \{\, \xi_j \,\}$.

To satisfy criterion (i) above, all that needs to be done, given a threshold η, is to ensure that all such minimum detection probability signals have, $Q_d(\eta;\, S_{\min}, (\xi_m + \xi_{m+1})/2) = Q_{d,\min}$. It follows from the location independence of $\mathcal{H}$ that only $\xi_{m+1} - \xi_m$ will enter into the calculation of the detection probability and not ξ_m and ξ_{m+1} separately. We call this quantity the *spacing* of the templates and denote it by δ. The whole template bank can now be constructed, using δ, as $\xi_k = \xi_{\min} + k\delta$ ($k = 0, 1, \ldots$) till $\xi_{\max}$ is reached.

The spacing of the templates depends on the value of $S_{\min}$. For instance, if $\eta = 7.9$ (a typical value), and $Q_{d,\min} = 0.95$, the observed strength required for a signal lying in the middle of two templates is $S_{\text{obs}} \simeq 8.6$. For a signal having $S = 8.8$, this would mean that the value of the intrinsic ambiguity function be $\approx 8.6/8.8 = 0.98$. From the profile of the ambiguity function one computes that the template chirp times would be placed $\simeq$ 0.030 sec apart. For a larger strength $S_{\min}$, the spacings obtained would be larger.

5. The Two Step Hierarchical Search Strategy

A one-step search would still be a computationally expensive proposition especially when post-Newtonian signals are considered. For the PN waveform there are two independent intrinsic parameters instead of just ξ of the Newtonian case, which are basically the masses of the two stars. Thus the dimension of the parameter space is increased by one, leading to an array of filters in place of each filter of the Newtonian case. A post$^{1.5}$-Newtonian template family, having spin parameters $\beta = \sigma = 0$, is adequate for the detection of signals up to (and possibly beyond) post^2-Newtonian order, even for maximally spinning systems [11], (except when the angle between the total angular momentum and orbital angular momentum is large which leads to a wobble). Estimates [10, 11] of the computational power that is required for an on-line one-step search using PN$^{1.5}$ templates turn out to be $\sim$ 200 Gflops (Giga flops) or higher. Therefore, it is desirable to have computationally less expensive detection strategies without, however, compromising too much on the performance afforded by matched filtering.

The basic idea behind a two step search is the use of two banks of templates B_1 and B_2, where the first stage bank B_1 is coarser than the second stage bank B_2. The maximum over a rectified output is first computed for each chirp time in a bank $B_1 = \{\, \xi_j \,\}$, which has a template spacing of $\delta^{(1)}$ and the number of templates is $n_t^{(1)} = (\xi_{\max} - \xi_{\min})/\delta^{(1)}$. If for some template in B_1 having a chirp time ξ_m, it so happens that the statistic crosses $\eta^{(1)}$ set at a lower level, then we call this event a *crossing* of $\eta^{(1)}$ produced by the chirp time ξ_m. Given a crossing, the next step involves using the template bank B_2 with a spacing $\delta^{(2)} < \delta^{(1)}$.

The set of chirp times used in B_2 will be located symmetrically around ξ_m (except when $\xi_m = \xi_{\rm min}$ or $\xi_{\rm max}$, but these can be ignored), as $\xi_p = \xi_m + p\delta^{(2)}$, where $-n+1 \le p \le n-1$. For every crossing of the first stage threshold we will employ a fixed number $M = 2(n-1)$ of templates from B_2.

The *two step template placement criteria* are : **(i)** Every signal with a strength greater than a given minimum strength $S_{\rm min}$ should produce, with a probability $Q_{d,\rm min}$, at least one crossing among the two templates which lie on either side of it. It should also be detected with a probability of $Q_{d,\rm min}$ when the second stage templates corresponding to the above crossings are employed. **(ii)** The false alarm should be less than a specified level $Q_{0,\rm max}$.

We quantify the computational requirement of a given two step search by the average of total number of templates, $n_t^{\rm av}$. We denote the average value of n_c in the absence of a signal by $n_c^{\rm av}$. The computational requirement can then be expressed as,

$$n_t^{\rm av} = n_c^{\rm av} \times M + n_t^{(1)}\,, \qquad\qquad n_c^{\rm av} = Q_0(\eta^{(1)}) \times n_t^{(1)}\,, \qquad (30)$$

where $Q_0(\eta^{(1)})$ is the probability of a crossing for a single template. Since the computational cost is proportional to $n_t^{\rm av}$, we minimise $n_t^{\rm av}$ as a function of the first stage threshold $\eta^{(1)}$. There does exist a minimum for the cost which makes the strategy worthwhile. Most of the cost goes in chasing up the false alarms. However, the strategy does reduce the cost remarkably especially for the PN case.

Fixing the false alarm rate to one per year and the detection probability at 0.95, we find that a two-step search brings about a significant reduction in computational requirements. For the Newtonian case the saving in the computational cost is by a factor of at most 8, falling to about 5 for a minimal match [10] of 0.97. For the post-Newtonian case the results are as follows: For the (i) initial LIGO noise p.s.d., a two-step search is $\sim$ 27.0 times faster than the corresponding one-step search and (ii) for the advanced LIGO noise p.s.d., a two-step search is $\sim$ 23.0 times faster than the corresponding one-step search. The range used for the masses m_1 and m_2 is $0.5 \le m_1, m_2 \le 30.0\,M_\odot$. This means an online detection of the PN^2 waveform can be performed with computing power of within 10 Gflops where 250 Gflops would be required for the corresponding one step search.

6. Concluding Remarks

About a decade ago the problem of detecting coalescing binary signals even with modest range of parameters as compared with today, was considered to be formidable. But the techniques developed over the years in data analysis have succeeded in slashing down the computational costs to the extent that the detection of the PN^2 waveform has become possible given today's computing capabilities. Needless to say that the leaps and bounds made by computer technology in the recent years have played a vital role in this endeavour.

The recently explored hierarchical methods of detection can be very useful for the case of coalescing binary signals and provide a strong motivation for more detailed investigations. Such methods would be indispensable if the number of signal parameters required becomes large. For instance, if the effects due to the spins of the binaries are included in the analysis, the dimension of the parameter space will increase leading to a large number of templates.

The subject is not closed as yet and any improvement in the computational costs in the detection process and for accurately determining the parameters of the signal is most welcome. Time frequency methods such as wavelet transforms have been recently explored. These methods may well prove to be useful in detection but as of now the question is still open. In this article I have not addressed the issue of parameter estimation on which a lot of work has also been done. But this is a more involved issue in which we are not just seeking for a yes/no answer but are interested in measuring the parameters of the signal. The noise in the data introduces errors in the measurement and reducing these errors by an appropriate analysis is an important question of data analysis.

References

1. A. Abramovici *et al* , *Science,*
2. C. Bradaschia *et al Nucl. Inst. A.,* **289**, 518 (1990).
3. K.S. Thorne, in *300 Years of Gravitation*, S.W. Hawking and W.Israel (eds.), (Cambridge Univ. Press, 1987).
4. B.F. Schutz, in *The Detection of Gravitational Radiation,* edited by D. Blair (Cambridge, 1989) pp 406-427.
5. R. Narayan, T. Piran and A. Shemi, *Astrophy. J.* **379**, L17 (1991); E.S.Phinney, *ibid.* **380**, L17 (1992).
6. B.S. Sathyaprakash and S.V. Dhurandhar, *Phys. Rev. D.*, **44**, 3819 (1991).
7. S.V. Dhurandhar and B.S. Sathyaprakash, *Phys. Rev. D.*, **49**, 1707 (1994).
8. S.D. Mohanty and S.V. Dhurandhar, *Phys. Rev. D.*, **54**, 7108 (1996).
9. S. D. Mohanty *Phys. Rev. D.*, **57**, 630 (1998).
10. B. Owen, *Phys. Rev. D.*, **53**, 6749 (1996).
11. T. A. Apostolatos, *Phys. Rev. D.*, **54**, 2421 (1996).
12. C.W. Helstrom, *Statistical Theory of Signal Detection,* 2nd. ed, (Pergamon Press, London, 1968).
13. L.S. Finn and D.F. Chernoff, *Phys. Rev. D.*, **47**, 2198 (1993).
14. J. P. A. Clarke and D. M. Eardley, *Astrophys. J.*, **215**, 311, (1977).
15. C. Cutler and E. E. Flanagan, *Phys. Rev. D.*, **49**, 2658 (1994).
16. S. V. Dhurandhar and M. Tinto, *Mon. Not. R. Astron. Soc.* **234**, 663 (1988).
17. P. Jaranowski and A. Krolak, *Phys. Rev. D.*, **49**, 1723 (1994).

29. PERTURBATIONS OF COSMOLOGICAL BACKGROUNDS

PETER K. S. DUNSBY AND GEORGE F. R. ELLIS
Department of Mathematics and Applied Mathematics, University of Cape Town, South Africa.

1. Introduction

The mathematical theory of cosmological perturbations has been a major topic in the literature ever since the pioneering work of Lifshitz and Khalatnikov [1, 2] and Bonnor [3]. Scalar perturbations have attracted particular attention, since they are directly related to density fluctuations in the early universe and therefore play an important role in understanding the formation of large-scale structure.

The original theory of Lifshitz does however suffer from the disadvantage that truly physical results can only be worked out after the correspondence Φ between the real (perturbed) universe and the background Friedmann-Robertson-Walker (FRW from now on) spacetime has been completely specified, that is a definite *gauge choice* has been made. If part of this correspondence is left arbitrary, we are left with some remaining *gauge freedom* and the solutions to the perturbed Einstein field equations will include unphysical *gauge modes.* This arbitrariness is inherent to the *gauge-invariance* of the linearized Einstein equations with respect to *gauge transformations* [4] (i.e. a change in the correspondence between the perturbed and the background spacetimes), simply because the equations are linear, so any linear combination of a physical solution and a gauge mode is also a mathematically acceptable solution. This problem was pointed out by Lifshitz and Khalatnikov in 1963 [2].

In 1966 Hawking attempted to circumvent the gauge issue by using covariant methods [5], and his work was extended by Olson [6]. These authors based their analysis on curvature variables rather than on the metric but, although for the most part they used gauge-invariant variables, their

B. R. Iyer and B. Bhawal (eds.), Black Holes, Gravitational Radiation and the Universe, 493–508.

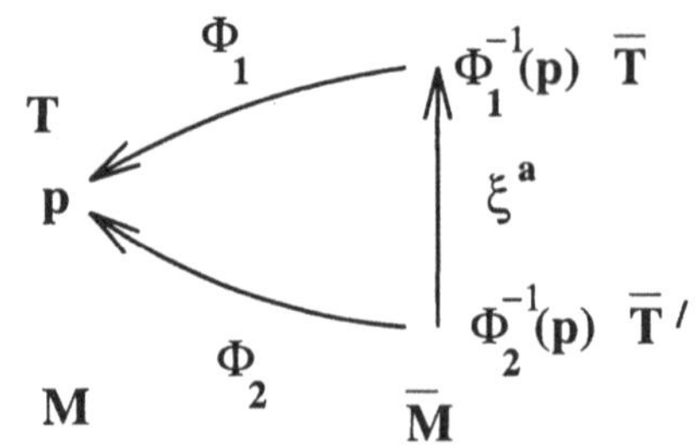

Figure 1. A gauge transformation changes the point in the background spacetime $\bar{M}$ corresponding to a point p in the physical spacetime M. This induces a change in any tensor $\bar{T}$ given by the Lie derivative of $\bar{T}$ along ξ.

analysis of density perturbations was still based on the gauge-dependent density contrast $\delta\mu/\mu$.

Following previous work by Gerlach and Sengupta [7], a fully gauge-invariant (GI from now on) theory of linear cosmological perturbations was first developed by Bardeen in 1980 [8]. In this paper GI variables are constructed in such a way that they completely specify the nature of the metric perturbations. There are however two problems with the Bardeen approach. Firstly, most of the variables in his theory do not have a transparent geometrical meaning, since they are defined with respect to a particular coordinate chart [9]. This may be seen as a consequence of the fact that in the standard approach to perturbations (also followed by Bardeen) any tensorial quantity T is usually split into a background part $\bar{T}$ and a perturbation δT: $T = \bar{T} + \delta T$, where the perturbation δT is treated as a propagating field on the background spacetime. Such a splitting is however only meaningful with respect to a given coordinate system [10], because δT is not a tensor field with respect to coordinate transformations in the background. Secondly, because Bardeen's variables are constructed as gauge-invariant linear combinations of gauge-dependent perturbations, the physical and geometrical meaning of the resulting quantities is often obscure, unless a particular hypersurface condition (i.e. a time gauge) is specified [11].

Gauge transformations can be regarded both from a geometrical and from a coordinate point of view. Stewart [9] gives a nice and clear presentation and comparison of both, and we refer the reader to his work for details. Here we recall only that the effect of a gauge transformation induced by an infinitesimal vector field ξ on a tensorial quantity $\bar{T}$ in the background model equals the Lie derivative of $\bar{T}$ along ξ (see figure 1):

$$\bar{T}' = \bar{T} + \mathcal{L}_\xi \bar{T} \quad \Rightarrow \quad \delta T' = \delta T + \mathcal{L}_\xi \bar{T}\ , \quad \mathcal{L}_\xi \bar{T} = 0 \quad \Rightarrow \quad \delta T' = \delta T. \qquad (1)$$

From this follows the Stewart and Walker Lemma [12]: perturbations to a geometrical background quantity $\bar{T}$ will be gauge-invariant iff $\bar{T}$ is (1) a constant scalar, (2) it vanishes, or (3) is a linear combination of products of Kronecker deltas with constant coefficients. From this point of view, what seems unsatisfactory regarding the gauge-invariant variables introduced by Bardeen and other authors[1] is that although they are first order gauge-invariant perturbations by construction, it is often not clear *to which quantity T they correspond.*

The quoted Lemma suggests a different approach, partially considered by Hawking [5] and extended in 1989 by Ellis and Bruni [15] to treat density perturbations gauge-invariantly. The basic idea is to introduce covariantly defined exact variables T (i.e. these variables are meaningful in any spacetime) such that their values $\bar{T}$ in a FRW universe vanish. In this way the quantity itself is a gauge-invariant perturbation in an *almost* FRW universe, and its physical significance is apparent through the covariant definition: from now on we shall refer to such quantities as *covariant gauge-invariant variables*[2]. There are in our view three main advantages in following this *geometrical approach*: *a)* it provides a unified treatment for the exact and the linearized theory; *b)* the Stewart and Walker Lemma is valid for any background spacetime, so one can often use the same GI variables in perturbing different universe models: in particular GI variables can be easily identified in perturbing homogeneous anisotropic spacetimes, and in this case the geometrical approach can be much simpler than the standard perturbative technique (see [16]); *c)* as pointed out by Stewart and Walker themselves and remarked by Mukhanov *et al.* [14] the use of exactly defined covariant variables has the advantage that if they vanish in the background FRW universe they are also GI under large gauge transformations.

The work by Ellis and Bruni [15] (which was applied to almost FRW universes with pressure-free matter) has been extended to treat density perturbation in a perfect fluid universe by Ellis, Bruni and Hwang [17, 18][3], to scalar fields with particular emphasis on *open inflationary models* and *conserved quantities* by Bruni, Ellis and Dunsby [23, 24] and to expanding Newtonian cosmologies by Ellis [25]. Hwang and Vishniac [26] and

[1]An alternative derivation of Bardeen's formalism based on a variational principle was provided by Brandenberger *et al.* [13], and Kodama and Sasaki [44] extended Bardeen's work to a multi-component fluid. This work and a recent paper by Mukhanov *et al.* [14] are both detailed reviews on the construction of the standard gauge-invariant formalism.

[2]We define an almost FRW universe as a spacetime in which the covariant GI variables do not vanish, but terms in the equations quadratic in these variables can be neglected (see section 3.2).

[3]A similar approach to that of Ellis Bruni and Hwang has been followed by Woszczyna and Kulak [19] with the aim of extending Olson's work to non flat universes and Jackson [20] has derived equations using variables with a different normalization. Lyth [21] and Zimdahl [22] have also used the covariant approach to derive equations in the comoving gauge.

Dunsby, Bruni and Ellis [27, 28] provided exact equations for density inhomogeneities in an imperfect fluid as well as linearized equations for multi-component fluids while Bruni, Dunsby and Ellis [29] provided a detailed comparison of the Bardeen and covariant approaches[4].

Dynamical systems methods have been used together with this approach to study perturbations of open FRW models [31], homogeneous and anisotropic backgrounds [16, 32] and dust-radiation FRW models [33].

More recently the covariant approach to cosmological perturbations has been applied to studies of dissipative fluids [34], cosmological magnetic fields [35] and generalized gravity theories [36].

In this review we shall not further discuss the gauge problem[5] but rather focus on the physical meaning of the covariant GI variables, the equations they obey and the overall consistency of the linear theory.

2. Covariant Cosmology

As a basis for the rest of the article, we present a summary of the covariant approach. We will restrict our analysis to perfect fluids with vanishing cosmological constant for simplicity and refer the reader to Ellis [38, 39] for details.

2.1. THE 4-VELOCITY AND KINEMATIC QUANTITIES

The 4-velocity tangent to the worldlines of fundamental observers in the universe is

$$u^a = \frac{dx^a}{d\tau}, \quad u^a u_a = -1 , \tag{2}$$

where τ is the proper time measured along the fundamental worldlines. We assume this 4-velocity is unique: that is, there is a preferred motion of matter at each spacetime event in any physically realistic cosmological model. At recent times this is taken to be the 4-velocity defined by the dipole of the Cosmic Blackbody Radiation: for there is precisely one 4-velocity which will set this dipole to zero. It is usually assumed that this is the same as the average 4-velocity of matter in a suitably sized volume [38].

Given u^a, there is defined the unique projection tensor

$$h_{ab} = g_{ab} + u_a u_b \quad \Rightarrow \quad h^a{}_b h^b{}_c = h^a{}_c,\ h^a{}_a = 3,\ h_{ab}u^b = 0 , \tag{3}$$

[4]This paper was inspired by earlier work due to Goode [30] in which he gave a GI characterization of scalar, vector and tensor perturbations based on Bardeen's variables.

[5]There is an entire literature on this subject, some authors emphasizing the physics [8, 15, 37], some the more mathematical aspects of the problem [9, 12].

which determines the metric properties of the instantaneous rest-space of observers moving with 4 - velocity u^a. There is also defined a volume element for these rest - spaces:

$$\eta^{abc} = \eta^{abcd}u_d \;\;\Rightarrow\;\; \eta^{abc} = \eta^{[abc]},\; \eta^{abc}u_c = 0\,, \tag{4}$$

where η^{abcd} is the 4 - dimensional volume element:

$$\eta^{abcd} = \eta^{[abcd]}\,, \quad \eta^{0123} = 1/\sqrt{|\det g_{ab}|}\,. \tag{5}$$

Two derivatives may also be defined: the time (dot) derivative along the fundamental world lines, where for any tensor T

$$\dot{T}^{ab}{}_{cd} = \nabla_e T^{ab}{}_{cd}u^e\,, \tag{6}$$

and the orthogonal spatial derivative D, where for any tensor T

$$\mathrm{D}_e T^{ab}{}_{cd} = h^a{}_s h^b{}_t h_c{}^v h_d{}^w h^p{}_e \nabla_p T^{st}{}_{vw}\,, \tag{7}$$

with total projection on all free indices.

The first covariant derivative of u_a can be uniquely decomposed into four parts:

$$\nabla_b u_a = \mathrm{D}_b u_a - \dot{u}_a u_b\,, \tag{8}$$

where

$$\mathrm{D}_b u_a = \sigma_{ab} + \omega_{ab} + \tfrac{1}{3}\Theta h_{ab}\,. \tag{9}$$

Here $\Theta = \nabla_a u^a$ is the volume expansion, $\sigma_{ab} = \sigma_{(ab)}$ is the shear tensor ($\sigma_{ab}u^a = \sigma^a{}_a = 0$), $\omega_{ab} = \omega_{[ab]}$ is the vorticity ($\omega_{ab}u^b = 0$) and $\dot{u}^a = \nabla_b u_a u^b$ is the acceleration. The vorticity vector is defined by

$$\omega^a = \tfrac{1}{2}\eta^{abcd}u_b\omega_{cd} \;\;\Rightarrow\;\; \omega^a u_a = 0,\; \omega^a\omega_{ab} = 0\,. \tag{10}$$

The average length scale ℓ is determined by

$$\dot{\ell}/\ell = \tfrac{1}{3}\Theta = H\,, \tag{11}$$

so the volume of a fluid element varies as ℓ^3; when the universe is an exact FRW spacetime and H is just the usual Hubble parameter.

2.2. MATER DESCRIPTION AND CURVATURE

The energy - momentum tensor of a perfect fluid is uniquely decomposed as follows:

$$T_{ab} = \mu u_a u_b + p h_{ab}\,, \tag{12}$$

where μ and p are the energy density and pressure of the fluid[6]. These quantities determine the space-time curvature through the Einstein field equations

$$R_{ab} = \kappa(T_{ab} - \tfrac{1}{2}Tg_{ab}) \tag{13}$$

for the Ricci tensor R_{ab} (here $T = h^b{}_a T^a{}_b$ and κ is the gravitational constant).

The non-gravitational physics of the situation is determined by the *equation of state* $p = p(\mu)$ relating the pressure and energy density, for example the simplest case $p = 0$ corresponds to pressure-free matter (often referred to as 'dust' or 'baryonic matter'). In addition we require that various *energy conditions* hold: one or all of

$$\mu > 0, \quad \mu + p > 0, \quad \mu + 3p > 0 \tag{14}$$

(the latter however being violated in inflationary universe models), and additionally demand the speed of sound in the fluid $c_s = (dp/d\mu)^{1/2}$ obeys

$$0 \le c_s^2 \le 1 \;\Leftrightarrow\; 0 \le dp/d\mu \le 1. \tag{15}$$

For later convenience we also define the ratio $w = p/\mu$.

The Weyl conformal curvature tensor C_{abcd} is split relative to u^a into *electric* and *magnetic* parts:

$$E_{ac} = C_{abcd}u^b u^d \quad\Rightarrow\quad E^a{}_a = 0,\; E_{ab} = E_{(ab)},\; E_{ab}u^b = 0\,, \tag{16}$$

$$H_{ac} = \tfrac{1}{2}\eta_{ab}{}^{ef}C_{efcd}u^b u^d \quad\Rightarrow\quad H^a{}_a = 0,\; H_{ab} = H_{(ab)},\; H_{ab}u^b = 0\,. \tag{17}$$

These represent the *free gravitational field*, enabling gravitational action at a distance (tidal forces, gravitational waves). Together with the Ricci tensor R_{ab} (determined locally at each point by the matter tensor through the Einstein field equations (13)), these quantities completely represent the spacetime Riemann curvature tensor R_{abcd}.

3. Cosmological perturbations

We consider first the background FRW (zero-order) model, and then linear perturbations (that is, first-order models).

3.1. CHARACTERIZATION OF FRW MODELS

An exact FRW can be covariantly characterized by the vanishing of the shear and the vorticity of u^a and by the vanishing of the spatial gradients

[6]If the fluid is imperfect, the fluid 4-velocity is not uniquely defined. It is however standard to choose u^a as either the *particle frame* or as the *energy frame*. For a detailed discussion of these choices see Ref.[29].

(i.e. orthogonal to u^a) of any scalar f:

$$\sigma = \omega = 0 \,, \quad \mathrm{D}_a f = 0 \,; \tag{18}$$

In particular the gradients of energy density, pressure and expansion

$$X_a \equiv \mathrm{D}_a\mu \,, \quad Y_a \equiv \mathrm{D}_a p \quad Z_a \equiv \mathrm{D}_a\Theta \,, \tag{19}$$

vanish, where $Y_a = 0 \Rightarrow a_a = 0$. Then $\mu = \mu(t)$, $p = p(t)$ and $\Theta = \Theta(t) = 3H(t)$ depend only on the cosmic time t defined (up to a constant) by the FRW fluid flow vector through $u_a = -t_{,a}$. It follows that these models are completely determined by an equation of state $p = p(\mu)$, the energy conservation equation

$$\dot{\mu} + 3(\mu + p)H = 0 \,, \tag{20}$$

and the Friedmann equations

$$3\dot{H} + 3H^2 + \tfrac{1}{2}\kappa(\mu + 3p) = 0 \,, \tag{21}$$

and

$$H^2 + \tfrac{K}{\ell^2} = \tfrac{1}{3}\kappa\mu \,, \tag{22}$$

where the latter is a first integral of the former (when $H \neq 0$).

3.2. GAUGE-INVARIANT VARIABLES

The approach to GI perturbations we shall follow is based on the Stewart and Walker lemma discussed in the introduction. We shall now consider a set of exactly defined covariant quantities which have significance in any spacetime and which, unless otherwise specified, vanish in a FRW model. It then follows by this lemma that these variables constitute a set of covariantly defined GI variables.

3.2.1. *Linearization procedure*

The variables that we will introduce in the next section are exactly defined covariant quantities, thus they have a physical and geometrical meaning in any spacetime. Those that vanish in a FRW model are GI. The exact covariant equations for them were derived by Ellis [38] and Ellis and Bruni [15]. These equations are linearized by assuming that terms quadratic in these quantities are negligible relative to those which are linear. Variables such μ, p and Θ *do not* vanish in a FRW model, but since they always appear in the exact equations as coefficients of GI variables, we need only their background value, i.e. they are treated as known functions in the equations. Thus the linearization procedure of the exact equations for GI variables we consider is trivial.

3.2.2. *Spatial gradients*
It follows from Eq.(18) that the shear and vorticity of u^a, together with the spatial gradients orthogonal to it are GI[7].

In particular the gradients (19) are GI and from these, we find it convenient to define two key variables

$$\mathcal{D}_a = \frac{\ell}{\mu} \mathrm{D}_a \mu \,, \qquad \mathcal{Z}_a = \ell \mathrm{D}_a \Theta \,, \tag{23}$$

the comoving fractional density gradient and the comoving gradient of expansion; these quantities have a direct link with observations [15, 40]. As we shall see, the analysis of density perturbations within the covariant approach is based on these two variables and their evolution equations.

3.2.3. *Curvature variables*
In the standard approach to perturbations [8] one looks at gravitational potentials, i.e at metric perturbations. Following Hawking [5], we focus instead on the curvature, i.e on the Riemann tensor $R^a{}_{bcd}$: this is made up of its trace $R_{ab} = R^c{}_{acb}$, the Ricci tensor, and its trace-free part, the Weyl tensor C_{abcd}, representing the free gravitational field determined non-locally by matter. FRW spacetimes are conformally flat: $C_{abcd} = 0$, which implies that the Weyl tensor is gauge-invariant. We note that in this case there is an interesting difference from the spatial gradients, shear and vorticity: while the latter are gauge-invariant only when orthogonal to a unit timelike vector that coincides with 4-velocity in the background, the Weyl tensor is gauge-invariant in any frame. This means that any possible decomposition of C_{abcd} gives gauge-invariant variables: in particular the electric E_{ab} and magnetic H_{ab} parts of the Weyl tensor are covariant and GI.

3.2.4. *Scalar variables*
Up to now we have been dealing with true tensors in the real physical almost-FRW universe. In treating cosmological perturbations it is however standard to split them into "scalar", "vector" and "tensor" parts through a non-local decomposition [9], where scalar density perturbations are solely responsible for the formation of structure in the universe.

We may however *locally* decompose the covariant derivative of any vector V_a in analogy with (8): for $\mathcal{D}_a$ we have

$$\ell \mathrm{D}_b \mathcal{D}_a \equiv \Delta_{ab} = \tfrac{1}{3} h_{ab} \Delta + \Sigma_{ab} + W_{ab} \,, \tag{24}$$

where $W_{ab} \equiv \Delta_{[ab]}$ and $\Sigma_{ab} \equiv \Delta_{(ab)} - \frac{1}{3} h_{ab} \Delta$ are respectively the skew symmetric and symmetric trace-free part of Δ_{ab}; $\Delta \equiv \Delta^a{}_a$ is the scalar

[7]In general, shear, vorticity and spatial gradients defined with respect to an arbitrary unit timelike 4-vector field n^a will only be GI iff n^a coincides with u^a in the background.

GI variable representing the clumping of matter[8]. More generally, a *locally* defined scalar variable is obtained from any tensor by taking its total divergence.

In particular in section 3.6 we shall derive equations for the following set of GI scalar variables:

$$\Delta = \ell \mathrm{D}_a \mathcal{D}^a \;, \quad \mathcal{Z} = \ell \mathrm{D}_a \mathcal{Z}^a \;, \tag{25}$$

that follow from the gradients previously introduced.

3.3. DYNAMIC EQUATIONS

There are three sets of equations to be considered, resulting from the Einstein field equations (13). We consider them in turn.

3.3.1. *The Ricci identity*

The first set arises from the *Ricci identity* applied to the 4-velocity field u^a, i.e.

$$\nabla_c \nabla_b u^a - \nabla_b \nabla_c u^a = R_d{}^a{}_{bc} u^d \,. \tag{26}$$

We obtain three propagation equations and three constraint equations. The *propagation equations* are:

1. The Raychaudhuri equation:

$$3\dot{H} + 3H^2 - \nabla_a \dot{u}^a + \tfrac{1}{2}\kappa\,(\mu + 3p) = 0 \;, \tag{27}$$

which is *the basic equation of gravitational attraction.*

2. The vorticity propagation equation:

$$\dot{\omega}_{ab} + \left(2 - 3c_s^2\right) H \omega_{ab} = 0 \;, \tag{28}$$

together with equation (34) showing how vorticity conservation follows if there is a perfect fluid with acceleration potential.

3. The shear propagation equation:

$$\dot{\sigma}_{ab} + E_{ab} + 2H\sigma_{ab} + \frac{c_s^2}{a(1+w)}\left[\mathrm{D}_{(a}\mathcal{D}_{b)} - \tfrac{1}{3}h_{ab}\mathrm{D}^c\mathcal{D}_c\right] = 0 \;, \tag{29}$$

showing how E_{ab} induce shear.

[8] Δ is equivalent to Bardeen's density perturbation measure ε_m.

In addition to these propagation equations we have three sets of *constraint equations*:

1. The $(0,\nu)$ equations:

$$\tfrac{2}{3}\mathcal{Z}_a + \ell \mathrm{D}^b\omega_{ab} - \ell \mathrm{D}^b\sigma_{ab} = 0\,, \tag{30}$$

2. The vorticity divergence identity:

$$\ell \mathrm{D}^a\omega_a = 0\,, \tag{31}$$

3. The H_{ab} equation:

$$H_{ab} = -\mathrm{D}^c\left[\omega_{(a}{}^d + \sigma_{(a}{}^d\right]\eta_{b)dc}\,. \tag{32}$$

3.3.2. *The contracted Bianchi identities*
The second set of equations arise from the *contracted Bianchi identities* $R^{ab}{}_{;b} - \frac{1}{2}R^{;a} = 0$, which by the field equations (13) imply conservation equations:

$$T^{ab}{}_{;b} = 0\,. \tag{33}$$

Projecting parallel and perpendicular to u^a, we obtain one propagation equation:

$$\dot{\mu} + (\mu + p)\Theta = 0\,, \tag{34}$$

the *energy conservation equation*, and one constraint equation:

$$(\mu + p)\dot{u}_a + \mathrm{D}_a p = 0\,, \tag{35}$$

the *momentum conservation equation*[9].

3.3.3. *The other Bianchi Identities*
The third set of equations arise from the *Bianchi identities*

$$R_{ab[cd;e]} = 0\,. \tag{36}$$

Double contraction gives (33), already considered. The full Bianchi identities give two further propagation equations and two constraint equations, which are similar in form to Maxwell's equations for the electromagnetic field [39].

The *propagation equations* are,

$$\dot{E}_{ab} + 3HE_{ab} + J_{ab} + \tfrac{1}{2}\kappa h\sigma_{ab} = 0\,, \tag{37}$$

[9] If imperfect fluids are considered, this is no longer a constraint equation, but rather is a time evolution equation for the energy flux q^a.

the '$\dot{E}$' equation, and

$$\dot{H}_{ab} + 3HH_{ab} - I_{ab} = 0 , \tag{38}$$

the '$\dot{H}$' equation, where we have defined

$$J_{ab} = \mathrm{D}^c H_{(a}{}^d \eta_{b)dc} = curl\ H , \tag{39}$$

$$I_{ab} = \mathrm{D}^c E_{(a}{}^d \eta_{b)dc} = curl\ E . \tag{40}$$

The *constraint equations* are the 'div E' equation:

$$\ell \mathrm{D}^b E_{ab} = \tfrac{1}{3}\kappa\mu\mathcal{D}_a , \tag{41}$$

and the 'div H' equation:

$$\mathrm{D}^b H_{ab} = \kappa(\mu + p)\omega_a . \tag{42}$$

3.4. THE STRUCTURE OF THE LINEAR EQUATIONS

The full set of equations given so far consists of two different kinds: propagation equations (involving time-derivatives of the kinematic or dynamic quantities) and constraint equations (involving only their spatial derivatives).

Because the equations have been obtained by linearization, we can regard their solutions as consisting additively of different parts that each themselves solve these linear equations. As is well known, Eq.(28) shows that rotational perturbations evolve independently of variables other than ω_a, which however in general affect the evolution of other quantities: since the equations are linear, this just means that in the general case all the vectorial and tensorial variables have a rotational (or vector) contribution. Thus the vorticity terms in the above equations can be regarded as known source terms. Gravitational waves are represented by those parts of the Weyl tensor components E_{ab} and H_{ab} which do not arise from rotational (vector) and density (scalar) perturbations, i.e. by their TT (transverse traceless) parts satisfying $\mathrm{D}^b E_{ab} = 0$ and $\mathrm{D}^b H_{ab} = 0$. Gravitational waves perturbations were discussed within the covariant approach by Hawking [5] and more recently by Hogan and Ellis [41] and Dunsby, Bassett and Ellis [42]. Density perturbations will be discussed in section 3.6.

3.5. CONSISTENCY OF ALMOST FRW MODELS

An important issue arising is the consistency of these equations. In order to prove that the linear equations are consistent we must show that the constraint equations are preserved in time along the fluid flow lines. Consider for example the constraint equation (30), say $\mathcal{C}_1$. We can take the time derivative of this equation, and then, using the commutation relations for time and space derivatives, substitute for all the time derivatives occurring, from the propagation equations. The result will be a new constraint equation, say $\mathcal{C}_1'$ (because we have eliminated all the time derivatives). Now it may be that $\mathcal{C}_1'$ is identically satisfied; then the constraint $\mathcal{C}_1$ is preserved in time. On the other hand $\mathcal{C}_1'$ could be a genuinely new equation; In this case we can take its time derivative to obtain a further constraint equation $\mathcal{C}_1''$. This in turn may be identically satisfied, or may be a further constraint equation that has to be satisfied in a non-trivial way. If too many non-trivial constraints arise in this way, we will have proved that the set of equations is inconsistent.

The constraint equations are:

$$\begin{aligned} \mathcal{C}^1_a &= \tfrac{2}{3}\mathcal{Z}_a + \ell \mathrm{D}^b\omega_{ab} - \ell \mathrm{D}^b\sigma_{ab} = 0 \, , \\ \mathcal{C}^2 &= \ell \mathrm{D}^a\omega_a = 0 \, , \\ \mathcal{C}^3_{ab} &= H_{ab} + \mathrm{D}^c\left[\omega_{(a}{}^d + \sigma_{(a}{}^d\right]\eta_{b)dc} \, , \end{aligned} \tag{43}$$

Taking the time derivative of the above constraints we obtain

$$\begin{aligned} \dot{\mathcal{C}}^1_a &= -2H\mathcal{C}^1_a \, , \\ \dot{\mathcal{C}}^2 &= \left(3c_s^2 - 2\right)H\mathcal{C}^2 \, , \\ \dot{\mathcal{C}}^3_{ab} &= -3H\mathcal{C}^3_{ab} \, , \end{aligned} \tag{44}$$

It follows that since

$$\mathcal{C}^1_a = \mathcal{C}^2 = \mathcal{C}^3_{ab} = 0 \, , \tag{45}$$

we end up with the identity $0 = 0$ for each constraint equation. We can therefore say that *the constraint equations are preserved in time so the complete set of linearized equations are consistent with each other*. If additional covariant conditions are imposed, which would occur, for example, if some of the non-linear terms were retained in the above equations, then this generally valid consistency may be disturbed and lead to further integrability conditions. This leads to the interesting possibility of a *linearization instability* in such models [43].

3.6. EVOLUTION EQUATIONS FOR THE GI VARIABLES

In order to investigate density perturbations we need to complete the above set of equations with an evolution equation for the GI density variable. In

Bardeen [8] and Kodama and Sasaki [44] this is coupled with an equation for a variable associated with the shear of matter. In Ellis and Bruni [15] the GI density variable is $\mathcal{D}_a$, and its evolution equation is coupled with an equation for the gradient of expansion. For a perfect fluid we obtain

$$\dot{\mathcal{D}}_a - 3Hw\mathcal{D}_a + (1+w)\,\mathcal{Z}_a = 0\,, \tag{46}$$

and

$$\dot{\mathcal{Z}}_a + 2H\mathcal{Z}_a + \tfrac{1}{2}\kappa\mu\mathcal{D}_a + \tfrac{c_s^2}{(1+w)}\left(\tfrac{K}{\ell^2} + \mathrm{D}^2\right)\mathcal{D}_a = -6aHc_s^{2\,(3)}\nabla^b\omega_{ab}\,, \tag{47}$$

on taking the gradient of the energy conservation (34) and Raychaudhuri equation (27).

It should be emphasized that these equations do not contain new dynamical information (they are implied by those already given), rather they extract the specific information we want (the propagation of the GI variables along the fluid flow lines) from the full set of equations. We can determine the behavior we are interested in from these new equations alone; they are of course consistent with the full set of equations.

As was shown in Ellis, Bruni and Hwang [18], it follows from these evolution equations for the density gradient $\mathcal{D}_a$ that although this is obviously not affected by gravitational wave (TT tensors) contributions, a rotational mode in the density gradient can be generated by the vorticity ω_{ab}.

3.7. EQUATIONS FOR THE SCALAR VARIABLES

True density perturbations related with the growth of matter clumping are represented by Δ. We could actually consider the scalar equivalent of the whole set of hydrodynamic and gravitational equations in the previous section on taking their divergences and using the following first-order identities for any vector V_a and tensor T_{ab}:

$$\ell\mathrm{D}_a\dot{V}_b = (\ell\mathrm{D}_aV_b)^{\cdot}\,, \qquad \ell\mathrm{D}_a\dot{T}_{bc} = (\ell\mathrm{D}_aT_{bc})^{\cdot}\,, \tag{48}$$

but this is not directly needed. Indeed it is possible to show that all the scalar variables we can obtain (on taking divergences of the vector and tensor variables previously used) are determined to first order by Δ through the given equations[10]; Thus we concentrate on the evolution equations for Δ.

Taking the divergence of Eq.(46) and Eq.(47), using (48), we obtain

$$\dot{\Delta} - 3Hw\Delta + (1+w)\,\mathcal{Z} = 0\,, \tag{49}$$

[10]The equivalent of this statement within the Bardeen formalism has been proved by Goode [30]: there, all the scalar perturbation variables are determined once ϵ_m is known.

$$\dot{\mathcal{Z}} + 2H\mathcal{Z} + \tfrac{1}{2}\kappa\mu\Delta + \tfrac{c_s^2}{(1+w)}\left(\mathrm{D}^2 + \tfrac{3K}{\ell^2}\right)\Delta = 0\,. \tag{50}$$

The evolution of Δ can be computed directly from a second order equation which follows from the above two first order equations:

$$\ddot{\Delta} + \mathcal{A}(t)\dot{\Delta} - \mathcal{B}(t)\Delta - c_s^2\mathrm{D}^2\Delta = 0\,, \tag{51}$$

where

$$\mathcal{A}(t) = \left(2 + 3c_s^2 - 6w\right)H\,, \tag{52}$$

and

$$\mathcal{B}(t) = \left[\left(\tfrac{1}{2} + 4w - \tfrac{3}{2}w^2 - 3c_s^2\right)\kappa\mu + \left(c_s^2 - w\right)\tfrac{12K}{\ell^2}\right]\,. \tag{53}$$

3.8. SOLUTIONS FOR BAROTROPIC PERFECT FLUIDS

When studying scalar (or density) perturbations it is useful to introduce a new variable $\Phi \equiv \mu\ell^3\Delta$. This simplifies the analysis considerably. In terms of this variable, for barotropic perfect fluids ($w = c_s^2$) equation (51) becomes:

$$\Phi'' + (1+3w)\tfrac{\ell'}{\ell}\Phi' - \left[3Kw + \tfrac{1}{2}(1+w)\mu\ell^2\right]\Phi - w^{(3)}\nabla^2\Phi = 0\,, \tag{54}$$

where the prime denotes differentiation with respect to conformal time η (i.e. $\frac{d\eta}{dt} \equiv \frac{1}{\ell}$). Harmonically analyzing this equation and setting $K = 0$ gives:

$$\Phi''_{(k)} + (1+3w)\tfrac{\ell'}{\ell}\Phi'_{(k)} + \left[(wk^2 - \tfrac{1}{2}(1+w)\mu\ell^2\right]\Phi = 0\,. \tag{55}$$

For a flat background, the scale factor ℓ, energy density μ and expansion Θ are given by:

$$\ell \propto \eta^{\frac{2}{3w+1}}\,,\quad \mu \propto \eta^{-\frac{6(1+w)}{3w+1}}\,,\quad \Theta \propto \eta^{-\frac{3(1+w)}{3w+1}}\,. \tag{56}$$

Substituting for the background parameters in Eq.(55) and integrating gives the following general solution:

$$\begin{aligned} \Phi_{(k)} &= (k\eta)^{2-\nu}\left[C_+J_\nu(wk\eta) + C_-N_\nu(wk\eta)\right]\,,\quad w \neq 0\,,\\ \Phi_{(k)} &= C_+\eta^2 + C_-\eta^{-3}\,,\quad w = 0\,, \end{aligned} \tag{57}$$

where C_+ and C_- are arbitrary constants and J_ν, N_ν denote the Bessel functions of the first and second kind, of order ν and k is the wave number. The growing mode in the solution for $w = 0$ (dust)[11] corresponds to the famous *gravitational instability* and has been used extensively in studies of the

[11] In terms of cosmic time, the growing mode $\sim t^{2/3}$.

onset of galaxy formation and Cosmic Microwave Background anisotropies [45, 46]. Dealing with that anisotropy in detail requires a kinetic theory description, which we do not have space to develop here; See however Stoeger Maartens and Ellis [47] for an important application. Challinor and Lasenby [48] and Gebbie, Ellis and Maartens [49] give a detailed development of a covariant GI to kinetic theory, based on the covariant harmonic formalism of Ellis, Matravers and Treciokas [50].

4. Conclusion

In this review paper, we considered the complete set of linearized covariant equations for a set of GI variables in a almost FRW universe as they follow from the Ricci identities (hydrodynamic equations) and the Bianchi identities (gravitational equations). In this approach Einstein's equations serve to establish algebraic relations between the Ricci tensor and the energy momentum tensor of matter, described as a fluid. The variables used are exactly defined in any spacetime, and are GI with respect to a FRW background as a consequence of the Stewart and Walker lemma. These linearized equations lead, in the case of scalar perturbations, to a second order equation governing the evolution of a locally defined scalar GI density perturbation variable Δ. In the case of pressure free matter (dust) the solution of this equation leads to the well known gravitational instability.

References

1. Lifshitz, E. M. 1946. *J. Phys.* USSR, **10**, 116.
2. Lifshitz, E. M. and Khalatnikov, I. M. 1963. *Adv. Phys.*, **12**, 185.
3. Bonnor, W. B. 1956. *Mon. Not. Roy. Ast. Soc.*, **117**, 104.
4. Sachs, R. K. 1964. Gravitational radiation, in *Relativity, Groups, and Topology*, eds. B. de Witt and C. de Witt (Gordon and Breach), 521.
5. Hawking, S. W. 1966. *Ap. J.*, **145**, 544.
6. Olson, D. W. 1976. *Phys. Rev. D.*, **14**, 327.
7. Gerlach, U. H. and Sengupta, U. K. 1978. *Phys. Rev. D.*, **18**, 1789.
8. Bardeen, J. M. 1980. *Phys. Rev. D.*, **22**, 1882.
9. Stewart, J. M. 1990. *Classical and Quantum Gravity*, **7**, 1169.
10. Faraoni, V. 1991, *Phys. Lett. A.*, **153**, 67.
11. Bardeen, J. M. 1991. Cosmological perturbations from quantum fluctuations to large scale structure, in *Particle Physics and Cosmology*, ed. A. Zee (Gordon and Breach).
12. Stewart, J. M. and Walker, M. 1974. *Proc. R. Soc. London*, **A341**, 49.
13. Brandenberger, R., Khan R. and Press, W. H. 1983. *Phys. Rev. D.*, **28**, 1809.
14. Mukhanov, V. F. Feldman, H. A. and Brandenberger, R. H. 1992. *Phys. Rep.*, 215, 203.
15. Ellis, G. F. R. and Bruni, M. 1989. *Phys. Rev. D.*, **40**, 1804.
16. Dunsby, P. K. S. 1993. *Phys. Rev. D.*, **48**, 3562.
17. Ellis, G. F. R., Hwang, J. and Bruni, M. 1989. *Phys. Rev. D.*, **40**, 1819.
18. Ellis, G. F. R., Bruni, M. and Hwang, J. 1990. *Phys. Rev. D.*, **42**, 1035.
19. Woszczyna, A. and Kulak, A. 1989. *Class. Quantum Grav.*, **6**, 1665.
20. Jackson, J. C. 1993. *Mon. Not. R. Astron. Soc.*, **264**, 729.

21. Lyth, D. H. 1985. *Phys. Rev. D.*, **31**, 1792.
22. Zimdahl, W. 1991. *Class. and Quantum Grav.*, **8**, 677.
23. Bruni, M., Ellis, G. F. R. and Dunsby, P. K. S. 1991. *Class. Quant. Gravity*, **9**, 921.
24. Dunsby, P. K. S. and Bruni, M. 1994. *Int. J. Mod. Phys. D.*, **3**, 443.
25. Ellis, G. F. R. 1989. *Mon. Not. Roy. Ast. Soc.*, **24**, 509.
26. Hwang, J. and Vishniac, E. 1990. Ap. J., **353**, 1.
27. Dunsby, P. K. S. 1991. *Classical and Quantum Gravity*, **8**, 1785.
28. Dunsby, P. K. S., Bruni, M. and Ellis, G. F. R. 1992. *Ap. J.*, **395**, 54.
29. Bruni, M., Dunsby, P. K. S. and Ellis, G. F. R. 1992.
30. Goode, S. W. 1989. *Phys. Rev. D.*, **39**, 2882.
31. Bruni, M. 1993. *Phys. Rev. D.*, **47**, 738.
32. Dunsby, P. K. S. 1997. Cosmological density perturbations, In *Dynamical Systems in Cosmology*, Ed. J. Wainwright and G. F. R. Ellis, *Cambridge University Press.*
33. Bruni, M. and Piotrkowska, K. 1994. *Mon. Not. R. Astron. Soc.*, **270**, 630.
34. Maartens, R. and Triginer, J. 1997. *Phys. Rev. D.*, **56** 4640.
35. Tsagas, C. G. and Barrow, J. D. 1997. *Class.Quantum Grav.*, **14**, 2539.
36. Hirai, T. and Maeda, K. 1994. *Ap J*, **431**, 6.
37. Press, W. H. and Vishniac, E. T. 1980. *Astrophys. J.*, **239**, 1.
38. Ellis, G. F. R. 1971. in *General Relativity and Cosmology*, Proceedings of XLVII Enrico Fermi Summer School, ed Sachs, R. K. (New York Academic Press).
39. Ellis, G. F. R. 1973. Relativistic cosmology, in *Cargese Lectures in Physics, Volume 6*, Ed. Schatzmann, E. (Gordon and Breach), 1.
40. Kristian, J. and Sachs, R. K. 1966. *Ap. J.*, **143**, 379.
41. Hogan, P. and Ellis, G. F. R. 1996. *Class. Quantum Grav.*, **14** A171.
42. Dunsby, P. K. S., Bassett, B. A. C. C., and Ellis, G. F. R. 1997. *Classical and Quantum Gravity*, **14**, 1215.
43. Maartens, R. 1997. *Phys. Rev. D.*, **55**, 463.
44. Kodama, H. and Sasaki, M. 1984, *Prog. Theor. Phys.*, **78**, 1.
45. Dunsby, P. K. S. 1997. A fully covariant description of CMB anisotropies, to appear in *Classical and Quantum Gravity.*
46. Sachs, R. K. and Wolfe, A. M. 1966. *Astrophys. J.*, **147**, 1, 73.
47. Stoeger, W., Maartens, R. and Ellis, G. F. R. 1995. *Ap J*, **443**, 1.
48. Challinor, A. and Lasenby, A. 1997. Recent developments in the calculation of CMB anisotropies, to appear in "Current Topics in Astrofundamental Physics (1997)", Ed. N. Sanchez, Kluwer Academic Press.
49. Gebbie, T. Ellis, G. F. R. and Maartens R. 1998. In preparation.
50. Ellis, G. F. R., Matravers, D. R. and Treciokas, R. 1983. *Annals of Physics*, (NY) **150**, 455.

30. MACH'S PRINCIPLE IN ELECTRODYNAMICS AND INERTIA

JAYANT V. NARLIKAR
Inter-University Centre for Astronomy and Astrophysics,
Post Bag 4, Ganeshkhind, Pune 411 007, India.

1. Introduction

The concept of inertia, which originated with Galileo, found a mathematical expression in Newton's laws of motion. According to the second law of motion, the force acting on a particle is proportional to the acceleration of the particle. The constant of proportionality measures the inertia of the particle and is called its inertial mass. Thus, if $\mathbf{P}$ is the force, $\mathbf{f}$ the acceleration, and m the inertial mass, we have

$$\mathbf{P} = m\mathbf{f}. \tag{1}$$

However, when formulating this law, Newton faced a fundamental difficulty : How to fix the reference frame relative to which $\mathbf{f}$ is measured? Clearly, Eq.(1) cannot hold in all reference frames. If it is valid in frame Σ, then in another frame Σ', with an acceleration $\mathbf{a}$ relative to Σ, the law of motion becomes

$$\mathbf{P}' \equiv \mathbf{P} - m\mathbf{a} = m\mathbf{f}', \tag{2}$$

where $\mathbf{f}'$ is the acceleration of the particle with respect to Σ', and $\mathbf{P}'$ is the force measured in it. Thus the force is modified from $\mathbf{P}$ to $\mathbf{P}$ $-m\mathbf{a}$. The extra term $-m\mathbf{a}$ that has to be added to $\mathbf{P}$ depends on the mass of the particle and hence is called the inertial force. The frames in which Eq.(1) holds without the necessity of such a modification are called inertial frames. All these frames (including Σ) are unaccelerated relative to one another.

What distinguishes inertial frames from noninertial ones? A priori Newton could see no physical reason to make such a distinction. That such a distinction exists in nature can be seen or demonstrated in numerous ways. Newton has discussed the so-called bucket experiment. The rotation of a

B. R. Iyer and B. Bhawal (eds.), Black Holes, Gravitational Radiation and the Universe, 509–524.

bucket is a relative term. If one observer sees the bucket rotating, another (e.g., one sitting on the bucket) can claim that it is nonrotating. In Newton's experiment the distinction can be made absolute. If the bucket is suspended by a string, the string is given a twist and then the bucket let go, it spins as the twisted string unwinds. If the bucket contains some water, its surface will become curved and dip toward the centre. This absolute effect demonstrates the presence of inertial forces and distinguishes between inertial and noninertial frames.

Why such a distinction should exist, Newton could not understand. However he used this distinction to postulate so-called absolute space. This is, in effect, a specific inertial frame in which Eq.(1) holds. All accelerations relative to this are detectable by the inertial forces.

Mach, in the last century, indicated that a clue to the identification of absolute space is indeed provided by the universe: a clue that was not available to Newton. Astronomical observations have shown that the frame of reference of the local observer, in which the distant parts of the universe appear to be nonrotating, is an inertial frame. This is a remarkable result! An observer can measure the rotation of the Earth relative to Newton's absolute space by observing the motion of a Foucault pendulum that is driven by the inertial forces. Or, he can measure the rotation relative to distant galaxies. In either case he gets the same answer! In other words, cosmology appears to provide a handle on the question of why certain frames enjoy a special status.

This was the point made by Mach in his critique of Newtonian mechanics. As one who believed in formulating all scientific deductions on directly observable quantities (and not on abstract theories), Mach strongly attacked the concept of absolute space; and sought to replace it by a background space of the remote parts of the universe. He went even further. The Newtonian concept of inertia and its measure in terms of mass were to him unsatisfactory. If mass is the quantity of matter in a body how does one set about measuring it? To Mach, mass and inertia were not the intrinsic properties of the body but the consequences of the existence of the body in a universe containing other matter. To measure mass one has to use Eq.(1): measure the force and divide it by the acceleration produced. But Eq.(1) itself depends on the use of absolute space, which has now been identified with the background space of distant matter. So, according to Mach's reasoning, mass is somehow determined by the distant matter.

Although initially impressed by Mach's ideas, Einstein later abandoned his efforts to incorporate them in his theories, for, Mach's principle and Newtonian gravitation were examples of an action-at-a-distance theories. Electromagnetism began as an action-at-a-distance theory but proved inadequate to describe all the observed phenomena including radiation. Einstein was impressed by Maxwell's field theory, not only because it successfully described electromagnetism but also because it had internal beauty. The theory already contained the Lorentz invariance required by special relativity. Action at a distance, on the other hand, was instantaneous and hence inconsistent with special relativity. Although, unlike electromagnetism, there was no experimental data against Newtonian gravitation (with the esoteric exception of the perihelion precession of Mercury!) its logical inconsistency was already apparent. And Mach's principle seemed to imply a similar idea of action at a distance.

Nevertheless, it is worth asking now, whether the Machian ideas are really irreconcilable with a theory of gravitation. Cannot the Newtonian action at a distance be reformulated to give expression to Mach's ideas without offending relativity? We first turn our attention to the action at a distance approach and see how this approach can be made to work in electromagnetism. We then formulate a theory of gravity on similar lines, which also incorporates Mach's ideas, and demonstrate how this theory approaches general relativity in the case of many particle systems.

2. The Absorber Theory of Radiation

In a letter to Weber on March 19, 1845, Gauss wrote:

I would doubtless have published my researches long since were it not that at the time I gave them up I had failed to find what I regarded as the keystone. Nil actum reputans si quid superesset agendum, namely, the derivation of the additional forces - to be added to the interaction of electrical charges at rest, when they are both in motion - from an action which is propagated not instantaneously but in time as is the case with light.

Gauss's attempts came some three decades before the Maxwellian field theory and six decades before special relativity. The success of these two theories shifted the emphasis from action at a distance to fields and it was not until well into the present century that the problem posed by Gauss was solved.

A beginning was made by Schwarzschild [1], Tetrode [2], and Fokker [3, 4, 5], who independently formulated the concept of delayed action at a distance. The action principle as formulated by Fokker may be written in the following form:

$$J = -\sum_a \int m_a da - \sum_{a<b}\sum \int\int e_a e_b \delta(s_{AB}^2)\eta_{ik} da^i db^k. \qquad (3)$$

In the above expression the charged particles are labeled, $a, b, \ldots$ with e_a and m_a the charge and mass of particle a. The worldline of a is given by the coordinate functions $a^i(a)$ of the proper time a. The spacetime is

Minkowskian, so that

$$da^2 = \eta_{ik} da^i da^k, \tag{4}$$

with $\eta_{ik}=$ diag $(-1,-1,-1,1)$. The first term of J therefore describes the inertial term. The second term describes the electromagnetic interaction between the worldlines of a typical pair of particles a, b. The delta function shows that the interaction is effective only when s^2_{AB} , the invariant square of distance between typical world points A, B on the worldlines of a and b, vanishes. This implies delayed action: $s^2_{AB}= 0$ means that world points A and B are connected by a light ray.

Although this formulation met the requirement of relativistic invariance, it gave rise to other difficulties. The major difficulty is as follows. For a typical point A on the worldline of a there are two points B_+ and B_- on the world line of b for which $s^2_{AB} = 0$. The effect of A is felt at B_+ (at a later time) and at B_- (at an earlier time). Similarly, since the action principle guarantees the equality of action and reaction, the reaction from B_+ and B_- is felt at A. Thus there are influences propagating with the speed of light, not only into the future but also into the past. This led to a conflict with the principle of causality, which seems to hold in everyday life. The other difficulties were of a less serious nature although not ignorable. For example, there was no "self-action" ($a = b$ is avoided in the double sum) and so there did not appear to be any obvious way of accounting for radiation damping.

These difficulties were removed by Wheeler and Feynman [6] by bringing into the discussion the important role of the absorber. In our above example, the reactions from B arrive at A instantaneously, whatever be the spatial separation of a and b. So it becomes necessary to take into account the reaction from the entire universe to A. Although the remote particles are expected to contribute less, their total number is large enough to make the calculation nontrivial. The essence of the argument given by Wheeler and Feynman is described below.

To begin with, define the 4-potential at X due to particle b by

$$A_i^{(b)}(X) = e_b \int \delta(s^2_{XB}) \eta_{ik} db^k, \tag{5}$$

and the corresponding direct-particle field by

$$F_{ik}^{(b)} = A_{k;i}^{(b)} - A_{i;k}^{(b)}. \tag{6}$$

A direct particle field is not an ordinary field, because it does not have any independent degrees of freedom. The 4-potential *identically* satisfies the relations

$$A^{(b)k}_{;k} \equiv 0, \quad \Box A^{(b)k} \equiv \eta^{mn} A^{(b)k}_{;mn} = 4\pi j^{(b)k}, \tag{7}$$

where $j^{(b)k}(X)$ is the current density vector of the particle b at a typical point X, defined in the usual way. Thus although Eq.(7) resembles the Maxwell wave equation (and the gauge condition) it represents identities.

The equation of motion of a typical charge a is obtained by varying its worldline and requiring $\delta J = 0$. We get the analogue of the Lorentz force equation in which the charge a is acted on by all other charges in the universe.

We now turn to the difficulty introduced by the time symmetry of this formulation. Instead of being the retarded solution of Eq.(7), Eq.(5) is the time-symmetric half-advanced and half-retarded solution. The same applies to the direct-particle fields. Suppressing the indices i, k, we may write

$$F^{(b)} = \frac{1}{2}[F^{(b)}_{\text{ret}} + F^{(b)}_{\text{adv}}]. \tag{8}$$

This field is present in the past as well as the future light cone of B.

Wheeler and Feynman argued in the following way. If we move the charge b, it generates a disturbance that affects all other charges in the universe. Their reaction arrives back instantaneously. Wheeler and Feynman showed how to calculate such a reaction in a universe of static Minkowski type with a uniform distribution of electric charges. They showed that the reaction to the motion of charge b can be calculated in a consistent fashion and comes out to be

$$R^{(b)} = \frac{1}{2}[F^{(b)}_{\text{ret}} - F^{(b))}_{\text{adv}}]. \tag{9}$$

Thus a test particle in the neighbourhood of charge b experiences a net total "field"

$$F^{(b)}_{\text{tot}} = F^{(b)} + R^{(b)} = F^{(b)}_{\text{ret}}. \tag{10}$$

This is the pure retarded field observed in real life! The self-consistency of the argument follows from the fact that the reaction $R^{(b)}$ has been calculated by adding the $\frac{1}{2}F^{(a)}$ fields of all particles $a \neq b$ that have been excited by this total field $F^{(b)}$. Thus only the future light cone of B comes into play. The reaction from the future cancels the advanced component of $F^{(b)}$ and doubles its retarded component.

Also according to the Lorentz force equation, $R^{(b)}$ is the force arising from all *other* particles in the universe experienced by the particle B. This is nothing but the radiative reaction to the motion of b as obtained earlier by Dirac [7] on empirical grounds. Earlier, Dirac's rule was difficult to understand within the context of the field theory, although it was known to give the right answer. Here the Dirac formula is understood as the consequence of a response of the universe to the local motion of the charge. Thus the theory not only gets round the problem of causality but it also accounts for the radiation damping formula.

Physically, what happens is the following. To the motion of b the future half of the universe acts as an absorber. It "absorbs" all the "energy" radiated by b, and in this process sends the reaction $R^{(b)}$, which does the trick! For this reason Wheeler and Feynman called this theory the *absorber theory of radiation.* The presence of the absorber is essential for the calculation to work. For example, it will not work in an empty universe surrounding the electric charge.

In the above self-consistent derivation there was still one defect: it was not unique. Another self-consistent picture was possible in which the nett field near every particle was the pure advanced field and the radiative reaction was of opposite sign to that of Eq.(9). The two solutions are compared thus. In (a) we have the retarded solution while in (b) we have the advanced solution. In (a) absorption in the future light cone is responsible while in (b) it is the absorption in the past that plays the crucial role. The important role of the absorbers is that they convert the time-symmetric situation of an isolated charge, to a time-asymmetric one of (a) or (b) type. It is, however, not possible to distinguish between (a) and (b) without reference to some other independent type of time asymmetry.

Wheeler and Feynman realized this and linked the choice of (a) to thermodynamics. Given the usual thermodynamic time asymmetry, they argued that the situation (b) would be highly unlikely (under the probability arguments of statistical mechanics) and that the usual asymmetry of initial conditions will favour (a) to (b).

It was, however, pointed out by Hogarth [8] that it is not necessary to bring thermodynamics into the picture at all. If one takes account of the fact that the universe is expanding, its past and future are naturally different. The reaction from the absorbing particles in the future light cone (designated by Hogarth collectively as the *future absorber*) does not automatically come out equal and opposite to the reaction from the past absorber. Thus the two pictures (a) and (b) do not always follow in an expanding universe. Hogarth found that for (a) to hold but not (b), the future absorber must be perfect and and past absorber imperfect; and vice versa for (b) to hold but not (a).

An absorber is perfect if it entirely absorbs the radiation emitted by a typical charge. In the static universe discussed by Wheeler and Feynman both the past and future absorbers are perfect; and this leads to the ambiguity mentioned earlier. However, Hogarth found that the ambiguity is resolved if the cosmological time asymmetry is taken into account. He found, for example, that in most big-bang models which expand forever, (b) is valid and not (a). In the steady-state model (Bondi and Gold [9]; Hoyle [10]), (a) is valid and not (b). In the big-bang models that expand and contract both the absorbers are perfect and the outcome is ambiguous.

Though interesting, Hogarth's work was incomplete in two aspects. First, he had not shown how to generalize the Fokker action to curved spacetimes needed to describe the expanding world models. Second, he had used collisional damping to decide upon the nature of absorbers, past and present; and this brought in thermodynamics by the back door!

Later Hoyle and Narlikar [11] completed the work by first rewriting the Fokker action Eq.(3) in curved space as is necessary for any cosmological discussion. They also deduced conclusions similar to Hogarth's but by using the radiative damping for producing absorption. This kept the asymmetry entirely within electrodynamics and cosmology.

Finally they extended the entire picture to quantum theory. Thus it is possible to discuss the entire range of phenomena of quantum electrodynamics without recourse to field theory (Hoyle and Narlikar [12, 13]). This therefore removes any possible objection to the concept of action at a distance in so far as it is applicable to electrodynamics.

The crucial role played in the whole calculation is that of the response of the universe. In the classical calculation the steady-state universe generates the "correct" response so that the local electric charges interact through retarded signals. The response from the big-bang models is of the wrong type. We are thus able to distinguish between the different cosmological models and decide on their validity or otherwise on the basis of the Wheeler-Feynman theory. We also see why charges interact through retarded signals: they do so because of the response of the universe. In the Maxwell field theory the choice of retarded solutions of Maxwell's equations is by an arbitrary fiat.

In the quantum calculation also it can be shown that the asymmetric phenomenon with respect to time, like the spontaneous downward transition of an atomic electron, is caused by the response of the universe. By contrast, the quantization of the Maxwell electromagnetic field ascribes these asymmetries to the so-called vacuum and to the rather abstract rules of quantization.

In a recent paper Hoyle and Narlikar [14] have shown that with suitable cosmological boundary conditions, like the de Sitter horizon, there is a cut off on high frequencies that lead to divergent integrals in the standard field theory. Thus the electron self energy problem and the various radiative corrections of quantum electrodynamics can be handled without subtraction of one infinity from another. In this sense the direct particle theory fares better than the field theory.

The direct-particle approach therefore achieves for electrodynamics what Mach sought to achieve for inertia. By bringing in the response of the universe to a local experiment in electrodynamics we have essentially incorporated Mach's principle into electromagnetic theory. Given the correct response of the universe, we can almost decouple our local system from it, although, strictly speaking, the theory would not be possible without the universe.

Can the same prescription be applied to inertia and gravitation? We discuss this problem in the following section.

3. Inertia as a Direct-Particle Field

We now return to the problem of achieving a "reconciliation" between general relativity and Mach's principle. To this end we shall look for a theory with the following properties:

(a) It has Mach's principle built into one of its postulates.
(b) It is conformally invariant.
(c) It does not have the conceptual difficulties associated with the case of a single particle in an otherwise empty universe.
(d) For a universe containing many particles the theory reduces to general relativity for most physical situations.

Some discussion is needed as to why the theory should be conformally invariant. The reasons are twofold. First, when we take note of the local Lorentz invariance of special relativity, the natural units to use are those in which the fundamental velocity c=1. The quantum theory, with which our theory should be consistent throws up another fundamental constant, the Planck constant related to the uncertainty principle. Thus it is natural to use units in which $\hbar$=1. This makes the classical action J, for example, dimensionless and the natural unit of mass the Planck mass which we shall quantify by $\sqrt{3c\hbar/4\pi G}$.

All masses are then expressible as numbers in units of this mass. With our choice of units, only one independent dimension out of the three length, mass and time, remains. Taking it as the dimension of mass, length goes as reciprocal of mass.

However, in a Machian theory, we expect the particle masses to be functions of space and time and as such not necessarily constant. Therefore, the standards of lengths and time intervals may also vary from one point to another. We therefore need laws of physics which are invariant with respect to this variation. Conformal invariance guarantees this.

Our second reason is based on the nature of action-at-a-distance. Given that, as in electrodynamics, the interaction propagates principally along null rays, we need an invariance that preserves the global structure of light cones. This again is guaranteed by the conformal invariance. Just as Lorentz invariance identifies the light cone as an invariant structure *locally*, so does conformal invariance identify it *globally*.

A theory following these guidelines was developed by Hoyle and Narlikar [15, 16] and its broad features are described next. We begin by a second look at the Fokker action for electrodynamics, this time rewritten in a curved Riemannian space-time:

$$J = -\sum_a \int m_a da - \sum\sum_{a<b} 4\pi e_a e_b \int\int \bar{G}_{i_A k_B} da^{i_A} db^{k_B}. \tag{11}$$

Here, in going from Eq.(3) to Eq.(11) the first term of J needs a trivial modification: da is now computed with a Riemannian metric. The modification of the second term of J requires considerable thought. The $\delta(s^2_{AB})\eta_{ik}$ is now replaced by $\bar{G}_{i_A k_B}$, a bivector propagator between A and B. It is the symmetric Green's function for the wave equation

$$\Box \bar{G}_{ik_B} + R^l_i \bar{G}_{lk_B} = [-\bar{g}(X,B)]^{-1/2} \bar{g}_{ik_B} \delta_4(X,B). \tag{12}$$

Here $\bar{G}_{ik_B}$ behaves as a vector at X and B, respectively, with the indices i and k_B (the subscript X on i is suppressed for the convenience of writing). $\bar{g}_{ik_B}$ is the parallel propagator between X and B [see Synge [17] for details] and $\bar{g}(X,B)$ its determinant. In the limit $g_{ik} \to \eta_{ik}$, $\bar{G}_{i_A k_B} \to \delta(s^2_{AB})\eta_{ik}$. The detailed structure of this propagator has been studied by DeWitt and Brehme [18].

The electromagnetic part of J is conformally invariant but the mechanical part (the first term) is not. We now compare Eq.(11) with the action for field theory of Maxwell and for general relativity. This action, denoted by $J^{(F)}$ is given by

$$J^{(F)} = \frac{1}{16\pi G}\int R(-g)^{1/2} d^4x - \sum_a \int m_a da - \frac{1}{16\pi}\int F^{lm} F_{lm} (-g)^{1/2} d^4x$$
$$-\sum_a \int A_i da^i. \tag{13}$$

The third and fourth term of $J^{(F)}$ represent, respectively, the free-field term and the field-particle interaction term.

In the direct-particle theory the second term of J replaces these two terms of the field theory. The fields as such lose their independent status and are replaced by propagators connecting particle world lines. What can we do about the first two terms of Eq.(13)? The second term already exists in Eq.(11) and it is tempting to simply insert the first term into Eq.(11) as representing gravity.

This procedure, however, is contrary to the spirit of the direct-particle picture. The first term of Eq.(13), although containing geometrical information, has also the character of a field. Hence it is out of place. We have already commented on the non-Machian character of the second term of Eq.(13). For these reasons the approach suggested above is not desirable.

The clue to the correct procedure that needs to be adopted is provided by a comparison of the last term of Eq.(13) with the second term of Eq.(11). If in the former we replace the potential A_i by a sum over the direct-particle potentials defined by a relation analogous to Eq.(5) for a curved space, we shall recover something that looks like the latter! In the same way we now replace the masses m_a by direct-particle fields defined in the following manner:

$$m^{(b)}(X) = \int \lambda_b G(X, B) db, \quad \lambda_b = \text{a coupling constant.} \tag{14}$$

$$m_a(A) = \lambda_a \sum_{b \neq a} m^{(b)}(A), \quad \lambda_a = \text{a coupling constant.} \tag{15}$$

The propagator $G(X, B)$ has to be a biscalar since masses are scalars and we wish to preserve a symmetry between X and B. The action Eq.(11) is now changed to

$$J = -\sum_{a<b}\sum \int\int \lambda_a \lambda_b G(A, B) da\, db - \sum_{a<b}\sum 4\pi e_a e_b \int\int \bar{G}_{i_A k_B} da^{i_A} db^{k_B}. \tag{16}$$

What should be the exact form of $G(A, B)$? Taking a clue from electromagnetism, we expect it to be a symmetric Green's function of a scalar wave equation. However, we also want the equation to be conformally invariant. These two requirements fix the form of the scalar propagator uniquely to within a multiplicative factor. We shall take $G(A, B)$ to satisfy the scalar wave equation

$$\Box G(X, B) + \frac{1}{6} R(X) G(X, B) = [-\bar{g}(X, B)]^{-1/2} \delta_4(X, B). \tag{17}$$

The wave operator is uniquely fixed by the requirement of conformal invariance.

Turning from these purely formal aspects to those of interpretation we note that Eq.(14) and Eq.(15) are essentially Machian ideas on inertia expressed mathematically. The mass of a particle a at its world point A is the sum of the contributions of all other particles in the universe. Thus requirement (a) has been met. Requirement (c) is also met, because for a single particle in an otherwise empty universe there is no action! The minimum number of particles required to define J is two. Thus for each of the two particles the other provides the "background" in the Machian sense. The requirement of conformal invariance is also met by our choice of the propagator. It therefore remains to examine requirement (d).

So far we have concentrated on inertia and ignored gravity. The action Eq.(16) does not contain the gravitational term $(1/16\pi G)\int R(-g)^{1/2}d^4x$ explicitly. Yet, as we shall see in the following section, the theory is fully capable of describing gravitational phenomena.

4. Conformal Gravity

Returning to the action Eq.(13) we note that when we try to derive the Einstein field equations by the Hilbert action principle, we get the Einstein tensor from the first term. This term does not exist any more in the direct particle action Eq.(16). Shall we get any gravitational term at all from Eq.(16) if we sought to perform the metric variation $g_{ik} \rightarrow g_{ik} + \delta g_{ik}$? A look at the electromagnetic part of Eq.(13) does not inspire confidence that the answer to this question should be in the affirmative. There it is the third rather than the fourth term that contributes the energy tensor of electrodynamics, and it is the fourth term that was used in going over to Eq.(16). Nevertheless, a closer examination shows that the terms in Eq.(13) do give nontrivial answers when the metric variation is performed.

The reason for this is understood as follows. Consider the electromagnetic propagator $\bar{G}_{i_A k_B}$ connecting A and B respectively, on the world lines of a and b. Suppose we perform a variation in the space-time metric of a compact region Ω. Since the propagator is a global property of space-time structure, it will change because of this change in structure of Ω. The change in the propagator is therefore expressible, in a first order calculation, as a functional of δg_{ik} over the volume Ω.

In the electromagnetic case the answer may be expressed in the following form

$$-\delta \sum_{a<b}\sum 4\pi e_a e_b \int\int \bar{G}_{i_A k_B} da^{i_A} db^{k_B} = -\frac{1}{2}\int T^{ik}\delta g_{ik}(-g)^{1/2}d^4x, \quad (18)$$

where

$$T^{ik} = \frac{1}{8\pi}\sum_{a<b}\sum [\frac{1}{2}g^{ik}F^{(a)mn}_{\text{ret}}F^{(b)}_{mn\ \text{adv}} - F^{(a)i}_{l\ \text{ret}}F^{(b)kl}_{\text{adv}} - F^{(b)i}_{l\ \text{ret}}F^{(a)kl}_{\text{adv}}]. \quad (19)$$

The details of this derivation are given by Narlikar [19].

It is interesting to note in passing that this derivation resolves an ambiguity about the energy tensors of direct-particle electrodynamics. Wheeler and Feynman [20] had discussed two tensors for this theory. Of these one was the canonical tensor given above by Eq.(19) and the other was the Frenkel tensor whose form differs from that given in Eq.(19) in having all the direct particle fields $F^{(a)lm}$ as the symmetric half-advanced-plus -half-retarded fields. Wheeler and Feynman had concluded:

From the standpoint of pure electrodynamics it is not possible to choose between the two tensors. The difference is of course significant for the general theory of relativity, where energy has associated with it a gravitational mass. So far we have not attempted to discriminate between the two possibilities by way of this higher standard.

As mentioned above, the usual prescription of metric variation uniquely yields the canonical tensor. The fact that we could get a nontrivial answer to the variational problem and that this resolves a long-standing ambiguity, reinforces our belief that we are proceeding along the correct path toward a theory of gravitation.

We now consider the variation of the first term of Eq.(16) as $g_{ik} \rightarrow g_{ik} + \delta g_{ik}$. We shall ignore the second term and concentrate on gravitation alone. Also for simplicity we begin by putting $\lambda_a = 1$ for all a. This does not alter the essential features of the theory.

The method is similar to that adopted for electromagnetism. We compute the change in the propagator $G(A, B)$ as the geometry changes in any compact region Ω. The details of this somewhat lengthy calculation are given elsewhere [see Hoyle and Narlikar [21]]. We simply quote the result. The field equations turn out to be

$$\frac{1}{2}\phi(R_{ik} - \frac{1}{2}g_{ik}R) = -T_{ik} + \frac{1}{6}[g_{ik}\Box\phi - \phi_{;ik}] + \frac{1}{2}[m_i^{\text{ret}}m_k^{\text{adv}} + m_k^{\text{ret}}m_i^{adv}$$
$$-g_{ik}m^{l\ \text{ret}}m_l^{\text{adv}}], \quad \text{where} \tag{20}$$

$$m(X) = \sum_a \int G(X, A)da, \quad m_i = \partial m/\partial x^i, \tag{21}$$

$$\phi(X) = m^{\text{adv}}(X)m^{\text{ret}}(X), \tag{22}$$

and m^{ret} and m^{adv} denote twice the retarded and advanced parts of $m(X)$, respectively. The energy tensor T_{ik} is the familiar energy tensor for a system of particles $a, b, ...$ with masses as defined by the Machian preciption, Eq.(14) and Eq.(15). Note that the masses are time symmetric. The function $m(X)$ satisfies the conformally invariant wave equation

$$\Box m + \frac{1}{6}Rm = N, \quad \text{where} \tag{23}$$

$$N(X) = \sum_a \int \delta_4(X, A)[-\bar{g}(X, A)]^{-1/2}da \tag{24}$$

is the invariant particle number density.

There are 10 equations in Eq.(20) and one equation Eq.(23) for the 11 unknowns g_{ik} and m. However, the divergence and trace of Eq.(20)

identically vanish, showing that there are in fact five fewer independent equations. This is hardly surprising since four of these five are due to the general coordinate invariance (as in general relativity) while the fifth identity (the vanishing of trace) is due to conformal invariance. It is easy to verify that if $[g_{ik}, m]$ is a solution of these equations then so is $[\zeta^2 g_{ik}, \zeta^{-1} m]$ for an arbitrary well-behaved (i.e., of type C^2) nonvanishing finite function ζ. This arbitrary function is nothing but the expression of the arbitrariness of mass-dependent units discussed in Section 3.

Suppose now that it is possible to choose ζ such that

$$m^{\text{ret}}\zeta^{-1} = m_0 = \text{constant}. \tag{25}$$

Suppose also that the response of the universe is such as to cancel all advanced components and double the retarded ones so that the effective mass function is m^{ret}. Then the field equations are simplified to

$$R_{ik} - \frac{1}{2} g_{ik} R = -\kappa T_{ik}, \quad \kappa \equiv \frac{8\pi G}{c^4} = 6/m_0^2. \tag{26}$$

We shall later identify m_0 with the mass of the Planck particle. However, as seen above, we have arrived at the familiar equations of general relativity! The conformal frame for which Eqs.(25) and (26) hold will be called the *Einstein frame*. We have thus completed the remaining part of the programme outlined at the beginning of Section 3.

The following points are worth emphasizing in the above derivation of Einstein's equations, which is so radically different from the standard ones (used for example, by Einstein in 1915 and by Hilbert later the same year):

(I) The approach to Einstein's equations is via the wider framework of a conformally invariant gravitation theory. Only in the limit of many particles in a suitably responding universe do we arrive at Einstein's equations. In the other limit of zero or no particles there is no theory! Thus it brings out the reason why the Machian paradox of one particle in an empty universe is not valid in the context of Einstein's equations. This reason does not emerge in the standard derivations of Einstein's equations.

(II) It is significant that the coupling constant κ is positive in this approach. This conclusion is unaffected by the change of sign of the coupling constants λ_a, λ_b, etc. (taken here as unity); nor it is affected by the choice of signature (i.e., $---+$ instead of $+++-$) of the spacetime metric. The choice of the conformally invariant scalar propagator leads to the coupling constant being positive, i.e. to gravity being "attractive". In the standard deviation the coupling constant is fixed (in sign as well as magnitude) by a comparison with Newtonian gravity.

(III) A considerable discussion has gone on regarding the admissibility of the so-called λ-term in Einstein's equations. This is because this term could be accommodated in Einstein's heuristic derivation or in Hilbert's action principle. It is worth emphasizing that the direct-particle approach to gravity given so far does not permit the λ-term. As we shall see later, this term does arise in a Machian way in the direct particle theory, *provided* we allow the wave equation (23) for inertia to be nonlinear. The present cosmological observations generally seem to require the cosmological constant (Bagla, et al, [22]).

(IV) The condition Eq.(25) that leads to Einstein's equations needs to be re-examined carefully under two special circumstances. Near a typical particle a, we expect the mass function $m^{(a)}(X)$ to "blow up" so that $m(x) \to \infty$, as $X \to A$, on the worldline of a. In order to make $\zeta^{-1}m(X)$ finite at A, we therefore require $\zeta \to \infty$, as $X \to A$. However, we have already ruled out such conformal functions by restricting ζ to finite values. Thus the transition to Einstein's equations is not valid as we tend to any typical source particle. The nature of the equations and their solutions near a particle in this theory have been discussed by Hoyle and Narlikar [16] and by Islam [23].

5. Cosmology and the Creation of matter

In recent years, this theory has been further generalized and applied to cosmology to include the cosmological constant as well as explicit description of creation of matter (see Hoyle *et al.* [24]).

Taking the cosmological constant into account first, we may ask whether Eq.(23) is the most general conformally invariant wave equation satisfied by a scalar function $m(X)$. The answer is *No*. The most general such equation is

$$\Box m + 1/6Rm + \Lambda m^3 = N, \tag{27}$$

where Λ is a constant. This, of course, makes the scalar Machian interaction non-linear and more difficult to handle. However, it very naturally leads to the cosmological constant of the right magnitude at the present epoch. For, if we assume that for a single particle, the value of this constant is unity, then in the sum Eq.(21) leading to $m(X)$, because of the presence of a large number n of particles within the cosmological horizon, the cube term is less effective by the factor n^{-2}. Thus, the factor Λ in Eq.(27) is of this order. With the identification of m_0, with the only possible fundamental constant, viz, the Planck mass

$$m_0 = \sqrt{\frac{3\hbar c}{4\pi G}}, \tag{28}$$

one can write the familiar cosmological constant in the Einstein equations as

$$\lambda = -3\Lambda m_0^2. \tag{29}$$

With around $2 X 10^{60}$ Planck particles in the horizon, we get the value of $\lambda \sim -2 X 10^{-56}\text{cm}^{-2}$. It is of the right order, but *negative.* However, it leads to interesting and physically relevant cosmological models.

Let us consider matter creation next. The standard relativity theory starts with the assumption that matter can neither be created nor destroyed, that is, the worldlines of particles are endless. (Even where matter is converted into radiation the above assumption holds in the sense that the worldline of a matter particle is converted to a worldline of particle of radiation.) However, in the big bang singularity, all worldlines are incomplete, thus leading to a contradiction with the basic assumption. Indeed the presence of the singularity signifies the break-down of the basic rules such as the action principle from which the equations of general relativity are obtained.

To account for the creation of matter in a nonsingular fashion, we introduce the additional input into the theory that the particle worldlines are with ends. The same action as before then describes this theory but the endpoints generate extra terms in the field equations.

For details of this work we refer the reader to Hoyle *et al.* [24]. The field equations in the "constant mass" conformal frame then take the form:

$$R_{ik} - \frac{1}{2}g_{ik}R + \lambda g_{ik} = -\kappa[T_{ik} - \frac{2}{3}(c_i c_k - \frac{1}{4}g_{ik}c^l c_l)]. \tag{30}$$

The scalar c-field arises from the contribution to inertia from ends of particle worldlines. These are the contributions of the Planck particles created which last a very short time scale$\sim 10^{-43}$ second.

Sachs et al [25] have solved these field equations and obtained a series of cosmological models which are a combination of two kinds: (i) models with creation of matter and (ii) models without creation of matter. The generic solution is known as the *Quasi Steady State Cosmology.* It is a nonsingular model which has a de Sitter type expansion with short term oscillations superposed on it. The former represents the creative and the latter the noncreative mode. The QSSC is being proposed as an alternative to the standard hot big bang cosmology (Narlikar [26]).

6. Conclusions

To summarize, the ideas which go under the name Mach's Principle are capable of wider applications than thought earlier by Ernst Mach. One can use the Machian concept in electrodynamics where the response of the universe can play a key role in both classical and quantum electrodynamics. The action at a distance framework used here is consistent with special relativity as well as with causality. The formalism can be used to give an expression to inertia as a direct long range effect from the distant parts of the universe. From inertia, one can arrive at a theory of gravity which is wider in its applications than general relativity. The theory can be extended to incorporate the cosmological constant and the concept of creation of matter without spacetime singularity. It leads to a viable cosmological model, known as the quasi-steady state cosmological model.

Acknowledgement: It is a pleasure to dedicate this article to the Festschrift of C.V. Vishveshwara with whom I have shared many common interests including the desire to understand this mysterious phenomenon called gravitation.

References

1. Schwarzschild, K. (1903), *Nachr. Ges. Wis. Gottingen*, **128**, 132.
2. Tetrode, H. (1922), *Z. Phys.*, **10**, 317.
3. Fokker, A.D. (1929a), *Z. Phys.*, **58**, 386.
4. Fokker, A.D. (1929b), *Physica*, **9**, 33.
5. Fokker, A.D. (1932), *Physica*, **12**, 145.
6. Wheeler, J.A. and Feynman, R.P. (1945), *Rev. Mod. Phys.*, **17**, 156.
7. Dirac, P.A.M. (1938), *Proc. Roy. Soc.*, **A167**, 148.
8. Hogarth, J.E. (1962), *Proc. Roy. Soc.*, **A267**, 365.
9. Bondi, H. and Gold, T. (1948), *M.N.R.A.S.*, **108**, 252.
10. Hoyle, F. (1948), *M.N.R.A.S.*, **108**, 372.
11. Hoyle, F. and Narlikar, J.V. (1963), *Proc. Roy. Soc.*, **A277**, 1.
12. Hoyle, F. and Narlikar, J.V. (1969), *Ann. Phys., New York*, **54**, 207.
13. Hoyle, F. and Narlikar, J.V. (1971), *Ann. Phys., New York*, **62**, 44.
14. Hoyle, F. and Narlikar, J.V. (1993), *Proc. Roy. Soc.*, **A442**, 469.
15. Hoyle, F. and Narlikar, J.V. (1964), *Proc. Roy. Soc.*, **A282**, 191.
16. Hoyle, F. and Narlikar, J.V. (1966), *Proc. Roy. Soc.*, **A294**, 138.
17. Synge, J.L. (1960) *Relativity, the General Theory*, North Holland, Amsterdam .
18. DeWitt, B.S. and Brehme, R.W. (1960), *Ann. Phys. (New York)*, **9**, 220.
19. Narlikar, J.V. (1974), *J. Phys.*, **A7**, 1274
20. Wheeler, J.A. and Feynman, R.P. (1949), *Rev. Mod. Phys.*, **21**, 424.
21. Hoyle, F. and Narlikar, J.V. (1974), *Action at a Distance in Physics and Cosmology*, Freeman, San Francisco.
22. Bagla, J.S., Padmanabhan, T. and Narlikar, J.V. (1996), *Comm. Astrophys.*, **18**, 275.
23. Islam, J.N. (1968), *Proc. Roy. Soc.*, **A306**, 487.
24. Hoyle, F., Burbidge and Narlikar, J.V. (1995), *Proc. Roy. Soc.*, **A448**, 191.
25. Sachs, R.,Narlikar, J.V. and Hoyle, F., *A & A*, **313**, 703.
26. Narlikar, J.V. (1998), in Proceedings of IAU Symposium 183 : *Cosmological Parameters and the Evolution of the Universe*, Ed. K. Sato, Dordrecht : Kluwer, to be published.

31. THE EARLY HISTORY OF QUANTUM GRAVITY (1916-1940)

JOHN STACHEL
Department of Physics and Center for Einstein Studies, Boston University, Boston, MA 02215, USA

It is with great affection that I dedicate this paper to Vishu, a dear friend and valued colleague for many years.

Most accounts of quantum gravity, insofar as they are concerned with its history at all, start with the post-World-War II period. Without attempting a full reconstruction of the pre-war history, I shall show that there was a lively discussion of some of the most important issues in the field, especially in the 1930s, which was cut off by the advent of the war and the untimely deaths of some of the main protagonists. When discussions of quantum gravity resumed after the war, they took place largely in ignorance of the fact that the main positions had been staked out earlier.

As so often the case in relativity, the story of quantum gravity begins with Einstein himself. Soon after the final formulation of general relativity, he pointed out the need for a quantum modification of the theory. In his first paper on gravitational radiation [1], Einstein argued that quantum effects must modify the general theory of relativity:

Due to the intra-atomic movement of electrons, atoms would have to radiate not only electromagnetic but also gravitational energy, if only in tiny amounts. As this is hardly true in nature, it appears that quantum theory would have to modify not only Maxwellian electrodynamics, but also the new theory of gravitation (p.209). Two years later, he reiterated this conclusion [2]:

As already emphasized in my previous paper, the final result of this argument, which demands a [gravitational] energy loss by a body due to its thermal agitation, must arouse doubts about the universal validity of the theory. It appears that a fully developed quantum theory must also bring about a modification of the theory of gravitation (p.164).

He soon began to speculate whether gravitation plays a role in the atomistic structure of matter [3]:

B. R. Iyer and B. Bhawal (eds.), Black Holes, Gravitational Radiation and the Universe, 525–534.

There are reasons for thinking that the elementary formations which go to make up the atom are held together by gravitational forces. ... The above reflexions show the possibility of a theoretical construction of matter out of the gravitational field and the electromagnetic field alone... In order to construct such a model of an "elementary particle," Einstein shows that it is necessary to modify the original gravitational field equations. The details of this paper are unimportant; the major interest of this paper is that his attention now shifted from possible quantum modifications of general relativity to the search for a unified theory of the electromagnetic and gravitational fields, on the basis of which he hoped to explain the structure of matter. Quantum effects are to be derived from such a theory, rather than postulated ad hoc. Einstein remained committed to this approach for the rest of his life: the search for a "natural" mathematical extension of the general theory in the hope that such a theory would somehow explain the quantization of matter and energy.

If he had lost interest in the possibility of applying quantization procedures –of either the old quantum theory or the new quantum mechanics– to the gravitational field equations [4], others took up this challenge. The intimate connection between space, time and gravitation in general relativity suggested the need for a quantum-based modification of general relativity to Oskar Klein [5]. In 1927 he argued that the quantum theory must ultimately modify the role of spatio-temporal concepts in fundamental physics:

As is well known, the concepts of space and time have lost their immediate sense as a result of the development of quantum theory. This is connected, as Bohr in particular has emphasized, with the fact that, according to the quantum theory, our measuring tools (light and free material particles) manifest a behavior that is characterized by the dilemma particle-wave, as a result of which a sharp definition of a spatio-temporal coincidence, which lies at the foundation of the usual concepts of space and time, becomes impossible. From this standpoint it is to be expected that the general theory of relativity stands in need of a revision in the sense of the quantum postulate, as also results from the fact that various of its consequences stand in contradiction to the requirements of the quantum theory.

Then follows a footnote:

For example, Einstein light-deflection by a collection of moving material bodies in a thermal equilibrium state would destroy the equilibrium of the radiation, which allows one to suspect a sort of gravitational Compton effect. One would expect a corresponding result in the case of statistical equilibrium between gravitational waves and light waves according to the general theory of relativity, due to the non-occurence of Planck's constant in the relevant equations.

Klein was cautious in his approach to the problem. After discussing the complementarity between the space-time description and the conservation laws for energy and momentum, he commented:

When one considers the apparently rather accidental role of the conservation laws in analytical mechanics as four of the many possible integrals of the equations of motion, then it could appear, in accord with the viewpoint just mentioned, as if the spatio-temporal magnitudes, which are of such central significance for our immediate experience, will retreat to the background in the mathematical structure of the future quantum theory. In spite of the great fruitfulness of quantum mechanics, many things speak for the view that the starting point for a general mathematical formulation of quantum theory is to be sought in the equations of classical field theory rather than in the equations of motion of mechanics. Thereby, the question of the role of spatio-temporal quantities (or rather of their quantum-mechanical analogues) appears in a different light, for a large part of the state variables characterizing field theory (the Einstein gravitational potentials) are indeed formally immediately tied to space-time concepts (pp.188-189).

Others were less cautious in approaching the problem of a quantum theory of gravity. In their first paper on quantum electrodynamics, Heisenberg and Pauli asserted [6]:

Quantization of the gravitational field, which appears to be necessary for physical reasons [in a footnote, they refer to the works of Einstein and Klein cited above], may be carried out without any new difficulties by means of a formalism fully analogous to that applied here (p.3).

Almost seventy years have elapsed since this optimistic prediction, yet we are still without a real quantum theory of gravity! The first attempt to fulfill Heisenberg and Pauli's prediction was made by Leon Rosenfeld [7], who was then working with Pauli. The latter had developed a method for the quantization of fields with gauge groups, the Hamiltonian formulations of which involve constraints. Rosenfeld applied Pauli's method to the linearized Einstein field equations. After setting up a formalism to deal with both the electromagnetic and linearized gravitational fields, he showed that the gravitational field– like the electromagnetic– can only be quantized using commutators (Bose quantization). In a second paper, he calculated the gravitational self-energy of a light quantum, which he found to diverge. He attributed this infinity to the fact "that one cannot attribute a finite radius to a light quantum. The analogy with the case of the electron need hardly be emphasized" (p.596) [8]. He then proceeded to discuss various possible transition processes involving light quanta and gravitational quanta, and concluded: "If we imagine a cavity initially filled with radiation (without dust particles!), then the first-approximation gravitational interactions between light quanta suffice to establish Planckian [i.e., thermal] equilibrium (with a speed proportional to $1/k$ [$k = 8\pi G/c^4$]" (p.598) [9]. In contrast to his post-World War II position, at this time Rosenfeld apparently saw no problems for a quantization of the gravitational field arising from the peculiar features of gravity.

During the thirties, the concept of a gravitational quantum analogous to the photon came to be generally (but as we shall see, not universally) accepted by theoretical physicists. (I exclude from this discussion the small group that followed Einstein in his continued search for a unified theory that would explain the quantum.) In the early thirties, there was some discussion of Bohr's idea that the neutrino might be related to the gravitational quantum [10]. The most elaborate such discussion that I know of appeared in a paper by Blokhintsev and Gal'perin [11], cited at length in Gorelik and Frenkel 1994 (pp.96-97), who maintain that the name "graviton" for the gravitational quantum here first appears in print.

In the late thirties another Pauli coworker, Markus Fierz, developed a formalism to handle free quantum fields of arbitrary spin and mass [12], soon generalized by Pauli and Fierz to include electromagnetic interactions [13]. They noted that the massless spin-two field obeys equations formally identical with the linearized Einstein equations. Pauli prepared a "Report on the general properties of elementary particles" for the projected 1939 Solvay Congress, which was never held due to the outbreak of World War II. At the urging of Heisenberg [14], Pauli's report included a section entitled "Remarks on Gravitational Waves and Gravitational Quanta (Spin 2)." Although the complete report was only published recently [15], most of its contents had already been published elsewhere. The section on gravitation was not, but Heisenberg– and no doubt many another leading physicist– was aware of its contents around 1939 [16].

In this section, Pauli develops a Lagrangian for the massless spin-two field formally identical with that for the linearized Einstein equations. He notes that the invariance of the theory under a group of what (by analogy with the gauge transformations of Maxwell's theory) he calls "gauge transformations of the second kind" corresponds to the invariance under infinitesimal coordinate transformations of the linearized Einstein theory. He quantizes the theory by a method analogous to Fermi's method for electrodynamics, i.e., by requiring the operators corresponding to the linearized harmonic coordinate conditions to annihilate the wave function. This method allows imposition of commutation relations on all components of the linearized metric using the invariant D-function, with which the harmonic constraints are compatible. Pauli concludes:

The gravitational quanta or "gravitons" so defined have the...spin 2 ... It is certainly a limitation of the quantum-mechanical side of this treatment, that one leaves it at that approximation, in which the general-relativistic field equations are linear. This limitation is most closely connected with the well-known divergence difficulties of field theory (p.901).

Apart from signaling these divergence difficulties, common to all quantum field theories at that time, Pauli, like Rosenfeld before him, seemed satisfied with linearized quantum gravity. At any rate, he did not offer any indication that quantizing the gravitational field might confront unique physical problems.

But in 1935, well before Pauli's work and apparently unknown to him, a young physicist working at the Leningrad Physico-Technical Institute, Matvei Petrovich Bronstein, had already applied Fermi's quantization technique to the linearized Einstein equations. If his technical approach was similar to Pauli's, Bronstein drew much more skeptical conclusions about the physical significance of the resulting formalism.

Bronstein was no newcomer to the problem of quantum gravity in 1935. He had been pondering the issue since at least 1930, when he opined: "It is a task of the nearest future to identify the ties between quantum mechanics and the theory of gravitation" (cited from Gorelik and Frenkel 1994, p.88). In 1931, he summed up an article on unified field theories by stating: "It seems that Einstein's program of unified field theory will prove impossible... what will be needed is a sort of marriage between relativity and quantum theory" (ibid.). By 1933 he was stressing the crucial significance of the three dimensional constants, c, G, and h, in defining the past and future development of physics (Bronstein 1933):

After relativistic quantum theory is formulated, the next task would be to realize the third step .... namely, to merge quantum theory (h constant) with special relativity (c constant) and the theory of gravitation (G constant) into a single whole (cited from Gorelik and Frenkel 1994, p.90) [17].

In 1935 Bronstein presented his Doctoral Thesis on "Quantizing Gravitational Waves" to a committee that included Vladimir Fock and Igor Tamm [18]. In 1936, the work was published in Russian [19], as well as in a condensed German version [20]. After carrying out the quantization of the linear theory by the Fermi method, developing the quantum analogue of Einstein's quadrupole radiation formula, and deducing the Newtonian law of attraction from the interchange of longitudinal gravitational quanta, Bronstein proceeds to some critical reflections on the physical significance of his results.

He carries out an analysis of the measurability of the (linearized) Christoffel symbols, which he takes to be the components of the gravitational field. By analogy with the then-recent Bohr-Rosenfeld analysis of the measurability of the electromagnetic field components, he shows that there are limitations on the measurability of the gravitational field components implied by the uncertaintly relations between position and momentum of a test body, the acceleration of which is used to measure the gravitational field. But he notes that there is an additional gravitational complication, which has no electromagnetic analogue: To measure the components of the electromagnetic field, it is permissible to introduce electrically neutral test bodies, which have no effect on the field being measured. But in the gravitational case, due to the universality of gravitational interactions, the effect of the energy-momentum of the test bodies on the gravitational field cannot be neglected– even in linear approximation. Bronstein derives an expression for the minimum uncertainty in a measurement of a component of the Christoffel symbols that depends inversely on the mass density of the test body, just as Bohr-Rosenfeld's corresponding result does on the charge density ρ of the test body. He then states what he sees as the crucial difference between the two cases:

Here we should take into account a circumstance that reveals the fundamental distinction between quantum electrodynamics and the quantum theory of the gravitational field. Formal quantum electrodynamics that ignores the structure of the elementary charge does not, in principle, limit the density of ρ. When it is large enough we can measure the electric field's components with arbitrary precision. In nature, there are probably limits to the density of electric charge ... but formal quantum electrodynamics does not take these into account... The quantum theory of gravitation represents a quite different case: it has to take into account the fact that the gravitational radius of the test body ($k\rho V$) must be less than its linear dimensions $k\rho V < V^{1/3}$ (Bronstein 1936b, p.217, transl. from Gorelik and Frenkel 1994, p.105 and Gorelik 1992, pp.376-377).

He acknowledges that "this was a rough approximation, because with the measuring device's relatively large mass departures from the superposition principle [i.e., the linearized approximation] will probably be noticeable;" but thinks "a similar result will survive in a more exact theory since it does not follow from the superposition principle. It merely corresponds to the fact that in General Relativity there cannot be bodies of arbitrarily large mass with the given volume" (Gorelik and Frenkel, 1994, pp.105-106). He concludes:

The elimination of the logical inconsistencies connected with this result requires a radical reconstruction of the theory, and in particular, the rejection of a Riemannian geometry dealing, as we have seen here, with values unobservable in principle, and perhaps also rejection of our ordinary concepts of space and time, modifying them by some much deeper and nonevident concepts. Wer's nicht glaubt, bezahlt einen Taler (Bronstein 1936b, transl. from Gorelik 1992, p.377) [21].

At least one physicist outside the Soviet Union acknowledged, and indeed extended, Bronstein's views. In 1938 the French physicist Jacques Solomon after summarizing Bronstein's argument concluded [22]:

In the case when the gravitational field is not weak, the very method of quantization based on the superposition principle fails, so that it is no longer possible to apply a relation such as [the equation setting a lower limit on the measurability of the linearized field strength] in an unambiguous way ... Such considerations are of a sort to put seriously in doubt the possibility of reconciling the present formalism of field quantization with the non-linear theory of gravitation (p.484).

In one of the many tragic ironies of history, both of these pre-war advocates of the need for a radically different approach to quantum gravity perished prematurely. On August 6, 1937, Bronstein was arrested by the Soviet State Security agency (NKVD); only twenty years later did his widow, the writer Lydia Chukovskaya, learn the exact date of his death: February 18, 1938 [23]. References to his name disappeared from the annals of Soviet physics for several decades. Jacques Solomon, a Communist militant active in the underground resistance to the German occupation of France, was arrested together with his wife, Hélène Langevin, in March 1942. He was killed by the Germans on May 23, 1942; she was sent to Auschwitz, but survived the war [24]. Between them, Stalin and Hitler saw to it that the post-World-War-II discussion of quantum gravity took place without what could have been two significant voices.

Bronstein was a student of Yakov Frenkel, who was quite skeptical of the whole project of quantizing the gravitational field equations. In an article prepared for the Schilpp volume (see footnote 4), he argued against the analogy between the gravitational and electromagnetic fields that is the basis of the graviton approach. Since this article is unavailable to me, I shall quote a summary that includes citations from Gorelik and Frenkel 1994:

[Frenkel] argued that "the electromagnetic field [was] matter while the gravitational field merely [determined] the metrical properties of the space-time continuum." He insisted that "strictly speaking there [were] no such things as gravitational energy or gravitational momentum since the corresponding values [did] not form a true tensor and [were] nothing more than a pseudotensor." It was his conviction that the attempts to quantize gravitation were seneseless since "the gravitational field [had] macroscopic, rather than microscopic meaning; it merely [supplied] a framework for physical events occuring in space and time while quantizing [could] be applied only to microscopic processes in material fields." These considerations were little affected by the developments in physics that occurred after November 1935-this and his remark during Bronstein's defense ... allows us to surmise that his position was the same in 1935 (p.85).

Another physicist who expressed doubts during the pre-war period about the need to quantize general relativity was van Dantzig. He considered general relativity to be a sort of thermodynamic limit of a deeper, underlying theory of interactions between particles [25].

Thus, by the mid-1930's the three positions that were to characterize post-World War II discussions of quantum gravity (among physicists not still wedded to Einstein's unified field theory program) had already been enunciated:

1) Quantum gravity should be formulated by analogy with quantum electrodynamics. In particular, one should start from quantization of the linearized gravitational field equations. Technical problems that arise will be similar to those arising in quantum electrodynamics and will presumably be solved pari passu with the problems of the latter (Rosenfeld, Pauli, Fierz) [26],

2) The unique features of gravitation will require special treatment. In particular, the full theory, with its non-linear field equations must be quantized. This implies that, while the general techniques of quantum field theory may be relevant, they must be generalized in such a way as to be applicable in the absence of a background metric (Bronstein, Solomon).

3) General relativity is essentially a macroscopic theory, to which the techniques of quantum field theory should not be (mis)applied (Frenkel, van Dantzig). It is sobering to realize how little real progress has been made on the problem of quantum gravity since these alternatives were posed sixty years ago, particularly when one recalls that sixty years is the time span that separates Maxwell's treatise from the mid-thirties!

I thank the Division of Humanities and Social Sciences, California Institute of Technology for its hospitality during the period in which the final version of this paper was prepared.

References

1. Albert Einstein, "Noherungsweise Integration der Feldgleichungen der Gravitation," Preussische Akademie der Wissenschaften (Berlin). Sitzungsberichte (1916): 688-696, translated as "Approximative Integration of the Field Equations of Gravitation," in The Collected Papers of Albert Einstein, vol. 6, The Berlin Years: Writings 1914-1917, English Translation of Selected Texts (Princeton University Press, 1997), Alfred Engel, transl., pp.201-210.
2. Albert Einstein, "Über Gravitationswellen," Preussische Akademie der Wissenschaften (Berlin). Sitzungsberichte (1918):154-167.
3. Albert Einstein, "Spielen Gravitationsfelder im Aufbau der materiellen Elementarteilchen eine wesentliche Rolle?," Preussische Akademie der Wissenschaften (Berlin). Sitzungberichte (1919): 349-356; translated as "Do Gravitational Fields Play an Essential Part in the Structure of the Elementary Particles of Matter?" in The Principle of Relativity, Otto Blumenthal, ed., W. Perret and J. B. Jeffery, transls. (Methuen, London, 1923), reprint 1952 (Dover, New York), pp.191-198.

4. There is an intriguing comment by Y. I. Frenkel, in a paper written for the Schilpp volume Albert Einstein: Philosopher- Scientist, but not submitted: "Einstein was probably the first to assimilate gravitational waves and the corresponding particles in a conversation with the author back in 1925" (quotation from Gennady E. Gorelik and Victor Y. Frenkel, "Matvei Petrovich Bronskin and Soviet Theoretical Physics in the Thirties", Birkhauser, Cambridge, MA, 1994, p.85, *cited hereafter*).
5. Oskar Klein, "Zur fiifdimensionalen Darstellung der Relativittstheorie," Zeitschrift für Physik 46 (1927): 188 -xxx. Klein was working with Bohr in Copenhagen at this time, and his comments may well reflect Bohr's views.
6. Werner Heisenberg and Wolfgang Pauli, "Zur Quantenelektrodynamik der Wellenfelder," Zeitschrift für Physik 56 (1929): 1-61.
7. Leon Rosenfeld, "Zur Quantelung der Wellenfelder," Annalen der Physik 5 (1930): 1113-152; "Über die Gravitationswirkungen des Lichtes," Zeitschrift für Physik 65 (1930): 589-599.
8. In response to a query from Pauli (see Pauli to Rosenfeld, 12 April 1931, in Wolfgang Pauli, "Scientific correspondence with Bohr, Einstein, Heisenberg and others", vol.3, 1940-1949, Karl von Meyenn, ed., Springer Verlag, New York, 1993, p.746) Rosenfeld added a supplement to his paper showing that the gravitational self-energy of any one-photon state is infinite. Solomon soon showed that this divergence was not due to the zero-point energy of the field. See Jacques Solomon, "Nullpunktsenergie der Strahlung und Quantentheorie der Gravitation," Zeitschrift für Physik 71 (1931): 162-170.
9. Rosenfeld himself later came to question the necessity of quantizing the gravitational field, and did not include his work on this topic in a collection of papers that he selected; instead he reprinted critical comments on field quantization, including arguments against the need for a quantum gravity. See Leon Rosenfeld, "On Quantization of Fields," in Selected Papers of Leon Rosenfeld, Robert S. Cohen and John Stachel, eds. (Reidel, Dordrecht/Boston/London, 1979)(hereafter Rosenfeld 1979), pp.442-445, and "Quantum Theory and Gravitation," ibid., pp.598-608.
10. See Wolfgang Pauli, Wissenschaftlicher Briefwechsel, vol. 2, 1930-1939, Karl von Meyenn, ed. (Springer Verlag, Berlin/Heidelberg/New York/Tokyo, 1985), Bohr to Pauli, 15 March 1934, p.308: "The idea was that the neutrino, for which one assumes a zero rest mass, could hardly be anything else than a gravitational wave with appropriate quantization" (transl. from Niels Bohr, Collected Works. vol. 7, Foundations of Quantum Phyics II (1933-1958), J. Kalcar, ed. (Elsevier, Amsterdam/Lausanne/New York/Oxford/Shannon/Tokyo, 1996), p.479). Fermi had evidently had a similar idea, but was aware of the problem of the different spins. See ibid., Pauli to Heisenberg 6 February 1934, p.277: "Fermi would prefer to make a connection between neutrinos and half gravitational quanta." As late as November 1934, Pauli cautiously stated: "While up to now it has been held almost certain that gravitational phenomena play practically no role in nuclear physics, it now seems that the possibility cannot be immediately rejected, that the phenomena of beta-radiation could be connected with the square root of kappa [the gravitational constant] ("Raum, Zeit, und Kausalität 1in der modernen Physik," Scientia 59 (1936): 65-76, p.76). This suggests that Pauli may have had in mind construction of a graviton from two neutrinos, along the lines of DeBroglie's neutrino theory of light.
11. Dmitri Ivanovich Blokhintsev and F. M. Gal'perin, "Gipoteza neitrino i zakon sokhraneniya energii," Pod znamenem marxisma (1934), no. 6, pp.147-157. As cited by Gorelik and Frenkel, they wrote: "The comparison displayed above indicates that the graviton and the neutrino have much in common, This probably testifies that in general the highly improbable process of graviton radiation becomes practically observable in beta-decay. If the neutrino turns out to be the graviton this would mean that contemporaray physics had approached the limits beyond which there would

be no present insurmountable barrier between gravitation and electromagnetism. Due to theoretical considerations it is hard to identify gravitons with the neutrino since it is hard to admit that they have the same spin 1/2 as the neutrino. In this respect gravitons have much more in common with light quanta. It is impossible, however, to totally rule out a theoretical possibility of their identification. So far it is much more correct to regard the neutrino as an independent type of particle" (Gorelik and Frenkel 1994, p.97).

12. Marcus Fierz, "Über die relativistische Theorie kräftefreier Teilchen mit beliebigem Spin," Helvetica Physica Acta 12 (1939): 3-37.
13. Wolfgang Pauli and Marcus Fierz, "Über relativistische Wellengleichungen von Teilchen mit beliebigem Spin im elektromagnetischen Feld," Helvetica Physica Acta 12 (1939): 297-300; ibid., "On relativistic wave equations for particles of arbitrary spin in an electromagnetic field," Royal Society (London). Proceedings A173 (1939): 211-232.
14. See Pauli to Heisenberg, 10 June 1939, in Wolfgang Pauli, Wissenschafliche Briefwechsel, vol 2, 1930-1939, Karl von Meyenn, ed. (Springer, Berlin/Heidelberg/New York/Tokyo, 1985), p.662; and Heisenberg to Pauli, 12 June 1939, ibid., p.665.
15. See Wolfgang Pauli, Wissenschafliche Briefwechsel, vol 2, 1930-1939, Karl von Meyenn, ed. (Springer, Berlin/Heidelberg/New York/Tokyo, 1985), pp.833-901; the section on gravitation is on pp.897-901.
16. See, for example, Pauli to Schrödinger, 5 November 1939, ibid., p.823-825.
17. In 1934 Pauli also discussed "the three fundamental natural constants," but added: "for the sake of simplicity we ignore gravitational phenomena for the present" (the article was not published until 1936 in Scientia; see the reference in note 10).
18. For Bronstein's life and work, see Gorelik and Frenkel 1994.
19. Matvei Petrovich Bronstein, "Kvantovanie gravitatsionnykh voln [Quantization of gravitational waves]," Zhurnal Eksperimentalnoy i Teoreticheskoy Fiziki 6 (1936): 195-236.
20. Matvei Petrovich Bronstein, "Quantentheorie schwacher Gravitationsfelder," Physikalische Zeitschrift der Sowjetunion 9 (1936): 140-157 (*hereafter Bronstein, 1936b*).
21. The German phrase–"Let him who doubts it pay a Thaler"– comes from the Grimm brother's tale, "Der tapfere Schneider."
22. Jacques Solomon, "Gravitation et Quanta," Journal de Physique et de Radium 9 (1938): 479-485.
23. See Gorelik and Frenkel 1994, pp.144-147; and Lydia Chukovskaya, The Akhmatova Journals/ Volume I 1938-41 (Farrar, Strauss and Giroux, New York 1994).
24. See Leon Rosenfeld, "Jacques Solomon," in Rosenfeld 1979, pp.297-301; Martha Cecilia Bustamente, "Jacques Solomon (1908-1942): Profil d'un physicien thèoricien dans la France des annèes trente," Revue d'histoire des sciences 50, 49-87, 1997.
25. See D. van Dantzig, "Some possibilities of the future development of the notions of space and time," Erkenntnis 7 (1938): 142-146; "On the Relation Between Geometry and Physics and the Concept of Space-time," in Füenfzig Jahre Relativitäetstheorie, P. Kervaire, ed., Helvetica Physica Acta Supplementum IV, (Birkhäuser, Basel 1956), pp.48-53.
26. Indeed, in 1949 Bryce DeWitt, using Schwinger's covariant technique, recalculated the gravitational self-energy of the photon and showed that it vanished identically. See Carl Bryce Seligman [DeWitt], "I. The Theory of Gravitational Interactions. II. The Interactions of Gravity With Light," Ph.D. Thesis, Harvard University, December 1949. Note that DeWitt emphasized the need to quantize the full, nonlinear theory, and never regarded quantization of the linearized equations as more than a preliminary exercise.

32. GEOMETRY IN COLOR PERCEPTION

ABHAY ASHTEKAR[1,3], ALEJANDRO CORICHI[1,2],
MONICA PIERRI[1]

[1] *Center for Gravitational Physics and Geometry*
Department of Physics, Penn State,
University Park, PA 16802-6300, USA

[2] *Instituto de Ciencias Nucleares*
Universidad Nacional Autónoma de México
A. Postal 70-543, México D.F. 04510, México

AND

[3] *Erwin Schrödinger International Institute for Mathematical Sciences*
Boltzmanngasse 9, A-1090 Vienna, Austria

1. Introduction

Of the five human senses, only two can be excited essentially instantaneously by signals from distant sources: hearing and sight. In both cases, signals propagate as waves. It is not an accident therefore that we have television but not its equivalent in the realm of smell, taste or touch. Smell for example, propagates by diffusion, a slow process which is difficult to control, while taste and touch require immediate contact with the source of the signal. There is, however, a rather fundamental difference also between hearing and sight. Hearing occurs because sound excites cells in the cochlear tube of the inner ear and each little longitudinal section of the tube is sensitive to a different frequency. Consequently, we are capable of distinguishing a very large class of 'sound profiles'. Thus, the space of sounds that human beings are capable of hearing has a very large dimension, which, for all practical purposes, can be taken to be *infinite*. This is in striking contrast with sight. It turns out that the space of colors that human beings are capable of seeing is only *three* dimensional!

B. R. Iyer and B. Bhawal (eds.), Black Holes, Gravitational Radiation and the Universe, 535–549.

The subject of color has a long and distinguished history and some of the most creative mathematicians, physicists and philosophers have devoted years of study to unravel its mysteries[1]. As is the case with so many fascinating discoveries, here too the road begins with Newton [1]. It was he who first realized that distinct intensity (or spectral) distributions $I(\lambda)$ of light are often indistinguishable to the human eye. In the modern terminology, he concluded that a *perceived color* is an *equivalence class* $\{I(\lambda)\}$ of spectral distributions and then went on to argue that, while the space of all spectral distributions $I(\lambda)$ is infinite dimensional, the space $\mathcal{C}$ of perceived colors $\{I(\lambda)\}$ is only three-dimensional. A century later, in 1802 Thomas Young [2] made the bold conjecture that the origin of this three-dimensionality may lie in the property that color sensation is caused by absorption of light by three different pigments in the human retina. It took another century and a half to definitively confirm this conjecture experimentally [3]!

In the meantime, in 1853 Grassmann [4, 5] gave a clean mathematical formulation of Newton's ideas. The following year, Riemann gave his epoch making inaugural address in Göttingen, in which he laid down the foundations of differential geometry. During this lecture, he singled out the physical space in which we live and the space of perceived colors as possible examples of 'curved spaces' of *three* dimensions [6]. The first of these has been quoted often in the context of general relativity. However, perhaps it is the second that constitutes a more immediate example, as far as our common experience is concerned. Maxwell sharpened these notions and then used them in 1861 to explicitly demonstrate that color photography was possible, despite the fact that the photographic emulsions themselves were black and white. The method of his demonstration supported Young's hypothesis. About the same time, Helmholtz was interested in our perception of both sound and color and made significant contributions to both areas. In particular, in 1891, he suggested [7] a form of the metric on the space $\mathcal{C}$ of perceived colors which determines the 'distance' between any two colors, i.e., how far or close an average human subject perceives the two colors to be. A little later, Schrödinger devoted several years to the study of colors. Indeed, this was one of his primary scientific pre-occupations during his Vienna years. In 1920, he proposed another, 'improved' form of the metric [8]. Since then, there have been several other attempts at improvements, perhaps the most notable being that of Stiles [9].

The aim of this contribution is to present a simple physical model of color perception and derive from it the geometric structure of $\mathcal{C}$ in a systematic, pedagogical fashion. In the spirit of this volume, the account is

[1]For example, the chemist John Dalton, the poet Johann Wolfgang von Goethe, and philosophers John Locke and Arthur Schopenhauer all had theories of color of their own. Ludwig Wittgenstein's last work was *Remarks on Colour*.

addressed to advanced undergraduate or beginning post-graduate students and assumes only a rudimentary knowledge of differential geometry.

Historically, the basic, geometric structure of C was discovered slowly, using simple experiments on human subjects. In the spirit of mathematical physics, one can therefore ask if this well-known structure can be derived systematically from first principles, knowing only adsorption coefficients of cones in the human eye that are now known. We will see that the answer is in the affirmative. This derivation is a simple example of a *phenomenological* application of differential geometry; one which is qualitatively similar to but *significantly* simpler than the more well-known applications to, e.g., cosmology where one uses the observed large scale homogeneity of the universe to severely constrain its geometry. We should emphasize that this is *not* meant to be a scholarly account. In particular, we will use terminology which should be intuitively obvious to mathematical physicists, even though it occasionally differs from that used in the scholarly literature, decreed by the International Commission on Illumination. (For example, what we call 'perceived color' is called 'psycho-physical color' in the standard lore.)

The structure of the paper is as follows. Section 2 contains the main mathematical discussion. The model we discuss is very simple, based only on physical –rather than psycho-physical– aspects of color perception. It ignores the intervention of the brain which in fact plays an important role in certain aspects of color perception. These limitations are discussed in section 3.

2. Geometry

This section is divided into three parts. In the first, we outline the procedure by which early researchers in the field arrived at a qualitative picture (figure 1) of the space C of perceived colors. In the second part, we will show how one can derive this structure in a straightforward fashion starting from the qualitative features of the experimentally observed adsorption curves. In the third, we discuss the Riemannian geometry of this space.

2.1. PHENOMENOLOGY

The key experimental technique used to unravel the structure of the space of perceived colors was quite simple: Mix and match. More precisely, mixtures of spectral distributions $I(\lambda)$ were exposed to the a large number of human subjects who were asked to comment on whether they perceived the same color or different, and, if different, how 'close' or 'far' these colors seemed to be. From the resulting data, qualitative features of the space C of perceived colors could be deduced.

First, let us consider the so-called *spectral colors*, i.e., those spectral distributions which can *not* be further decomposed by a prism or a grating. In the idealized description, in this case, we have

$$I(\lambda) = k\delta(\lambda, \lambda_0) \tag{1}$$

for some constant k and a fixed wave-length λ_0, and, in a more realistic description,

$$I(\lambda) = k \begin{cases} \frac{1}{\epsilon}, & \text{if } \lambda_0 - \frac{\epsilon}{2} < \lambda < \lambda_0 + \frac{\epsilon}{2} \\ 0, & \text{otherwise} \end{cases} \tag{2}$$

for some $\epsilon << 1$. It is an observed fact that each spectral color is perceived as distinct and a mixture of two spectral colors produces a new color; spectral colors can not be mixed to obtain another spectral color. Furthermore, binary mixtures of spectral colors suffice to produce all observed colors. That is, any color produced by mixing three or more spectral colors can also be produced by mixing a (different, in general) pair of spectral colors. Finally, the space of spectral colors can be parametrized by two numbers, the wave-length λ_0 which takes values in the visual range and the intensity k. It is thus a two-dimensional space.

The second important feature of $\mathcal{C}$ is that it has an underlying convex-linear structure. To see this, let us first introduce some notation. Denote by $\mathcal{I}$ the space of all spectral distributions,

$$\mathcal{I} := \{I(\lambda) : \quad (-1, 1) \longrightarrow R; \quad I(\lambda) \geq 0.\} \tag{3}$$

where for notational simplicity we have normalized the spectrum of visible light to lie in the interval $(-1, 1)$. Because $I(\lambda)$ are required to be non-negative, $\mathcal{I}$ is not a vector space. However, if $I_n(\lambda)$ are elements of $\mathcal{I}$, then for all constants $a_n \geq 0$, $\sum a_n I_n(\lambda)$ also belong to $\mathcal{I}$ if n runs over a finite range. (In particular, if $I_1(\lambda)$ and $I_2(\lambda)$ belong to $\mathcal{I}$, so does the 'line' joining them, $aI_1(\lambda) + (1 - a)I_2(\lambda)$ where $0 \leq a \leq 1$.) That is, $\mathcal{I}$ is an infinite-dimensional, convex sub-space of the vector space consisting of all functions on the interval $(-1, 1)$. Since distinct spectral intensities can be perceived as the same color, there is a (in fact, huge!) projection P:

$$P: \ \mathcal{I} \longrightarrow \mathcal{C}\,. \tag{4}$$

Let us denote by c the image of the spectral distribution $I(\lambda)$. There is an obvious equivalence relation $\sim$ defined on $\mathcal{I}$ given by: $I(\lambda) \sim I'(\lambda) \iff P(I) = P(I')$. Thus, we have a $1-1$ relation between these equivalence classes and perceived colors: $\{I(\lambda)\} \equiv c$. Spectral distributions $I(\lambda)$ which belong to the same equivalence class, c, are called *metamers* (of the color c). Now, observations show that the projection map P has a key property: it

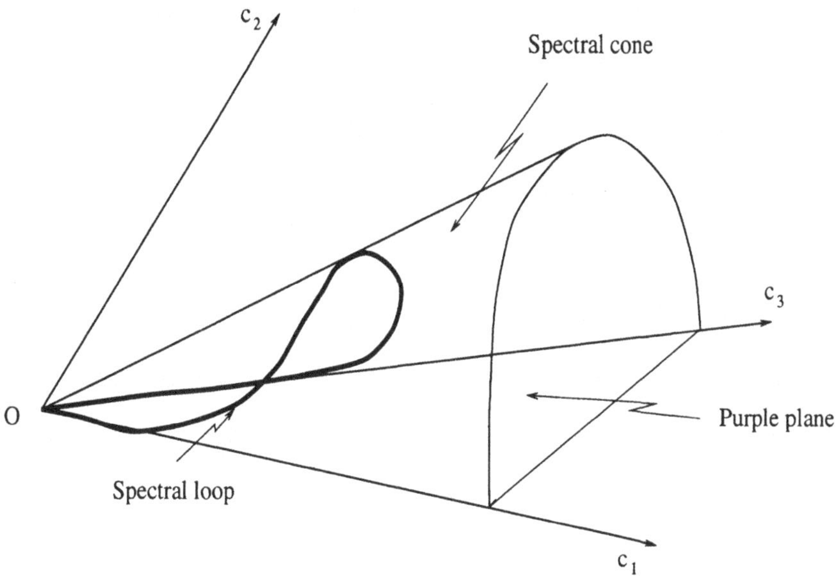

Figure 1. The general structure of the space $\mathcal{C}$ is shown. The Spectral cone together with the Purple plane are the boundary of the convex set $\mathcal{C}$. The axes are not drawn at 'right angles' because there is no metric to define the notion of orthogonality.

preserves linearity. More precisely, for each integer n, suppose that the spectral distributions $I_n(\lambda)$ and $I'_n(\lambda)$ are metamers, i.e., $\{I_n(\lambda)\} = \{I'_n(\lambda)\}$. Then, their linear combinations are also metamers (of a different color):

$$\{I_n(\lambda)\} = \{I'_n(\lambda)\} \Longrightarrow \{\sum_n a_n I_n(\lambda)\} = \{\sum_n a_n I'_n(\lambda)\} \tag{5}$$

for all non-negative coefficients a_n. Therefore, the space $\mathcal{C}$ of perceived colors inherits from $\mathcal{I}$ the structure of a convex-linear space.

By putting together the observations of the last two paragraphs, one can conclude that the spectral colors constitute a boundary of $\mathcal{C}$. Since the boundary is two-dimensional, *the space $\mathcal{C}$ is a 3-dimensional convex-linear space.* This underlying mathematical structure is reminiscent of the structure of states (i.e. density matrices) in quantum mechanics. The space of all states is also convex-linear, the boundary consists of pure states and any mixed state can be obtained by a superposition of pure states. In the present case, the spectral colors are the analogs of pure states.

Finally, let us consider the structure of the part of the boundary of $\mathcal{C}$ formed by spectral colors. The boundary is parametrized by λ_0, the wave length of the spectral color and k the intensity. Hence, it is of interest to examine the one-dimensional curve made of all colors of unit intensity

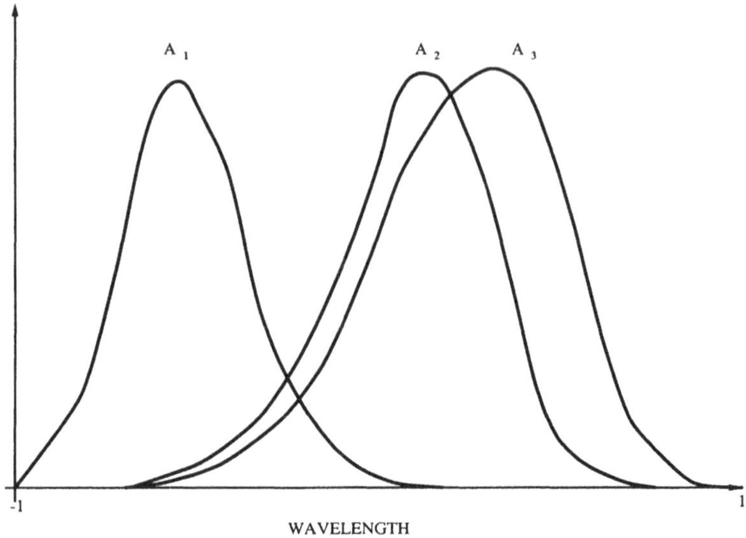

Figure 2. The adsorption functions for the three eye cones. The actual domain, (380nm, 780nm) is normalized to the interval $(-1, 1)$ for simplicity.

$(k = 1)$. It turns out that this curve is a closed loop; it is called the *spectral loop*.

Figure 1 summarizes the qualitative structure of $\mathcal{C}$ that had been arrived at phenomenologically. The space $\mathcal{C}$ is the manifold with boundary that resembles a solid cone. It turns out that the complete boundary consists of a *spectral cone* and the *purple plane*. As one moves along the spectral loop, one runs through distinct spectral colors while motion along 'radial' direction corresponds only to change in the intensity of these colors. The purple plane consists of those colors which can be obtained by combining extreme violet and extreme red.

2.2. DERIVATION OF THE STRUCTURE OF $\mathcal{C}$

Let us now see, in the spirit of mathematical physics, how the structure of $\mathcal{C}$ can be derived from first principles.

The experimental input will now consist only of the observed adsorption curves of the three cones in the human eye, which are qualitatively sketched in figure 2. All results of this sub-section will follow from this information.

Denote the three adsorption functions by $A_i(\lambda)$, $i = 1, 2, 3$. From physical considerations, it is obvious that, given an incident spectral distribution $I(\lambda)$, there are three 'output' signals c_i, one for each cone, given by

$$c_i := \int_{-1}^{1} \mathrm{d}\lambda \, I(\lambda) \, A_i(\lambda) \tag{6}$$

sent to the brain. Therefore, if we ignore potential complications in the brain itself –these are discussed in Sec. III– we can characterize each perceived color by the three numbers, c_i. Thus, in the simplest mathematical model, we have a precise formula for the projection map P (of (4)) which maps the infinite dimensional space of spectral distributions to the three dimensional space of perceived colors:

$$P[I(\lambda)] = c_i \, . \tag{7}$$

We will use this map to recover the structure of the space of colors depicted in figure 1, obtained in Sec II.A from several phenomenological inputs.

First of all, it is obvious that the space $\mathcal{C}$ is three dimensional. Next, since we now have the *explicit* form of the projection map, we study its properties. It is obvious that the map is convex-linear, i.e. satisfies (5). Note that this is now a *derived property*, not an independent phenomenological input that it was in Sec. 2.1. Thus, we have shown that $\mathcal{C}$ has a natural convex-linear structure.

Next, the space is endowed with a natural 'radial' vector field. To see this, note that, since the space of spectral distributions, $\mathcal{I}$, is naturally embedded in the vector space of all functions on the interval $(-1, 1)$, it has a natural radial vector field. Motions along the integral curves of this vector field correspond to spectral distributions which have the same profile except for an overall multiplicative factor. This vector field naturally projects to a vector field B^a on $\mathcal{C}$. This is the *brightness vector field* and its components in the natural chart c_i are given by $B^i = c_i$. The integral curves of this vector field are rays through the origin and, as the name suggests, points that lie on any one ray represent colors which are all the same except for brightness.

Finally, let us look at spectral colors, i.e., (idealized) spectral distributions (1). Because of the brightness vector field, the structure of this sub-set can be understood entirely in terms of the structure of the spectral curve, i.e., colors corresponding to incident intensities $I(\lambda) = \delta(\lambda, \lambda_0)$. Since these colors correspond to coordinates $c_i = A_i(\lambda_0)$, the spectral curve $\mathbf{c}(\lambda_0)$, parametrized by λ_0 is given simply by

$$c_i(\lambda_o) = A_i(\lambda_0) \tag{8}$$

in the natural chart. Some of the elementary properties of the spectral curve now follow immediately from the form of the adsorption functions (i.e., figure 2). To see this, let us embed $\mathcal{C}$ in R^3 (without a specific metric) as in figure 1. Let us begin at the origin, $c_i = 0$, the totally black color corresponding to the incident intensity $I(\lambda) = 0$, and follow the spectral curve as λ increases from -1 to 1. If we move a little bit, only the first set of cones is excited, whence only A_1 is non-zero, i.e., for a finite interval, the spectral curve $\mathbf{c}(\lambda_0)$ is tangential to the c_1-axis. Physically, this part of the spectral curve corresponds to the red end of the spectrum. As we continue to move, we reach the value of λ at which the function A_2 starts taking values different from zero. The curve $\mathbf{c}(\lambda_0)$ then lies in the (c_1, c_2) plane. The curve $\mathbf{c}(\lambda_0)$ departs that plane when A_3 stops being zero, and it goes in to the 'octant' defined by the vectors $\partial/\partial c_i$ with positive coefficients. Thus, $\mathbf{c}(\lambda_0)$ traces a convex curve and then hits the (c_2, c_3) plane; it then goes to the axis c_3 and finally reaches back the point O. Since the final segment lies in the (c_2, c_3) plane, the spectral curve has a discontinuity at the origin. Thus, we conclude:

1. The starting point and the end point of the curve coincide: $\mathbf{c}(\lambda_0 = -1) = \mathbf{c}(\lambda_0 = 1) = O$. Thus, the spectral curve is indeed a loop.
2. The derivative of the curve at the origin O is discontinuous:

$$\left.\frac{\mathrm{d}\mathbf{c}(\lambda_0)}{\mathrm{d}\lambda_0}\right|_{-1} \neq \left.\frac{\mathrm{d}\mathbf{c}(\lambda_0)}{\mathrm{d}\lambda_0}\right|_{1} \tag{9}$$

3. The two-surface generated by rays starting from O and passing through points on the spectral loop –the *spectral cone*– is a boundary of a 3-dimensional convex set.

Thus, detailed properties of the spectral curve that had previously been arrived at phenomenologically from a large number of facts can be readily derived in our model. (Incidentally, Newton did not realize that the spectral curve passes through the origin. This fact as well as the discontinuity at the origin was however well-known by Schrödinger's time.)

Finally, let us 'explain' the origin of the purple plane. Consider incident intensity profiles $I(\lambda)$ whose support is the union of the extreme red part of the spectrum (i.e., on wave-lengths where only $A_1 \neq 0$) and the extreme violet (i.e., wave-lengths where only $A_3 \neq 0$). The corresponding colors lie in the c_1-c_2 plane and fill it. *This is the purple plane which constitutes a part of the boundary of $\mathcal{C}$.* Note that the c_2-c_3 plane or the c_1-c_2 plane does not constitute an analogous boundary because the form of the overlap of the A_2 adsorption curve with the A_1 and the A_3 curves implies that most points on these planes do not lie in the image of the map P of (7).

Let us summarize. The basic assumption of the model is simply that the output signal from the retina to the brain be given by the overlap integrals of the incident intensity and the three adsorption functions. This assumption is well-motivated since the spectral distribution $I(\lambda)$ represents the flux of energy per unit wave-length, per unit solid angle that is incident on the eye and the adsorption coefficients $A_i(\lambda)$ physically represent the fraction of the incident energy that is absorbed by the ith set of cones at the wavelength λ. The model is 'compact' in the sense that this is the *only* assumption. From it, one can systematically derive that the structure of the space of perceived colors is that of figure 1, i.e., is the one that had been put together slowly from diverse phenomenological inputs. The model is of course a drastically simplified version of reality since it ignores the role played by the rods in the retina and, more importantly, mechanisms in the brain. However, such idealizations are in the best tradition of theoretical physics and it is pleasing that the model is capable of providing complete insight in to the qualitative features of the manifold structure of $\mathcal{C}$.

2.3. RIEMANNIAN STRUCTURE

We can now turn to the next key question: that of finding a physical metric on the space $\mathcal{C}$. We can all tell if the two colors are close to one another or far apart and the mathematical physics problem is that of introducing a positive definite metric g_{ab} on $\mathcal{C}$ which correctly captures this perception. Thus, what we mean by a physical metric is a metric tensor which properly encodes the notion of 'distance between colors' as perceived by an average human subject. The underlying (but often implicit) assumption of researchers in this field, going back at least to Helmholtz [7] and Schrödinger [8], appears to be that there is indeed a unique physical metric on $\mathcal{C}$. Whether this is indeed the case is not entirely clear to us. For, it may not be possible to make a *sufficient number* of experiments to pin down the metric tensor uniquely. Certainly, an average human subject does not have a *direct* perception that the distance between a pair of colors c_1 and c_2 is, say, twice the distance between another pair, c_2 and c_3, even when the three colors are reasonably close to one another (so that issues such as 'geodesic crossings' do not arise). Therefore, if the metric can indeed be singled out uniquely, it would have to be via a series of a sufficient number of indirect experiments. What *is* clear is that experiments can be made to constrain the *form* of the possible metrics. In this section, we will provide an illustration. More precisely, we will see that there do exist experiments that show that $\mathcal{C}$ is equipped with an additional structure, which in turn provides a handle on the *form* of allowable metrics. Note that this requires the use a new experimental input i.e., in addition to the form of the three adsorption coefficients used so far.

Recall that there is a well-defined vector field B^a on $\mathcal{C}$, the brightness vector field whose integral curves are the 'radial lines' in $\mathcal{C}$. The idea is to experimentally probe its properties which are tied to the physical metric on $\mathcal{C}$. More precisely, let us try to gain control on some of the properties of the unknown metric by seeking directions which are *orthogonal* to B^a. Fix an arbitrary color c in $\mathcal{C}$. It lies on an integral curve ℓ of B^a. Look at a 'nearby' integral curve ℓ', consider colors which lie on it and compare them to the original color c. Choose from all points on ℓ' the point c' that seems 'closest' to c. Then, one can say that c and c' have the same brightness. Repeat this procedure for all integral curves that are close to ℓ. When carried out by a large sample of human subjects with 'normal' vision, this experiment provides a two-dimensional 'flat' passing through c, *along which brightness is constant* for the average human subject. Along each integral curve of B^a, by contrast, it is only the brightness that changes. Thus, through each point c in $\mathcal{C}$, there is a two-flat which is *orthogonal* to the brightness vector field. Within experimental errors, these flats are 'integrable'. That is, through each point c of $\mathcal{C}$ there passes a two-dimensional surface which is everywhere orthogonal to the brightness vector field B^a. Thus we have a foliation of $\mathcal{C}$ and hence we can introduce a *brightness function* B whose level surfaces are given by the foliation. A priori, the function B is not unique since we can replace it by any function $f(B)$ of B. However, we can remove this arbitrariness by tying the function B with the vector field B^a. We will make the simplest choice $B^a\partial_a B = 1$, i.e., that the brightness function B is the affine parameter along the integral curves of the brightness vector field B^a.

Let us restate the argument in a slightly different way. The manifold $\mathcal{C}$ is equipped with a vector field B^a. The experiments indicate that the allowable metrics g_{ab} should be such that the co-vector field $B_a := g_{ab}B^a$ is hypersurface orthogonal: $B_{[a}\,\partial_b\, B_{c]} = 0$. Note that the metric is essential in the formulation of this condition because the notion of hypersurface orthogonality refers to co-vector fields and *not* to vector fields. This condition is thus a restriction on the permissible metrics.

This new structure tells us that the metric g_{ab} has the following form:

$$g_{ab} = q_{ab} + N^2\, \partial_a B\, \partial_b B \tag{10}$$

where q_{ab} is the intrinsic metric (or projection operator) on the two-manifolds of constant brightness and $N^2 = g_{ab}B^aB^b$ is the squared-norm of the vector field B^a. Thus, while a general metric on $\mathcal{C}$ has six arbitrary components, the allowable metrics have only four; three in q_{ab} and one in N. (In the relativity terminology, the 'shift vector' is zero.) Furthermore, since the constant brightness surfaces are two-dimensional and have trivial topology, q_{ab} is in fact conformally flat.

There *is* a mathematically natural metric on $\mathcal{C}$ with respect to which B^a is hypersurface orthogonal as required. However, it is not clear to us that the brightness level surfaces determined by that metric are physically correct. Nonetheless, as an illustration, we will now sketch how this metric arises. This discussion will also bring out an interesting feature of the manifold $\mathcal{C}$.

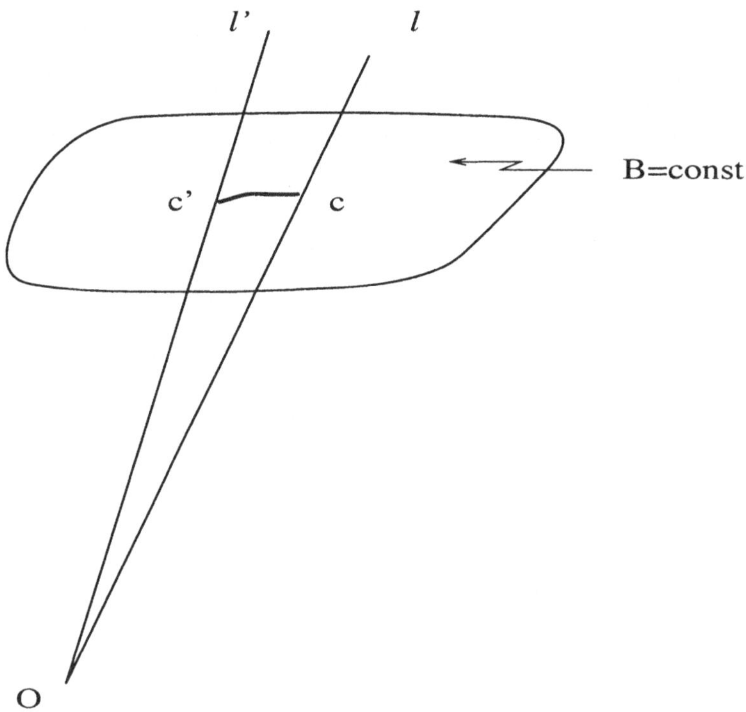

Figure 3. The color c' is the point on ℓ' that seems closest to the fixed c. The surface of constant brightness $B = \text{const}$ is formed from such colors.

In Sec. 2.2 we provided the explicit expression (7) of the projection operator P from the space $\mathcal{I}$ of intensities to the space $\mathcal{C}$ of perceived colors. Let us consider $\mathcal{I}$ as a convex-linear sub-space of the Hilbert space $\mathcal{H} = L^2([-1,1], d\lambda)$, i.e., space of (real-valued) square-integrable functions on the interval $(-1,1)$. This space is obviously equipped with a positive definite metric, the inner-products being given by:

$$\langle f_1, f_2 \rangle = \int_{-1}^{1} d\lambda \, f_1(\lambda) f_2(\lambda) \,. \tag{11}$$

The absorption functions $A_i(\lambda)$, in particular, belong to $\mathcal{H}$. Denote by $\mathcal{S}$ the three dimensional subspace of $\mathcal{H}$ spanned by the $A_i(\lambda)$. The projection map (7) can be trivially extended to all of $\mathcal{H}$. For notational simplicity, we will denote the extension also by P. It is obvious from the explicit expression of this map that the image of $\mathcal{S}$ under it is a three-dimensional vector space, which we will denote by V. It is straightforward to verify that the map P is an isomorphism between $\mathcal{S}$ and V. Note that $\mathcal{C}$ is a convex-linear sub-space of V. Denote by $\mathcal{S}^+$ the convex-linear sub-space of $\mathcal{S}$ obtained by projecting to $\mathcal{S}$ the convex-linear space $\mathcal{I}$ of all spectral distributions.

(This is a Hilbert-space projection within $\mathcal{H}$.) Then, the image under P of $\mathcal{S}^+$ is precisely $\mathcal{C}$. Since P is an isomorphism between $\mathcal{S}$ and V, it is in particular an isomorphism between $\mathcal{S}^+$ and $\mathcal{C}$. We will use this isomorphism to identify the two spaces. Thus, *not only is there a projection from $\mathcal{I}$ to $\mathcal{C}$, there is also an embedding of* $\mathcal{C}$ (on to $\mathcal{S}^+$ and therefore) *in to $\mathcal{H}$*. But since $\mathcal{S}^+$ is a real, smooth sub-manifold of $\mathcal{H}$, it inherits from $\mathcal{H}$ a positive definite, smooth metric. Hence $\mathcal{C}$ is equipped with a metric g^o_{ab}. This is a natural metric at least from a mathematical point of view because the L^2-norm features also in the projection map (7).

Finally, let us discuss hypersurface orthogonality of B^a. The radial vector field is hypersurface orthogonal everywhere on $\mathcal{H}$ (except at the origin), and hence also on any finite dimensional sub-space, such as $\mathcal{S}$, thereof. Restricted to $\mathcal{S}^+$, this is precisely the brightness vector field B^a. Thus, the positive-definite metric g^o_{ab} induced on $\mathcal{C}$ from $\mathcal{H}$ does have the required property that $g^o_{ab}B^b$ is hypersurface orthogonal.

This particular metric g^o_{ab} on $\mathcal{C}$ is flat. However, the natural coordinates, c_i do not provide a Cartesian chart for this flat metric because the adsorption functions $A_i(\lambda)$ are not mutually orthogonal in the L^2-norm. But the components of g^o_{ab} in the chart c_i can be written down readily once the L^2-inner products of the $A_i(\lambda)$ are known. The latter can be estimated analytically rather easily by, for example, approximating the $A_i(\lambda)$ by Gaussians. Indeed, in this approximation, $\mathcal{S}$ can be thought of as the linear span of three quantum mechanical coherent states and calculations simplify considerably because of the available mathematical machinery associated with coherent states.

Finally, as explained in the Introduction, several other metrics have been proposed in the literature. However, they are invariably written in coordinates which are rather unnatural from the point of adsorption coefficients (and hence our simple model). It would take too long to explain the meaning of these coordinates here. Furthermore, none of the available metrics have a definitive status. Therefore, we will refrain from discussing them here. We wish to emphasize, however, that the problem of finding the complete description of the physical metric(s) on $\mathcal{C}$ is extremely interesting from a practical viewpoint; its solution would be regarded as a major breakthrough with applications to a variety of industries including film, photography and color television.

3. Discussion

In mathematical physics, one generally begins by constructing simplified models in which the actual physical complications of the system under consideration are stripped down to a bare minimum —the oft-quoted example is that of a 'spherical cow'. Such a strategy is generally essential because the model has to be sufficiently simple to be amenable to mathematical analysis. Furthermore, if these assumptions are physically well-motivated, the model generally leads to results which are correct at least qualitatively. Further improvements can be often made in increments by treating the ignored complications as perturbations.

We have proceeded in this spirit. Our basic assumption is that the essential features of perceived colors are all captured in the three adsorption coefficients $A_i(\lambda)$, whence the perceived colors are completely characterized by the three numbers c_i of (6). In the first approximation, this assumption is physically reasonable because c_i are the overlap integrals of the function $I(\lambda)$ representing the incident energy flux per unit wavelength and functions $A_i(\lambda)$ representing the fraction of the incident energy absorbed by the i-th cones. That is, c_i are a good measure of the signal that is sent to the brain.

However, as indicated already, our model fails to incorporate two important features: (i) role of the rods in the retina which also contributes to the signal; and, more importantly, (ii) signal processing in the brain itself. The role of rods is to enhance the shades of grey and they are therefore important in processing information that is received when the illumination is poor, e.g. in the evening and at night. In color perception which occurs in brightly illuminated environments, therefore, it seems safe to ignore their role. The role played by the brain, of course, is *much* more complex. However, these complexities seem not to be critical for 'normal human subjects' and under 'normal viewing conditions'. That is, in this case, there appears to be a strong correlation between the final perception of the color and the input signal to the brain, coded in the three numbers c_i. This assertion is supported by the fact that our simple mathematical model reproduced all the general features of $\mathcal{C}$ described in figure 1. Further support for the model comes from color blindness. Most color blind people can see some colors but not all. In terms of the model, they have only two types of cones rather than all three. Their space of perceived colors is thus only two dimensional. A very small fraction perceive no colors at all. They have only one type of cones and so their space of colors is only one dimensional. All this is in agreement with observations.

Nonetheless, it is important to remember that color perception is in fact a 'psycho-physical' phenomenon while the model treats it simply as a physical phenomenon. Thus, it is quite incomplete. What does it miss? Here is a simple example. Let us consider an apple. Under normal illuminations –e.g., in reasonably bright sunlight– the model works well. Under these 'normal' conditions, the spectral distribution $I(\lambda)$ reflected by the apple is highly peaked around red, whence, according to the model only c_1 is significant and we do indeed perceive it as being red. However, the apple can be illuminated by artificial light and then the reflected intensity $I(\lambda)$ can be quite different. So, our model would now predict that all three c_i should be non-zero and hence we should perceive quite a different color. This does *not* happen for a large set of illuminations; an average human subject continues to perceive a red apple! This is because the brain now performs further calculations to re-interpret the incoming signal. Such re-interpretations are essential for our survival which requires that we should be capable of recognizing the 'sameness' of the object even when it is illuminated quite differently. There are dozens of day to day examples in which the brain interferes non-trivially to make the model completely inadequate. Snow, in particular, looks white even when we are walking in a pine forest, although now it in fact reflects predominantly green light. Similarly, for a large class of illuminations, a patch color can seem to be green in certain surroundings but pale grey when looked at in isolation. A sheet of paper appears white in both bright day light and in a windowless office even though the amount of light reflected from a dark grey sheet in daylight is much more than that reflected by the 'white' sheet in a windowless office. In each case, the brain has absorbed not just the three inputs c_i but a whole range of other 'relevant' information to produce a final color perception. By now, neurologists even know the place in the cortex which responds to color as opposed to wavelengths. It is a pea-sized region, called V_4, which is distinct from another region V_1 that responds to specific wave-lengths, or, in terms of our model, to the three c_i.

A striking example of this inadequacy of our purely physical model was seen recently. It is described by Oliver Sacks in his poignant essay *The case of the colorblind painter* [10]. This is a story of a successful artist who had been painting for over 40 years. In 1986, he had a minor car accident. He suffered a concussion, but recovered rather quickly except for one thing: he lost his color vision! Detailed examination showed that his retina –with all its three types of cones– were intact and functioning normally. So, his brain was receiving the color signals as before; the space of colors *received* in the brain continued to be three-dimensional. However, the space of *actually perceived* colors was now only one dimensional! Examination showed that the whole brunt of damage to the cortex from the accident was borne precisely by the V_4 area.

To summarize, our simple model works well when we study colors in isolation, e.g., when a uniform field of color is presented to the average eye on a black background. In the 'real world' however factors such as illumination, the state of adaptation of the visual system and neighboring stimuli in the visual field cause the brain to intervene and remove correlations between the input signal (the three c_i) and the actual color perception. These interventions are incredibly fascinating processes. But they also seem to be among the most complex ones. It will probably take a long time before such processes can be successfully coded in the mathematical models of color perception.

There is another fascinating problem that has attracted considerable attention: the role of color perception in the evolution of species. One might have naively thought that color perception becomes more and more sophisticated as the species become more advanced in evolutionary terms. Surprisingly, this appears *not* to be the case! The space of perceived colors of a frog, for example, is *four* dimensional —their eyes have four types of cones. Could it be that there is actually an anti-correlation between the number of colors perceived and place of the species in the evolutionary ladder? Again, the answer is in the negative! While the dimension is four for frogs and three for humans, it is one for most mammals. Thus, most mammals, including the big cats which are such excellent hunters, are color blind. Naively, one might have thought that, because they have to hunt in diverse environments, evolutionary forces would have endowed them with a large dimensional color space. This is simply not the case. While there do exist interesting attempts to understand the role of evolution and natural selection in color perception (see, e.g., [11]), this is still a largely open issue.

Acknowledgments

This work was supported in part by the NSF grant PHY 95-14240 and INT 97-22514, by the Eberly Research funds of The Pennsylvania State University and by DGAPA of the Universidad Nacional Autónoma de México.

References

1. I. Newton, *Phil. Trans. Roy. Soc. London*, **80**, 3075 (1671); I. Newton, *Opticks*, Smith and Walford, London, (1704).
2. T. Young, *Phil. Trans. Roy. Soc. London*, **92**, 20 (1802).
3. D. Mitchell and W. Rushton, *Vision Research*, **11**, 1033 (1971).
4. H. Grassmann, *Ann. Phys. Chem.*, (peggendorf), **89**, 69 (1853); translation in [5].
5. D.L. MacAdam, *Sources of Color Science*, MIT Press, Cambridge Mass (1970).
6. B. Riemann, *Ueber die Hypothesen, Welche der Geometrie Zu Grunde liegen* Gött. Nachr. **13**, 133 (1868).
7. H. Helmholz, *Sitzungber. d. Berl. Akad.*, Dec 17, 1071 (1891).
8. E. Schrödinger, *Annalen del Physik*, **63**, 397 (1920); ibid, 481 (1920).
9. G. Wyszecki and W.S. Stiles, *Color Science*, Wiley, New York (1982).
10. O. Sachs, *An Anthropologist on Mars*, (Vintage Books, New York, 1996) pages 3-41.
11. J.W. Weinberg, *Gen. Rel. Grav.*, **7**, 135 (1976).

Henri Poincaré once remarked that real problems can never be classified as solved or unsolved ones, but that they are always *more or less solved.* This remark applies particularly well to the problem of motion which has had a chequered history. Even the Newtonian problem of motion, which appeared to be well understood after the development of the powerful methods of celestial classical mechanics, embarked on an entirely new career after the work of Poincaré which has led to many furthur developments. In my opinion the Einsteinian problem of motion has not even reached a classical stage where the basic problems appear (provisionally) as 'well understood'...
One thing is certain: the Einsteinian problem of motion is no longer a purely theoretical problem, thanks to the dramatic improvement in the precision of position measurements in the solar system, and to the discovery of the binary pulsar 1913+16 which is a marvellous relativistic laboratory; the Einsteinian problem of motion has become an important tool of modern astrophysics. It is therefore of some urgency, not only to complete and unify the work already done but also develop new approaches which will keep the best aspects of existing methods while freeing themselves from their conceptual and/or technical drawbacks. These new approaches should aim at both formal and conceptual clarification of the basic issues, and at securing more accurate explicit results. Let us hope that such methods will be developed before the (first) centenary of the publication of Einstein's classic paper on 'the foundation of the general theory of relativity' ('Die Grundlage der allgemeinen Relativitätstheorie') which is comparable with Newton's *Philosophiae Naturalis Principia Mathematica* both in its importance for physics and its structure which consists, after a conceptual introduction, of a formal part followed by a physical part where connection is made with the natural world.

— THIBAULT DAMOUR

33. C.V. VISHVESHWARA - A PROFILE

N. PANCHAPAKESAN

Department of Physics and Astrophysics,
Delhi University,
Delhi- 110 007, India.

Vishveshwara's name will be associated with that of Black Holes in the history of science. He took up the study of black holes when very few persons took them seriously. He proved their stability and discussed the scattering of radiation from black holes to be able to identify them. His pioneering study was followed by more detailed studies of increasing sophistication later on. (See for example Mathematical Theory of Black Holes, S. Chandrasekhar).

Vishveshwara was born on March 6,1938 in Bangalore, a city where he has spent most of his life. He had his schooling there and then went on to Mysore University for further studies. He obtained the B.Sc.(Hons) degree in 1958 and the M.Sc. degree in 1959. He then went for higher studies to U.S.A. After getting his A.M. from Columbia University, New York in 1964 he moved to University of Maryland from where he got his Ph.D. in 1968. His thesis advisor for Ph.D. was Prof. C.W. Misner and his thesis subject was "Stability of Schwarzschild Metric".

After working in a post doctoral position and serving as a visiting faculty member, he returned to India (i.e., Bangalore) in 1976 and joined the Raman Research Institute. He moved from there, in December 1992, to his present position, as Senior Professor, at Indian Institute of Astrophysics, Bangalore.

For more than a decade now he has also been associated with the Jawaharlal Nehru Planetarium and Science Centre. He was the Director from November 1987 to March 1990 and since then has been the Honorary Director. The Planetarium, apart from its usual activities, also provides for non-formal science education and some amount of research.

B. R. Iyer and B. Bhawal (eds.), Black Holes, Gravitational Radiation and the Universe, 551–558.

1. Early Research

The first major work of C.V. Vishveshwara comprised one of the earliest studies of black hole geometry utilising spacetime symmetries. It provided a coordinate invariant characterisation of the stationary limit and event horizon for Kerr black hole thereby establishing the existence of the ergosphere. His work on the stability of black holes was done in the late 1960s. The problem had been formulated and studied earlier in 1957 by T. Regge and J.A. Wheeler. At that time there was no systematic way of handling the co-ordinate singularity at the event horizon in the Schwarzschild metric. A satisfactory solution was given only in 1960 by M. Kruskal (and independently by Szekeres). Vishveshwara used the Kruskal co-ordinates to examine the behaviour of small perturbations. He found that if they are growing perturbations (with imaginary frequency) then they cannot be small at the event horizon at the initial time. Thus such perturbations are disallowed and the metric is stable. His detailed analysis published in 1970 established the stability of the Schwarzschild metric. More general approaches have been formulated later (see S. Chandrasekhar's, Mathematical Theory of Black holes for details).

The normal modes of the black hole are necessarily damped and hence go under the name of Quasi-Normal modes and were first studied by Vishveshwara and published in Nature in 1970. They are the subject of intense study as the gravitational radiation at these frequencies, so called 'ringing modes', would signal the collapse of a star to a black hole. A significant remark heard at GR15 was that all the intense calculations still reproduce the same figure given in Vishveshwara's publication in Nature in 1970. There is however no truth in the rumour that this figure would be featured on the crest of Vishveshwara, if and when he is knighted!

2. Sense Of Humour

Anyone who has known him at a personal level is immediately impressed by his other qualities. To quote Alberto Chamorro: "I was always struck and pleased by Vishu's keen sense of humour, amusing subtle irony and by his fine writing". He is also full of ideas about creating cartoons to present the scientific and human situation in the character of research and scientific activity. Many proceedings of conferences e.g. *Highlights in gravitation and cosmology (1988); Advances and gravitation and cosmology (1992)* have very striking cartoons due to him. Many conferences have featured him as the after-dinner speaker. His parody of well known scientific events like Galileo's feather and penny experiment have regaled many diners at conferences. We give a small example from his after dinner talk given at GR15 held in Pune, India:

All this is well known. But what is not known and only gedanken research has revealed is the hitherto unknown existence of Galileo's Hungarian graduate student by the name Zolta Zbznzky, a name with four z's. Like all good thesis advisors, Galileo asked Zolta to work on a problem he himself could not solve. He asked Zolta to compute the radiation reaction as the quadrupole moment of the tower of Pisa changed suddenly when it leaned to one side. Zolta wanted to do this using the 4.37th order, or it could be 4.73rd, PPN formalism. That is Plagiarized Post-Nostradamus formalism which had tremendous predictive power having been formulated by Nostradamus himself. The orders were irrational because the formalism itself was a bit irrational. This was a formidable problem indeed. Zolta's motto was: Solve a solvable problem rather than unsolve an unsolvable one. Therefore he decided to construct a two-dimensional toy model. The two-dimensional analogue of the tower of Pisa, which is a cylinder in three space, is a disk. Zolta made a disk and whirled it to simulate gravity. But no takers. Then Zolta made his first discovery. He realized that his toy model could become more palatable if he baked it, along with his half-baked ideas, with tomato sauce and cheese. He named it after the town in which it was discovered replacing the s with two z's from his name and called it Pizza. Now his toy model sold like, what else, pizza. The demand was so great that Zolta had to sell it by slices. Now he made his second discovery. When he had sold one slice, he realized that he could not go round the disk through an angle of two pi. People were puzzled wondering why Zolta was talking about two pies when he had only one. Anyway, Zolta claimed that this represented, for some unknown reason, an erotic cosmic g-string, 'g' standing for gravity and 'erotic' being the abbreviation of 'heterotic'. Some people said Zolta was into string theory. Some said he had gone mad. And some unkind ones said what is the difference. Seeking greener pastures or what is popularly known as the green card, he migrated to Las Vegas, along with his toy model and g-string and all, and became godfather to generations of general relativists.

3. Research work

A summary of his research work over the last thirty years would be as follows: (The references are to the numbers in the list of publications of technical papers included elsewhere in this volume).

1. **Theory of Black Holes: Refs: 2-16, 19, 22, 33, 37.**
   One of the earliest geometric characterisations of stationary limits and event horizons associated with static and rotating black hole spacetimes was formulated. Stability of the Scwarzschild black hole was proved. Quasi-normal modes of the Schwarzschild black hole were discovered. Various phenomena in black hole spacetimes, such as gravitational synchrotron radiation, charged particle motion, electromagnetic fields of charges, effects of electromagnetic fields on accretion disks, have been studied.
2. **Exact Solutions: Refs: 18, 20, 21, 31, 39.**
   Three exact solutions to the Einstein field equations involving rotating dust sources have been obtained. One of them, however, turned out to be the well known Gödel solution in disguise. Magnetic versions of cylindrically symmetric solutions were obtained.
3. **Perturbations of specific space times: Refs: 23, 25, 30, 34, 36.**
   Electromagnetic, gravitational and neutrino perturbations of a number of spacetimes with local rotational symmetry have been investigated using generalised Hertz potential methods. Separation of massive Dirac equation has been studied in this class of geometries.
4. **Neutrinos in strong gravitational fields: Refs: 24, 26-30**
   Detailed study of neutrinos in gravitational fields, such as those associated with compact objects and spherical gravitational collapse, have been carried out. Both geodesic and Dirac formalisms have been used in these investigations.
5. **Neutron Star Models, Ultracompact Objects: Refs: 35**
   The unusual behaviour of objects with sizes $R < 3GM/c^2$, was noted and the existence of these 'ultracompact objects' investigated using the core-envelope paradigm of Sabbadini-Hartle-Chitre. The stability of these configurations was proved. General relativistic interior solutions were studied for various equations of state to study the most compact configuration the equation of state would allow.
6. **Higher Dimensional spacetimes, Higher Derivative Gravity: Refs: 40-43.**
   Higher dimensional spacetimes have come to be of interest following the advent of new theories such as superstring theory and generalised Kaluza-Klein theory. In this context some problems involving higher dimensions have been investigated, such as Frenet-Serret formalism in higher dimensional black hole spacetimes, scalar waves in Boulware-Deser black hole geometries and generalisations of the Vaidya metric.
7. **Gyroscopic Precession and inertial forces: Refs: 32, 45, 50-52**
   One of the phenomena that carries the imprint of spacetime curvature and rotation is gyroscopic precession. A study of gyroscopic precession

has been made using Frenet-Serret formalism in a completely covariant and geometric manner. There has been considerable interest in the general relativistic analogues of inertial forces. The relation between gyroscopic precession and inertial forces has been studied and covariant connections made between them.

8. **Balance between Gravitational and Electromagnetic forces: Refs: 46-48.**
   In collaboration with the relativity group at Bilbao, Spain, the study of the equilibrium of a charged test particle in the background a Reissner Nordstrom black hole has been extended to investigate spinning particles in the Kerr Newman background.
9. **Quasinormal Modes of Black Holes: Ref: 5,49.**
   One of the most important dynamic attributes of a black hole is its quasi-normal modes of oscillation and these were discovered in 1970 by C.V. Vishveshwara. They offer a method of observing both black holes and gravitational radiation. Quasi-normal modes may be modified by changes in the effective potential of the black hole: While the fundamental mode remains essentially the same, the higher modes change considerably. Thus the fundamental mode would carry the imprint of the black hole, while the higher ones may reveal the nature of the perturbing influences.
10. **Miscellaneous Problems: Ref: 10, 11, 13, 17, 38.**
   Diverse types of problems studied include motion of charges in homogeneous electromagnetic fields, Thomas precession in general relativity, superluminal velocities associated with quasars and linearisation stability in general relativity.

4. Fest at GR15, IUCAA, Pune

Vishveshwara has been involved in the organising committees of GR's for quite some time now. Along with Profs. Vaidya, Raychaudhari, Narlikar, Prasanna, Dadhich, and other younger colleagues, he has also been involved in the organisation of workshops and instructional conferences for young relativists as also the by now successful series of International conferences on Gravitation and Cosmology (ICGC) in India. His large circle of friends and well wishers was evident in the warmth and enthusiasm shown at the Fest held in Pune, India during GR15 in December '97. The general sentiment at the Fest is best captured in the following three messages from Charles Misner, Alberto Chamorro and Biplab Bhawal who unfortunately could not participate in the festivities.

Congratulations and all the best wishes on your soon-to-arrive 60th birthday.
Our days together when you were a student at Maryland were great times as you worked on crucial problems contributing to the developing understanding of black holes, even before they had been named. Your subsequent work in this and related fields are the sort that warm the heart of any thesis advisor. I have also come to know and admire your broader interests in art, literature, and history which add to your stature as a most valued figure in the world-wide relativity community and in your country. Those of us who occasionally have the opportunity to interact more informally with you have also benefitted from your hospitality and been entertained by your sense of humor.
Although circumstances intervened to cancel my planned participation in GR15 and this celebration there for you, I look forward to future visits to Bangalore which now has been made even more attractive since it is home to my daughter-in-law's family as well as to you and Saru.

Charles Misner

Although I had known Vishu for many years from his important contributions to General Relativity, my friendship with him dates from 1988 when I had the fortune of meeting him at Queen Mary and Westfield College, London. A natural empathy immediately developed between us and very soon I expressed to him the deep cultural interest I had always had for India. As it was, Vishu turned out to be my gate to India. Through him I first visited his country for two months in 1991, and on a couple of occassions more in the following years. Likewise, he visited our Department -in the University of the Basque Country, Bilbao, Spain- several times so that at present we have established a firm scientific collaboration and a warm and close friendship. By the way of Vishu today India -a country that may leave indifferent no one- is one of my loved nations. There I made other good friends and acquaintances, both professional and human. We have both travelled together in India and Spain, and talked many hours on scientific, cultural, psychological, personal, gastronomic and a wide variety of other matters. I think that we both always felt at ease with each other and enjoyed a great deal of placid talking. For my part, I was always struck and pleased by Vishu's keen sense of humour, amusing subtle irony and by his fine writing. In fact I have been encouraging him to write about some personal experiences, something for which I believe he is also endowed with a literary talent. Were it not for the obligatory brevity of these lines I could tell of quite a number of interesting recollections of my relationship with Vishu. Let me finish by wishing him a very happy sixtieth birthday and many more years to come of good research, humour, art and succesful fatherhood -the latter is an extremelly meritory quality I had the oportunity to see in what high degree is present in Vishu's personality. I also hope for many more years of fruitful collaboration and enriching friendship with him. I consider a privilege for me to be counted among his friends.

Alberto Chamorro

My most sincere congratulations to you!
Any person who grows up amidst the all-pervading uncertainties in all aspects of our society to complete 60 years of his life should be congratulated. In Japan they celebrate this occasion by making this lucky person wear a red vest - they call this occasion *Kan-reki*, *Kan* means circle and *reki* means history – because the lucky one completes five phases, each of twelve years, of his life to receive a great chance of a transition to another life – a new one. I hope this transition brings to you more happy moments and lets you to explore various paths of creativity which you always wished to walk in.
I'm sorry Vishu that I'll not be present in your 60th birthday party. At present I'm also in a transition state – transition in many aspects. I'm moving from one project to the other, from one country to another and trying to follow a great team of people in a search operation to look into ripples in the most intricate structure of our background spacetime which you so beautifully taught many of us to incorporate in our culture and to love and to carry to our beds.
I guess, unlike a scientist in a transition state in his thirties like presently I am, you'll find sufficient time in your new life to make it to my 60th birthday party. I am looking forward to that day to tell you and to hear from you all those stories ... how people, amidst their fear and anguish, amidst the all-pervading uncertainties in all aspects of our society, dared to write equations for the truth hidden behind horizons unknown, how they dared to perturb supermassive blackholes and to look into the most difficult intricacies of our spacetime background on which we live our life and care for our friends and breathe and wonder, how they strove valiantly and how they erred and came short again and again and how they finally performed that last transition to have the knowledge of the triumph of high achievement... And how they could hear those ripples sent by the drums of the heaven! ... For this and much more, please wait till then.
Cheers to you, your family and to all your friends who shared your joy and anguish and happiness and uncertainties! After all, no one is a failure who has friends and I'm sure, looking back at the first phase of your life, you would certainly find how greatly successful you have been!

Biplab

GRAVITATIONAL RADIATION

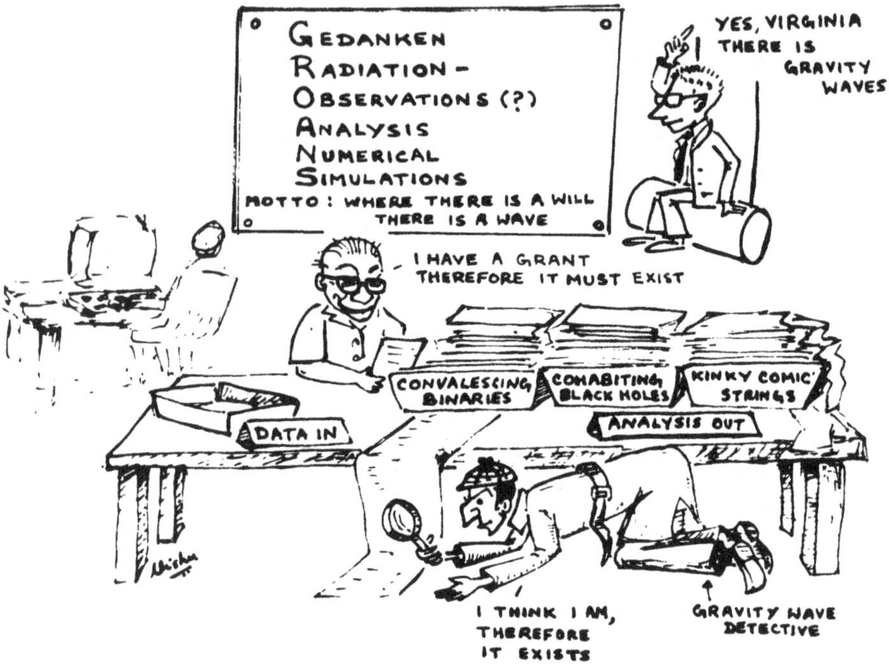

This cartoon originally appeared in *Advances in Gravitation and Cosmology*, B.R. Iyer, A. R. Prasanna and C.V. Vishveshwara (eds.), Wiley Eastern Limited, New Delhi, India (1993) and is reproduced with kind permission of the original publishers.

34. PUBLICATIONS OF C. V. VISHVESHWARA

1. Ph.D Dissertion

- Title : *Stability of the Schwarzschild Metric*
- Advisor : Professor Charles W. Misner
- University : University of Maryland
- Year : 1968

2. Technical papers

1. *Gravitational Radiation from Binary Systems*, C. V. Vishveshwara, University of Maryland Technical Report (1965),
2. *Generalization of the "Schwarzschild Surface" to Arbitrary Static and Stationary Metrics*, C. V. Vishveshwara, J Math. Phys., 9, 1319 (1968),
3. *Differential Equations for Perturbations on the Schwarzschild Metric*, L. A. Edelstein and C. V. Vishveshwara, Phys. Rev. D, 1, 3514 (1970),
4. *Stability of the Schwarzschild Metric*, C. V. Vishveshwara, Phys. Rev. D, 1, 2870 (1970),
5. *Scattering of Gravitational Radiation by a Schwarzschild Black-Hole*, C. V. Vishveshwara, Nature, 227, 936 (1970),
6. *Electromagnetic Field of a Particle Moving in a Spherically Symmetric Black Hole Background*, R. Ruffini, J. Tiomno and C. V. Vishveshwara, Nuovo Cimento Letters, 3, 211 (1971),
7. *Polarization of Gravitational Synchrotron Radiation*, R. A. Breuer, J. Tiomno and C. V. Vishveshwara, Nuovo Cimento Letters, 4, 857 (1972),
8. *Vector and Tensor Radiation from Schwarzschild Relativistic Circular Geodesics*, R. A. Breuer, R. Ruffini, J. Tiomno and C. V. Vishveshwara, Phys. Rev. D, 7, 1002 (1973),
9. *Polarization of Synchrotron Radiation from Relativistic Schwarzschild Circular Geodesics*, R. A. Breuer and C. V. Vishveshwara, Phys. Rev. D, 7, 1008 (1973),

10. *Thomas Pression and the Motion of Charged Particles*, E. Honig, E. L. Schucking and C. V. Vishveshwara, Bull. Am. Phys. Soc., 18, 644 (1973),
11. *Thomas Pression and the Motion of Charged Particles*, E. Honig, E. L. Schucking and C. V. Vishveshwara, unpublished preprint,
12. *Point Charge in the Vicinity of a Kerr Black Hole*, J. M. Cohen, L. Kegeles, C. V. Vishveshwara and R. M. Wald, Annals of Physics, 82, 597 (1974),
13. *Motion of Charged Particles in Homogeneous Electromagnetic Fields*, E. Honig, E. L. Schucking and C. V. Vishveshwara, J. Math. Phys., 15, 774 (1974),
14. *The Rest Frame in Stationary Space-times with Axial Symmetry*, R. D. Greene, E. L. Schucking and C. V. Vishveshwara, J. Math. Phys. 16, 153 (1975),
15. *Polarization of Gravitational Geodesic Synchrotron Radiation*, R. A. Breuer, J. Tiomno and C. V. Vishveshwara, Nuovo Cimento, 25B, 851 (1975),
16. *Electromagnetic Field of a Current Loop around a Kerr Black Hole*, D. M. Chitre and C. V. Vishveshwara, Physical Review D, 12, 1538 (1975),
17. *Kinematics of Relativistic Ejection*, C. Behr, E. L. Schucking, C. V. Vishveshwara and W. Wallace, Astronom. Journ., 81, 147 (1976),
18. *Relativistically Rotating Dust Cylinders*, C. V. Vishveshwara and J. Winicour, J. Math. Phys., 18, 1280 (1977),
19. *Charged Particle Motion in an Electromagnetic Field on Kerr Background Geometry*, A. R. Prasanna and C. V. Vishveshwara, Pramana, 11, 359 (1978),
20. *Interiors with Relativistic Dust Flow*, C. Hoenselaers and C. V. Vishveshwara, Journal of Physics A, 12, 209 (1979),
21. *A Relativistically Rotating Fluid Cylinder*, C. Hoenselaers and C. V. Vishveshwara, GRG, 10, 43 (1979),
22. *Trajectories of Charged Particles in the Static Ernst Space-time*, N. Dadhich, C. Hoenselaers and C. V. Vishveshwara, Journal of Physics A, 12, 215 (1979),
23. *Electromagnetic Fields in the Godel Universe*, J. M. Cohen, C. V. Vishveshwara and S. V. Dhurandhar, Journal of Physics A, 13, 933 (1980),

24. *Neutrinos in Compact Objects*, A. K. Kembhavi and C. V. Vishveshwara, Phys. Rev. D, 22, 2349 (1980),

25. *Electromagnetic Fields in Space-times with Local Rotational Symmetry*, S. V. Dhurandhar, C. V. Vishveshwara and J. M. Cohen, Phys. Rev. D, 21, 2794 (1980),

26. *Neutrinos and Gravitational Collapse-I: Analysis of Trajectories*, S. V. Dhurandhar and C. V. Vishveshwara, Astrophysical Journal, 245, 1094 (1981),

27. *Neutrinos in Gravitational Collapse: Analysis of the Flux Profiles*, S. V. Dhurandhar and C. V. Vishveshwara, Pramana, 22, 159 (1984),

28. *Neutrinos in the Kerr and Robertson-Walker Geometries*, S. V. Dhurandhar, C. V. Vishveshwara and J. M. Cohen, Journal of Physics A, 15, 1643 (1982),

29. *Neutrinos in Gravitational Collapse : The Dirac Formalism*, B. R. Iyer, S. V. Dhurandhar and C. V. Vishveshwara, Phys. Rev. D, 25, 2053 (1982),

30. *Neutrinos in Perfect Fluid Space-times with Local Rotational Symmetry*, S. V. Dhurandhar, C. V. Vishveshwara and J. M. Cohen, Phys. Rev. D, 26, 2598 (1982),

31. *Magnetization of all Stationary Cylindrically Symmetric Vacuum Metrics*, B. R. Iyer and C. V. Vishveshwara, J. Math. Phys., 24, 1568 (1983),

32. *An Example of Induced Centrifugal Force in General Relativity*, J. M. Cohen, W. J. Sarill and C. V. Vishveshwara, Nature, 298, 829 (1982),

33. *Magnetic Fields and Accretion Discs Around Kerr Black Holes*, P. J. Wiita, C. V. Vishveshwara, M. J. Siah and B. R. Iyer, J. Phys., A16, 2077 (1983),

34. *Electromagnetic, Neutrino and Gravitational Fields in the Kasner Space-time with Rotational Symmetry*, S. V. Dhurandhar, C. V. Vishveshwara and J. M. Cohen, Class. Quantum Grav., 1, 61, (1984),

35. *Ultracompact ($R < 3M$) Objects in General Relativity*, B. R. Iyer, C. V. Vishveshwara and S. V. Dhurandhar, Class. Quantum Grav., 2, 219 (1985),

36. *Separability of the Dirac Equation in a Class of Perfect Fluid Space-times with Local Rotational Symmetry*, B. R. Iyer and C. V. Vishveshwara, J. Math. Phys., 26, 1034 (1985),

37. *Accretion onto a Kerr Black Hole in the Presence of a Dipole Magnetic Field*, B. R. Iyer, C. V. Vishveshwara, P. J. Wiita and J. J. Goldstein, Pramana, 25,135 (1985),
38. *Joint Linerarization Instabilities in General Relativity*, D. Brill and C. V. Vishveshwara, J. Math Phys., 27, 1813 (1986),
39. *Exact Solutions for Space-times with Local Rotational Symmetry in which the Dirac Equation Separates*, B. R. Iyer and C. V. Vishveshwara, J. Math. Phys., 28, 1377 (1987),
40. *The Frenet-Serret Formalism and Black Holes in Higher Dimensions*, B. R. Iyer and C. V. Vishveshwara, Class. Quantum Gravity, 5, 961 (1988),
41. *The Vaidya Solution in Higher Dimensions*, B. R. Iyer and C. V. Vishveshwara, Pramana, 32, 749 (1989),
42. *Scalar Waves in Boulware-Deser Black Hole background*, B. R. Iyer, Sai Iyer and C. V. Vishveshwara, Class. Quantum Gravity, 6, 1627 (1989),
43. *Scalar Waves in the Witten Bubble Spacetime*, B. Bhawal and C. V. Vishveshwara, Phys. Rev. D, 42, 1996 (1990),
44. *Modified Photon Equation of motion as a Test for the Principle of Equivalence*, B. Bhawal, H. S. Mani and C. V. Vishveshwara, Phys. Rev. D, 44, 1323 (1991),
45. *Frenet-Serret Description of Gyroscopic Precession*, B. R. Iyer and C. V. Vishveshwara, Phys. Rev. D, 48, 5706 (1993),
46. *Equilibrium of a charged test particle in the Kerr-Newman Spacetime*, J. M. Aguirregabiria, A. Chamorro, J. Suinaga and C. V. Vishveshwara, Class.Quantum Grav., 12, 699 (1995),
47. *Equilibrium of a charged test particle with spin in the Kerr-Newman Spacetime*, J. M. Aguirregabiria, A. Chamorro, J. Suinaga and C. V. Vishveshwara, Class. Quantum Grav., 13, 417 (1996),
48. *Equilibrium of charged test particle in the Kerr-Newman spacetime: force analysis*, J. M. Aguirregabiria, A. Chamorro, K. Rajesh Nayak, K. Suinaga
and C. V. Vishveshwara, Class. Quantum Grav., 13, 2179 (1996),
49. *Scattering by Black Holes: a simulated potential approach*, J. M. Aguirregabiria and C. V. Vishveshwara, Phys. Lett. A, 210, 251 (1996),
50. *Gyroscopic Precession and Centrigugal Forces in the Kerr-Newman spacetime*, K. R. Nayak and C. V. Vishveshwara, Class. Quantum Grav., 13, 1783 (1996),

51. *Gyroscopic Precession and Centrifugal Force in the Ernst Spacetime*, K. R. Nayak and C. V. Vishveshwara, GRG, 29, 291 (1997).
52. *Gyroscopic Precession and Inertial Forces in Axially Symmetric Stationary Spacetimes*, K. R. Nayak and C. V. Vishveshwara, GRG, 1998 (in press)

3. Articles in Books

1. *Black Holes for Bedtime*, C. V. Vishveshwara in 'Gravitation, Quanta and the Universe', eds. A. R. Prasanna, J. V. Narlikar and C. V. Vishveshwara (Wiley Eastern, 1980),
2. *General Relativity and Gravitational Collapse*, C. V. Vishveshwara in 'Gravitation and Relativistic Astrophysics', eds. A. R. Prasanna, J. V. Narlikar and C. V. Vishveshwara (Indian Academy of Sciences/World Scientific, 1984),
3. *General Relativistic Aspects of Neutron Star Models*, B. R. Iyer and C. V. Vishveshwara in 'A Random Walk through General Relativity and Cosmology', eds. N. Dadhich, J. Krishna Rao, J. V. Narlikar and C. V. Vishveshwara (Wiley Eastern, 1985),
4. *Some Recent Developments in General Relativity*, C. V. Vishveshwara in 'Recent Developments in Theoretical Physics', eds. E. C. G. Sudarshan, K. Srinivasa Rao and R. Sridhar (World Scientific, 1987),
5. *Black Holes : A Slanted Overview*, C. V. Vishveshwara in 'Highlights in Gravitation and Cosmology', eds. B. R. Iyer, A. Kembhavi, J. V. Narlikar and C. V. Vishveshwara (Cambridge University Press, 1988),
6. *Introduction to Black Holes*, C. V. Vishveshwara in 'Gravitation, Gauge Theories and the Early Universe', eds. B. R. Iyer, N. Mukunda and C. V. Vishveshwara (Kluwer Academic Publishers, 1989),
7. *Introduction to Relativistic Cosmology*, C.V. Vishveshwara in 'Gravitation, Gauge Theories and the Early Universe', eds. B. R. Iyer, N. Mukunda and C. V. Vishveshwara (Kluwer Academic Publishers, 1989),
8. *Geometry and the Universe*, C. V. Vishveshwara in 'Cosmic Perspectives', eds. S. K. Biswas, D. C. V. Mallik and C. V. Vishveshwara (Cambridge University Press, 1989),
9. *Black Holes, Bubbles and Radiation: An Excursion into Higher Dimensions*, in 'Frontier Physics', eds. S. MacDowell, H. M. Nussenzveig and R. A. Salmeron (World Scientific, 1991),

10. *Relativity and Rotation*, in 'Directions in General Relativity', eds. Bei-Lok Hu, Michael Ryan and C. V. Vishveshwara (Cambridge University Press, 1994),
11. *Gyroscopic Precession in General Relativity*, C. V. Vishveshwara in 'Inhomogeneous Cosmological Models', eds. A. Molina and J.M.M. Senovilla (World Scientific, 1996).
12. *A Simulated Potential Approach to the Scattering of Black Holes*, J. M. Aguirregabiria and C. V. Vishveshwara, in 'Relativistic Astrophysics and Cosmology', eds. J. Buitrigo, E. Mediavilla and A. Oscot (World Scientific, 1997).
13. *Gyroscopic Precession, Inertial Forces and Gravi-electromagnetism: A Covariant Approach*, K. R. Nayak and C. V. Vishveshwara in 'Relativistic Astrophysics and Cosmology', eds. J. Buitrago, E. Mediavilla and A. Oscot (World Scientific, 1997).
14. *Black Holes and Rotation*, C. V. Vishveshwara in 'Physics of Black Holes', eds. N. Mukunda, J. Pasupathy and C. V. Vishveshwara (Springer-Verlag, 1998).
15. *The Engelbert Experience: Pathways fvrom the past*, C. V. Vishveshwara in 'On Einstein's Path: Essays in Honor of Engelbert Schucking', eds. Alex Harvey (Springer-Verlag, 1998).
16. *After the Fall: From Adam and Eve to Apples and Elevators*, C. V. Vishveshwara in 'Gravitational and Cosmology: At the Turn of the Millenium', eds. N. K. Dadhich and J. V. Narlikar (IUCAA, Pune, 1998).

4. Books Edited

1. *Gravitation, Quanta and the Universe*, A. R. Prasanna, J. V. Narlikar, and C. V. Vishveshwara (New Delhi, Wiley Eastern, 1980),
2. *Gravitation and Relativistic Astrophysics*, A. R. Prasanna, J. V. Narlikar and C. V. Vishveshwara (Bangalore, Indian Academy of Sciences / World Scientific, 1984),
3. *A Random Walk in Relativity and Cosmology*, N. K. Dadhich, J. Krishna Rao, J. V. Narlikar and C. V. Vishveshwara (New Delhi, Wiley Eastern, 1985),
4. *Highlights in Gravitation and Cosmology*, B. R. Iyer, A. K. Kembhavi, J. V. Narlikar and C. V. Vishveshwara (Cambridge, Cambridge University Press, 1988),
5. *Gravitation, Gauge Theories and the Early Universe*, B. R. Iyer, N. Mukunda and C. V. Vishveshwara (Dordrecht, Kluwer, 1989),

6. *Cosmic Perspectives*, S. K. Biswas, D. C. V. Mallik and C. V. Vishveshwara (Cambridge, Cambridge University Press, 1989),
7. *Advances in Gravitation and Cosmology*, B. R. Iyer, A. R. Prasanna, R. K. Varma, C. V. Vishveshwara (New Delhi, Wiley Eastern, 1992),
8. *Directions in General Relativity*, Bei-Lok Hu, Michael Ryan and C. V. Vishveshwara (Cambridge, Cambridge University Press, 1994),
9. *Geometry, Fields and Cosmology: Techniques and Applications*, B. R. Iyer and C. V. Vishveshwara, (Dordrecht, Kluwer, 1997).
10. *Physics of Black Holes*, eds. N. Mukunda, J. Pasupathy and C. V. Vishveshwara, (Springer-Verlag, 1998).

5. Nontechnical Articles

1. *How will the World End?*, Akashavani, All India Radio Publication (May 1982),
2. *The Expanding Universe : A Reappraisal*, Akashavani, All India Radio publication (May 1983),
3. *Leelavathi - The Mathematical Poetry of Bhaskara*, Bulletin of Sciences (1985),
4. *From Cadavers to Computers*, Bulletin of Sciences (Oct-Dec 1991),
5. *The Apple and the Fall*, Bulletin of Sciences (Jan-Mar 1992),
6. *Food for Thought*, Bulletin of Sciences (Apr-Jun 1992),
7. *Conservation Laws*, Unpublished Manuscript,
8. *On the Black Hole Trail ... : A Personal Journey*, Current Science, 71, 824 (1996).

Fundamental Theories of Physics

Series Editor: Alwyn van der Merwe, *University of Denver, USA*

1. M. Sachs: *General Relativity and Matter.* A Spinor Field Theory from Fermis to Light-Years. With a Foreword by C. Kilmister. 1982 ISBN 90-277-1381-2
2. G.H. Duffey: *A Development of Quantum Mechanics.* Based on Symmetry Considerations. 1985 ISBN 90-277-1587-4
3. S. Diner, D. Fargue, G. Lochak and F. Selleri (eds.): *The Wave-Particle Dualism.* A Tribute to Louis de Broglie on his 90th Birthday. 1984 ISBN 90-277-1664-1
4. E. Prugovečki: *Stochastic Quantum Mechanics and Quantum Spacetime.* A Consistent Unification of Relativity and Quantum Theory based on Stochastic Spaces. 1984; 2nd printing 1986 ISBN 90-277-1617-X
5. D. Hestenes and G. Sobczyk: *Clifford Algebra to Geometric Calculus.* A Unified Language for Mathematics and Physics. 1984 ISBN 90-277-1673-0; Pb (1987) 90-277-2561-6
6. P. Exner: *Open Quantum Systems and Feynman Integrals.* 1985 ISBN 90-277-1678-1
7. L. Mayants: *The Enigma of Probability and Physics.* 1984 ISBN 90-277-1674-9
8. E. Tocaci: *Relativistic Mechanics, Time and Inertia.* Translated from Romanian. Edited and with a Foreword by C.W. Kilmister. 1985 ISBN 90-277-1769-9
9. B. Bertotti, F. de Felice and A. Pascolini (eds.): *General Relativity and Gravitation.* Proceedings of the 10th International Conference (Padova, Italy, 1983). 1984 ISBN 90-277-1819-9
10. G. Tarozzi and A. van der Merwe (eds.): *Open Questions in Quantum Physics.* 1985 ISBN 90-277-1853-9
11. J.V. Narlikar and T. Padmanabhan: *Gravity, Gauge Theories and Quantum Cosmology.* 1986 ISBN 90-277-1948-9
12. G.S. Asanov: *Finsler Geometry, Relativity and Gauge Theories.* 1985 ISBN 90-277-1960-8
13. K. Namsrai: *Nonlocal Quantum Field Theory and Stochastic Quantum Mechanics.* 1986 ISBN 90-277-2001-0
14. C. Ray Smith and W.T. Grandy, Jr. (eds.): *Maximum-Entropy and Bayesian Methods in Inverse Problems.* Proceedings of the 1st and 2nd International Workshop (Laramie, Wyoming, USA). 1985 ISBN 90-277-2074-6
15. D. Hestenes: *New Foundations for Classical Mechanics.* 1986 ISBN 90-277-2090-8; Pb (1987) 90-277-2526-8
16. S.J. Prokhovnik: *Light in Einstein's Universe.* The Role of Energy in Cosmology and Relativity. 1985 ISBN 90-277-2093-2
17. Y.S. Kim and M.E. Noz: *Theory and Applications of the Poincaré Group.* 1986 ISBN 90-277-2141-6
18. M. Sachs: *Quantum Mechanics from General Relativity.* An Approximation for a Theory of Inertia. 1986 ISBN 90-277-2247-1
19. W.T. Grandy, Jr.: *Foundations of Statistical Mechanics.* Vol. I: *Equilibrium Theory.* 1987 ISBN 90-277-2489-X
20. H.-H von Borzeszkowski and H.-J. Treder: *The Meaning of Quantum Gravity.* 1988 ISBN 90-277-2518-7
21. C. Ray Smith and G.J. Erickson (eds.): *Maximum-Entropy and Bayesian Spectral Analysis and Estimation Problems.* Proceedings of the 3rd International Workshop (Laramie, Wyoming, USA, 1983). 1987 ISBN 90-277-2579-9
22. A.O. Barut and A. van der Merwe (eds.): *Selected Scientific Papers of Alfred Landé. [1888-1975].* 1988 ISBN 90-277-2594-2

23. W.T. Grandy, Jr.: *Foundations of Statistical Mechanics.* Vol. II: *Nonequilibrium Phenomena.* 1988 ISBN 90-277-2649-3
24. E.I. Bitsakis and C.A. Nicolaides (eds.): *The Concept of Probability.* Proceedings of the Delphi Conference (Delphi, Greece, 1987). 1989 ISBN 90-277-2679-5
25. A. van der Merwe, F. Selleri and G. Tarozzi (eds.): *Microphysical Reality and Quantum Formalism, Vol. 1.* Proceedings of the International Conference (Urbino, Italy, 1985). 1988 ISBN 90-277-2683-3
26. A. van der Merwe, F. Selleri and G. Tarozzi (eds.): *Microphysical Reality and Quantum Formalism, Vol. 2.* Proceedings of the International Conference (Urbino, Italy, 1985). 1988 ISBN 90-277-2684-1
27. I.D. Novikov and V.P. Frolov: *Physics of Black Holes.* 1989 ISBN 90-277-2685-X
28. G. Tarozzi and A. van der Merwe (eds.): *The Nature of Quantum Paradoxes.* Italian Studies in the Foundations and Philosophy of Modern Physics. 1988 ISBN 90-277-2703-1
29. B.R. Iyer, N. Mukunda and C.V. Vishveshwara (eds.): *Gravitation, Gauge Theories and the Early Universe.* 1989 ISBN 90-277-2710-4
30. H. Mark and L. Wood (eds.): *Energy in Physics, War and Peace.* A Festschrift celebrating Edward Teller's 80th Birthday. 1988 ISBN 90-277-2775-9
31. G.J. Erickson and C.R. Smith (eds.): *Maximum-Entropy and Bayesian Methods in Science and Engineering.* Vol. I: *Foundations.* 1988 ISBN 90-277-2793-7
32. G.J. Erickson and C.R. Smith (eds.): *Maximum-Entropy and Bayesian Methods in Science and Engineering.* Vol. II: *Applications.* 1988 ISBN 90-277-2794-5
33. M.E. Noz and Y.S. Kim (eds.): *Special Relativity and Quantum Theory.* A Collection of Papers on the Poincaré Group. 1988 ISBN 90-277-2799-6
34. I.Yu. Kobzarev and Yu.I. Manin: *Elementary Particles. Mathematics, Physics and Philosophy.* 1989 ISBN 0-7923-0098-X
35. F. Selleri: *Quantum Paradoxes and Physical Reality.* 1990 ISBN 0-7923-0253-2
36. J. Skilling (ed.): *Maximum-Entropy and Bayesian Methods.* Proceedings of the 8th International Workshop (Cambridge, UK, 1988). 1989 ISBN 0-7923-0224-9
37. M. Kafatos (ed.): *Bell's Theorem, Quantum Theory and Conceptions of the Universe.* 1989 ISBN 0-7923-0496-9
38. Yu.A. Izyumov and V.N. Syromyatnikov: *Phase Transitions and Crystal Symmetry.* 1990 ISBN 0-7923-0542-6
39. P.F. Fougère (ed.): *Maximum-Entropy and Bayesian Methods.* Proceedings of the 9th International Workshop (Dartmouth, Massachusetts, USA, 1989). 1990 ISBN 0-7923-0928-6
40. L. de Broglie: *Heisenberg's Uncertainties and the Probabilistic Interpretation of Wave Mechanics.* With Critical Notes of the Author. 1990 ISBN 0-7923-0929-4
41. W.T. Grandy, Jr.: *Relativistic Quantum Mechanics of Leptons and Fields.* 1991 ISBN 0-7923-1049-7
42. Yu.L. Klimontovich: *Turbulent Motion and the Structure of Chaos.* A New Approach to the Statistical Theory of Open Systems. 1991 ISBN 0-7923-1114-0
43. W.T. Grandy, Jr. and L.H. Schick (eds.): *Maximum-Entropy and Bayesian Methods.* Proceedings of the 10th International Workshop (Laramie, Wyoming, USA, 1990). 1991 ISBN 0-7923-1140-X

Fundamental Theories of Physics

44. P.Pták and S. Pulmannová: *Orthomodular Structures as Quantum Logics*. Intrinsic Properties, State Space and Probabilistic Topics. 1991 ISBN 0-7923-1207-4
45. D. Hestenes and A. Weingartshofer (eds.): *The Electron*. New Theory and Experiment. 1991 ISBN 0-7923-1356-9
46. P.P.J.M. Schram: *Kinetic Theory of Gases and Plasmas*. 1991 ISBN 0-7923-1392-5
47. A. Micali, R. Boudet and J. Helmstetter (eds.): *Clifford Algebras and their Applications in Mathematical Physics*. 1992 ISBN 0-7923-1623-1
48. E. Prugovečki: *Quantum Geometry*. A Framework for Quantum General Relativity. 1992 ISBN 0-7923-1640-1
49. M.H. Mac Gregor: *The Enigmatic Electron*. 1992 ISBN 0-7923-1982-6
50. C.R. Smith, G.J. Erickson and P.O. Neudorfer (eds.): *Maximum Entropy and Bayesian Methods*. Proceedings of the 11th International Workshop (Seattle, 1991). 1993 ISBN 0-7923-2031-X
51. D.J. Hoekzema: *The Quantum Labyrinth*. 1993 ISBN 0-7923-2066-2
52. Z. Oziewicz, B. Jancewicz and A. Borowiec (eds.): *Spinors, Twistors, Clifford Algebras and Quantum Deformations*. Proceedings of the Second Max Born Symposium (Wrocław, Poland, 1992). 1993 ISBN 0-7923-2251-7
53. A. Mohammad-Djafari and G. Demoment (eds.): *Maximum Entropy and Bayesian Methods*. Proceedings of the 12th International Workshop (Paris, France, 1992). 1993 ISBN 0-7923-2280-0
54. M. Riesz: *Clifford Numbers and Spinors* with Riesz' Private Lectures to E. Folke Bolinder and a Historical Review by Pertti Lounesto. E.F. Bolinder and P. Lounesto (eds.). 1993 ISBN 0-7923-2299-1
55. F. Brackx, R. Delanghe and H. Serras (eds.): *Clifford Algebras and their Applications in Mathematical Physics*. Proceedings of the Third Conference (Deinze, 1993) 1993 ISBN 0-7923-2347-5
56. J.R. Fanchi: *Parametrized Relativistic Quantum Theory*. 1993 ISBN 0-7923-2376-9
57. A. Peres: *Quantum Theory: Concepts and Methods*. 1993 ISBN 0-7923-2549-4
58. P.L. Antonelli, R.S. Ingarden and M. Matsumoto: *The Theory of Sprays and Finsler Spaces with Applications in Physics and Biology*. 1993 ISBN 0-7923-2577-X
59. R. Miron and M. Anastasiei: *The Geometry of Lagrange Spaces: Theory and Applications*. 1994 ISBN 0-7923-2591-5
60. G. Adomian: *Solving Frontier Problems of Physics: The Decomposition Method*. 1994 ISBN 0-7923-2644-X
61 B.S. Kerner and V.V. Osipov: *Autosolitons*. A New Approach to Problems of Self-Organization and Turbulence. 1994 ISBN 0-7923-2816-7
62. G.R. Heidbreder (ed.): *Maximum Entropy and Bayesian Methods*. Proceedings of the 13th International Workshop (Santa Barbara, USA, 1993) 1996 ISBN 0-7923-2851-5
63. J. Peřina, Z. Hradil and B. Jurčo: *Quantum Optics and Fundamentals of Physics*. 1994 ISBN 0-7923-3000-5
64. M. Evans and J.-P. Vigier: *The Enigmatic Photon*. Volume 1: The Field $\boldsymbol{B}^{(3)}$. 1994 ISBN 0-7923-3049-8
65. C.K. Raju: *Time: Towards a Constistent Theory*. 1994 ISBN 0-7923-3103-6
66. A.K.T. Assis: *Weber's Electrodynamics*. 1994 ISBN 0-7923-3137-0
67. Yu. L. Klimontovich: *Statistical Theory of Open Systems*. Volume 1: A Unified Approach to Kinetic Description of Processes in Active Systems. 1995 ISBN 0-7923-3199-0; Pb: ISBN 0-7923-3242-3

Fundamental Theories of Physics

68. M. Evans and J.-P. Vigier: *The Enigmatic Photon.* Volume 2: Non-Abelian Electrodynamics. 1995 ISBN 0-7923-3288-1
69. G. Esposito: *Complex General Relativity.* 1995 ISBN 0-7923-3340-3
70. J. Skilling and S. Sibisi (eds.): *Maximum Entropy and Bayesian Methods.* Proceedings of the Fourteenth International Workshop on Maximum Entropy and Bayesian Methods. 1996 ISBN 0-7923-3452-3
71. C. Garola and A. Rossi (eds.): *The Foundations of Quantum Mechanics – Historical Analysis and Open Questions.* 1995 ISBN 0-7923-3480-9
72. A. Peres: *Quantum Theory: Concepts and Methods.* 1995 (see for hardback edition, Vol. 57) ISBN Pb 0-7923-3632-1
73. M. Ferrero and A. van der Merwe (eds.): *Fundamental Problems in Quantum Physics.* 1995 ISBN 0-7923-3670-4
74. F.E. Schroeck, Jr.: *Quantum Mechanics on Phase Space.* 1996 ISBN 0-7923-3794-8
75. L. de la Peña and A.M. Cetto: *The Quantum Dice.* An Introduction to Stochastic Electrodynamics. 1996 ISBN 0-7923-3818-9
76. P.L. Antonelli and R. Miron (eds.): *Lagrange and Finsler Geometry.* Applications to Physics and Biology. 1996 ISBN 0-7923-3873-1
77. M.W. Evans, J.-P. Vigier, S. Roy and S. Jeffers: *The Enigmatic Photon.* Volume 3: Theory and Practice of the $\boldsymbol{B}^{(3)}$ Field. 1996 ISBN 0-7923-4044-2
78. W.G.V. Rosser: *Interpretation of Classical Electromagnetism.* 1996 ISBN 0-7923-4187-2
79. K.M. Hanson and R.N. Silver (eds.): *Maximum Entropy and Bayesian Methods.* 1996 ISBN 0-7923-4311-5
80. S. Jeffers, S. Roy, J.-P. Vigier and G. Hunter (eds.): *The Present Status of the Quantum Theory of Light.* Proceedings of a Symposium in Honour of Jean-Pierre Vigier. 1997 ISBN 0-7923-4337-9
81. M. Ferrero and A. van der Merwe (eds.): *New Developments on Fundamental Problems in Quantum Physics.* 1997 ISBN 0-7923-4374-3
82. R. Miron: *The Geometry of Higher-Order Lagrange Spaces.* Applications to Mechanics and Physics. 1997 ISBN 0-7923-4393-X
83. T. Hakioğlu and A.S. Shumovsky (eds.): *Quantum Optics and the Spectroscopy of Solids.* Concepts and Advances. 1997 ISBN 0-7923-4414-6
84. A. Sitenko and V. Tartakovskii: *Theory of Nucleus.* Nuclear Structure and Nuclear Interaction. 1997 ISBN 0-7923-4423-5
85. G. Esposito, A.Yu. Kamenshchik and G. Pollifrone: *Euclidean Quantum Gravity on Manifolds with Boundary.* 1997 ISBN 0-7923-4472-3
86. R.S. Ingarden, A. Kossakowski and M. Ohya: *Information Dynamics and Open Systems.* Classical and Quantum Approach. 1997 ISBN 0-7923-4473-1
87. K. Nakamura: *Quantum versus Chaos.* Questions Emerging from Mesoscopic Cosmos. 1997 ISBN 0-7923-4557-6
88. B.R. Iyer and C.V. Vishveshwara (eds.): *Geometry, Fields and Cosmology.* Techniques and Applications. 1997 ISBN 0-7923-4725-0
89. G.A. Martynov: *Classical Statistical Mechanics.* 1997 ISBN 0-7923-4774-9
90. M.W. Evans, J.-P. Vigier, S. Roy and G. Hunter (eds.): *The Enigmatic Photon.* Volume 4: New Directions. 1998 ISBN 0-7923-4826-5
91. M. Rédei: *Quantum Logic in Algebraic Approach.* 1998 ISBN 0-7923-4903-2
92. S. Roy: *Statistical Geometry and Applications to Microphysics and Cosmology.* 1998 ISBN 0-7923-4907-5

The manufacturer's authorised representative in the EU is Springer Nature Customer Service Centre GmbH, Europaplatz 3, 69115 Heidelberg, Germany. If you have any concerns regarding our products, please contact ProductSafety@springernature.com

Printed and bound by CPI Group (UK) Ltd, Croydon, CR0 4YY

22/09/2026

02220988-0009